AF812469

Studies in Systems, Decision and Control

Volume 383

Series Editor

Janusz Kacprzyk, Systems Research Institute, Polish Academy of Sciences, Warsaw, Poland

The series "Studies in Systems, Decision and Control" (SSDC) covers both new developments and advances, as well as the state of the art, in the various areas of broadly perceived systems, decision making and control–quickly, up to date and with a high quality. The intent is to cover the theory, applications, and perspectives on the state of the art and future developments relevant to systems, decision making, control, complex processes and related areas, as embedded in the fields of engineering, computer science, physics, economics, social and life sciences, as well as the paradigms and methodologies behind them. The series contains monographs, textbooks, lecture notes and edited volumes in systems, decision making and control spanning the areas of Cyber-Physical Systems, Autonomous Systems, Sensor Networks, Control Systems, Energy Systems, Automotive Systems, Biological Systems, Vehicular Networking and Connected Vehicles, Aerospace Systems, Automation, Manufacturing, Smart Grids, Nonlinear Systems, Power Systems, Robotics, Social Systems, Economic Systems and other. Of particular value to both the contributors and the readership are the short publication timeframe and the world-wide distribution and exposure which enable both a wide and rapid dissemination of research output.

Indexed by SCOPUS, DBLP, WTI Frankfurt eG, zbMATH, SCImago.

All books published in the series are submitted for consideration in Web of Science.

More information about this series at http://www.springer.com/series/13304

Samsul Ariffin Abdul Karim · Afza Shafie
Editors

Towards Intelligent Systems Modeling and Simulation

With Applications to Energy, Epidemiology and Risk Assessment

Springer

Editors
Samsul Ariffin Abdul Karim
Fundamental and Applied Sciences Department and Centre for Systems Engineering (CSE)
Institute of Autonomous System
Universiti Teknologi PETRONAS
Seri Iskandar, Perak Darul Ridzuan
Malaysia

Afza Shafie
Fundamental and Applied Sciences Department
Universiti Teknologi PETRONAS
Seri Iskandar, Perak Darul Ridzuan
Malaysia

ISSN 2198-4182 ISSN 2198-4190 (electronic)
Studies in Systems, Decision and Control
ISBN 978-3-030-79605-1 ISBN 978-3-030-79606-8 (eBook)
https://doi.org/10.1007/978-3-030-79606-8

© Institute of Technology PETRONAS Sdn Bhd 2022
This work is subject to copyright. All rights are solely and exclusively licensed by the Publisher, whether the whole or part of the material is concerned, specifically the rights of translation, reprinting, reuse of illustrations, recitation, broadcasting, reproduction on microfilms or in any other physical way, and transmission or information storage and retrieval, electronic adaptation, computer software, or by similar or dissimilar methodology now known or hereafter developed.
The use of general descriptive names, registered names, trademarks, service marks, etc. in this publication does not imply, even in the absence of a specific statement, that such names are exempt from the relevant protective laws and regulations and therefore free for general use.
The publisher, the authors and the editors are safe to assume that the advice and information in this book are believed to be true and accurate at the date of publication. Neither the publisher nor the authors or the editors give a warranty, expressed or implied, with respect to the material contained herein or for any errors or omissions that may have been made. The publisher remains neutral with regard to jurisdictional claims in published maps and institutional affiliations.

This Springer imprint is published by the registered company Springer Nature Switzerland AG
The registered company address is: Gewerbestrasse 11, 6330 Cham, Switzerland

Preface

This book is a collection of the works that have been conducted by researchers at various universities in the world. The emergence of disruptive technologies has led to a significant change in the role of mathematics and statistics for problem solving, with the use of sophisticated software and hardware in solving complex systems and process. In the era of digital technology, mathematics and statistics need to be highly relevant to be able to cater for the needs of Industrial Revolution 4.0 (IR 4.0) such as big data analytics, simulation, autonomous system and cloud computing. Motivated by this development, a total of 26 chapters are contributed by respectable experts that actively involve in many researches and grants. The main scope of the book is to conduct a new system of modelling and simulations on solving differential equations, nonlinear equations, energy, epidemiology and risk assessment. This book will be of interest for postgraduate students, researchers as well as other scientists who are working in numerical modelling and simulations based on efficient mathematical and statistical techniques.

The editors would like to express their gratitude to all the contributing authors for their great efforts and full dedication in preparing the manuscripts for the book. We would like to thank all reviewers for reviewing all manuscripts and providing very constructive feedback. The first editor is supported by Universiti Teknologi PETRONAS (UTP) and the Ministry of Education, Malaysia, through a research grant **FRGS/1/2018/STG06/UTP/03/1015MA0-020 (New rational quartic spline for image refinement)** and **YUTP: 015LC0-315** (Uncertainty estimation based on Quasi-Newton methods for Full Waveform Inversion (FWI)).

The reader from this book will know about current trend of technologies, and this book could be a good reference for the postgraduate studies and researchers.

Seri Iskandar, Malaysia
September 2021

Samsul Ariffin Abdul Karim
Afza Shafie

Contents

Convexity Preservation of the Ternary 6-point Interpolating Subdivision Scheme 1
Mudassar Iqbal, Samsul Ariffin Abdul Karim, Afza Shafie, and Muhammad Sarfraz

Pandemic Measures for the COVID19 Outbreak Control in Malaysia: A Correlative Comparison with China 25
Chockalingam Aravind Vaithilingam, S. Gayathri, R. Lalitha, R. R. Rajalakshmi, and Arthanari Jagadeeshwaran

Efficient Numerical Schemes for MHD Convection in Porous Cavity 39
Zaihar Yaacob, Mohammad Khatim Hasan, and Bahari Idrus

The Impact of the New Ordering Strategy on Over—relaxation Variants for Two Dimensional Poisson's problems 47
Mohammad Khatim Hasan, Samsul Ariffin Abdul Karim, and Jumat Sulaiman

High—Order Hosoya Polynomials with Collocation Approach for the Solution of Two—Point Boundary Value Problems 63
Shazliyanna Ussin, Jumat Sulaiman, Mohammad Khatim Hasan, and Samsul Ariffin Abdul Karim

The Quaternion Fourier Transform of Finite Measure and Its Properties 79
Mawardi Bahri, Amran Rahim, Muh. Nur, and Amir Kamal Amir

On Relationship Between Ambiguity Function and Linear Canonical Ambiguity Function 93
Mawardi Bahri and Samsul Ariffin Abdul Karim

Numerical Solution of Oil Spill Weathering with Two Points Block Backward Differentiation Formula 107
Nooraini Zainuddin, Hira Soomro, Noraini Jamaludin, Siti Ainor Mohd Yatim, and Zarina Bibi Ibrahim

Visualization of Electric Field Response for a Semi Circle Transmitter in Seabed Logging 123
Elsayed Dahy and Afza Shafie

Component Analysis of Electromagnetic Wave Propagation in a Layered Model for CSEM Hydrocarbon Surveying 135
Amir Rostami, Afza Shafie, Hassan Soleimani, and Surajudeen Olalekan Sikiru

A Review on the Important Key Properties of Mathematical Models Describing Photovoltaic/Thermal (PV/T) Solar Collectors System 149
Saiful Azmi Husain, Mahmod Othman, and Noran Nur Wahida Khalili

Trisection and Pentasection Method: A Modification of the Bisection Method for Solving Non-linear Algebraic Equations 157
Saratha Sathasivam and Shehab Abdulhabib Saeed Alzaeemi

Predicting and Assessing Road Accidents Using Autoregressive Model and Value at Risk Approach 163
Teh Raihana Nazirah Roslan, Chee Keong Ch'ng, Masnita Misiran, and Nattakorn Phewchean

Global Stability Analysis of Dengue Model with Awareness, Vector Control and Time Delays 177
Jay Michael R. Macalalag, Timothy Robin Y. Teng, and Elvira P. de Lara-Tuprio

A Copula Based-Approach for Image Splicing Detection 197
Aniq A. Rohmawati and Rimba W. Ciptasari

Artificial Intelligence Forecasting for Transmission Line Ampacity 217
Yaman Hamed, Muhamad Safwan Abd Rahman, Mohd Zainal Abidin Ab Kadir, Miszaina Osman, Azrul Mohd Ariffin, and Nur Fadilah Ab Aziz

Estimating Marine CSEM Responses Using Gaussian Process Regression Based on Synthetic Models 235
Muhammad Naeim Mohd Aris, Hanita Daud, Khairul Arifin Mohd Noh, and Sarat Chandra Dass

Stochastic Process Algebra Model of Amyloidogenic Processing in Alzheimer's Disease 249
Angelyn Lao, Michelle Andrea Dela Cruz, Grazielle Joy Tenio, Polly Sy, and Eduardo Mendoza

Soliton, Rogue Wave and Double Layer in an Unmagnetized Collisionless Plasma 265
Md. Golam Hafez and Samsul Ariffin Abdul Karim

Sparse Closed-Loop Surface Related Multiple Elimination 283
Abdul Halim Abdul Latiff and Hazwan Hashim

Hybrid Illumination Analysis for Complex Overburden Geological Structure 329
Abdul Halim Abdul Latiff

Sol-gel Synthesis of Silicadioxide/Organopolymer Nanocomposite for Potential Applications in EOR Nanotechnology 357
Hassan Soleimani, Mohammed Falalu Hamza, Zulkifli Merican Aljunid Merican, and Karl Dunbar Stephen

Empirical and Neural Network Modelling of Oil Recovery in Nano Assisted Enhanced Oil Recovery 367
Afza Shafie, Sayed Ameenuddin Irfan, and Norsyahirah Aripin

Mathematical Study of a Fractional-Order General Pathogen Dynamic Model with Immune Impairment 379
Shaimaa Abdelhamed Azoz and Fatima Hussien

Efficiency Evaluation of Half-Sweep Newton-EGSOR Method to Solve 1D Nonlinear Porous Medium Equations 399
Jackel Vui Lung Chew, Elayaraja Aruchunan, and Jumat Sulaiman

Examination of Half-Sweep Closed Newton–Cotes Quadrature Schemes in Solving Dense System 413
Elayaraja Aruchunan, Mohana Sundaram Muthuvalu, Zailan Siri, Sachin Sharma Ashok Kumar, Jumat Sulaiman, Jackel Vui Lung Chew, and Majid Khan Majahar Ali

Index 431

Editors and Contributors

About the Editors

Samsul Ariffin Abdul Karim is Senior Lecturer at Fundamental and Applied Sciences Department, Universiti Teknologi PETRONAS (UTP), Malaysia. He has been at the department for more than twelve years. He obtained his B.App.Sc., M.Sc. and Ph.D. in Computational Mathematics and Computer-Aided Geometric Design (CAGD) from Universiti Sains Malaysia (USM). He is a Professional Technologists registered with Malaysia Board of Technologists (MBOT). He had 20 years of experience using mathematica and MATLAB software for teaching and research activities. His research interests include curves and surfaces designing, geometric modelling and wavelets applications in image compression and statistics. He has published more than 140 papers in journals and conferences as well as eight books including two research monographs and three edited conferences volume and 40 chapters. He was the recipient of Effective Education Delivery Award and Publication Award (Journal and Conference Paper), UTP Quality Day 2010, 2011 and 2012, respectively. He was Certified WOLFRAM Technology Associate, Mathematica Student Level. He has published eight books with Springer.

Dr. Afza Shafie is currently Senior Lecturer with the Fundamental and Applied Science Department (FASD) at Univerisiti Teknologi PETRONAS. She holds Bachelor's and Master's degrees in Mathematics and Statistics, respectively. In addition, she has a Ph.D. in Systems Technology. Her interest is in mathematical and statistical modelling. Her research includes the application of these knowledge in the field of electromagmetics for hydrocarbon detection in Malaysian waters, application of electromagnetic and nanofluids in enhanced oil recovery. Some of her research has been funded by Petroleum Research Fund (PRF), Yayasan—UTP grant, the Prototype Research Grant (PRGS) and the Fundamental Research Grant (FRGS) from the Ministry of Education Malaysia. In addition, she is actively involved with the University initiatives to promote STEM to the schools in Malaysia. Her work has been presented in international conferences and published in ISI and high-impact journals. Her current area of interest is the application of artificial intelligence for electromagnetic and nanofluids enhanced oil recovery.

Contributors

Nur Fadilah Ab Aziz Institute of Power Engineering, Universiti Tenaga Nasional (UNITEN), Kajang, Malaysia

Mohd Zainal Abidin Ab Kadir Institute of Power Engineering, Universiti Tenaga Nasional (UNITEN), Kajang, Malaysia

Muhamad Safwan Abd Rahman Institute of Power Engineering, Universiti Tenaga Nasional (UNITEN), Kajang, Malaysia

Samsul Ariffin Abdul Karim Fundamental and Applied Sciences Department and Centre for Systems Engineering (CSE), Institute of Autonomous System, Universiti Teknologi PETRONAS, Seri Iskandar, Perak Darul Ridzuan, Malaysia

Abdul Halim Abdul Latiff Centre for Subsurface Imaging, Universiti Teknologi PETRONAS, Seri Iskandar, Perak Darul Ridzuan, Malaysia

Majid Khan Majahar Ali School of Mathematical Sciences, Universiti Sains Malaysia, Penang, Malaysia

Shehab Abdulhabib Saeed Alzaeemi School of Mathematical Sciences, Universiti Sains Malaysia, Penang, Malaysia

Amir Kamal Amir Department of Mathematics, Hasanuddin University, Makassar, Indonesia

Azrul Mohd Ariffin Institute of Power Engineering, Universiti Tenaga Nasional (UNITEN), Kajang, Malaysia

Norsyahirah Aripin Faculty of Science, Universiti Teknologi Malaysia, Johor Bahru, Malaysia

Elayaraja Aruchunan Faculty of Science, Institute of Mathematical Sciences, University of Malaya, Kuala Lumpur, Malaysia

Shaimaa Abdelhamed Azoz Faculty of Science, Department of Mathematics, Assiut University, Assiut, Egypt

Mawardi Bahri Department of Mathematics, Hasanuddin University, Makassar, Indonesia

Jackel Vui Lung Chew Faculty of Computing and Informatics, Universiti Malaysia Sabah Labuan International Campus, Labuan F.T., Malaysia

Chee Keong Ch'ng School of Quantitative Sciences, Universiti Utara Malaysia, Sintok, Kedah, Malaysia

Rimba W. Ciptasari School of Computing, Telkom University, Bandung, Indonesia

Elsayed Dahy Department of Mathematics, Faculty of Science, University Al-Azhar, Assuit, Egypt

Sarat Chandra Dass School of Mathematical and Computer Sciences, Heriot-Watt University Malaysia, Putrajaya, Malaysia

Hanita Daud Department of Fundamental and Applied Sciences, Universiti Teknologi PETRONAS, Seri Iskandar, Perak, Malaysia

Elvira P. de Lara-Tuprio Department of Mathematics, School of Science and Engineering, Ateneo de Manila University, Quezon City, Metro Manila, Philippines

Michelle Andrea Dela Cruz Mathematics and Statistics Department, De La Salle University, Manila, Philippines

S. Gayathri Xavier Institute of Management & Entrepreneurship, Chennai, India

Md. Golam Hafez Department of Mathematics, Chittagong University of Engineering and Technology, Chattogram, Bangladesh

Yaman Hamed Institute of Power Engineering, Universiti Tenaga Nasional (UNITEN), Kajang, Malaysia

Mohammed Falalu Hamza Department of Pure & Industrial Chemistry, Bayero University Kano, Kano, Nigeria

Mohammad Khatim Hasan Centre for Artificial Intelligence Technology, Faculty of Information Science and Technology, Universiti Kebangsaan Malaysia, Bangi, Malaysia

Hazwan Hashim Centre for Subsurface Imaging, Universiti Teknologi PETRONAS, Seri Iskandar, Perak Darul Ridzuan, Malaysia

Saiful Azmi Husain Mathematical Sciences Department, Faculty of Science, Universiti Brunei Darussalam, Gadong, BE, Brunei Darussalam

Fatima Hussien Faculty of Science, Department of Mathematics, Assiut University, Assiut, Egypt

Zarina Bibi Ibrahim Department of Mathematics, Faculty of Science, Universiti Putra Malaysia (UPM), Selangor, Malaysia

Bahari Idrus Centre for Artificial Intelligence Technology, Faculty of Information Science and Technology, Universiti Kebangsaan Malaysia, Bangi, Malaysia

Mudassar Iqbal Department of Fundamental and Applied Sciences, Universiti Teknologi PETRONAS, Seri Iskandar, Perak Darul Ridzuan, Malaysia;
Department of Mathematical Sciences, Balochistan University of Information Technology, Engineering and Management Sciences (BUITEMS), Quetta, Pakistan

Sayed Ameenuddin Irfan Shale Gas Research Group, Universiti Teknologi PETRONAS, Seri Iskandar, Perak, Malaysia

Arthanari Jagadeeshwaran Sona College of Technology, Salem, India

Noraini Jamaludin Centre for Foundation Studies, Universiti Teknologi PETRONAS, Bandar Seri Iskandar, Perak Darul Ridzuan, Malaysia

Noran Nur Wahida Khalili Fundamental and Applied Sciences Department, Universiti Teknologi PETRONAS, Bandar Seri Iskandar, Perak Darul Ridzuan, Malaysia

Sachin Sharma Ashok Kumar Department of Physics, Faculty of Science, Centre for Ionic Universiti Malaya, Kuala Lumpur, Malaysia

R. Lalitha Computer Science and Engineering, Rajalakshmi Institute of Technology, Chennai, India

Angelyn Lao Mathematics and Statistics Department, De La Salle University, Manila, Philippines;
Systems and Computational Biology Research Unit, CENSER, De La Salle University, Manila, Philippines

Jay Michael R. Macalalag College of Mathematics and Natural Sciences, Caraga State University, Butuan City, Agusan Del Norte, Philippines

Eduardo Mendoza Mathematics and Statistics Department, De La Salle University, Manila, Philippines;

Systems and Computational Biology Research Unit, CENSER, De La Salle University, Manila, Philippines

Zulkifli Merican Aljunid Merican Fundamental and Applied Sciences Department, Universiti Teknologi PETRONAS, Bandar Seri Iskandar, Perak Darul Ridzuan, Malaysia

Masnita Misiran Institute of Strategic Industrial Decision Science Modeling, School of Quantitative Sciences, Universiti Utara Malaysia, Sintok, Kedah, Malaysia

Muhammad Naeim Mohd Aris Department of Fundamental and Applied Sciences, Universiti Teknologi PETRONAS, Seri Iskandar, Perak, Malaysia

Khairul Arifin Mohd Noh Department of Geosciences, Universiti Teknologi PETRONAS, Seri Iskandar, Perak, Malaysia

Mohana Sundaram Muthuvalu Fundamental and Applied Sciences Department, Universiti Teknologi PETRONAS, Seri Iskandar Perak, Malaysia

Muh. Nur Department of Mathematics, Hasanuddin University, Makassar, Indonesia

Miszaina Osman Institute of Power Engineering, Universiti Tenaga Nasional (UNITEN), Kajang, Malaysia

Mahmod Othman Fundamental and Applied Sciences Department, Universiti Teknologi PETRONAS, Bandar Seri Iskandar, Perak Darul Ridzuan, Malaysia

Nattakorn Phewchean Department of Mathematics, Faculty of Science, Mahidol University, Bangkok, Thailand;
Centre of Excellence in Mathematics, Perdo, Thailand

Amran Rahim Department of Mathematics, Hasanuddin University, Makassar, Indonesia

R. R. Rajalakshmi Computer Sciences and Engineering, Kongu Engineering College, Erode, India

Aniq A. Rohmawati School of Computing, Telkom University, Bandung, Indonesia

Teh Raihana Nazirah Roslan Othman Yeop Abdullah Graduate School of Business, Universiti Utara Malaysia, Kuala Lumpur, Malaysia;
Institute of Strategic Industrial Decision Science Modeling, School of Quantitative Sciences, Universiti Utara Malaysia, Sintok, Kedah, Malaysia

Amir Rostami Shale Gas Research Group, Universiti Teknologi PETRONAS, Bandar Seri Iskandar, Perak Darul Ridzuan, Malaysia

Muhammad Sarfraz Department of Information Science, College of Life Sciences, Kuwait University, Sabah AlSalem University City, Safat, Kuwait

Saratha Sathasivam School of Mathematical Sciences, Universiti Sains Malaysia, Penang, Malaysia

Afza Shafie Department of Fundamental and Applied Sciences, Universiti Teknologi PETRONAS, Bandar Seri Iskandar, Perak Darul Ridzuan, Malaysia

Surajudeen Olalekan Sikiru Fundamental and Applied Sciences Department, Universiti Teknologi PETRONAS, Bandar Seri Iskandar, Perak Darul Ridzuan, Malaysia

Zailan Siri Institute of Mathematical Sciences, Faculty of Science, University of Malaya, Kuala Lumpur, Malaysia

Hassan Soleimani Fundamental and Applied Sciences Department, Universiti Teknologi PETRONAS, Bandar Seri Iskandar, Perak Darul Ridzuan, Malaysia

Hira Soomro Fundamental and Applied Sciences Department, Universiti Teknologi PETRONAS, Bandar Seri Iskandar, Perak Darul Ridzuan, Malaysia

Karl Dunbar Stephen Institute of Petroleum Engineering, Heriot-Watt University, Edinburgh, UK

Jumat Sulaiman Mathematics with Economics Programme, Faculty of Science and Natural Resources, Universiti Malaysia Sabah, Kota Kinabalu, Sabah, Malaysia

Polly Sy Institute of Mathematics, University of the Philippines Diliman, Diliman, Quezon City, Philippines

Timothy Robin Y. Teng Department of Mathematics, School of Science and Engineering, Ateneo de Manila University, Quezon City, Metro Manila, Philippines

Grazielle Joy Tenio Mathematics and Statistics Department, De La Salle University, Manila, Philippines

Shazliyanna Ussin Mathematics with Economics Programme, Faculty of Science and Natural Resources, Universiti Malaysia Sabah, Kota Kinabalu, Malaysia

Chockalingam Aravind Vaithilingam High Impact Research Laboratories, Taylor's University, Subang Jaya, Malaysia

Zaihar Yaacob Polymer Engineering Technology, Universiti Kuala Lumpur Branch Campus Malaysian Institute of Chemical and Bio-Engineering Technology, Kuala Lumpur, Malaysia

Siti Ainor Mohd Yatim School of Distance Education, Universiti Sains Malaysia, USM, Penang, Malaysia

Nooraini Zainuddin Fundamental and Applied Sciences Department, Universiti Teknologi PETRONAS, Bandar Seri Iskandar, Perak Darul Ridzuan, Malaysia

Convexity Preservation of the Ternary 6-point Interpolating Subdivision Scheme

Mudassar Iqbal, Samsul Ariffin Abdul Karim, Afza Shafie, and Muhammad Sarfraz

Abstract The frequent use of subdivision schemes (SSs) for curve sketching can be seen in various papers published during recent years. These SSs provides efficient method for producing curves and surfaces. Further, these SSs can be divide in various categories but we focused only one of the type of SS called interpolating SS. Here, we have discussed some properties of the "ternary 6-point interpolating subdivision scheme" introduced by Faheem and Mustafa [1] in 2008. The aim of this paper is to discuss about a geometric property called "Convexity" with the tension parameter μ and its effect on curve sketching. The scheme is analyzed and its convexity preserving property is proved using mathematical induction. Strictly convex initial data is used for the evaluation of the condition on the tension parameter μ. This condition shows that our scheme is more efficient than other schemes in terms of lower value of tension parameter μ. The significance of the condition is displayed through numerical applications.

M. Iqbal · A. Shafie
Department of Fundamental and Applied Sciences, Universiti Teknologi PETRONAS, 32610 Seri Iskandar, Perak Darul Ridzuan, Malaysia
e-mail: mudassar.iqbal@buitms.edu.pk; mudassar_20000083@utp.edu.my

A. Shafie
e-mail: afza@utp.edu.my

M. Iqbal
Department of Mathematical Sciences, Balochistan University of Information Technology, Engineering and Management Sciences (BUITEMS), Quetta, Pakistan

S. A. Abdul Karim (✉)
Fundamental and Applied Sciences Department and Centre for Systems Engineering (CSE), Institute of Autonomous System, Universiti Teknologi PETRONAS, 32610 Seri Iskandar, Perak Darul Ridzuan, Malaysia
e-mail: samsul_ariffin@utp.edu.my

M. Sarfraz
Department of Information Science, College of Life Sciences, Kuwait University, Sabah AlSalem University City, BShadadiya, 5969, Safat 13060, Kuwait
e-mail: prof.m.sarfraz@gmail.com

© Institute of Technology PETRONAS Sdn Bhd 2022

S. A. Abdul Karim and A. Shafie (eds.), *Towards Intelligent Systems Modeling and Simulation*, Studies in Systems, Decision and Control 383,
https://doi.org/10.1007/978-3-030-79606-8_1

1 Introduction

Subdivision schemes (SSs) can generate curves and surfaces more efficiently, these are frequently used in sketching the curves and surfaces. Subdivision schemes and shape-preserving properties together are studied in recent years for the development of shaping the desire curves and surfaces. This side of CAGD has become very famous for research areas like CAGD, computer animation, computer graphics (CG), reverse engineering, robotics, CAD/CAM, and many other areas with so many interesting applications in industry. Convexity preserving property has shown great significance in controlling the shape of desire control points. This has been done in the industrial design sector e.g. imposing convexity in controlling the shape of cars, airplanes or ships. In addition, in graphing smooth limiting curves and surface designs, despite of properties like convergence, stability, smoothness, approximation order, error estimation, and convexity preservation is also a very new and important subject. Several review papers on this subject have been published in the last few years. Convexity protection is a fundamental concept of geometry, but it's also used in other fields of mathematics such as calculus of variations, complex analysis, partial differential equations, graph theory,probability theory, discrete mathematics, functional analysis, and coding theory. Other divisions of science, such as economics, physics, chemistry, and biology, all rely on convexity preservation.

1.1 Literature Review

Yadshalom [2] suggested a class of SS with finite curve modeling support that maintained monotonicity. Carnicer and Dahmen [3] investigated a characterization of strict convexity-preserving SS. Dyn et al. [4] studied an interpolatory 4-point SS that preserving convexity and was implemented for curve construction. Kuijt and Damme [5] investigated interpolating SS on shape preserving for non-uniform data. They also provided adequate conditions for the preservation of convexity. Cai [6] formed a four-point ternary interpolating SS with a C2 limit curve that kept convexity. A binary six-point approximating SS that preserves convexity was introduced by Hao et al. [7]. Mustafa et al. [8] presented n-ary interpolating SS, which also has the scheme's convexity preservation. Floater et al. [9] provided input data monotonicity preservation with conditions that ensure differentiability to the SS's limit function With the bell-shaped mask, the shape-preserving property of ternary SS was investigated by Pitolli [10]. Mustafa and Ashraf [11] proposed a six-point interpolating ternary SS in which smoothness and differentiability of limit curves were addressed. Siddiqi and Noreen [12] investigated to maintain the convexity of 6-point ternary interpolating SS. Akram et al. [13] suggested a non-stationary binary 4-point interpolating scheme that retained monotonicity, positivity, and convexity. Mustafa and Bashir [14] treated the univariate scheme and its bivariate SS non-tensor product generalisation. The suggested schemes maintained the initial data's monotonicity.

Novara and Romani [15] presented the constraints that the interpolating 5-point ternary SS's free parameter and the vertices of a strictly convex initial polygon must meet in order for the limit curve's convexity to be preserved. Asghar et al. [16] tested SS on a probability distribution with a higher continuity and convexity was retained in the SS. Ghaffar et al. [17] introduced a class of non-stationary 2 m-point binary scheme and also discussed curvature and torsion of the limit curves of the proposed schemes. Zulkifli et al. [18] addressed the use of a new rational bi-cubic Ball function with six parameters in image interpolation, especially for grayscale images.

In 2019, Kanwal et al. [19] produced the curves, a numerical approximating collocation algorithm based on a binary 6-point approximating SS was created. The scheme is shown to generate more smooth continuous solutions to the problems. To explain the algorithm with its graphic representation, a numerical illustration was given. Also Ghaffar et al. [20] to produce the curves, a new class of $2q$-point non-stationary SSs was proposed. Again, Ghaffar et al. [21] to produce smooth curves and surfaces, a tensor product of nine-tic B-spline approximating subdivision scheme was developed. To illustrate the surface representations, numerical examples were given. For more recent work on SS one may refer to References [22–33].

We have found some benefits of the proposed schemes as compare to other schemes such as:

1. The proposed scheme preserve strict convexity inside the parametric range. Some numerical examples are given to verify and examine the conditions of convexity preserving for the 6-point ternary scheme 2.1.
2. The suggested scheme would yield results of convex interpolating curve everywhere inside the range $\mu = (0.01152, 0.01183)$. Meanwhile in the scheme proposed by Ashraf et al. [24] having parametric range $[-2, +\infty[-\{-1\}$ convexity might not be preserved on some sub-intervals such that along the whole interval. On the other hand, in the scheme of Siddiqui and Noreen (2015) [12] having parametric range $\omega = (0.0072016, 0.0090535$ does show some convexity preserving phenomena but their parameter range is less than our discovered range.
3. For convex data interpolation, the proposed scheme produces interpolating curve that lies above a straight line, but comparing with Ashraf et al. [24], the proposed scheme gives more visually pleasing results (Table 1).

Table 1 List of abbreviations and symbols

Abbreviations	Meaning
SSs	Subdivision schemes
CAGD	Computer aided geometric designs
CAD	Computer aided designs

2 Preliminaries

Here we discussed a ternary 6-point interpolating SS and its smoothness and Hölder exponent, presented by Faheem and Mustafa [1] in 2008.

2.1 Construction of Scheme

Definition 2.1 Review three recurrent relations that enhance the provided kth level polygon $f^k = \{f_i^k\}_{i\in\mathbb{Z}}$ to the (k+*1*)th level polygon

$$f^{k+1} = \{f_i^{k+1}\}_{i\in\mathbb{Z}}$$

$$\begin{cases} f_{3i}^{k+1} = f_i^k, \\ f_{3i+1}^{k+1} = a_1 f_{i-2}^k + a_2 f_{i-1}^k + a_3 f_i^k + a_4 f_{i+1}^k + a_5 f_{i+2}^k + a_6 f_{i+3}^k, \\ f_{3i+2}^{k+1} = a_6 f_{i-2}^k + a_5 f_{i-1}^k + a_4 f_i^k + a_3 f_{i+1}^k + a_2 f_{i+2}^k + a_1 f_{i+3}^k, \end{cases} \tag{2.1}$$

We get following mask from above recursive relations

$$\mathbf{a} = \{..., a_6, a_1, 0, a_5.a_2, 0, a_4, a_3, 1, a_3, a_4, 0, a_2, a_5, 0, a_1, a_6, 0, ...\}$$

This mask's Laurent Polynomial is

$$\mathbf{a}(y) = a_6 y^{-8} + a_1 y^{-7} + a_5 y^{-5} + a_2 y^{-4} + a_4 y^{-2} + a_3 y^{-1} + 1 + a_3 y + a_4 y^2 + a_2 y^4 + a_5 y^5 + a_1 y^7 + a_6 y^8 \tag{2.2}$$

To produce C^0 functions, we need $\mathbf{a}(1) = 3$, which gives us the following condition:

$$a_1 + a_2 + a_3 + a_4 + a_5 + a_6 = 1 \tag{2.3}$$

To find the mask of our scheme, we first find the Laurent Polynomial $\mathbf{a}^n(y)$ corresponding to the SS S_n, subject to the condition that the mask of $a^{(n)}(y)$. Using $n = 1, 2, 3, 4$ in [34] we get

$$
\begin{aligned}
a^{(1)} &= 3\{a_6, a_1 - a_6, -a_1, a_5 + a_6, a_2 + a_1 - a_5 - a_6, -a_2 - a_1, a_4 + a_5 + a_6, \\
&\quad 1 - 2a_4 - 2a_5 - 2a_6, a_4 + a_5 + a_6, -a_2 - a_1, a_2 + a_1 - a_5 - a_6, a_5 + a_6, \\
&\quad -a_1, a_1 - a_6, a_6\} \\
a^{(2)} &= 9\{a_6, a_1 - 2a_6, a_6 - 2a_1, a_5 + a_1 + 2a_6, a_2 - 2a_5 + 2a_1 - 4a_6, \\
&\quad -2a_2 + a_5 - 4a_1 + 2a_6, \frac{1}{3} + 2a_2 + 4a_1, -2a_2 + a_5 - 4a_1 + 2a_6, \\
&\quad a_2 - 2a_5 + 2a_1 - 4a_6, a_5 + a_1 + 2a_6, a_6 - 2a_1, a_1 - 2a_6, a_6\} \\
a^{(3)} &= 27\{a_6, a_1 - 3a_6, 3a_6 - 3a_1, a_5 + 3a_1 + 2a_6, a_2 - 3a_5 + 2a_1 - 9a_6, \\
&\quad -3a_2 + 3a_5 - 9a_1 + 9a_6, a_2 - 3a_5 + 2a_1 - 9a_6, \\
&\quad a_5 + 3a_1 + 2a_6, 3a_6 - 3a_1, a_1 - 3a_6, a_6\} \\
a^{(4)} &= 81\{a_6, a_1 - 4a_6, 6a_6 - 4a_1, a_5 + 6a_1, a_2 - 4a_5 - 15a_6, \\
&\quad a_5 + 6a_1, 6a_6 - 4a_1, a_1 - 4a_6, a_6\}
\end{aligned}
$$

where

$$
\begin{cases}
-6a_1 - 3a_2 + 3a_4 + 6a_5 + 9a_6 = 1 \\
3a_1 + a_2 + a_5 + 3a_6 = \frac{-1}{9} \\
21a_1 + 4a_2 - 5a_5 - 24a_6 = 0 \\
9a_1 + a_2 - 4a_5 - 30a_6 = 0
\end{cases}
\tag{2.4}
$$

Solving (2.3) and (2.4) by taking $a_6 = \mu$, The ternary six-point interpolating SS 's, mask (2.1) is:

$$
a_1 = -\frac{11}{81} + 13\mu, \quad a_2 = \frac{13}{27} - 51\mu, \quad a_3 = -\frac{2}{27} + 74\mu,
$$
$$
a_4 = \frac{74}{81} - 46\mu, \quad a_5 = -\frac{5}{27} + 9\mu, \quad a_6 = \mu
$$

2.2 *Smoothness of the Scheme*

Here we presents the analysis of ternary 6-point interpolating SS using Laurent polynomial method.

Theorem 2.1 *The ternary 6−point interpolating ss (2.1) is C^2 continuous on $\mu \in (\frac{1152}{100{,}000}, \frac{1183}{100{,}000})$.*

Proof The Laurent polynomial (2.2) for the mask of the SS can also be rewritten as:

$$\begin{aligned} a(y) =& \mu y^8 + \left(-\frac{11}{81} + 13\mu\right) y^7 + \left(-\frac{5}{27} + 9\mu\right) y^5 + \left(\frac{13}{27} - 51\mu\right) y^4 \\ &+ \left(\frac{74}{81} - 46\mu\right) y^2 + \left(-\frac{2}{27} + 74\mu\right) y \\ &+ 1 + \left(-\frac{2}{27} + 74\mu\right) y^{-1} + \left(\frac{74}{81} - 46\mu\right) y^{-2} \\ &+ \left(\frac{13}{27} - 51\mu\right) y^{-4} + \left(-\frac{5}{27} + 9\mu\right) y^{-5} + \left(-\frac{11}{81} + 13\mu\right) y^{-7} + \mu y^{-8}. \end{aligned}$$

From [34] the Laurent polynomial of scheme is

$$a(y) = y^{-2}(1 + y + y^2)\xi_1(y), \tag{2.5}$$

where

$$\begin{aligned} \xi_1(y) =& \mu y^8 + \left(-\frac{11}{81} + 12\mu\right) y^7 + \left(\frac{11}{81} - 13\mu\right) y^6 + \left(-\frac{5}{27} + 10\mu\right) y^5 \\ &+ \left(\frac{43}{81} - 48\mu\right) y^4 + \left(-\frac{28}{81} + 38\mu\right) y^3 \\ &+ \left(\frac{59}{81} - 36\mu\right) y^2 + \left(-\frac{37}{81} + 72\mu\right) y \\ &+ \left(\frac{59}{81} - 36\mu\right) + \left(-\frac{28}{81} + 38\mu\right) y^{-1} + \left(\frac{43}{81} - 48\mu\right) y^{-2} \\ &+ \left(-\frac{5}{27} + 10\mu\right) y^{-3} \\ &+ \left(\frac{11}{81} - 13\mu\right) y^{-4} + \left(-\frac{11}{81} + 12\mu\right) y^{-5} + \mu y^{-6}. \end{aligned}$$

From [34] using Laurent polynomial for $L = n = 1$ and (2.5), we get

$$b^{[1,1]}(y) = \frac{1}{3} a^{(1)}(y) = \frac{y^2}{1 + y + y^2} a(y) = \xi_1(y). \tag{2.6}$$

We need the Laurent polynomial $a(y)$ to maintain C^0 continuity of S and $\|\frac{1}{3}S_1\|_\infty < 1$. Then

$$\left\|\frac{1}{3}S_1\right\|_\infty = max\left\{\sum_{j\in\mathbb{Z}} |b^{[1,1]}_{i+3j}| : i = 0, 1, 2\right\},$$

This implies for $0.00411 < \mu < 0.01453$

$$\left\|\frac{1}{3}S_1\right\|_\infty = max\{\aleph_1, \aleph_2\} < 1,$$

where

$$\aleph_1 = |\mu| + \left|-\frac{5}{27} + 10\mu\right| + \left|\frac{59}{81} - 36\mu\right| + \left|-\frac{28}{81} + 38\mu\right| + \left|\frac{11}{81} - 13\mu\right|,$$
$$\aleph_2 = \left|-\frac{11}{81} + 12\mu\right| + \left|\frac{43}{81} - 48\mu\right| + \left|-\frac{37}{81} + 72\mu\right| + \left|\frac{43}{81} - 48\mu\right| + \left|-\frac{11}{81} + 12\mu\right|.$$

Therefore $\left(\frac{1}{3}S_1\right)$ converges uniformly. Hence, using Theorem from [34], S is C^0. By (2.6), the S_1 can be rewritten as

$$a^{(1)}(y) = 3y^{-2}(1 + y + y^2)\xi_2(y), \tag{2.7}$$

where

$$\begin{aligned}\xi_2(y) =& \mu y^8 + \left(-\frac{11}{81} + 11\mu\right)y^7 + \left(\frac{22}{81} - 25\mu\right)y^6 \\ &+ \left(-\frac{26}{81} + 24\mu\right)y^5 + \left(\frac{47}{81} - 47\mu\right)y^4 \\ &+ \left(-\frac{49}{81} + 61\mu\right)y^3 + \left(\frac{61}{81} - 50\mu\right)y^2 \\ &+ \left(-\frac{49}{81} + 61\mu\right)y + \left(\frac{47}{81} - 47\mu\right) \\ &+ \left(-\frac{26}{81} + 24\mu\right)y^{-1} + \left(\frac{22}{81} - 25\mu\right)y^{-2} + \left(-\frac{11}{81} + 11\mu\right)y^{-3} + \mu y^{-4}.\end{aligned}$$

From Laurent polynomial [34] for $L = 1, n = 2$ and (2.7), we have

$$b^{[2,1]}(y) = \frac{1}{3}a^{(2)}(y) = \frac{y^2}{1 + y + y^2}\, a^{(1)}(y) = 3\xi_2(y). \tag{2.8}$$

For C^1 continuity of S, we require that the Laurent polynomial $a^{(1)}(y)$ and $\|\frac{1}{3}S_2\|_\infty < 1$. The norm of scheme $\frac{1}{3}S_2$is

$$\left\|\frac{1}{3}S_2\right\|_\infty = max\left\{\sum_{j\in\mathbb{Z}} |b^{[2,1]}_{i+3j}| : i = 0, 1, 2\right\},$$

This shows that $0.01105 < \mu < 0.01337$

$$\left\|\frac{1}{3}S_2\right\|_\infty = max\,\{\aleph_3, \aleph_4\} < 1,$$

where

$$\aleph_3 = 3\left(\left|-\frac{11}{81} + 11\mu\right| + \left|\frac{47}{81} - 47\mu\right| + \left|-\frac{49}{81} + 61\mu\right| + \left|\frac{22}{81} - 25\mu\right|\right),$$
$$\aleph_4 = 3\left(2\,|\mu| + 2\left|-\frac{26}{81} + 24\mu\right| + \left|\frac{61}{81} - 50\mu\right|\right).$$

Therefore $\left(\frac{1}{3}S_2\right)$ converges uniformly. Hence, using Theorem from [34], S is C^1. By (2.7), the S_2 can again be rewritten as

$$a^{(2)}(y) = 9y^{-2}(1 + y + y^2)\xi_3(y), \tag{2.9}$$

where

$$\xi_3(y) = \mu y^8 + \left(-\frac{11}{81} + 10\mu\right)y^7 + \left(\frac{33}{81} - 36\mu\right)y^6 + \left(-\frac{48}{81} + 50\mu\right)y^5$$
$$+ \left(\frac{62}{81} - 61\mu\right)y^4 + \left(-\frac{63}{81} + 72\mu\right)y^3 + \left(\frac{62}{81} - 61\mu\right)y^2$$
$$+ \left(-\frac{48}{81} + 50\mu\right)y + \left(\frac{33}{81} - 36\mu\right) + \left(-\frac{11}{81} + 10\mu\right)y^{-1} + \mu y^{-2}.$$

And using $n = 3$, $L = 1$ and (2.9), we get

$$b^{3,1}(y) = \frac{1}{3}a^{(3)}(y) = \frac{y^2}{1 + y + y^2}\,a^{(2)}(y) = 9\xi_3(y). \tag{2.10}$$

To maintain C^2 continuity of S, we require the Laurent polynomial $a^{(2)}(y)$ and $\|\frac{1}{3}S_3\|_\infty < 1$. The norm of scheme $\frac{1}{3}S_3$is

$$\left\|\frac{1}{3}S_3\right\|_\infty = max\left\{\sum_{j\in\mathbb{Z}} |b^{3,1}_{i+3j}| : i = 0, 1, 2\right\},$$

This implies for $0.01152 < \mu < 0.01183$

$$\left\|\frac{1}{3}S_3\right\|_\infty = max\,\{\aleph_5, \aleph_6\} < 1, \tag{2.11}$$

where

$$\aleph_5 = 9\left(|\mu| + \left|-\frac{48}{81} + 50\mu\right| + \left|\frac{62}{81} - 61\mu\right| + \left|-\frac{11}{81} + 10\mu\right|\right),$$
$$\aleph_2 = 9\left(2\left|\frac{33}{81} - 36\mu\right| + \left|-\frac{63}{81} + 72\mu\right|\right).$$

Therefore $\left(\frac{1}{3}S_3\right)$ converges uniformly. Hence, using Theorem from [34], S is C^2. By (2.10), the S_3 be rewritten as

$$a^{(3)}(y) = 27y^{-2}(1 + y + y^2)\xi_4(y), \tag{2.12}$$

where

$$\begin{aligned}\xi_4(y) =& \mu y^8 + \left(-\frac{11}{81} + 9\mu\right)y^7 + \left(\frac{44}{81} - 46\mu\right)y^6 + (-1 + 87\mu)y^5 \\ &+ \left(\frac{33}{27} - 102\mu\right)y^4 + (-1 + 87\mu)y^3 + \left(\frac{44}{81} - 46\mu\right)y^2 \\ &+ \left(-\frac{11}{81} + 9\mu\right)y + \mu.\end{aligned}$$

For C^3 continuity, we require that $a^{(3)}(y)$, but there is no common range of μ for which $\|\frac{1}{3}S_4\|_\infty < 1$ is true. Therefore scheme S is not C^3 continuous.

2.3 Hölder Exponent

Here, we generalize the smoothness based on Hassan et al. [34] in generalize form. We conclude that

$$R_H = 3 + \vartheta^\eta, \quad \forall \eta \geq 1, \tag{2.13}$$

whereas ϑ^η is written as

$$3^{-\eta\vartheta^\eta} = \left\|\left(\frac{1}{3}S_3\right)^\eta\right\|_\infty. \tag{2.14}$$

For the sake of computing efficiency, we take $\eta = 1$. Then using (2.11), we get

$$\left\|\frac{1}{3}S_3\right\|_\infty = \begin{cases}\aleph_5, & 0.01152 < \mu \leqslant \frac{125}{10692}, \\ \aleph_6, & \frac{125}{10692} \leqslant \mu < 0.01183.\end{cases} \tag{2.15}$$

we obtain that Hölder regularity against μ of the ternary 6-point scheme is

$$R(\mu) = \begin{cases} 2 - \log_3(\aleph_5), \ 0.01152 < \mu \leqslant \frac{125}{10692}, \\ 2 - \log_3(\aleph_6), \ \frac{125}{10692} \leqslant \mu < 0.01183. \end{cases} \tag{2.16}$$

We can observe that the ternary 6-point scheme achieves the maximum smoothness at $\mu = \frac{125}{10,692}$, and its Holder exponent is $R_H = R(\frac{125}{10,692}) = 2.1827$.

3 Basic Features

The symmetry of the limit function, approximation order and support of our scheme are all discussed in this section.

3.1 Approximation Order

Here,we prove that the proposed scheme's approximation order is four.

Theorem 3.1 *The proposed 6-point ternary SS reproduces all the cubic polynomials for $\mu \in (0.01152, 0.01183)$ and quartic at $\mu = \frac{125}{10,692}$.*

Theorem 3.2 *The SS introduced in (2.1) is exact for cubic.*

Proof It is sufficient to check that the scheme is exact for the monomials x^j, $j = 0, 1, 2, 3$ using its linearity.

$$Let \ \ f_i^k = \left(\frac{i}{3^k}\right)^0, \quad i \in \mathbb{Z}, then$$

$$f_{3i}^{k+1} = \left(\frac{3i}{3^{k+1}}\right)^0, \ f_{3i+1}^{k+1} = \left(\frac{3i+1}{3^{k+1}}\right)^0, \ f_{3i+2}^{k+1} = \left(\frac{3i+2}{3^{k+1}}\right)^0$$

This means that the scheme is exact in terms of constants.

$$Consider \ \ f_i^k = \left(\frac{i}{3^k}\right)^1, \quad i \in \mathbb{Z}, then$$

$$
\begin{aligned}
f_{3i}^{k+1} &= \left(\frac{3i}{3^{k+1}}\right)^1, \\
f_{3i+1}^{k+1} &= \mu_1\left(\frac{i-2}{3^k}\right)^1 + \mu_2\left(\frac{i-1}{3^k}\right)^1 + \mu_3\left(\frac{i}{3^k}\right)^1 \\
&\quad + \mu_4\left(\frac{i+1}{3^k}\right)^1 + \mu_5\left(\frac{i+2}{3^k}\right)^1 + \mu_6\left(\frac{i+3}{3^k}\right)^1 \\
&= \left(\frac{3i+1}{3^{k+1}}\right)^1, \\
f_{3i+2}^{k+1} &= \mu_6\left(\frac{i-2}{3^k}\right)^1 + \mu_5\left(\frac{i-1}{3^k}\right)^1 + \mu_4\left(\frac{i}{3^k}\right)^1 \\
&\quad + \mu_3\left(\frac{i+1}{3^k}\right)^1 + \mu_2\left(\frac{i+2}{3^k}\right)^1 + \mu_1\left(\frac{i+3}{3^k}\right)^1 \\
&= \left(\frac{3i+2}{3^{k+1}}\right)^1.
\end{aligned}
$$

As a result, for linear, the scheme is exact. And, we can observe that the scheme is both quadratic and cubic.

Theorem 3.3 *A SS (2.1) has approximation order 4.*

3.2 Basic Limit Function

The basic limit function of a SS for the following data at $m = 0$,

$$
f_i^0 = \begin{cases} 1, \ i = m, \\ 0, \ i \neq m \end{cases} \tag{3.1}
$$

By Theorem (2.1) it shows that the basis function of the scheme (2.1) belongs to $C^2(\mathbb{R})$.

We can now prove that it is symmetric around the Y-axis and has a compact support over the $[-4, 4]$ interval.

Theorem 3.4 *Around the Y axis, the basic limit function is symmetric.*

Proof Let

$$
D_n = \left\{\frac{j}{3^n}, j \in \mathbb{Z}, n \in \mathbb{Z}_+\right\}.
$$

The simple limit function F is then restricted to Dn, which is satisfying

$$F\left(\frac{j}{3^n}\right) = f_j^n, \tag{3.2}$$

We'll show that symmetry appears as

$$F\left(\frac{j}{3^n}\right) = F\left(\frac{-j}{3^n}\right), \quad \forall n \in \mathbb{Z}_+.$$

We will show it using induction on n. For $n = 0$, using the interpolation property

$$F(j) = F(-j), \quad \forall j \in \mathbb{Z},$$

suppose that

$$F\left(\frac{j}{3^n}\right) = F\left(\frac{-3j}{3^n}\right), \quad \forall j \in \mathbb{Z}. \tag{3.3}$$

By (2.1), we have

$$F\left(\frac{3j+1}{3^{n+1}}\right) = f_{3j+1}^{n+1} = \mu_1 F\left(\frac{j-2}{3^n}\right) + \mu_2 F\left(\frac{j-1}{3^n}\right) + \mu_3 F\left(\frac{j}{3^n}\right)$$
$$+\mu_4 F\left(\frac{j+1}{3^n}\right) + \mu_5 F\left(\frac{j+2}{3^n}\right) + \mu_6 F\left(\frac{j+3}{3^n}\right).$$

By (2.1) and (3.3), we get

$$F\left(\frac{3j+1}{3^{n+1}}\right) = f_{-3j-1}^{n+1} = F\left(-\frac{3j+2}{3^{n+1}}\right)$$

In same way,

$$F\left(\frac{3j+2}{3^{n+1}}\right) = f_{-3j-2}^{n+1} = F\left(-\frac{3j+2}{3^{n+1}}\right)$$

Hence,

$$F\left(\frac{j}{3^n}\right) = F\left(-\frac{j}{3^n}\right), \quad \forall j \in \mathbb{Z} \ \ and \ \ \forall n \in \mathbb{Z}_+.$$

From the smoothness of F, we get

$$F(x) = F(-x), \quad \forall x \in \mathbb{R}.$$

Hence proved.

4 Convexity Preservation of the Scheme

Convexity preserving is also an important geometric property in curve shaping. Here, we will prove the convexity preservation of the SS (2.1).

A set of initial control points is given $\{P_i^0\}_{i\in\mathbb{Z}}, P_i^0 = \{(i, f_i^0) : i \in [-2, 3^{-k}n + 2]\}$, which are strictly convex that is $d_i^0 > 0$, and here $i \in [-2, n+2]$. The notation for second-order divided differences is $d_i^k = 3^{2k}(f_{i-1}^k - 2f_i^k + f_{i+1}^k)$.

As a result, the SS (2.1) can be rewritten as follows in terms of second order divided differences

$$d_{3i}^{k+1} = 9\mu d_{i-2}^k + \left(\frac{-26}{9} + 216\mu\right) d_{i-1}^k + \left(\frac{61}{9} - 450\mu\right) d_i^k + \left(\frac{-26}{9} + 216\mu\right) d_{i+1}^k + 9\mu d_{i+2}^k \tag{4.1}$$

$$d_{3i+1}^{k+1} = \left(\frac{22}{9} - 225\mu\right) d_{i-1}^k + \left(\frac{-49}{9} + 549\mu\right) d_i^k + \left(\frac{47}{9} - 423\mu\right) d_{i+1}^k + \left(\frac{-11}{9} + 99\mu\right) d_{i+2}^k \tag{4.2}$$

$$d_{3i+2}^{k+1} = \left(\frac{-11}{9} + 99\mu\right) d_{i-1}^k + \left(\frac{47}{9} - 423\mu\right) d_i^k + \left(\frac{-49}{9} + 549\mu\right) d_{i+1}^k + \left(\frac{22}{9} - 225\mu\right) d_{i+2}^k \tag{4.3}$$

The following are the convexity-preserving conditions for the SS (2.1).

Theorem 4.1 *Let $r^k = max_i\{p_i^k, q_i^k\}$, where $p_i^k = \frac{d_{i+1}^k}{d_i^k}$ and $q_i^k = \frac{1}{p_i^k} = \frac{d_i^k}{d_{i+1}^k}$. If the initial control points $P^0 = \{(i, f_i) : i \in [-2, n+2]\}$ are strictly convex that is $d_i^k > 0$, then the parameter μ satisfies $\mu \in (0.01152, 0.01183)$ and $r^0 < \lambda$ where*

$$1 < \lambda < \frac{-49 + 4941\mu}{52 - 4050\mu},$$

then for $k \geq 1$ and $r^k < \lambda$ that is the 6-point ternary interpolating SS (2.1) will be convexity preserving.

Proof From the definition of p_i^k and q_i^k and condition (5) we have immediately

$$\frac{1}{\lambda} < r^0 < \lambda \tag{4.4}$$

Since

$$1 < \lambda < \frac{-49 + 4941\mu}{52 - 4050\mu},$$

This implies that $\lambda > 1$, in order for inequality (4.4) to hold, it must satisfy this condition.

Mathematical induction is employed to prove $di^k > 0$ and $r^k < \lambda$. Of course, for $k = 0$ this statement is valid. For any $k \geq 0$, let $di^k > 0$ and $r^k < \lambda$, for the proof that an inductive part exists. Using the Eq. (4.1)

$$\begin{aligned}
d_{3i}^{k+1} &= 9\mu d_{i-2}^k + \left(\frac{-26}{9} + 216\mu\right) d_{i-1}^k + \left(\frac{61}{9} - 450\mu\right) d_i^k \\
&+ \left(\frac{-26}{9} + 216\mu\right) d_{i+1}^k + 9\mu d_{i+2}^k \\
&= d_i^k \left[9\mu \frac{d_{i-2}^k}{d_i^k} + \left(\frac{-26}{9} + 216\mu\right) \frac{d_{i-1}^k}{d_i^k} + \left(\frac{61}{9} - 450\mu\right) + \left(\frac{-26}{9} + 216\mu\right) \frac{d_{i+1}^k}{d_i^k} + 9\mu \frac{d_{i+2}^k}{d_i^k}\right] \\
&= d_i^k \left[9\mu q_{i-2}^k + \left(\frac{-26}{9} + 216\mu\right) q_{i-1}^k + \left(\frac{61}{9} - 450\mu\right) + \left(\frac{-26}{9} + 216\mu\right) p_i^k + 9\mu p_{i+1}^k\right] \\
&> d_i^k \left[9\mu\lambda + \left(\frac{-26}{9} + 216\mu\right) \lambda + \left(\frac{61}{9} - 450\mu\right) + \left(\frac{-26}{9} + 216\mu\right) \lambda + 9\mu\lambda\right] \\
&> d_i^k \left[2\left(\frac{-26}{9} + 225\mu\right) \lambda + \left(\frac{61}{9} - 450\mu\right)\right] \\
&> d_i^k \left[\frac{110}{9} - 999\mu\right] > 0
\end{aligned} \tag{4.5}$$

Which shows that for $1 < \lambda < \frac{-49+4941\mu}{52-4050\mu}$, $d_{3i}^{k+1} > 0$.
Similarly it can be shown that $d_{3i+1}^{k+1} > 0$ and $d_{3i+2}^{k+1} > 0$ for $1 < \lambda < \frac{-49+4941\mu}{52-4050\mu}$.
Thus, $d_i^{k+1} > 0, \forall k > 0$.
Now we show $r^{k+1} < \lambda$.
Since

$$\begin{aligned}
p_{3i}^{k+1} &= \frac{d_{3i+1}^{k+1}}{d_{3i}^{k+1}} \\
&= \frac{(\frac{22}{9} - 225\mu)d_{i-1}^k + (\frac{-49}{9} + 549\mu)d_i^k + (\frac{47}{9} - 423\mu)d_{i+1}^k + (\frac{-11}{9} + 99\mu)d_{i+2}^k}{9\mu d_{i-2}^k + (\frac{-26}{9} + 216\mu)d_{i-1}^k + (\frac{61}{9} - 450\mu)d_i^k + (\frac{-26}{9} + 216\mu)d_{i+1}^k + 9\mu d_{i+2}^k} \\
&= \frac{(\frac{22}{9} - 225\mu)q_{i-1}^k + (\frac{-49}{9} + 549\mu) + (\frac{47}{9} - 423\mu)p_i^k + (\frac{-11}{9} + 99\mu)p_{i+1}^k}{9\mu q_{i-2}^k + (\frac{-26}{9} + 216\mu)q_{i-1}^k + (\frac{61}{9} - 450\mu) + (\frac{-26}{9} + 216\mu)p_i^k + 9\mu p_{i+1}^k}
\end{aligned}$$

we have

$$\begin{aligned}
&p_{3i}^{k+1} - \lambda \\
&= \frac{1}{9\mu q_{i-2}^k + \left(\frac{-26}{9} + 216\mu\right) q_{i-1}^k + \left(\frac{61}{9} - 450\mu\right) + \left(\frac{-26}{9} + 216\mu\right) p_i^k + 9\mu p_{i+1}^k} \\
&\times \left[\left(\frac{22}{9} - 225\mu\right) q_{i-1}^k + \left(\frac{-49}{9} + 549\mu\right) + \left(\frac{47}{9} - 423\mu\right) p_i^k\right.
\end{aligned}$$

$$+\left(\frac{-11}{9}+99\mu\right)p_{i+1}^k-9\mu\lambda q_{i-2}^k-\left(\frac{-26}{9}+216\mu\right)\lambda q_{i-1}^k$$
$$-\left(\frac{61}{9}-450\mu\right)\lambda-\left(\frac{-26}{9}+216\mu\right)\lambda p_i^k-9\mu\lambda p_{i-1}^k\Big]$$

$$p_{3i}^{k+1}-\lambda=\frac{N}{D} \tag{4.6}$$

As *Denomenator* $= D = d_{3i}^{k+1} > 0$, we have to show that $N < 0$, for this consider

$$\begin{aligned}
N <&\left(\frac{22}{9}-225\mu\right)q_{i-1}^k+\left(\frac{-49}{9}+549\mu\right)+\left(\frac{47}{9}-423\mu\right)p_i^k+\left(\frac{-11}{9}+99\mu\right)p_{i+1}^k\\
&-9\mu\lambda q_{i-2}^k-\left(\frac{-26}{9}+216\mu\right)\lambda q_{i-1}^k-\left(\frac{61}{9}-450\mu\right)\lambda-\left(\frac{-26}{9}+216\mu\right)\lambda p_i^k-9\mu\lambda p_{i-1}^k\\
=&\left(\frac{22}{9}-225\mu\right)\lambda+\left(\frac{-49}{9}+549\mu\right)+\left(\frac{47}{9}-423\mu\right)\lambda+\left(\frac{-11}{9}+99\mu\right)\lambda\\
&-9\mu\lambda^2-\left(\frac{-26}{9}+216\mu\right)\lambda^2-\left(\frac{61}{9}-450\mu\right)\lambda-\left(\frac{-26}{9}+216\mu\right)\lambda^2-9\mu\lambda^2\\
=&-2\left(\frac{-26}{9}+225\mu\right)\lambda^2+\left(\frac{-3}{9}-99\mu\right)\lambda+\left(\frac{-49}{9}+549\mu\right)\\
=&[\lambda-1]\left[\lambda+\frac{49-4941\mu}{52-4050\mu}\right]=0
\end{aligned}$$

that proves that $N < 0$, for $1 < \lambda < \frac{-49+4941\mu}{52-4050\mu}$.
Therefore, $p_{3i}^{k+1}-\lambda=\frac{N}{D}<0$
i.e.,$p_{3i}^{k+1} < \lambda$.
Similarly, it can be shown that $p_{3i+1}^{k+1} < \lambda$ and $p_{3i+2}^{k+1} < \lambda$ for $1 < \lambda < \frac{-49+4941\mu}{52-4050\mu}$.
Thus,

$$p_i^{k+1} < \lambda, \forall k \geq 0. \tag{4.7}$$

To show $q_{3i}^{k+1} < \lambda$, consider

$$\begin{aligned}
q_{3i}^{k+1} &= \frac{d_{3i}^{k+1}}{d_{3i+1}^{k+1}}\\
&=\frac{9\mu d_{i-2}^k+\left(\frac{-26}{9}+216\mu\right)d_{i-1}^k+\left(\frac{61}{9}-450\mu\right)d_i^k+\left(\frac{-26}{9}+216\mu\right)d_{i+1}^k+9\mu d_{i+2}^k}{\left(\frac{22}{9}-225\mu\right)d_{i-1}^k+\left(\frac{-49}{9}+549\mu\right)d_i^k+\left(\frac{47}{9}-423\mu\right)d_{i+1}^k+\left(\frac{-11}{9}+99\mu\right)d_{i+2}^k}\\
&=\frac{9\mu q_{i-2}^k+\left(\frac{-26}{9}+216\mu\right)q_{i-1}^k+\left(\frac{61}{9}-450\mu\right)+\left(\frac{-26}{9}+216\mu\right)p_i^k+9\mu p_{i+1}^k}{\left(\frac{22}{9}-225\mu\right)q_{i-1}^k+\left(\frac{-49}{9}+549\mu\right)+\left(\frac{47}{9}-423\mu\right)p_i^k+\left(\frac{-11}{9}+99\mu\right)p_{i+1}^k}
\end{aligned}$$

we have

$$\begin{aligned}
&q_{3i}^{k+1}-\lambda\\
&=\frac{1}{\left(\frac{22}{9}-225\mu\right)q_{i-1}^k+\left(\frac{-49}{9}+549\mu\right)+\left(\frac{47}{9}-423\mu\right)p_i^k+\left(\frac{-11}{9}+99\mu\right)p_{i+1}^k}
\end{aligned}$$

$$\times\left[9\mu q_{i-2}^k+\left(\frac{-26}{9}+216\mu\right)q_{i-1}^k+\left(\frac{61}{9}-450\mu\right)+\left(\frac{-26}{9}+216\mu\right)p_i^k+9\mu p_{i+1}^k\right.$$
$$-\left(\frac{22}{9}-225\mu\right)\lambda q_{i-1}^k-\left(\frac{-49}{9}+549\mu\right)\lambda-\left(\frac{47}{9}-423\mu\right)\lambda p_i^k$$
$$\left.-\left(\frac{-11}{9}+99\mu\right)\lambda p_{i+1}^k\right]$$
$$q_{3i}^{k+1}-\lambda=\frac{N}{D} \tag{4.8}$$

As *Denominator* $= D = d_{3i+1}^{k+1} > 0$, we have to show that $N < 0$, for this consider

$$\begin{aligned}
N <&\frac{9\mu}{\lambda}+\frac{(\frac{-26}{9}+216\mu)}{\lambda}+\left(\frac{61}{9}-450\mu\right)+\frac{(\frac{-26}{9}+216\mu)}{\lambda}+\frac{9\mu}{\lambda}\\
&-\left(\frac{22}{9}-225\mu\right)-\left(\frac{-49}{9}+549\mu\right)\lambda-\left(\frac{47}{9}-423\mu\right)-\left(\frac{-11}{9}+99\mu\right)\\
&=-\frac{1}{\lambda}\left[\left(\frac{-49}{9}+549\mu\right)\lambda^2-\left(\frac{3}{9}+99\mu\right)\lambda-\left(\frac{-52}{9}+450\mu\right)\right]\\
&=-\frac{1}{\lambda}\left[\lambda-1\right]\left[\lambda+\frac{52-4050\mu}{49-4941\mu}\right]=0
\end{aligned}$$

that proves that $N < 0$, for $1 < \lambda < \frac{-49+4941\mu}{52-4050\mu}$.
Therefore, $p_{3i}^{k+1}-\lambda=\frac{N}{D}<0$
i.e. $q_{3i}^{k+1} < \lambda$.
Similarly, it can be shown that $q_{3i+1}^{k+1} < \lambda$ and $q_{3i+2}^{k+1} < \lambda$ for $1 < \lambda < \frac{-49+4941\mu}{52-4050\mu}$.
Thus,

$$q_i^{k+1} < \lambda, \forall k \geq 0. \tag{4.9}$$

From Eqs. (4.7) and (4.9), it can be combined as

$$r_i^{k+1} < \lambda$$

This completes the proof (Table 2).

5 Numerical Examples

Here, we demonstrate the efficiency of the projected scheme (2.1) on various open and closed polygons. Figures 1 and 2a–d represents the initial polygons and their refined shapes for open curves that we obtain after applying subdivision rules to the

Table 2 Comparison of proposed schemes with other interpolating schemes

Subdivision scheme	Type	Convexity type	Approximation order	Support
4-point ternary[25]	Interpolating	No	3	5
4-point ternary [27]	Interpolating	Yes	3	5
6-point binary [30]	Interpolating	No	5	10
6-point ternary [28]	Interpolating	Yes	5	3
6-point (Proposed) [1]	Interpolating	Yes	3	8

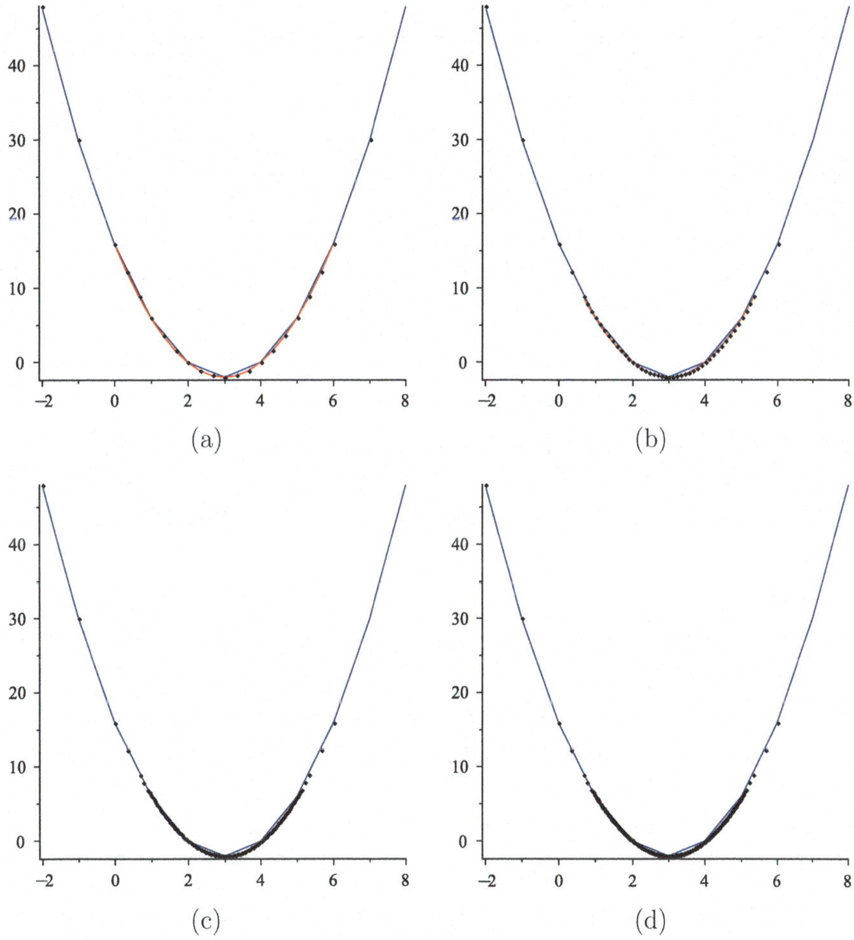

Fig. 1 **a** Demonstrate initial polygon. After applying first and 2nd subdivision step we have (**b**) and (**c**). While result of 3rd subdivision step is illustrated in (**d**)

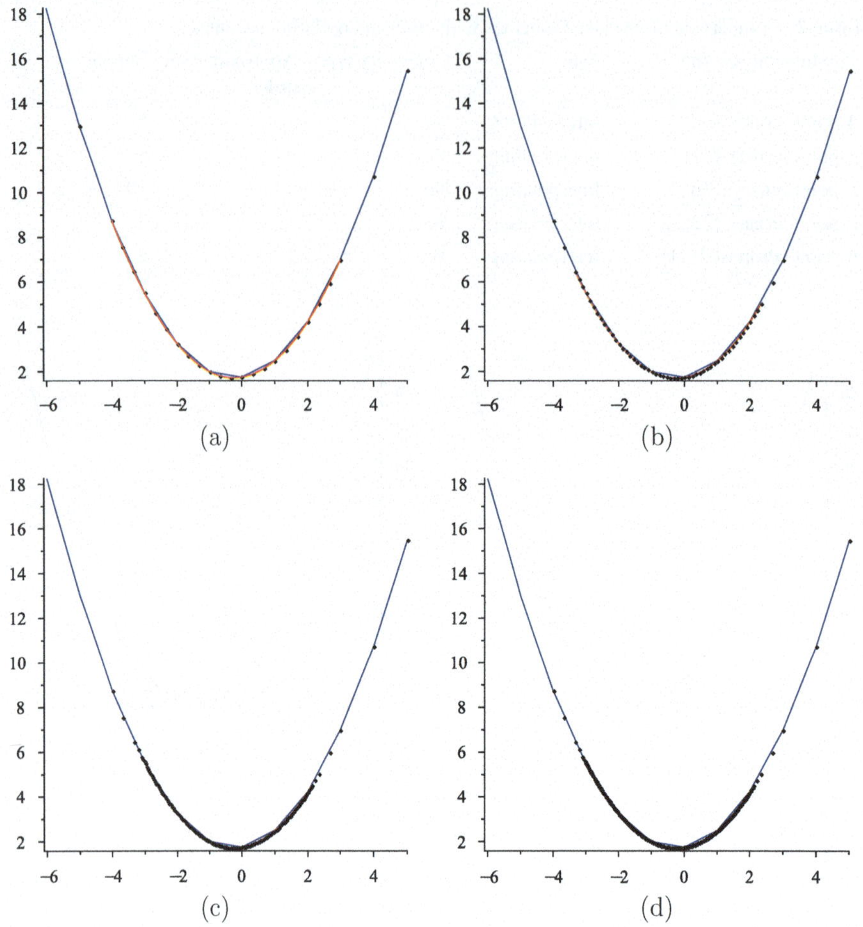

Fig. 2 **a** Demonstrate initial polygon. After applying first and 2nd subdivision step we have (**b**) and (−). While result of 3rd subdivision step is illustrated in (**d**)

scheme (2.1), having parametric value $\mu = 1152/100,000$ using strictly convex data sets (see Tables 3 and 4). But in Fig. 3a–d, we illustrate the initial polygons and their refined form,and the curve produced with parametric value $\mu = 1190/100,000$ using convex data set (Not strictly convex curve) (see Table 5). The initial closed polygon and the limit curve created by our proposed scheme are shown in Fig. 4 choosing $\mu = 1152/100,000$, that we obtain after applying refined subdivision rules to the Scheme (2.1) (Table 6).

Table 3 Strictly convex data set used for Fig. 1

x	−2	−1	0	1	2	3	4	5	6	7	8
y	48	30	16	6	0	−2	0	6	16	30	48

Table 4 Strictly convex data set used for Fig. 1

x	5	4	3	2	1	0	−1	−2	−3	−4	−5	−6
y	15.5	10.75	7	4.25	2.5	1.75	2	3.25	5.5	8.75	13	18.25

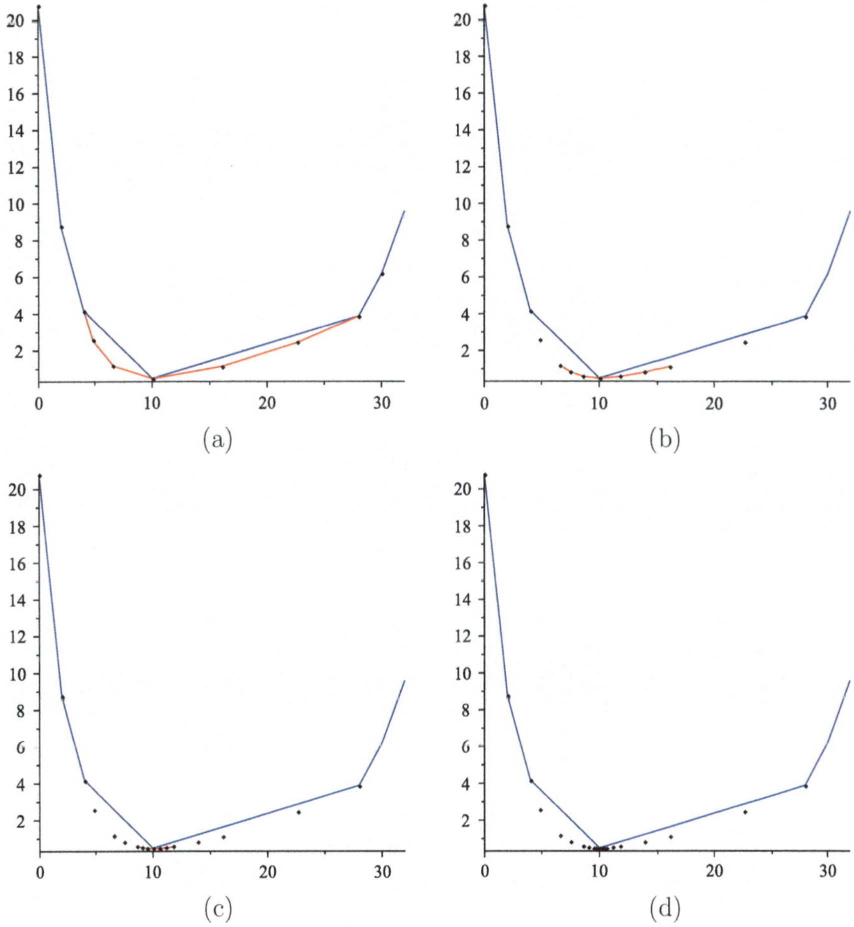

Fig. 3 **a** Demonstrate initial polygon. After applying first and 2nd subdivision step we have (**b**) and (**c**). While result of 3rd subdivision step is illustrated in (**d**)

Table 5 Convex data set used for Fig. 3

x	0	2	4	10	28	30	32
y	20.8	8.8	4.2	0.5	3.9	6.2	9.6

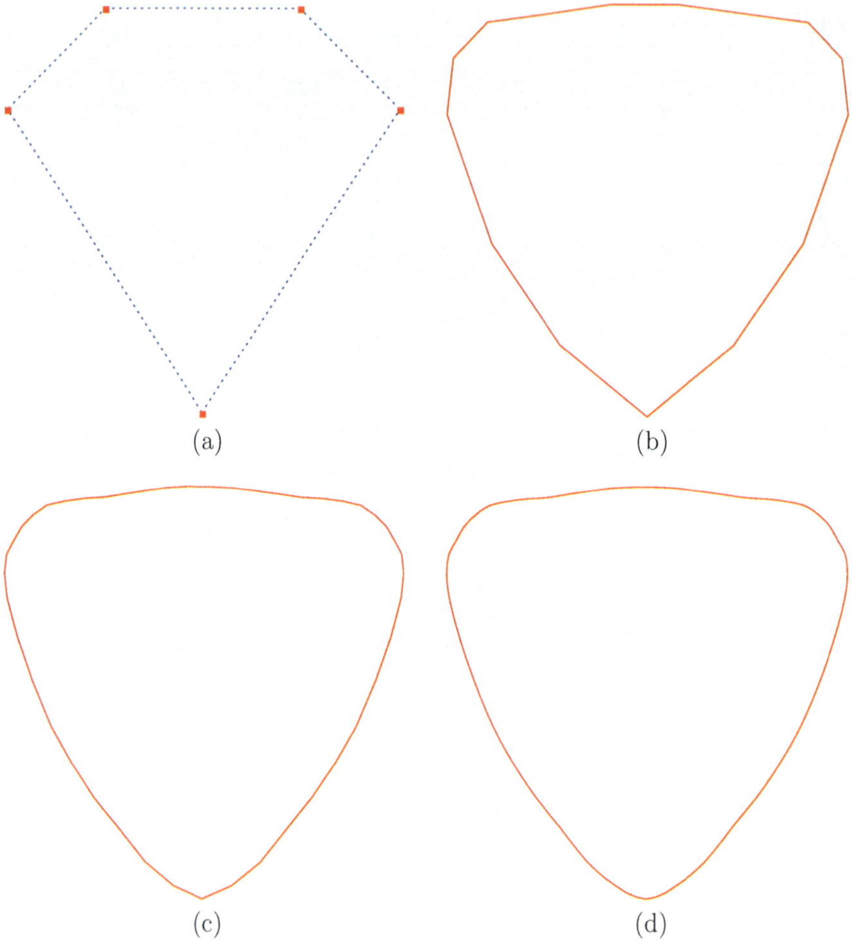

Fig. 4 **a** Demonstrate initial polygon. After applying first and 2nd subdivision step we have (**b**) and (**c**). While result of 3rd subdivision step is illustrated in (**d**)

Table 6 Data set used for Fig. 4

x	3	5	4	2	1	3
y	0	3	4	4	3	0

6 Conclusions

In this chapter, we have presented some fundamental properties of the 6-point ternary interpolating SS, which improves on the scheme in various ways that meet different requirements. The generated figures displays that the convexity-preservation of the limit curve is a useful method for modifying the proposed scheme. We have shown that by taking initial convex data, the limit curves created by the proposed scheme are also strictly convex inside the parametric range $\mu = (0.01152, 0.01183)$. As can be seen in Fig. 3a–d the scheme can not produced strictly convex curves if parametric value $\mu = 1190/100, 000$ lies outside the parametric range. As shown, the proposed scheme for various choice of shape control parameters can be considered more universal. Several examples are given which support our findings. Finally, considering scattered data and applying shape preserving properties on SS is a good idea.

Funding This work is fully supported by Universiti Teknologi PETRONAS (UTP) and Ministry of Education, Malaysia through FRGS/1/2018/STG06/UTP/03/1/015MA0020 (New rational quartic spline for image refinement) and UTP through a research grant YUTP: 015LC0-315 [Uncertainty estimation based on Quasi-Newton methods for Full Waveform Inversion (FWI)]. The first author is supported by Graduate Research Assistant Scheme (GRA) to pursue his Ph.D. study at Universiti Teknologi PETRONAS (UTP).

References

1. Faheem, K., Mustafa, G.: Ternary six-point interpolating SS. Lobachevskii J. Math. **29**(3), 153–163 (2008)
2. Yadshalom, I.: Monotonicity preserving SSs. J. Approx. Theory **74**, 41–58 (1993)
3. Carnicer, J.M., Dahmen, W.: Characterization of local strict convexity preserving interpolation methods by C1 functions. J. Approx. Theory **77**, 2–30 (1994)
4. Dyn, N., Kuijt, F., Levin, D., Van Damme, R.: Convexity preservation of the four-point interpolatory SS. Comput. Aided Geom. Des. **16**, 789–792 (1999)
5. Kuijt, F., Van Damme, R.: Shape preserving interpolatory SSs for non-uniform data. J. Approx. Theory **114**, 1–32 (2002)
6. Cai, Z.: Convexity preservation of the interpolating four-point C2 ternary stationary SS. Comput. Aided Geom. Des. **26**, 560–565 (2009)
7. Hao, Y.X., Wang, R.H., Li, C.J.: Analysis of a 6-point binary SS. Appl. Math. Comput. **218**, 3209–3216 (2011)
8. Mustafa, G., Deng, J., Ashraf, P., Rehman, N.A.: The mask of odd points n-ary interpolating SS. J. Appl. Math. (2012)

9. Floater, M., Beccari, C., Cashman, T., Romani, L.: A smoothness criterion for monotonicity preserving subdivision. Adv. Comput. Math. **39**, 193–204 (2013)
10. Pitolli, F.: Ternary shape-preserving SSs. Math. Comput. Simul. **106**, 185–194 (2014)
11. Mustafa, G., Ashraf, P.: A new 6-point ternary interpolating SS and its differentiability. J. Inf. Comput. Sci. **5**, 199–210 (2010)
12. Siddiqi, S.S., Noreen, T.: Convexity preservation of six point C2 interpolating SS. Appl. Math. Comput. **265**, 936–944 (2015)
13. Akram, G., Bibi, K., Rehan, K., Siddiqi, S.S.: Shape preservation of 4-point interpolating non-stationary SS. J. Comput. Appl. Math. **319**, 480–492 (2017)
14. Mustafa, G., Bashir, R.: Univariate approximating schemes and their non-tensor product generalization. Open Math. **16**, 1501–1518 (2018)
15. Novara, P., Romani, L.: On the interpolating 5-point ternary SS: a revised proof of convexity-preservation and an application-oriented extension. Math. Comput. Simul. **147**, 194–209 (2018)
16. Asghar, M., Iqbal, M.J., Mustafa, G.: A family of high continuity SSs based on probability distribution. Mehran Univ. Res. J. Eng. Technol. **38**, 389 (2019)
17. Ghaffar, A., Ullah, Z., Bari, M., Nisar, K.S., Al-Qurashi, M.M., Baleanu, D.: A new class of 2m-point binary non-stationary SSs. Adv. Differ. Equ. **2019**, 325 (2019)
18. Zulkifli, N.A.B., Karim, S.A.A., Sarfraz, M., Ghaffar, A., Nisar, K.S.: Image interpolation using a rational bi-cubic ball. Mathematics **7**, 1045 (2011)
19. Kanwal, G., Ghaffar, A., Hafeezullah, M.M., Manan, S.A., Rizwan, M., Rahman, G.: Numerical solution of 2-point boundary value problem by subdivision scheme. Commun. Math. Appl. **10**(2), 1–11 (2019)
20. Ghaffar, A., Bari, M., Ullah, Z., Iqbal, M., Nisar, K.S., Baleanu, D.: A new class of 2q-point nonstationary subdivision schemes and their applications. Mathematics **7**(7), 639 (2019)
21. Ghaffar, A., Iqbal, M., Bari, M., Muhammad Hussain, S., Manzoor, R., Sooppy Nisar, K., Baleanu, D.: Construction and application of nine-Tic B-spline tensor product SS. Mathematics **7**(8), 675 (2019)
22. Ashraf, P., Sabir, M., Ghaffar, A., Nisar, K.S., Khan, I.: Shape-preservation of ternary four-point interpolating non-stationary SS. Front. Phys. **7**, 241 (2020)
23. Iqbal M., Karim S.A.A., Sarfraz M.: Construction and application of Septic B-spline tensor product scheme. In: Abdul Karim S.A., Saad N., Kannan R. (eds) Advanced Methods for Processing and Visualizing the Renewable Energy. Studies in Systems, Decision and Control, vol 320. Springer, Singapore (2021)
24. Ashraf, P., Nawaz, B., Baleanu, D., Sooppy Nisar, K., Ghaffar, A., Khan, M.A.A., Akram, S.: Analysis of geometric properties of ternary four-point rational interpolating subdivision scheme. Mathematics **8**(3), 338 (2020)
25. Beccari, C., Casciola, G., Romani, L.: An interpolating 4-point C2 ternary non-stationary SS with tension control. Comput. Aided Geometr. Des. **24**(4), 210–219 (2007)
26. Ghaffar, A., Ullah, Z., Bari, M., Nisar, K.S., Baleanu, D.: Family of odd point non-stationary SSs and their applications. Adv. Differ. Equ. **171**, 1–20 (2019)
27. Siddiqi, S.S., Ahmad, N., Rehan, K.: Ternary 4-point interpolating scheme for curve sketching. Res. J. Appl. Sci. Eng. Tech **6**, 1556–1561 (2013)
28. Siddiqi, S.S., Noreen, T.: Convexity preservation of six point C^2 interpolating subdivision scheme. Appl. Math. Comput. **265**, 936–944 (2015)
29. Shahzad, A., Khan, F., Ghaffar, A., Mustafa, G., Nisar, K.S., Baleanu, D.: A novel numerical algorithm to estimate the subdivision depth of binary SSs. Symmetry **12**, 66 (2020)
30. Weissman, A.: A 6-Point Interpolatory SS for Curve Design. University of Tel-Aviv (1989)
31. Zou, L., Song, L., Wang, X., Chen, Y., Zhang, C., Tang, C.: Bivariate Thiele-like rational interpolation continued fractions with parameters based on virtual points. Mathematics **8**, 71 (2020)
32. Peng, K., Tan, J., Li, Z., Zhang, L.: Fractal behavior of a ternary 4-point rational interpolation SS. Math. Comput. Appl. **23**, 65 (2018)

33. Shang, Y.: Lack of Gromov-hyperbolicity in small-world networks. Cent. Eur. J. Math. **10**, 1152–1158 (2011)
34. Hassan Mohamed, F., Ivrissimitzis, I.P., Dodgson, N.A. , Sabin, M.A.: An interpolating 4-point C2 ternary stationary SS. Comput. Aided Geometr. Des. **19**(1), 1–18 (2002)

Pandemic Measures for the COVID19 Outbreak Control in Malaysia: A Correlative Comparison with China

Chockalingam Aravind Vaithilingam, S. Gayathri, R. Lalitha, R. R. Rajalakshmi, and Arthanari Jagadeeshwaran

Abstract The 2019 novel coronavirus (2019-nCoV) outbreak is declared as a pandemic by the World Health Organization. This chapter presents a simplified approach of the 2019-nCoV outbreak in Malaysia, based on a simple mathematical model and limited reference data. The profound model predictions is based on the actual data on the date of confirmation excluding deaths, considering the recovered will have the possibility to get infected again. The 14 days incubation characteristics are used in the computations as pronounced by CDC to improve the prediction characteristics. This includes the four stages of recovery characteristics in any pandemic cases towards cluster segregation, contact tracing towards flattening the growth curve. The model from china was taken as reference and the Malaysian recovery phase analyses and compared with the measures in place. The computational approach for the dataset available is presented and the similarity measure is a good reference point in handling the pandemic of this size in future.

Keywords Pandemic · COVID19 · nCOVID · Growth factor · Recovery factor · Flatten-curve

---

C. A. Vaithilingam (✉)
High Impact Research Laboratories, Taylor's University, Subang Jaya, Malaysia
e-mail: aravindcv@ieee.org

S. Gayathri
Xavier Institute of Management & Entrepreneurship, Chennai, India

R. Lalitha
Computer Science and Engineering, Rajalakshmi Institute of Technology, Chennai, India
e-mail: lalitha.r@ritchennai.edu.in

R. R. Rajalakshmi
Computer Sciences and Engineering, Kongu Engineering College, Erode, India
e-mail: rrr@kongu.ac.in

A. Jagadeeshwaran
Sona College of Technology, Salem, India

© Institute of Technology PETRONAS Sdn Bhd 2022

S. A. Abdul Karim and A. Shafie (eds.), *Towards Intelligent Systems Modeling and Simulation*, Studies in Systems, Decision and Control 383,
https://doi.org/10.1007/978-3-030-79606-8_2

1 Introduction

1.1 Escalating the COVID19 Scenario

Animal to human transmissions through a host or a series of host transfer, lead to several outbreaks with the earlier pandemics from the last century [1]. The new type of Severe Acute Respiratory Syndrome (SARS) family novel coronavirus (2019-nCoV) that out broke earlier in South China in December 2019, is challenging the world, since the reporting of the first case by the World Health Organization on 7 Jan 2020 and announcing it as a pandemic [2, 3]. The endemic that started from Wuhan, Hubei province from the Republic of China, slowly became an agent between the host population genetically mutating and introducing new pathogens. The rapid flow of the condition is significantly visible only when the growth is already inside the host, until then it shows ordinary flu symptoms blocking early detection like in some other cases. It is estimated that the virus can travel around 8 m and can survive up to 3 h on the surface and spreads through sneezing from the host person [4].

Figure 2.1 shows the way the outbreak is significantly classified, and Table 2.1 shows the description onto the pandemic phases of approach due to an outbreak. Figure 2.2 shows the outbreak condition and the relation with endemic, epidemic,

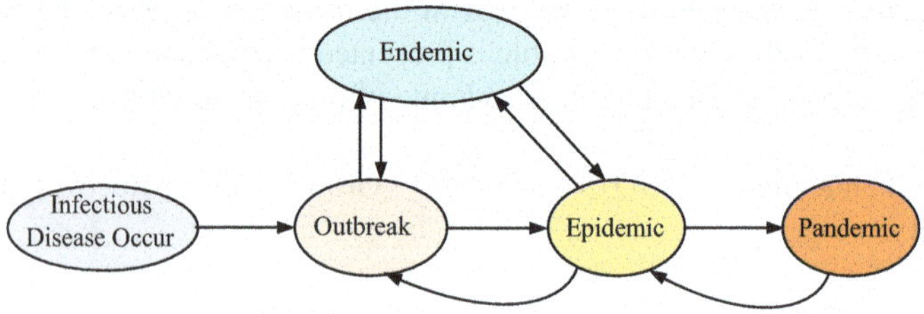

Fig. 2.1 Towards pandemic outbreak through the animal-host transmission

Table 2.1 Endemic towards pandemic characteristics

Phase	Description	Type	Ref
Phase 1	Virus is seen in animals (coronavirus is very common in animals such as bats, pangolin due to preying) and does not infect human	–	[5]
Phase 2	Known virus infects human (through bite, or as extort meat)	Endemic	[6]
Phase 3	Possible human to human transmission (within close contacts, accidental touch on contact surface, sexual act)	Endemic	[7]
Phase 4	Possible human to human transmission in communities outbreak (larger gatherings, community work)	Epidemic	[8]
Phase 5	Spread among humans in more than a country (due to travel, meetings)	Pandemic	[9]
Phase 6	Community outbreak in multiple country	Pandemic	[9]

Fig. 2.2 Phases in disease primarily through the animal-host transmission

and pandemic conditions. It grows in six phases with the final one bringing to the notice of the drastic efforts required to control. As the clusters are globally community bounded, stricter initiatives are needed to tackle it, or else the fatality will be the worst as happened during the outbreak of H1N1 Spanish flu in 1918, with the number of deaths estimated to be at least 50 million worldwide [4]. Table 2.2 shows the classification of the number of significant outbreaks and their characteristics. It is noteworthy here, that the pandemic is repeating since the occurrence of first Spanish flu in 1918, every 100 years.

The spread of the 2019-nCoV is bound with multiple causes with the majority towards lack of information at the early outbreak situation. Relevant corroborated information could have prohibited this widespread, making it to remain epidemic than allowing it to be pandemic [12]. The early release of the epidemic information in a way is highly critical for the medic response on any outbreak [13–15]. The scientific diagnostic criterion is another challenge, at the early stage, as until some point of time, it is seen as pneumonia that could be treated with the most common drug available in the market. The foremost criteria in the outbreak spread is the unavailability of the prediction system, making it tough to realize the actual impact the pandemic would relate to the public. Hence, it is imperative for the epidemiological survey to trace the infectious pathways, with specific interests on the subsequent generation spreaders. But, the outbreak is just coincident with population migration, with the globalization impact, throughout the world where six million people travel everyday through air, making the pandemic to grow exponentially within three

Table 2.2 Endemic towards pandemic characteristics

Year	Name	Originated	Classification	Type of the source	Fatality rate (Estimated)
1918	Spanish Flu H1N1	Unknown but rampant in Germany, Britain, US and France	Pandemic	Bird	25–30% of world population
1950	Hemorrhagic fever (Dengue)	Philippines and Thailand	Epidemic	Mosquito (Aedes Aegypti)	–
1976	Ebola	Democratic Republic of Congo	Epidemic	Bat	88%
1999	Niba	Malaysia	Epidemic	Pig	40–70%
2002	SARS	China	Epidemic	Bat-Pangolin	14–15%
2003	H5N1 Bird Flu Avian Flu	China	Epidemic	Chicken	56%
2009	A(H1N1)pdm09 Swine Flu	US/Mexico	Epidemic	Pig	80%
2012	MERS COV	Saudi	Epidemic	Camel	34.4%
2013	H7N9 Avian Flu	China	Epidemic	Chicken	40%
2015	MERS COV	Korea	Epidemic	Camel	40%
2015	Zika	Brazil	Epidemic	Mosquito	10.5%
2016	Yellow Fever	Angola	Epidemic	Mosquito	20–50%
2019	nCOV19	China	Pandemic	Pangolin/Bat	3.4%*

*as of 12 April 2020
Sources WHO, CDC [10, 11]

months duration. In a way, this makes a challenging aspect towards the epidemiological investigation but simultaneously monitoring and prediction becomes highly meaningful for decision making against the outbreak, which created the prevailing public health crisis.

2 Data and Methods

2.1 COVID19 Travel from Wuhan to the First Case in Malaysia

The first nCOVID virus stain infected person was identified on 25 Jan 2020, 15 days after its first observation in south china. The sample data were drafted to find the significance in the growth rate through the initial variation for 30 days as in Fig. 2.3.

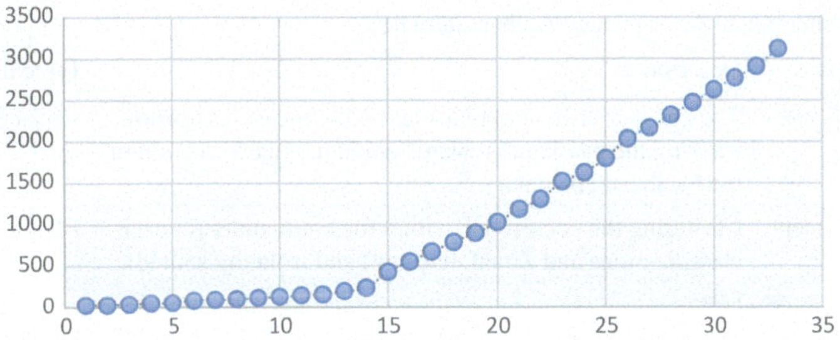

Fig. 2.3 Growth of the cases reported for the 30 days since the first case reported

Since the first case reported, normalcy is challenged, with an exponential growth within the next 20 days and this mooted towards the implement of Movement Control Order (MCO) by the Government of Malaysia [16, 17]. Reported cases were relatively low with cases spiking around mid-March 2020, most of them linked to a religious event held in Kuala Lumpur in late February [18]. The first few cases from the Tablighi cluster were reported in 14 March 2020, within the few days of reporting, the government invested interest on the growing population of cases in relation to the size of the gathering. As seen in Fig. 2.4, the major cluster of the outbreak is towards it and since then, through contact tracing, the links are established and the testing is continued still, at the time of this paper being written [19]. It is equally interesting to note that there are 69 cases (sporadic) without any link to this cluster [19] and makes it more challenging to identify the source, allowing a further delay towards moving to the recovery and restoration of the pandemic cycle.

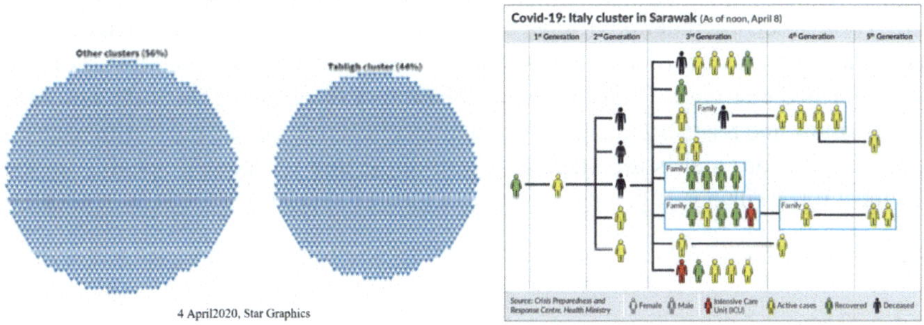

(a) Clustering (b) Contact Tracing

Fig. 2.4 Clustering towards contact tracking to control the epidemic

Table 2.3 Endemic towards pandemic characteristics

Stage	Class	Description	Typical Timeline
Stage 1	Monitor	The phase is critical in phasing out the measures towards tracking the pandemic towards epidemic (identification on the localized epicenter)	10–28 days
Stage 2	Prevent	Clustering the occurrence within the community (cluster classifications and Zonal marking) and isolating sporadic cases	14 days
Stage 3	Recover	Track onto the growth rate to the recovery rate considering the death rate (same type of pattern on the recovery rate Vs growth rate)	14–28 days
Stage 4	Restore	Where the growth rate is very less compared to recovery rate (start disinfect the cluster/zones marked epidemic)	14 days
		Continue social distancing with personal protection gear on	14–28 days

*all classification is of the slab of 14 days as typical incubation period of the virus is 10–14 days
Source CDC)[20]

2.2 *Relational Characterization of China on COVID19 Through the Stages of Recovery*

Any pandemic and localized epidemic should critically go through five stages of the eradication invoking different stages as in Table 2.3. Figure 2.5 shows the Chronological classification for the Wuhan case used in this study model. Stage 1(22–27 November 2019) is reported for the monitoring to weigh the class of severity of the outbreak, China reported pneumonia cluster cases in Wuhan, Hubei Province (31 Dec 2019) (classified as epidemic) (Stage 2), From there, several measures including the lockdown of the Hubei province makes the recovery picking up with downsizing of the confirmed new cases (Stage3) and as the ratio of the growth rate of the outbreak is hitting low and recovery rate faster (including clustering out the most critical cases pronounced dead) makes the Stage 4 of restoration and subsequent opening up the province for normalcy.

2.3 *Data*

The 2019-nCoV data used are of inclusive (1) Ministry of Health, Malaysia (http://www.moh.gov.my/index.php/pages/view/2019-ncov-wuhan), providing data in from January 25, 2020 to April 07, 2020 providing the latest epidemiological data, including cumulative confirmed cases, deaths, and recovered case; (2) Coronatracker (https://www.coronatracker.com/), providing the latest global data from January 20, 2020 up to the present. (3)The data for other epidemics the information are from WHO [21].

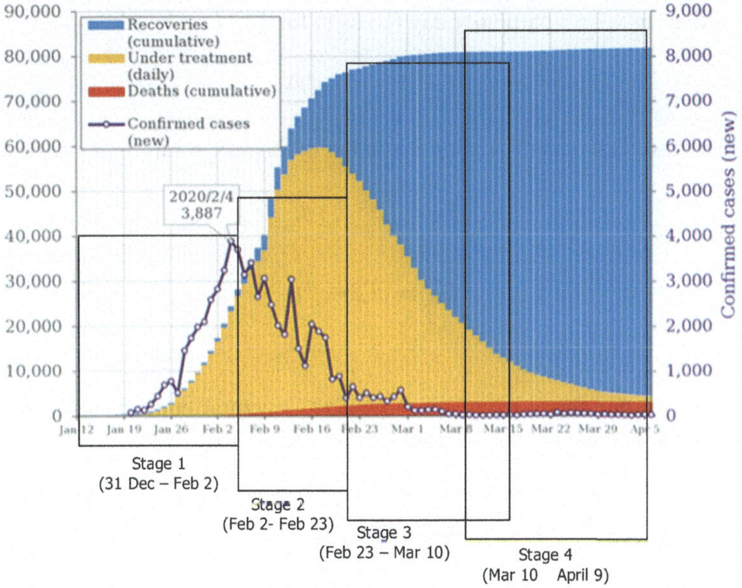

Fig. 2.5 Chronological classification for the Wuhan case used in this study model (https://commons.wikimedia.org/w/index.php?curid=87342125)

2.4 *Simplified Model*

We consider simple mathematical correlation equations built in as a model in particular with the scenario of the Malaysian data. However, we verified the model with respect to the other data available for countries and the results shows similar trend making our proposed model an easier one to seek the behavior of the virus inhaled pandemic. Assuming population unchanged due to other causes and considering a disease, immunized after recovery, we consider the following formulations:

T: total number of cases upto the day of computations.

A: susceptible cases on the day of computations.

A_1: cases registered prior to the day of computations.

A_x: cases registered prior to the incubation period (14 days earlier or after).

ε: growth rate of the pandemic(daily).

C: number of fatality on the day.

D: recovered numbers on the day.

B: forecast days after the first data of reference.

E_B: forecast actual cases after B number of days.

ε_1: growth rate of the previous day pandemic.

β: recovered growth rate.

α: Active case growth ratio.

α_y : data on the date of calculations (same as ε).

α_x : data before/after n days of considerations.

The active cases on a particular day (E) as in Eq. (2.1).

$$E = T - C - D \tag{2.1}$$

The Forecast total data (T_B) after B number of days as in Eqs. (2.2), (2.3).

$$T_B = \left(B^{\varepsilon - \varepsilon_1}\right) T \tag{2.2}$$

$$E_B = (T_B * \beta) \tag{2.3}$$

where $\varepsilon = \frac{A_1}{A}$.

We correlate the above using mathematical structural equation as to find the required forecast pandemic downsizing support unit (N) after B days as in Eqs. (2.4, 2.5).

$$N = T_B - E_B \tag{2.4}$$

$$n = \frac{\alpha_y}{\alpha_x} \tag{2.5}$$

According to the CDC, the incubation period for the novel coronavirus is somewhere between 2 days to two weeks after exposure [20]. To seek the effectiveness on the formulations we constructed the data using the correlation for few countries and compared with the actual data to validate. The days of incubation (n) is as in Eq. (2.6).

$$n = \frac{\varepsilon_y}{\varepsilon_x} \tag{2.6}$$

where

ε_y : *growth rate on the date of calculations (same as* ε*).*

ε_x : *growth rate before/after n days of considerations.*

The fatality growth rate (∂) (susceptible to death) is given as in Eq. (2.7)

$$\partial = \frac{\partial_y}{\partial_x} \tag{2.7}$$

where

∂_y :*fatality rate on the date of calculations (same as* ε*).*

∂_x : *fatality rate before/after n days of considerations.*

The downsizing support (N_x) considering the incubation and the fatality is given in Eq. (2.8).

$$N_x = N * n * \tag{2.8}$$

3 Results and Inferences

3.1 *Impact and Preliminary Screening on the Movement Controlled Order*

Within few weeks Malaysia become the country with the highest number of confirmed cases regionally [22]. Any outbreak of an infectious disease with a shoot-up period, followed by peak of intensity and when the margin between the new case and the recovered ones are larger (Stage 1–Stage 2) as seen in Fig. 2.6. Malaysian pandemic control measures relatively followed the way china handled when the outbreak was in place. Malaysia is presently at the ramping-up stage (Stage 2) and with the peak is yet to close. However, before that happens, we must do everything at our capacity to "flatten the curve" (Stage 3–Stage 4) with the objective to prevent sharper and significant growth, that would then make the healthcare system overwhelming. Even without the current pandemic situation, the public and private hospitals are already stretched with ill patients, needing ventilators across the globe. The number of breeding infections may not be reduced overall significantly, and there is even the possibility of the pandemic lasting slightly longer. This is evident and this is the point at which the measures to contain are to be in place, for which Malaysia has reacted well by placing the Movement Control Order primarily. With the numbers peaking towards exponential on March 16, 2020, with a single day hike

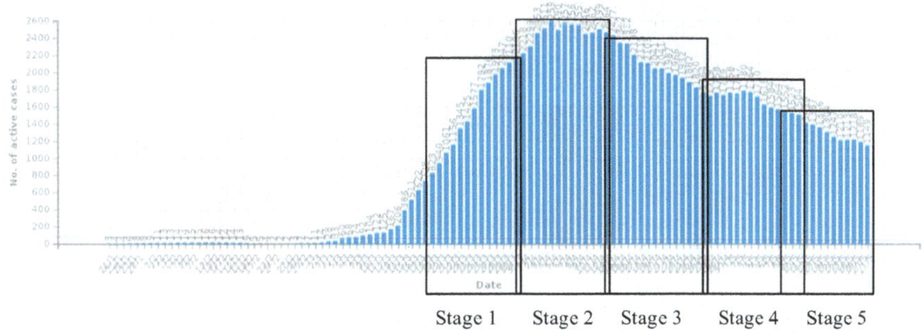

Fig. 2.6 Growth before the start of the MCO period

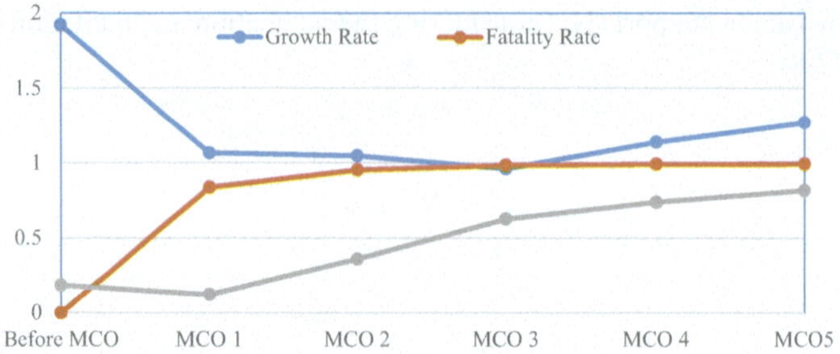

Fig. 2.7 Growth to recovery during the MCO period

from March 14, 2020 to 190 in March 15, the government issued the Movement Control Order (MCO). Figure 2.7 shows the trend between the first few cases noted till the MCO is applied in Malaysia. The similar trend on the characteristics at which the growth rate of the reported cases and the recovery cases deviate much (number of recovered is smaller in comparison with the new cases reported) is the point at which severe pandemic measures are to be initiated (Stage 2). The current drawn measures by the government on the restricted movement (18 March 2020–31 March 2020) showed significant force of closing the gap between the total cases to the recovered cases. But still the breakeven or the containment is not significant and this forces the government towards enhanced restricted movement (1 April 2020–14 April 2020). This extended period helped to segregate the clusters, thus making the situation under complete control in comparison with the available resources. Once the clusters are located for the hot spots, the next step is to lockdown the area with designated zone of red and orange as shown in Fig. 2.8. The third MCO was placed by the government (15 April 2020–28 April 2020) primarily due to the new cluster reported on [23] and the non-tracking of the sporadic cases. While this is being in continuation, the

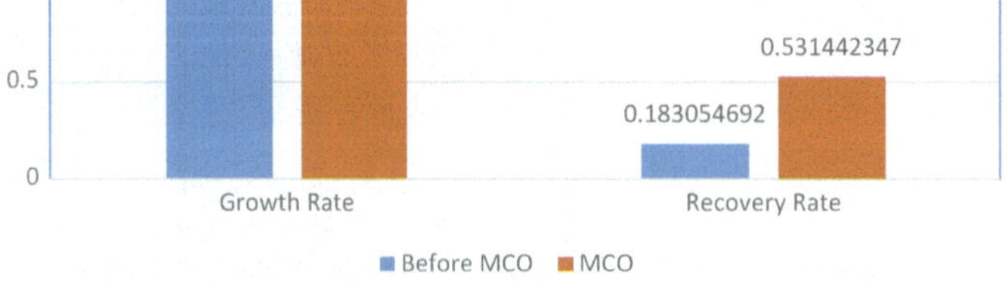

Fig. 2.8 Growth to recovery during the MCO period

incubation of the passive cases of the sporadic ones since reported, need to subsidize (either recover on its own or as new reported ones), once that is in control, then the country may enter the Stage 4 of restoration while still prevailing the enhanced MCO (29 April 2020–12 May 2020). The enhanced MCO is further extended (13 May 2020 till 9 June 2020) the major reasons aligned to avoid long travel due festive and the clustering within the enhanced area. Most of the cases reported to say are from the primary known source with less to no sporadic new cases. Because of the stress on the government due to the severe loss on the economic point the move by the government the opening of the industrial and most of the sectors with the new normal in place.

Malaysia's preventive measure is ranked at 51st among the 122 countries that implemented strict preventive measures, according to Government Response Tracker [24], and it stands 4th in the South East Asian Countries as on 12/04/2020. The countries with most stringent reaction include India, New Zealand, and Qatar. In reflection to the above and referring to the Fig. 2.5 on the stage on containment and referring to Fig. 2.7, the recovery rate is higher but still the growth rate is stable and hence the epidemic moving towards Stage 3 as seen in Fig. 2.8. Realizing this, the Government of Malaysia laid an extension till 28th of April 2020 with the expectation of flattening the curve first and then moving to Stage 3, which is the start of recovery (Stage 4). The marginal gap between the recovery factor and the growth factor is with larger stride in recovery factor. With the effort on the MCO standard operating procedure extensions, it is expected that the Stage 3 of recovery is expected to happen within the next 14 days. Post MCO, the restoration or the opening up to return to normalcy would take another 14–28 days to be seen through.

4 Summary

As it moves towards Stage 4 on the recovery/restoration phases, while combating the COVID-19, as on record, a mere normalcy is highly near to possible for restoration in the commercial, business, education, manufacturing, aviation and tourism industry. The government strategy towards making 2020 the year of tourism and later having to withdraw it, due to the pandemic, makes Malaysia losing out a larger share on GDP, as it is the second highest revenue for the country [25]. To balance out the loss, with an understanding that the tourism is to revive in most likely after third quarter, the government is looking into other revenue generation for the nation. To parallel out the GDP and also the health safety it is becoming a new norm to live with a new norm until a new drug to counter the virus is available in the market. The last century witnessed a number of newer types of viruses and this is becoming more common of the future. With the underlying pandemic and the parallel lessons of china and Malaysia, the standard operating procedure to tackle the secondary or repeated occurrence can be drafted towards knowing to live with viruses.

References

1. Kamikawa, T.L., et al.: Pandemic influenza detection by electrically active magnetic nanoparticles and surface plasmon resonance. IEEE Trans. Nanotechnol. **11**(1), 88–96 (2012)
2. Kwok, K.O., Tang, A., Wei, V.W., Park, W.H., Yeoh, E.K., Riley, S.: Epidemic models of contact tracing: systematic review of transmission studies of severe acute respiratory syndrome and middle east respiratory syndrome. J. Comput. Struct. Biotechnol. **17**, 186–194 (2019)
3. Wuhan Municipal Health Commission Infection Data (2020) [Online] Available: http://wjw.wuhan.gov.cn/front/web/list2nd/no/710 (accessed April 2, 2020).
4. Luo, Y., et al.: Using multi-source web data for epidemic surveillance: a case study of the 2009 Influenza A (H1N1) pandemic in Beijing. Proceedings of 2010 IEEE International Conference on Service Operations and Logistics, and Informatics, pp. 76–81, Qingdao, Shandong (2010)
5. American Chemical Society. Missing link in coronavirus jump from bats to humans could be pangolins, not snake. ScienceDaily (2020). [Online] www.sciencedaily.com/releases/2020/03/200326144342.htm (accessed April 2, 2020)
6. [Online] Available: https://www.doh.wa.gov/YouandYourFamily/IllnessandDisease/AnimalTransmittedDiseases (accessed April 2, 2020)
7. Stoltey, J.E., Cohen, S.E.: Syphilis transmission: a review of the current evidence. Sex Health. **12**(2), 103–109 (2015). https://doi.org/10.1071/SH14174
8. Parrish, C.R., Holmes, E.C., Morens, D.M., et al.: Cross-species virus transmission and the emergence of new epidemic diseases. Microbiol Mol Biol Rev. **72**(3), 457–470 (2008). https://doi.org/10.1128/MMBR.00004-08
9. Institute of Medicine (US) Forum on Microbial Threats. Microbial Evolution and Co-Adaptation: A Tribute to the Life and Scientific Legacies of Joshua Lederberg: Workshop Summary. Washington (DC), National Academies Press (US) (2009). 5, Infectious Disease Emergence: Past, Present, and Future. [Online] Available https://www.ncbi.nlm.nih.gov/books/NBK45714/ (accessed April 7, 2020).
10. World Health Organization. WHO The Cumulative Number of Reported Probable Cases of Severe Acute Respi ratory Syndrome (SARS). Geneva, Swiss (May 6, 2003). [Online]. Available: https://www.who.int/csr/sarscountry/2003_05_06/en/ (accessed April 7, 2020)
11. [Online] Available: https://www.cdc.gov/flu/pandemic-resources/1918-pandemic-h1n1.html (accessed April 2, 2020)
12. Jiang, C., Dong, M.: Optimal measures for SARS epidemics outbreaks. 2006 6th World Congress on Intelligent Control and Automation, pp. 9331–9336, Dalian (2006)
13. Chen, P., Chen, K.: Optimal Control of Epidemic Information Dissemination in Mobile Ad Hoc Networks 2011 IEEE Global Telecommunications Conference—GLOBECOM 2011, pp. 1–5, Houston, TX, USA (2011)
14. Aung, C.Y., Ali, G.G.M.N., Zhao, M., Cai, R., Chong, P.H.J.: Information epidemics control for data delivery in opportunistic networks. 2016 IEEE International Conference on Communications (ICC), pp. 1–6, Kuala Lumpur (2016)
15. [Online] https://economictimes.indiatimes.com/news/politics-and-nation/coronavirus-cases-in-india-live-news-latest-updates-april10/liveblog/75074919.cms (accessed April 12, 2020)
16. [Online] https://www.pmo.gov.my/2020/03/movement-control-order/ (accessed 25 March 2020)
17. [Online] https://theaseanpost.com/article/covid-19-asean-preparing-worst (accessed 20 March 2020)
18. [Online] https://www.straitstimes.com/asia/se-asia/malaysia-reports-190-new-coronavirus-cases-most-linked-to-mosque-event (accessed 25 March 2020)
19. https://www.therakyatpost.com/2020/04/09/these-are-malaysias-5-biggest-covid-19-clusters-that-you-should-know/ (accessed 9 April 2020)
20. [Online] https://www.cdc.gov/coronavirus/2019-ncov/symptoms-testing/symptoms.html (accessed April 12, 2020)

21. World Health Organization. WHO The Cumulative Number of Reported Probable Cases of Severe Acute Respi ratory Syndrome (SARS). Geneva, Swiss (May 6, 2003). [Online]. Available: https://www.who.int/csr/sarscountry/2003_05_06/en/
22. Nesteruk, I.: Estimations of the coronavirus epidemic dynamics in South Korea with the use of SIR model (2020). https://doi.org/10.13140/RG.2.2.15489.40807 (preprint)
23. [Online] https://www.thestar.com.my/news/nation/2020/04/10/covid-19-79-cases-found-in-the-three-kl-locations-under-enhanced-mco-says-ft-minister (accessed 10 April 2020)
24. [Online] https://www.bsg.ox.ac.uk/research/research-projects/oxford-covid-19-government-response-tracker (accessed 12 April 2020)
25. [Online] https://www.thestar.com.my/news/nation/2020/04/12/malaysia039s-response-to-covid-19-ranked-fourth-strictest-in-sea (accessed 14 April 2020)

Efficient Numerical Schemes for MHD Convection in Porous Cavity

Zaihar Yaacob, Mohammad Khatim Hasan, and Bahari Idrus

Abstract Natural convective heat transfer occurs when there is a temperature difference between the surface and the surrounding fluid. These physical problems often occur in real -life applications involving porous mediums such as microprocessor cooling in computers, underground oil and gas movement, food processing and more. The theory of convective heat transfer in a porous medium is a complex phenomenon that relies on interconnected partial differential equations that have no analytical solution except for simplified problems, resulting in numerical solutions having to be used. The standard finite difference method (FDM) is a popular method because of its straightforward mathematical formulation and programming. But FDM requires many grid points to get a precise and meaningful answer. This causes the program to take a long time to converge which also affects the storage and memory of the computer. This chapter examines the effectiveness of non -standard finite difference method (NSFDM) in solving MHD convection problem in rectangular porous medium.

Keywords Nonstandard finite difference method · Convection heat transfer · Porous medium · Magnetohydrodynamics

Z. Yaacob (✉)
Polymer Engineering Technology, Universiti Kuala Lumpur Branch Campus Malaysian Institute of Chemical and Bio-Engineering Technology, Kuala Lumpur, Malaysia
e-mail: zaihar@unikl.edu.my

M. K. Hasan · B. Idrus
Centre for Artificial Intelligence Technology, Faculty of Information Science and Technology, Universiti Kebangsaan Malaysia, UKM, 43600 Bangi, Malaysia
e-mail: mkh@ukm.edu.my

B. Idrus
e-mail: bahari@ukm.edu.my

© Institute of Technology PETRONAS Sdn Bhd 2022

S. A. Abdul Karim and A. Shafie (eds.), *Towards Intelligent Systems Modeling and Simulation*, Studies in Systems, Decision and Control 383,
https://doi.org/10.1007/978-3-030-79606-8_3

1 Introduction

A common numerical method used to solve conquering equations in unstable natural convection problems in porous mediums is the finite difference method by Alchaar et al. [1]; Chamkha [2]; Hossain et al. [3]; Pakdee and Rattanadecho [4]; Poulikakos and Bejan [5]; Revnic et al. [6]; Saleh et al. [7] and Saleh and Hashim [8].

This chapter will examine the problem of magnetohydrodynamics (MHD) convection in porous media. A new NSFD schemes for unsteady state are introduced. Our main objective is to propose a new scheme, to show the effectiveness of the new scheme and to estimate the heat transfer rate along the porous medium walls.

2 Magnetohydrodynamics

The study of magnetohydrodynamics was first introduced by Hannes Alfven that received Nobel Prize in 1970 for his contributions in physics. It is a study of magnetic properties and deportment of electrically conducting fluids. The effect of magnetic fields on natural convection in porous media is also receiving widespread attention due to the variety of engineering and technological applications such as solar energy collection, nuclear reactor insulation, geothermal energy generation and electronic equipment cooling Bian et al.[9]. This problem also arises in geophysics when the earth's crust saturated with fluids is exposed to the effects of magnetic fields Khanafer and Chamkha [10].

3 Non—steady State Mathematical Model

The governing equations for a non-steady state problem are:

$$\nabla \cdot \boldsymbol{V^*} = 0 \tag{1}$$

$$\boldsymbol{V^*} = \frac{K}{\mu}\left(-\nabla P^* + \rho \mathbf{g} + \boldsymbol{J} \times \boldsymbol{B}\right) \tag{2}$$

$$\frac{\partial T^*}{\partial t^*} + \left(\boldsymbol{V^*} \cdot \nabla\right)T^* = \alpha_m \nabla^2 T^* + \frac{q_o}{\rho_0 c_p}\left(T^* - T_c^*\right)^p \tag{3}$$

$$\nabla \cdot \boldsymbol{J} = 0 \tag{4}$$

$$\boldsymbol{J} = \sigma\left(-\nabla \varphi + \boldsymbol{V^*} \times \boldsymbol{B}\right) \tag{5}$$

$$\rho = \rho_0\left[1 - \beta\left(T^* - T_c^*\right)\right] \tag{6}$$

Here $\boldsymbol{V}^*$ fluid vector velocity, T^* fluid temperature, P pressure, $\boldsymbol{B}$ external magnetic field, $\boldsymbol{J}$ electric current vector, φ electrical potential, $\mathbf{g}$ gravity acceleration vector, K porous medium permeability, α_m effective heat permeability, ρ fluid density, μ dynamic viscosity, β coefficient of thermal expansion, T_c^* reference temperature, q_o internal heat generator, p internal heating exponent, c_p specific heat at constant pressure, σ electrical conductivity, ρ_0 reference density, $-\nabla\varphi$ electric field and t^* is time.

Equations (1)—(6) are reduced to Eqs. (7) and (8) after some simplifications and substitutions of nondimensional parameters.

$$\frac{\partial\theta}{\partial t} + \frac{\partial\psi}{\partial y}\frac{\partial\theta}{\partial x} - \frac{\partial\psi}{\partial x}\frac{\partial\theta}{\partial y} = \frac{\partial^2\theta}{\partial x^2} + \frac{\partial^2\theta}{\partial y^2} + G\theta^p \tag{7}$$

$$\frac{\partial^2\psi}{\partial x^2} + \frac{\partial^2\psi}{\partial y^2} = -Ra\frac{\partial\theta}{\partial x} - Ha^2\left(\frac{\partial^2\psi}{\partial y^2}\sin^2\phi + 2\frac{\partial^2\psi}{\partial x\partial y}\sin\phi\cos\phi + \frac{\partial^2\psi}{\partial x^2}\cos^2\phi\right) \tag{8}$$

where; $Ra = -K\beta\mathbf{g}\mathrm{L}\Delta\mathrm{T}/\alpha_m\nu$ is Rayleigh number, $G = q_o L^2(\Delta T)^{p-1}$ is internal heat generator constant, and $(Ha)^2 = \sigma K B_0^2/\mu$ is Hartmann number that represent magnetic field effect. Value of $\phi = 0$ means horizontal magnetic field and $\phi = \pi/2$ means vertical. Solutions of (7) and (8) gives the temperature distribution inside the porous medium and average Nusselt number at the walls can then be calculated.

4 Hybrid Non—standard Scheme

Nonsteady equation (7) is solve using alternating direct implicit method, while Eq. (8) is solve using Gauss–Seidel iteration. Discrete representation for both equations are shown next. Equations (9)–(12) are from (7), meanwhile (13)–(15) are from (8).

$$r_x\theta_{i+1,j}^{k+\frac{1}{2}} - (1+2r_x)\theta_{i,j}^{k+\frac{1}{2}} + r_x\theta_{i-1,j}^{k+\frac{1}{2}} = -r_y\theta_{i,j+1}^{k} - \left(1-2r_y\right)\theta_{i,j}^{k} - r_y\theta_{i,j-1}^{k} + (\Delta t)(S_\theta)_{i,j} \tag{9}$$

$$r_y\theta_{i,j+1}^{k+1} - \left(1+2r_y\right)\theta_{i,j}^{k+1} + r_y\theta_{i,j-1}^{k+1} = -r_x\theta_{i+1,j}^{k+\frac{1}{2}} - (1-2r_x)\theta_{i,j}^{k+\frac{1}{2}} - r_x\theta_{i-1,j}^{k+\frac{1}{2}} + (\Delta t)(S_\theta)_{i,j} \tag{10}$$

$$r_x = \frac{\Delta t}{(\Delta x)^2},\ r_y = \frac{\Delta t}{(\Delta y)^2} \tag{11}$$

$$(S_\theta)_{i,j} = \left(\frac{\partial\psi}{\partial y}\frac{\partial\theta}{\partial x}\right)_{i,j} - \left(\frac{\partial\psi}{\partial x}\frac{\partial\theta}{\partial y}\right)_{i,j} - G(\theta^p)_{i,j} \tag{12}$$

$$\frac{\psi_{i+1,j} - 2\psi_{i,j} + \psi_{i-1,j}}{(\Delta x)^2} + \frac{\psi_{i,j+1} - 2\psi_{i,j} + \psi_{i,j-1}}{(\Delta y)^2} = (S_\psi)_{i,j} \tag{13}$$

$$(S_\psi)_{i,j} = -Ra\left(\frac{\partial\theta}{\partial x}\right)_{i,j} - (SA_\psi)_{i,j} \tag{14}$$

$$(SA_\psi)_{i,j} = Ha^2\left(\left(\frac{\partial^2\psi}{\partial y^2}\right)_{i,j}\sin^2\phi + 2\left(\frac{\partial^2\psi}{\partial x\partial y}\right)_{i,j}\sin\phi\cos\phi + \left(\frac{\partial^2\psi}{\partial x^2}\right)_{i,j}\cos^2\phi\right) \tag{15}$$

The standard finite difference method (FDM) requires an interval size of h, in the x-direction (Δx) and the y-direction (Δy) small enough (h $\ll$ 0.01) to obtain good results. This causes the program to take quite a long time with a huge number of iterations to get meaningful results. For example, in this problem, when a program is run with a grid size of 1/h = 200, the convergence time exceeds 20 min, and the number of iterations approaches 100,000. To overcome this, we proposed a hybrid nonstandard scheme base on the observations by R.E. Mickens [10–12] that mentioned, a finite difference scheme at best must have a relationship between the space interval and the time interval ($\Delta t \equiv \Delta x$).

In this chapter we proposed two new nonstandard hybrid schemes (NSFD-H) as following:

Scheme	Space (Δx & Δy)	Time (Δt)
NSFD-H1	$\Delta x \to e^{\Delta x} - 1$; $\Delta y = \Delta x$	$\Delta t = \Delta x$
NSFD-H2	$\Delta x \to 2\sinh(\frac{\Delta x}{2})$; $\Delta y = \Delta x$	$\Delta t = \Delta x$

5 Numerical Results and Discussion

The results shown here compare between standard FDM and NSFD (H1 and H2) in terms of convergence time and number of iterations. All comparions are for the case $Ra = 10,\ 10^2,\ 10^3,\ \ Ha = 5,\ G = 3,\ p = 3,\ \ \phi = 0$ and range of Nx.

A comparison of convergence time and number of iterations can be referenced in Tables 1 and 2. It is calculated by dividing the value from FDM by the value from NSFD-H, and the result we call as the multiple difference. Here NSFD-H have faster convergence time up to 225 times faster compared to FDM as shown in Table 1. NSFD-H also have a smaller number of iterations up to 506 times less than FDM as depicted in Table 2. All this have been achieved while still maintaining the accuracy of the results, as in Table 3 for Nusselt number value $\overline{Nu_h}$.

Table 1 Convergence time difference between FDM and NSFD-H

FDM / NSFD-H			
Ra	*Nx*	H1	H2
10	4	40.34622	34.66879
	6	178.3068	165.2156
	8	94.1423	124.2407
	10	143.8129	131.7302
	20	56.71737	53.45134
	30	28.94423	30.9454
	40	15.54858	19.92512
	50	16.89483	14.94922
10^2	4	26.78065	26.003
	6	180.3656	90.09618
	8	121.849	164.6477
	10	**225.2718**	187.4582
	20	85.54737	67.00831
	30	39.44281	32.81738
	40	26.06475	20.81433
	50	14.94377	14.46225
10^3	4	32.3387	34.36835
	6	176.9254	173.7936
	8	130.3712	174.8612
	10	108.9441	103.3307
	20	29.98366	33.61304
	30	16.23907	16.88848
	40	20.50993	13.60933
	50	7.202614	7.573697

Bold value indicates NSFD-H have faster convergence time up to 225 times faster compared to FDM

6 Summary

In this chapter, we have manage to show that the proposed NSFD schemes are superior than FDM in terms of convergence time and number of iterations. Thus comfirmed the effectiveness of our numerical schemes in solving non-steady MHD convection problem. In the future, we will propose some new schemes and apply to some new problem.

Table 2 Number of iterations difference between FDM and NSFD-H

FDM / NSFD-H			
Ra	*Nx*	H1	H2
10	4	324.314	303.163
	6	282.4422	271.366
	8	163.1327	158.2871
	10	164.7564	161.2384
	20	61.75215	60.49038
	30	31.44884	33.73855
	40	16.68327	21.60588
	50	18.0422	15.87766
10^2	4	309.7912	299.9043
	6	291.1842	276.625
	8	218.6192	207.3413
	10	248.3271	215.1406
	20	89.06134	73.52572
	30	40.72325	35.93433
	40	26.39494	22.35361
	50	14.93902	15.47264
10^3	4	**506.3529**	461.1429
	6	341.8125	317.9651
	8	256.6138	235.5
	10	136.2388	137.608
	20	32.30588	36.88156
	30	17.93287	18.63721
	40	21.58742	14.67673
	50	7.641594	8.129702

Bold value indicates NSFD-H also have a smaller number of iterations up to 506 times less than FDM

Table 3 Absolute difference for $\overline{Nu_h}$ between FDM and NSFD-H

|FDM—NSFD-H|			
Ra	*Nx*	H1	H2
10	4	2.85E-05	0.001203
	6	0.002216	0.000357
	8	0.001832	0.001007
	10	0.0025	0.000279
	20	0.001792	0.000194
	30	0.001407	7.01E-05
	40	0.001143	2.64E-05
	50	0.000961	4.89E-06
10^2	4	0.028722	0.014424
	6	0.011307	0.003895
	8	0.006513	0.002025
	10	0.003099	7.06E-05
	20	0.001012	7.06E-05
	30	0.000643	0.000146
	40	0.000361	4.17E-05
	50	0.000264	3.31E-05
10^3	4	0.296398	0.036296
	6	0.260143	0.017734
	8	0.222348	0.003409
	10	0.205459	0.011841
	20	0.11172	0.004879
	30	0.07184	0.001167
	40	0.052642	0.000109
	50	0.041995	0.000145

Acknowledgements Authors gratefully acknowledge the financial support received from UNIKL and MARA.

References

1. Alchaar, S., Vasseur, P., Bilgen, E.: Hydromagnetic natural convection in a tilted rectangular porous enclosure. Numer. Heat Transf. Part A: Appl. **27**(1), 107–127 (1995)
2. Chamkha, A.J.: Hydromagnetic combined convection flow in a vertical lid driven cavity with internal heat generation or absorption. Numer. Heat Transf. Part A: Appl. **41**(5), 529–546 (2002)
3. Hossain, M.A., Hafiz, M.Z., Rees, D.A.S.: Buoyancy and thermocapillary driven convection flow of an electrically conducting fluid in an enclosure with heat generation. Int. J. Therm. Sci. **44**(7), 676–684 (2005)

4. Pakdee, W., Rattanadecho, P.: Numerical Analysis of Natural Convection in Porous Cavities with Partial Convective Cooling Conditions. J. Porous Media **12**(11), 1083–1100 (2009)
5. Poulikakos, D., Bejan, A.: Unsteady natural convection in a porous layer. Phys. Fluids **26**(5), 1183–1191 (1983)
6. Revnic, C., Grosan, T., Pop, I., Ingham, D.B.: Magnetic field effect on the unsteady free convection flow in a square cavity filled with a porous medium with a constant heat generation. Int. J. Heat Mass Transf. **54**(9–10), 1734–1742 (2011)
7. Saleh, H., Roslan, R., Hashim, I.: Natural convection in a porous trapezoidal enclosure with an inclined magnetic field. Comput. Fluids **47**(1), 155–164 (2011)
8. Saleh, H., Hashim, I.: Hydromagnetic natural convection flow in a fluid-saturated porous medium with nonuniform heat generation. Adv. Mech. Eng., (2012)
9. Bian, W., Vasseur, P., Bilgen, E.: Effect of an electromagnetic field on natural convection in an inclined porous layer. Int. J. Heat Fluid Flow **17**(1), 36–44 (1996)
10. Khanafer, K.M., Chamkha, A.J.: Hydromagnetic Natural Convection from an Inclined Porous Square Enclosure with Heat Generation. Numer. Heat Transf. Part A: Appl. **33**(8), 891–910 (1998)
11. Mickens, R.E.: A best finite difference scheme for the fisher equation. Numer. Methods Partial Diff. Equat. **585**(January), 581–585 (1994)
12. Mickens, R.E.: Analysis of a finite-difference scheme for a linear advection-diffusion reaction equation. J. Sound Vib. **236**(5), 901–903 (2000)

The Impact of the New Ordering Strategy on Over—relaxation Variants for Two Dimensional Poisson's problems

Mohammad Khatim Hasan, Samsul Ariffin Abdul Karim, and Jumat Sulaiman

Abstract The development of a numerical system for the solution of the partial differential equation is very crucial since it can simulate the behavior of many scientific phenomena. The developed system can be used by researchers, educators, and students for research and education purposes. The system was developed using five over-relaxation schemes to solve two-dimensional Poisson's equations. Besides, the results generated by the system can be exported to Microsoft Excel for further analysis. This system was developed using C # programming languages and ASP.NET. Microsoft SQL Server 2005 is used as a database to store data. In this research, the successive over-relaxation (SOR) method was used as the control of comparison. From the research, the Alternating Top-Down Successive Over Relaxation (AUD-AOR) gives the best performance as compared to others.

Keywords Ordering strategy · Successive overrelaxation scheme · Accelerative overrelaxation scheme · Grid size impact; ordering strategy impact

M. K. Hasan (✉)
Centre for Artificial Intelligence Technology, Faculty of Information Science and Technology, Universiti Kebangsaan Malaysia, UKM, 43600 Bangi, Malaysia
e-mail: mkh@ukm.edu.my

S. A. Abdul Karim
Fundamental and Applied Sciences Department, Centre for Systems Engineering (CSE), Institute of Autonomous System, Universiti Teknologi PETRONAS, Bandar Seri Iskandar, 32610 Seri Iskandar, Perak Darul Ridzuan, Malaysia
e-mail: samsul_ariffin@utp.edu.my

J. Sulaiman
Mathematics with Economics Programme, Faculty of Science and Natural Resources, Universiti Malaysia Sabah, Jalan UMS, Kota Kinabalu 88400, Sabah, Malaysia
e-mail: jumat@ums.edu.my

© Institute of Technology PETRONAS Sdn Bhd 2022

S. A. Abdul Karim and A. Shafie (eds.), *Towards Intelligent Systems Modeling and Simulation*, Studies in Systems, Decision and Control 383,
https://doi.org/10.1007/978-3-030-79606-8_4

1 Introduction

In the era of Industrial Revolution 4.0, information technology plays an important role in industrial activity, academics, research, and system development. Rapid progress in microelectronics development has resulted in the advancement of various information technology facilities such as personal computers, telecommunication tools, multimedia, Internet of Things, transportation, healthcare facilities, business, and many others. This technology allows information sharing, simulations, and many other activities processed or managed via computers or even used to solve complex problems that can be mimicked by mathematical formulations.

Elliptic partial differential equations (PDEs) are very important in most scientific and engineering simulation activities. The equations can be used to simulate various problems that arise in many branches of science and engineering, such as heat and magnetic distribution in convection problems, the steady temperature distribution in a two-dimensional field of constant conductivity, temperature distributions in a product, traveling wave in guiding conductors and electrostatics problems. For instance, the distribution of the electrostatic potential can be determined by solving the Poisson equation, if there is a charge density in the problem domain, otherwise, if the simulation domain is free from electric charge, the governing equation is known as the Laplace equation.

Numerical solutions have been used by researchers to solve the Elliptic problem [1–11] for so many years. Various approaches have been proposed by researchers such as the Successive over Relaxation (SOR) method [1], Accelerative over Relaxation (AOR) method [2, 3], Explicit Decoupled Group (EDG) method [4–9] and Modified Explicit Group (MEG) method [10, 11] to solve the elliptic problem. In 1950, Young proposed the SOR method [1]. The method updated the Gauss-Seidel method to produce a more accurate solver. Recently [12] apply SOR and few other finite difference methods to solve the system of linear equations in the bipolar fuzzy environment with only 10 iterations to gather the distance of Hausdorff metric significantly small. However, the chosen relaxation parameter is more than the interval (0,1). Yaacob et al. [13] have modified the SOR method to propose a new hybrid nonstandard iterative method that increases the accuracy of approximate solution of natural convection in porous cavity significantly.

In 1978, Hadjidimos [2, 3] modify the SOR method by adding another coefficient to the SOR method and the new method known as the AOR method. The method was shown to be efficient compared to the SOR method.

The EDG utilized the rotated scheme which only solved half of the solution domain, while the MEG utilized the $2h$ mesh size and solved only a quarter of the solution domain. These approaches have been extended to develop other methods such as the Half Sweep Iterative Alternating Decomposition Explicit (HSIADE) scheme [14], Quarter Sweep Arithmetic Mean (QSAM) method to solve fourth-order parabolic equation [15], and Quarter Sweep Iterative Alternating Decomposition Explicit (QSIADE) scheme to solve diffusion equation [16]. The concept has been further extended to produce complexity reduction approach methods such as the

High-Speed Low Order Finite Difference Time Domain [17], High-Speed High Order Finite Difference Time Domain [18, 19]. Some literature on this complexity reduction approach can be read in [20]. The extension to the parallel algorithm has been done to further increase the speed of the High-Speed Low Order Finite Difference Time Domain [21]. These methods have shown a very good performance, primarily on the computational aspect to solve the free space wave propagation problem.

Another technique to increase the speed of computation is by rearranging the ordering of computation. Sulaiman et al. [22] apply a red-black strategy to the Quarter-Sweep Successive Over Relaxation (QSSOR) method. The method improved the computational time of the ordinary QSSOR method. In [23], Sulaiman et al. had applied the red-black ordering strategy to Explicit Decoupled Group Successive Over Relaxation (EDGSOR) method and again the ordering strategy improved the computational time of the ordinary EDGSOR method. A year later, [24] apply the red-black strategy to Quarter-sweep Gauss-Seidel and again the method succeeded to improve the computational time of the ordinary Gauss-Seidel method.

In this chapter, we discuss the development of various complexity reduction relaxation schemes for solving two-dimensional Poisson equations given by Eq. (1). We also discuss a new ordering strategy, which is also improves the computation time for SOR and AOR methods.

$$\frac{\partial^2 U}{\partial x^2} + \frac{\partial^2 U}{\partial y^2} = f(x, y), \quad (x, y) \in \Omega = [a, b] \times [a, b] \tag{1}$$

with boundary conditions given by Eq. (2).

$$U(x, y) = g(x, y), \quad (x, y) \in d\Omega \tag{2}$$

This chapter is organized as follows. In Sect. 1, we give some basic introduction as well as related literature review. Section 2 is devoted to the over-relaxation schemes. Section 3, discuss the strategy proposed to speed-up the over-relaxation scheme. Section 4 is dedicated for the performance analysis of the proposed scheme. Then conclusion will be given in the final section.

2 Over-relaxation Schemes

Solving Eq. (1) using the finite difference approximation in a rectangular grid in-plane with equal grid spacing in both x and y directions with $x_i = ih, \ y_j = jh \, (i, j = 0, 1, 2, \cdots, N)$. In this chapter, the point $u(x_i, y_j)$ is denoted by u_{ij}. Since in this chapter, an equal square mesh is used, thus the mesh size, $h = \frac{1}{N}$. The construction of methods discussed in this chapter started by discretizing Eq. (1) using central difference approximation that arises from the Taylor series given in Eq. (3) and Eq. (4). Since we apply rectangular grid (x, y) plane, this imply $\Delta x = \Delta y = h$.

$$\frac{\partial^2 u}{\partial x^2} = \frac{u_{i+1,j} - 2u_{i,j} + u_{i-1,j}}{h^2}, \tag{3}$$

and

$$\frac{\partial^2 u}{\partial y^2} = \frac{u_{i,j+1} - 2u_{i,j} + u_{i,j-1}}{h^2}. \tag{4}$$

By applying Eqs. (3) and (4) into Eq. (1), the discrete equation for Eq. (1) is given by

$$u_{i+1,j} - 2u_{i,j} + u_{i-1,j} + u_{i,j+1} - 2u_{i,j} + u_{i,j-1} = h^2 f(x_i, y_i),$$

which can be simplifies to

$$u_{i+1,j} + u_{i,j+1} - 4u_{i,j} + u_{i-1,j} + u_{i,j-1} = h^2 f(x_i, y_i). \tag{5}$$

A linear system of equation in matrix form can be constructed from Eq. (5) by constructing a simultaneous discrete equation for $i, j = 1, 2, 3, \cdots, n-1$. The discrete linear equation for $i, j = 1, 2, 3, \cdots, n-1$ is given as below.

For $i = 1, j = 1$,

$$u_{2,1} + u_{1,2} - 4u_{1,1} = f(x_1, y_1) - u_{0,1} - u_{1,0}.$$

For $i = 2, j = 1$,

$$u_{3,1} + u_{2,2} - 4u_{2,1} + u_{1,1} = f(x_2, y_1) - u_{2,0}.$$

For $i = 2, j = 2$,

$$u_{3,2} + u_{2,3} - 4u_{2,2} + u_{1,2} + u_{2,1} = f(x_2, y_2).$$

For $i = n-2, j = n-2$,

$$u_{n-1,n-2} + u_{n-2,n-1} - 4u_{n-2,n-2} + u_{n-3,n-2} + u_{n-2,n-3} = f(x_{n-2}, y_{n-2}).$$

For $i = n-1, j = n-2$,

$$u_{n-1,n-1} - 4u_{n-1,n-2} + u_{n-2,n-2} + u_{n-1,n-3} = f(x_{n-1}, y_{n-2}) - u_{n,n-2}.$$

For $i = n-1, j = n-1$

$$-4u_{n-1,n-1} + u_{n-2,n-1} + u_{n-1,n-2} = f(x_{n-1}, y_{n-2}) - u_{n,n-1} - u_{n-1,n}.$$

These discrete equations can be re-written in matrix form, $A\tilde{u} = \tilde{b}$. For simplication of displaying the matrix, we only construct the matrix for $i, j = 1, 2, 3$ as given in Eq. (6).

$$\begin{bmatrix} -4 & 1 & 0 & 1 & 0 & 0 & 0 & 0 & 0 \\ 1 & -4 & 1 & 0 & 1 & 0 & 0 & 0 & 0 \\ 0 & 1 & -4 & 0 & 0 & 1 & 0 & 0 & 0 \\ 1 & 0 & 0 & -4 & 1 & 0 & 1 & 0 & 0 \\ 0 & 1 & 0 & 1 & -4 & 1 & 0 & 1 & 0 \\ 0 & 0 & 1 & 0 & 1 & -4 & 0 & 0 & 1 \\ 0 & 0 & 0 & 1 & 0 & 0 & -4 & 1 & 0 \\ 0 & 0 & 0 & 0 & 1 & 0 & 1 & -4 & 1 \\ 0 & 0 & 0 & 0 & 0 & 1 & 0 & 1 & -4 \end{bmatrix}
\begin{bmatrix} u_{1,1} \\ u_{2,1} \\ u_{3,1} \\ u_{1,2} \\ u_{2,2} \\ u_{3,2} \\ u_{1,3} \\ u_{2,3} \\ u_{3,3} \end{bmatrix} = \begin{bmatrix} h^2 f(x_1, y_1) - u_{0,1} - u_{1,0} \\ h^2 f(x_2, y_1) - u_{2,0} \\ h^2 f(x_3, y_1) - u_{3,0} - u_{4,1} \\ h^2 f(x_1, y_2) - u_{0,2} \\ h^2 f(x_2, y_2) \\ h^2 f(x_3, y_2) - u_{4,2} \\ h^2 f(x_1, y_3) - u_{0,3} - u_{1,4} \\ h^2 f(x_2, y_3) - u_{2,4} \\ h^2 f(x_3, y_3) - u_{4,3} - u_{3,4} \end{bmatrix} \tag{6}$$

The coefficient matrix, A is then decomposed into three separate matrices; i.e. block diagonal, D, lower triangular, L, and upper triangular, U, using relation given in Eq. (7).

$$A = D - L - U \tag{7}$$

Hadjidimos [2] utilized Eq. (7) to propose a method called Accelerative over-relaxation (AOR). The method is defined as Eq. (8).

$$\begin{aligned} u^{(k+1)} &= L_{r,w} u^{(k)} + w(D - rL)^{-1} b, \\ L_{r,w} &= (I - rD^{-1}L)^{-1}\big[(1-w)I + (w-r)L + wD^{-1}U\big]. \end{aligned} \tag{8}$$

Equation (8) can be re-written as Eq. (9).

$$(D - rL)\tilde{u}^{(k+1)} = (w - r)L\tilde{u}^{(k)} + wU\tilde{u}^{(k)} + w\tilde{b} + (1 - w)D\tilde{u}^{(k)}. \tag{9}$$

Re-arranging Eq. (9), we arrive at

$$D\tilde{u}^{(k+1)} = rL\tilde{u}^{(k+1)} + wU\tilde{u}^{(k)} + wb + (1 - w)D\tilde{u}^{(k)}. \tag{10}$$

Another relaxation method can be constructed, which is called the Successive over-relaxation (SOR) iterative scheme for Eq. (1) is gathered by introducing a weighted parameter, w to Eq. (5). The method can be written as Eq. (11).

$$u_{i,j}^{(k+1)} = w\left(\frac{u_{i-1,j}^{(k+1)} + u_{i+1,j}^{(k)} + u_{i,j-1}^{(+1)} + u_{i,j+1}^{(k)} - h^2 f_{i,j}}{4}\right) + (1-w)u_{i,j}^{(k)}. \quad (11)$$

The AOR iterative scheme given by Eq. (10) can also be constructed by modifying Eq. (11). It can be constructed by introducing a parameter, r and replace $u_{i-1,j}^{(k+1)}$ with $u_{i-1,j}^{(k)}$ and $u_{i,j-1}^{(k+1)}$ with $u_{i,j-1}^{(k)}$, and adding $\frac{r\left(u_{i-1,j}^{(k+1)}-u_{i-1,j}^{(k)}\right)}{4}$ and $\frac{r\left(u_{i,j-1}^{(k+1)}-u_{i,j-1}^{(k)}\right)}{4}$ to the SOR scheme and shown as Eq. (12).

$$u_{i,j}^{(k+1)} = r\left(\frac{u_{i-1,j}^{(k+1)} + u_{i-1,j}^{(k)} + u_{i,j-1}^{(k+1)} + u_{i,j-1}^{(k)}}{4}\right) + w\left(\frac{u_{i-1,j}^{(k)} + u_{i+1}^{(k)} + u_{i,j-1}^{(k)} + u_{i,j+1}^{(k)} - h^2 f_{i,j}}{4}\right) + (1-w)u_{i,j}^{(k)} \quad (12)$$

Another variant of AOR schemes can be constructed by rotating the x axis and the y axis by 45^0 clockwise [1, 7–9]. The location of nodes used in Eq. (11) transform as below.

$i, j \pm 1 \to i \pm 1,\ j \pm 1,$
$i \pm 1, j \to i \pm 1,\ j \mp 1,$
$\Delta x, \Delta y \to \sqrt{(\Delta x)^{2+}(\Delta x)^2} = \sqrt{2}h, \quad \Delta x = \Delta y,$
Thus, Eq. (11) transform into Eq. (13).

$$u_{i+1,j+1} + u_{i-1,j-1} + u_{i+1,j-1} + u_{i-1,j+1} - 4u_{ij} = 2h^2 f_{ij}. \quad (13)$$

A rotated variants of Eq. (11) can be constructed, we arrive at Eq. (14).

$$u_{i,j}^{(k+1)} = w\left(\frac{u_{i-1,j-1}^{(k+1)} + u_{i+1,j+1}^{(k)} + u_{i+1,j-1}^{(k+1)} + u_{i-1,j+1}^{(k)} - 2h^2 f_{i,j}}{4}\right) + (1-w)u_{i,j}^{(k)}. \quad (14)$$

The Rotated Accelerated Over-relaxation (RAOR) iterative scheme for Eq. (10) can be developed by introducing a parameter, r and replace $u_{i-1,j-1}^{(k+1)}$ with $u_{i-1,j-1}^{(k)}$ and $u_{i+1,j-1}^{(k+1)}$ with $u_{i+1,j-1}^{(k)}$, and adding $\frac{r\left(u_{i-1,j-1}^{(k+1)}-u_{i-1,j-1}^{(k)}\right)}{4}$ and $\frac{r\left(u_{i+1,j-1}^{(k+1)}-u_{+1i,j-1}^{(k)}\right)}{4}$ in Eq. (14). Thus gives

$$u_{i,j}^{(k+1)} = r\left(\frac{u_{i-1,j-1}^{(k+1)} + u_{i-1,j-1}^{(k)} + u_{i+1,j-1}^{(k+1)} + u_{i+1,j-1}^{(k)}}{4}\right)$$

$$+ w\left(\frac{u_{i-1,j-1}^{(k)} + u_{i+1,j-1}^{(k)} + u_{i-1,j+1}^{(k)} + u_{i+1,j+1}^{(k)} - 2h^2 f_{i,j}}{4}\right)$$
$$+ (1-w)u_{i,j}^{(k)}. \tag{15}$$

The third AOR scheme variant can be gathered by applying 2h steps instead of h steps in discretizing Eq. (1), thus produce Eq. (16).

$$u_{i,j+2} + u_{i,j-2} + u_{i+2,j} + u_{i-2,j} - 4u_{ij} = 4h^2 f_{ij}. \tag{16}$$

By introducing a weighted parameter w, into a SOR-like approach, we arrive at Eq. (17).

$$u_{i,j}^{(k+1)} = w\left(\frac{u_{i-2,j}^{(k+1)} + u_{i+2,j}^{(k)} + u_{i,j-2}^{(+1)} + u_{i,j+2}^{(k)} - 4h^2 f_{i,j}}{4}\right) + (1-w)u_{i,j}^{(k)}. \tag{17}$$

By introducing a new weighted parameter, r and replacing $u_{i-2,j}^{(k+1)}$ with $u_{i-2,j}^{(k)}$ and $u_{i,j-2}^{(k+1)}$ with $u_{i,j-2}^{(k)}$, and adding $\frac{r\left(u_{i-2,j}^{(k+1)} - u_{i-2,j}^{(k)}\right)}{4}$ and $\frac{r\left(u_{i,j-2}^{(k+1)} - u_{i,j-2}^{(k)}\right)}{4}$ in Eq. (17), we arrive at Eq. (18). Equation (18) will be known as Quarter sweep accelerated over relaxation (QSAOR) method.

$$u_{i,j}^{(k+1)} = r\left(\frac{u_{i-2,j}^{(k+1)} + u_{i-2,j}^{(k)} + u_{i,j-2}^{(k+1)} + u_{i,j-2}^{(k)}}{4}\right)$$
$$+ w\left(\frac{u_{i-2,j}^{(k)} + u_{i+2,j}^{(k)} + u_{i,j-2}^{(k)} + u_{i,j+2}^{(k)} - 4h^2 f_{i,j}}{4}\right)$$
$$+ (1-w)u_{i,j}^{(k)} \tag{18}$$

3 Alternating Top-Down Strategy

The implementation of the SOR alternating top-down ordering strategy (AUD-SOR) was done by modifying the index in Eq. (12). The Eq. (12) was re-written in three equations with difference indices as shown in Eqs. (19)–(21) for $\boldsymbol{u}_{i,n-j}^{(k+1)}$, $\boldsymbol{u}_{n-i,j}^{(k+1)}$ and $\boldsymbol{u}_{n-i,n-j}^{(k+1)}$.

$$u_{n-i,j}^{(k+1)} = w\left(\frac{(g)^2 u_{n-i-1,j}^{(k+1)} + (g)^2 u_{n-i+1,j}^{(k)} + (h)^2 u_{n-i,j-1}^{(k+1)} + (h)^2 u_{n-i,j+1}^{(k)} - (g)^2 (h)^2 f_{n-i,j}}{2\left((h)^2 + (g)^2\right)}\right)$$
$$+ (1-w)u_{n-i,j}^{(k)}. \tag{19}$$

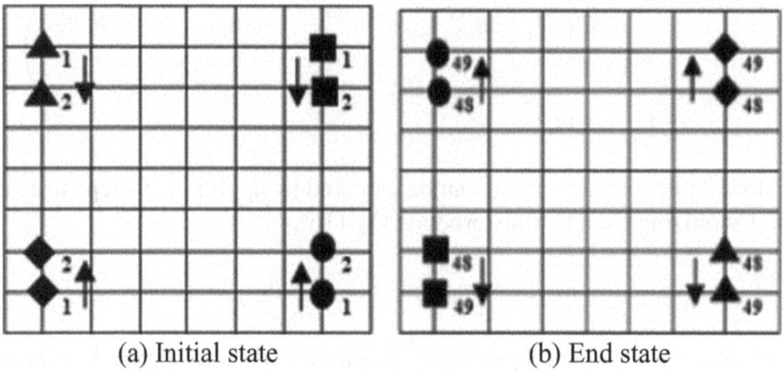

Fig. 1 Computing flow for each syimbols in AUD-SOR

$$u_{i,n-j}^{(k+1)} = w\left(\frac{(g)^2 u_{i-1,n-j}^{(k+1)} + (g)^2 u_{i+1,n-j}^{(k)} + (h)^2 u_{i,n-j-1}^{(k+1)} + (h)^2 u_{i,n-j+1}^{(k)} - (g)^2 (h)^2 f_{i,n-j}}{2\left((h)^2 + (g)^2\right)}\right) + (1-w)u_{i,n-j}^{(k)}. \quad (20)$$

$$u_{n-i,n-j}^{(k+1)} = w\left(\frac{(g)^2 u_{n-i-1,n-j}^{(k+1)} + (g)^2 u_{n-i+1,n-j}^{(k)} + (h)^2 u_{n-i,n-j-1}^{(k+1)} + (h)^2 u_{n-i,n-j+1}^{(k)} - (g)^2 (h)^2 f_{n-i,n-j}}{2\left((h)^2 + (g)^2\right)}\right) + (1-w)u_{n-i,n-j}^{(k)}. \quad (21)$$

The discrete node in the solution domain is denoted by four symbols(♦,●, ▲ dan ■). The illustration for the node location for $n = 8$ and the computing flow for each syimbols are displayed in Figs. 1 and 2 shows the complete node ordering.

The AUD-SOR need to be implemented by the ordering symbols (♦,●, ▲ dan ■). The node with (♦) symbol need to be solved first using Eq. (12), then solving node with (▲) symbol using Eq. (20), then node with (●) symbol using Eq. (19) and finally solving node with (■) symbol using Eq. (21) alternatingly until the convergence criteria is met.

4 Performance Analysis

To show the performance of the Accelerative over-relaxation (AOR), Rotated accelerative over- relaxation (RAOR), Quarter sweep accelerative over-relaxation (QSAOR) methods, and Successive over-relaxation with alternating top-down strategy (AUD-SOR) we conducted an experiment and recorded the performance parameter such as accuracy, number of iteration and computational time and compared it to the famous Successive over-relaxation (SOR) method.

In this chapter, we consider the problem of two-dimensional Poisson problems in Eq. (22) and its boundaries condition in Eq. (23).

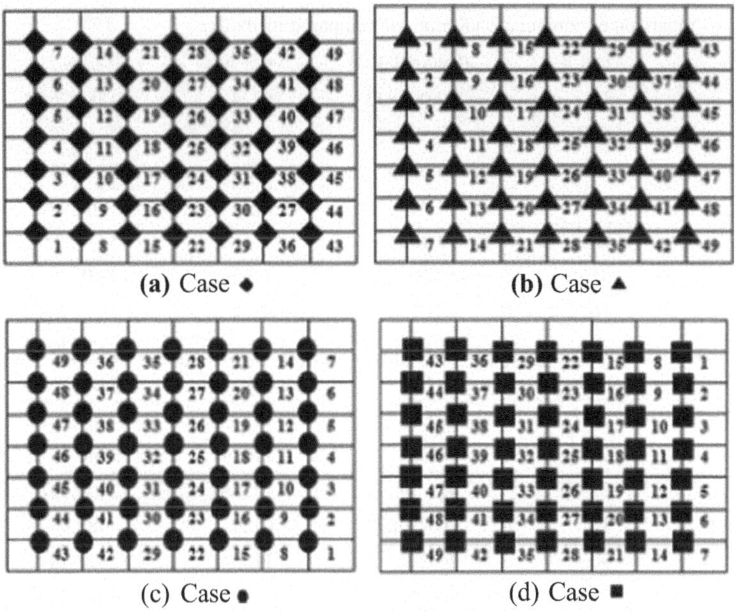

Fig. 2 Complete node ordering for AUD-SOR

$$\frac{\partial^2 u}{\partial x^2} + \frac{\partial^2 u}{\partial y^2} = f(x, y) = \left(x^2 + y^2\right)e^{xy}, \tag{22}$$

in R, (x,y) ∈ (0, 1)×(0, 1), with boundary conditions as in Eq. (23).

$$u(x, 0) = u(0, y) = 1, \ u(x, 1) = e^x, u(1, y) = e^y. \tag{23}$$

The exact solution for Eqs. (22) and (23) is given in Eq. (24).

$$u(x, y) = e^{xy}. \tag{24}$$

The convergence criteria used in this chapter was,

$$\text{maks}\left\{\left|u_{i,j}^{(k+1)} - u_{i,j}^{(k)}\right|\right\} < 10^{-6}, \tag{25}$$

as the stopping criterion.

In every experiment, the r values chosen are close to the optimum value of w. The selected parameter w should give the minimum number of iteration. Every experiment was conducted and recorded in Tables 1, 2, and 3.

The numerical experiment results (accuracy, computational time, and the number of iterations) were given in Tables 1, 2, and 3. These tables clearly show that using

Table 1 Comparison of computational time for compared methods

Methods	$n \times n$	t(seconds)	$n \times n$	$t(seconds)$
SOR	16	0.020833	6724	1382.766
	900	28.57292	11664	4034.49
	3136	326.6563	17956	9350.495
AOR	16	0.024	6724	1965.992
	900	39.985	11664	5439.289
	3136	449.203	17956	13248.59
RAOR	16	0.018	6724	832.763
	900	27.845	11664	2179.392
	3136	216.845	17956	5481.563
QSAOR	16	0.016	6724	282.952
	900	8.869	11664	840.131
	3136	92.779	17956	2099.621
AUD-SOR	16	0.020833	6724	988.9375
	900	26.0625	11664	2597.281
	3136	252.3177	17956	5710.547

Table 2 Comparison of iteration for compared methods

Methods	$n \times n$	$iteration$	$n \times n$	$iteration$
SOR	16	22	6724	6974
	900	1036	11664	11723
	3136	3391	17956	17596
AOR	16	22	6724	6820
	900	1013	11664	11466
	3136	3316	17956	17211
RAOR	16	15	6724	4527
	900	671	11664	7615
	3136	2199	17956	11438
QSAOR	16	14	6724	2348
	900	345	11664	3957
	3136	1138	17956	5949
AUD-SOR	16	33	6724	4608
	900	896	11664	7139
	3136	2488	17956	9995

Table 3 Comparison of accuracy for compared methods

Methods	$n \times n$	max *error*	$n \times n$	max *error*
SOR	16	1.56E-04	6724	4.48E-07
	900	3.48E-06	11664	2.36E-07
	3136	9.94E-07	17956	1.29E-07
AOR	16	1.91E-01	6724	4.89E-04
	900	3.66E-03	11664	2.82E-04
	3136	1.05E-03	17956	1.83E-04
RAOR	16	3.77E-01	6724	9.80E-04
	900	7.34E-03	11664	5.65E-04
	3136	2.10E-03	17956	3.67E-04
QSAOR	16	5.86E-01	6724	1.96E-03
	900	1.44E-02	11664	1.13E-03
	3136	4.19E-03	17956	7.33E-04
AUD-SOR	16	1.56E-04	6724	2.63E-05
	900	3.21E-06	11664	7.98E-05
	3136	5.08E-06	17956	1.89E-04

more node points to discretize the solution domain will result in higher accuracy for all numerical methods.

We analyzed the impact of grid size ratio to computational time ratio for all AOR, RAOR, QSAOR, and AUD-SOR methods show that more computational time is needed for a higher number of node points. For the AOR variant with grid size 900, the computational time needed 16.67 times compared to grid size 16 (grid size ratio 56.25). For grid size 3136 (grid size ratio 196), the computational time needed is 187.17 times compared to grid size 16. For grid size 6724 (grid size ratio 420.25), the computational time needed is 819.16 times compared to grid size 16. For grid size 11664 (grid size ratio 729), the computational time needed is 2266.37 times compared to grid size 16. For grid size 17956 (grid size ratio 1122.25), the computational time needed is 5520.25 times compared to grid size 16.

For the RAOR variant with grid size 900, the computational time needed 15.47 times compared to grid size 16 (grid size ratio 56.25). For grid size 3136 (grid size ratio 196), the computational time needed is 120.47 times compared to grid size 16. For grid size 6724 (grid size ratio 420.25), the computational time needed is 462.65 times compared to grid size 16. For grid size 11664 (grid size ratio 729), the computational time needed is 1210.77 times compared to grid size 16. For grid size 17956 (grid size ratio 1122.25), the computational time needed is 3045.31 times compared to grid size 16.

For the QSAOR variant with grid size 900, the computational time needed 5.54 times compared to grid size 16 (grid size ratio 56.25). For grid size 3136 (grid size ratio 196), the computational time needed is 57.99 times compared to grid size 16. For grid size 6724 (grid size ratio 420.25), the computational time needed is 176.85 times compared to grid size 16. For grid size 11664 (grid size ratio 729), the computational time needed is 525.08 times compared to grid size 16. For grid size 17956 (grid size ratio 1122.25), the computational time needed is 1312.26 times compared to grid size 16.

For the AUD-SOR variant with grid size 900, the computational time needed 12.51 times compared to grid size 16 (grid size ratio 56.25). For grid size 3136 (grid size ratio 196), the computational time needed is 121.11 times compared to grid size 16. For grid size 6724 (grid size ratio 420.25), the computational time needed is 474.70 times compared to grid size 16. For grid size 11664 (grid size ratio 729), the computational time needed is 1246.72 times compared to grid size 16. For grid size 17956 (grid size ratio 1122.25), the computational time needed is 2741.11 times compared to grid size 16.

We also analyze how much speed-up gain by comparing AOR, ROAR, QSAOR, and AUD-SOR to SOR. We compare the respective grid size for each method to the respective grid size of SOR. For AOR, 0.86 speed-up gain for grid size 16, 0.71 for grid size 900, 0.72 for grid size 3136, 0.70 for grid size 6724, 0.74 for grid size 11664, and 0.70 for grid size 17956. This shows that the AOR method is slower than the SOR method. However, the iteration number to converge is almost similar with 0–2.2% of the reduced number of iteration for AOR compared to the SOR method.

For RAOR, 1.16 speed-up gain for grid size 16, 1,02 for grid size 900, 1.51 for grid size 3136, 1.66 for grid size 6724, 1.85 for grid size 11664, and 1.71 for grid size 17956. This shows that the RAOR method is at least has the same speed as the SOR method and is almost 1.8 times faster with a higher grid size. The ROAR method has iteration number lesser by 31%-35% to converge compared to the SOR method.

For QSAOR, 1.30 speed-up gain for grid size 16, 3.22 for grid size 900, 3.52 for grid size 3136, 4.89 for grid size 6724, 4.80 for grid size 11664, and 4.45 for grid size 17956. This shows that the QSAOR method is faster than the SOR method with at least 1.30 time faster. The speed-up increases to about 4.5 times much faster than the SOR method. The QSOAR method has iteration number lesser by 36%-66% to converge compared to the SOR method.

For AUD-SOR, the method has a similar speed for grid size 16, 1.10 speed-up gain for grid size 900, 1.29 for grid size 3136, 1.40 for grid size 6724, 1.55 for grid size 11664, and 1.64 for grid size 17956. This shows that the QSAOR method is at least as faster as the SOR method. The speed-up increases to about 1.64 times much faster than the SOR method. This method shows to be at par with the ROAR method and better than the AOR method. The AUD-SOR method needs more iteration by 50% for grid size 16 but has less iteration number for other grid sizes by 13%-43% to converge compared to the SOR method.

We also analyze the impact grid size to computing time to the smallest grid size for all methods compared. We do experiments for every method as displayed in Table 1 with the smallest grid size is 16. The ratio of other grid sizes to the smallest grid size is between 56.25 to 1122.25. We calculate the computing time ratio to grid size 16 computing time for all methods compared. The computing time ratio for the SOR is between 13.72 to 4488.31. For the AOR method, the computing time ratio is between 16.66 to 5520.25 times, For the RAOR method, the computing time ratio is between

15.47 to 3045.31. For the QSAOR method, the computing time ratio is between 5.54 to 1312.26. For the AUD-SOR method, the computing time ratio is between 12.51 to 2741.11. This comparison shows that the QSAOR has the least impact from the increasing grid size, then followed by AUD-SOR. The AOR method has the most impact from increasing grid size.

The minimum iteration for SOR is 22 (for grid size 16), and the maximum iteration is 17596 (for grid size 17956), The minimum iteration for AOR is 22 (for grid size 16), and the maximum iteration is 17211 (for grid size 17956). The minimum iteration for RAOR is 15 (for grid size 16), and the maximum iteration is 11438 (for grid size 17956). The minimum iteration for QSAOR is 14 (for grid size 16), and the maximum iteration is 5949(for grid size 17956), The minimum iteration for AUD-SOR is 33 (for grid size 16), and the maximum iteration is 9995 (for grid size 17956),

The minimum of maximum error for SOR is 1.29E-07 (for grid size 17956), and the maximum of maximum error is 1.56E-04 (for grid size 16), The minimum value of maximum error for AOR is 1.83E-04 (for grid size 17956), and the maximum of maximum error is 1.91E-01(for grid size 16). The minimum value of maximum error for RAOR is 3.67E-04 (for grid size 17956), and the maximum value of maximum error is 3.77E-01 (for grid size 16). The minimum value of maximum error for QSAOR is 7.33E-04 (for grid size 17956), and the maximum value of maximum error is 5.86E-01(for grid size 16), The minimum value of maximum error for AUD-SOR is 3.21E-06 (for grid size 900), and the maximum of maximum error is .89E-04 (for grid size 17956). The result shows that the AUD-SOR has slightly different behavior on the maximum error. It seems that the optimum grid size for accuracy is 900. Having more grid size maybe not accuracy optimal anymore for the AUD-SOR method.

5 Summary

Five over relaxation methods, which are SOR, AUD-SOR, AOR, RAOR, and QSAOR methods have been demonstrated in this chapter. The SOR and AOR methods have been used for control of comparison. All methods show to converge to the exact solution. Using the concept of quarter sweep, the QSAOR has successfully reduced the complexity of the AOR method and resulted in faster computing time. However, by applying the rotated strategy, a method call RAOR is created and has successfully reduced the computational time of AOR by 25%-60% and reduced the computational time to SOR by 14%-46%. This finding is agreed by theoretically reduce complexity by 50%. In addition to that, this method needs a fewer number of iteration than the AOR method by 66%-68% and 65%-68% less than the SOR method.

When the quarter sweep reduction was applied, the method call QSAOR was created and have successfully reduced the computational time of AOR by 34–85%. This finding also agreed by theoretically reduce complexity by 75%. The finding reduced the computational time to SOR by 24–80%. In addition to that, this method needs a fewer number of iteration than the AOR method by 34–64% and 33–64%

less than the SOR method. When the alternating Up-Down ordering strategy was applied, the method call AUD-SOR was created and have successfully reduced the computational time of AOR by 14–57%. The finding reduced the computational time to SOR by 0–23%. In addition to that, this method needs a fewer number of iteration than the AOR method by 12–42% and 15–43% less than the SOR method for grid size from 900 to 17956.

In the conclusion, the ROAR and QSAOR have successfully increased the speed of AOR. The AUD-SOR also has successfully increased the speed of the SOR method. The AUD-SOR has the same performance as RAOR compared to SOR and AOR. The AUD-SOR also has better accuracy compared to AOR, RAOR, and QSAOR methods.

Acknowledgements We acknowledge the development of source code using C# by Ng Yit Hoe and the financial support from GGP-2017-023 grant.

References

1. Young, D.M.: Iterative methods for solving partial differential equations of elliptic type. A Ph.D. Thesis of Historical Importance, Harvard University Cambridge. Massachusets (1950)
2. Hadjidimos, A.: Accelerated overrelaxation method. Math. Comput. **32**(141), 149–157 (1978)
3. Hadjidimos, A.: Successive overrelaxation (SOR) and related methods. J. Comput. Appl. Math. **123**, 177–199 (2000)
4. Abdullah, A.R.: The four point explicit decoupled group (EDG) method: a fast Poisson solver. Int. J. Comput. Math. **38**, 61–70 (1991)
5. Sulaiman, J., Othman, M., Hasan, M.K.: Nine point-EDGSOR iterative method for the finite element solution of 2D poisson equations. Lect. Notes Comput. Sci. **5592** (Part 1), 764–774 (2009)
6. Sulaiman, J., Othman, M., Hasan, M.K.: Red-black half-sweep iterative method using triangle finite element approximation for 2D poisson equations. Lect. Notes Comput. Sci. **4487**, 326–333 (2007)
7. Sulaiman, J., Othman, M., Hasan, M.K.: Red-black EDGSOR iterative method using triangle element approximation for 2D poisson equations. Lect. Notes Comput. Sci. **4707**(Part 3), 326–333 (2007)
8. Ibrahim, A., Abdullah, A.R.: Solving the two dimensional diffusion equation by the four point explicit decoupled group (EDG) iterative method. Int. J. Comput. Math. **58**, 253–256 (1996)
9. Yousif, W., Evans, D.J.: Explicit de-coupled group iterative methods and their implementations. Parallel Algorithms Appl. **7**, 53–71 (1995)
10. Sulaiman, J., Hasan, M.K., Othman, M., Karim, S.A.A.: MEGSOR iterative method for the triangle element solution of 2D poisson equations. Proc. Comput. Sci. **1**(1), 377–385 (2010)
11. Sulaiman, J., Othman, M., Hasan, M.K.: MEGSOR iterative scheme for the solution of 2D elliptic pde's. Proc. World Acad. Sci. Eng. Technol. **62**, 471–477 (2010)
12. Akram, M., Muhammad, G., Kaom, A.N.A., Hussain, N.: Iterative methods for solving a system of linear equations in a bipolar fuzzy environment. Mathematics **7**(8), 728 (2019). https://doi.org/10.3390/math7080728
13. Yaacob, Z., Hasan, M.K., Idrus, B.: NSFD schemes for MHD convection in porous cavity heated and cooled on adjacent walls with non-uniform heat generation. Int. J. Adv. Sci. Technol. **29**(3), 1064–1082 (2020)

14. Sulaiman, J., Hasan, M.K., Othman, M.: The half-sweep iterative alternating decomposition explicit (HSIADE) method for diffusion equation. Lect. Notes Comput. Sci. **3314**, 57–63 (2004)
15. Sulaiman, J., Hasan, M.K., Othman, M., Yaacob, Z.: Quarter-sweep arithmetic mean iterative algorithm to solve fourth-order parabolic equations. Proceedings of the 2009 International Conference on Electrical Engineering and Informatics, vol. 1, pp. 194–198 (2009)
16. Sulaiman, J., Othman, M., Hasan, M.K.: Quarter-sweep iterative alternating decomposition explicit algorithm applied to diffusion equations. Int. J. Comput. Math. **81**(12), 1559–1565 (2004)
17. Hasan, M.K., Othman, M., Johari, R., Abbas, Z., Sulaiman, J.: The hslo(3)-fdtd with direct domain and temporary domain approaches on infinite space wave propagation. 2005 13th IEEE International Conference on Computers, Communications and Signal Processing with Special Track on Biomedical Engineering, pp. 65–68 (2005)
18. Hasan, M.K., Sulaiman, J., Karim, S.A.A.: A new high order algorithm with low computational complexity for electric field simulation. J. Comput. Sci. **6**(7), 769–774 (2010)
19. Hasan, M.K., Othman, M., Abbas, Z., Sulaiman, J., Ahmad, F.: Fourth order approximation with complexity reduction approach for the solution of time domain Maxwell equations in free space. Lect. Notes Comput. Sci. **5592**(Part 1), 752–763 (2009)
20. Hasan, M.K., Sulaiman, J., Karim, S.A.A., Othman, M.: Development of some numerical methods applying complexity reduction approach for solving scientific problem. J. Appl. Sci. **11**(7), 1255–1260 (2011)
21. Hasan, M.K., Othman, M., Abbas, Z., Sulaiman, J., Ahmad, F.: Parallel solution of high speed low order fdtd on 2D free space wave propagation. Lect. Notes Comput. Sci. **4706**(Part 2), 13–24 (2007)
22. Sulaiman, J., Othman, M., Hasan, M.K.: Red black QSSOR iterative method for solving 2D Helmholtz equations. Integrating Mathematical Sciences within Society, 28&29 November 2007, Bangi-Putrajaya Hotel
23. Sulaiman, J., Othman, M., Hasan, M.K.: Red-black EDG-SOR iterative method using triangle finite element approximation for 2D Poisson equations. In: Gervasi and Gavrilova, M. (Eds): ICCSA 2007, LNCS 4707, Part III, pp. 298–308. Springer, Berlin Heidelberg (2007)
24. Hasan, M.K., Sulaiman, J., Othman, M.: Implementation of red black strategy to quarter sweep iteration for solving first order hyperbolic equations. 2008 International Symposium on Information Technology, pp. 1–5, Kuala Lumpur (2008). https://doi.org/10.1109/ITSIM.2008.4632033

High—Order Hosoya Polynomials with Collocation Approach for the Solution of Two—Point Boundary Value Problems

Shazliyanna Ussin, Jumat Sulaiman, Mohammad Khatim Hasan, and Samsul Ariffin Abdul Karim

Abstract In this study, the two-point boundary value problem has been discretized using the high-order Hosoya polynomial discretization schemes to produce their corresponding Hosoya approximation equations. Next, each approximation equation can generate a system of linear equations. To get the numerical solution of the proposed problem, the direct method has been used in this study to solve the system of linear equations in various orders, namely 6, 8, 10 and 12. To validate the application of the high-order Hosoya polynomial, three examples of proposed problems have also been considered to study the accuracy of approximate solutions based on three orders of Hosoya approximation equations. Based on the results of numerical experiments, it can be pointed out that the eighth-order solution of Hosoya polynomial gives the higher accuracy as compared to sixth, tenth and twelfth order solutions.

Keywords High-order Hosoya polynomial · Boundary value problems · Direct method

---

S. Ussin · J. Sulaiman (✉)
Mathematics with Economics Programme, Faculty of Science and Natural Resources, Universiti Malaysia Sabah, Jalan UMS, Kota Kinabalu 88400, Sabah, Malaysia
e-mail: jumat@ums.edu.my

M. K. Hasan
Centre for Artificial Intelligence Technology, Faculty of Information Science and Technology, Universiti Kebangsaan Malaysia, UKM, 43600 Bangi, Malaysia
e-mail: mkh@ukm.edu.my

S. A. Abdul Karim
Fundamental and Applied Sciences Department, Centre for Systems Engineering (CSE), Institute of Autonomous System, Universiti Teknologi PETRONAS, Bandar Seri Iskandar, 32610 Seri Iskandar, Perak Darul Ridzuan, Malaysia
e-mail: samsul_arriffin@utp.edu.my

© Institute of Technology PETRONAS Sdn Bhd 2022

S. A. Abdul Karim and A. Shafie (eds.), *Towards Intelligent Systems Modeling and Simulation*, Studies in Systems, Decision and Control 383,
https://doi.org/10.1007/978-3-030-79606-8_5

1 Introduction

Graphs for Hosoya polynomials were first introduced in a seminar paper by Hosoya named "Calculation of several polynomials in chemistry" [1]. Then, the Hosoya polynomial has the benefit of including a lot of detail about the distance on a normal graph. Identifying the Hosoya polynomial graph, for example, allows determining the Weiner index graph as the polynomial's first derivative at point = 1 simple [2].

To study the feasibility of the Hosoya polynomial, let us start with $H(G, \lambda)$, which represents the Hosoya polynomial as follows

$$H(G, \lambda) = \sum_{k \geq 0} d(G, k)\lambda^k \quad (1)$$

with $d(G, k)$ is the number of vertex pairs of graphs G having distances at k and λ as parameters [3]. Furthermore, the polynomial addition of Hosoya can be defined as a function of the path i.e.

$$f(x) = \sum_{i=1}^{n} c_i H(P_i, x) = C^T H_P(x) \quad (2)$$

where C is the coefficient of the Hosoya polynomial and $H_P(x)$ is the matrix n × 1. Equation (2) is known as a Hosoya approximation function of order n with c_i, i = 1,2,..., n is the coefficient to be determined.

2 Boundary Value Problem and Recent Methods

In the last few decades, computational strategies for boundary value problems have been increasingly effective. Many scientists pay more attention to this problem numerically and show the capabilities of their computational methods used to solve it, especially TPBVPs; this issue can be found in many fields of chemistry, engineering, and physics, particularly optimal power, beam displacements, and heat flow [4–7]. Many scientists sought to attain higher precision by utilising different computational approaches in their early works on achieving the numerical solutions of TPBVPs. See Nazan and Hikmet [8]; Lin et al. [9]; Hamid et al. [10]; Rashidina and Sharif [11]; and Suardi et al. [12] for cases of spline methods for tackling second-order two-point boundary value problems. The fundamental idea behind these methods is to split the interval into subintervals and use a spline family such as spline, B-spline, and extended B-spline functions on every subinterval at the same time. These methods are required to solve the linear system of spline approximation equations with greater accuracy. TPBVPs can also be solved numerically by generating an ingenious approach based on the Galerkin method (see El-Gamel [13], Mohsen and Gamel [14], and Rahman et al. [15]. To solve boundary value problems, a number of computational approaches are used, including TPBVPs for getting the approximate solution [16–20].

Many studies were adopted through the concept of the finite difference approach, in addition to the use of the aforementioned numerical methods for solving the proposed problem (1) as indicated over the first paragraph (FDM). As a consequence, a new finite difference discretization method has been proposed by combining many numerical discretization schemes from different finite difference families. For instance, when solving TPBVPs, various finite difference approaches such as the classical finite difference [21], Rational Finite Difference [22] and Chebyshev finite difference [23] are introduced. The convergence of the regular finite difference principle together with the Chebyshev and rational approximation functions, respectively, has aided the growth of Chebyshev finite difference and Rational Finite Difference schemes. Pandey [24] suggested a different finite difference scheme through exponential formulae alone with finite difference discretization schemes identified as exponential finite difference discretization schemes that can solve TPBVPs in accordance with these combinations. Apart from rational, exponential and Chebyshev approximation functions, Suardi and Sulaiman [25] also evaluated the effectiveness of the Redlich–Kister polynomial as a numerical method used to solve TPBVPs, and they discovered that the solution of using the Redlich–Kister approximate function is more precise than Cubic Trigonometry B-Spline interpolation method.

We present a feasibility analysis of the Hosoya polynomial as a computational tool for solving TPBVPs, inspired by the high precision of various polynomial functions centred on high-order approximation functions. To solve the problem, first we discretize TPBVPs using the Hosoya polynomial function to form the Hosoya approximation equation. After that, we will use the approximate equation to construct a linear system. Since the generated linear system has a large and dense matrix, the approximate solution is obtained using the direct approach. Consider the two-point boundary value problems described as to explain the efficiency for the feasibility of the Hosoya polynomial function.

$$y'' + p(x)y' + q(x)y = f(x), \quad x \in [a, b] \tag{3}$$

subject to boundary conditions

$$y(a) = \alpha, \quad y(b) = \beta \tag{4}$$

where p(x), q(x) and f(x) are functions known to x or constants and a, b, α and β are constants (Abdullah, 1990).

3 High—Order Hosoya Polynomial with Collocation Approach

As mentioned in the introduction, our main purpose is to get the numerical solution of two-point boundary value problems. In this study, the solution process can be mainly divided into two steps, the first step is to apply high-order Hosoya polynomials with the collocation approach to discretize the proposed problem and to get the corresponding approximation equation. Then this approximation equation leads a system of Hosoya approximation equations. The second step is to implement the direct method in solving the linear system and then obtain the numerical solution of the approximation equation for problem (1). Before we introduce the rational finite difference-quadrature discretization scheme, the following discussion attempts to give the explanation of these describe these formulas.

3.1 Hosoya's Approximation Equation Towards the Proposed Problems

As it is known that in graph theory, the path in a graph is a finite or infinite side network that connects a network of vertices with all of them being different from each other. The path of P_n is a graph with the vertex n, i.e. $v_1, v_2, \ldots, v_n$, where v_i is adjacent to $v_{(i+1)}$, i = 1,2,..., n-1. Therefore, the path's length is the number of sides in it. Paths, P_1, P_2 and P_3 are illustrated in Fig. 1.

Because of that, the Hosoya polynomial, $H(P_n, \lambda)$ of the path, P_n is given as (Ramane et al. 2017)

$$H(P_n, \lambda) = n + (n-1)\lambda + (n-2)\lambda^2 + \ldots + [n-(n-2)]\lambda^{n-2} + [n-(n-1)]\lambda^{n-1} \tag{5}$$

In particular, Eq. (5) can be described in detail. For example, Eq. (6) is the first three lists for the Hosoya polynomial functions as follows.

$$\left.\begin{aligned} H(P_1, \lambda) &= 1 \\ H(P_2, \lambda) &= \lambda + 2 \\ H(P_3, \lambda) &= \lambda^2 + 2\lambda + 3 \\ &\vdots \end{aligned}\right\} \tag{6}$$

Fig. 1 Path of graphs P_1, P_2 and P_3

Therefore, consider the proposed problem, i.e. the two-point boundary value problems defined as

$$y'' + p(x)y' + q(x)y = f(x), \quad x \in [a, b] \tag{7}$$

subject to boundary conditions

$$y(a) = \alpha, \quad y(b) = \beta.$$

Before obtaining an approximate solution to the problem (7), then reconsider the definition of the Hosoya approximation function in Eq. (2) and redefine it as

$$u(x) = \sum_{i=1}^{n} c_i H(P_i,\ x) \tag{8}$$

where $C_i, i = 1, 2, 3, \ldots, n$ are unknown coefficients of the Hosoya approximation function that need to be determined numerically.

To illustrate the application of the approximation function (8) in this study, then consider the value of n = 10 in Eq. (8) and write as

$$u(x) = \sum_{i=1}^{10} c_i H(P_i, x) \tag{9}$$

$$\begin{aligned} \therefore u(x) = {} & C_1 H(P_1, x) + C_2 H(P_2, x) + C_3 H(P_3, x) + C_4 H(P_4, x) \\ & + \ldots + C_9 H(P_9, x) + C_{10} H(P_{10}, x) \end{aligned}$$

Then, to get the first and second derivatives of Eq. (9), we can get them as

$$\begin{aligned} u'(x) = {} & C_1 H'^{(P_1,x)} + C_2 H'^{(P_2,x)} + C_3 H'(P_3, x) \\ & + C_4 H'^{(P_4,x)} + \ldots + C_9 H'^{(P_9,x)} + C_{10} H'(P_{10}, x) \end{aligned} \tag{10}$$

$$\begin{aligned} u''(x) = {} & C_1 H''^{(P_1,x)} + C_2 H''^{(P_2,x)} + C_3 H''(P_3, x) + C_4 H''^{(P_4,x)} \\ & + \ldots + C_9 H^{(P_9,x)} + C_{10} H''(P_{10}, x). \end{aligned} \tag{11}$$

To derive the approximation equation of Hosoya, we need to substitute Eqs. (9), (10) and (11) on the proposed problem (7) and then it can be obtained as follows

$$\begin{aligned} & [H''(P_1, x) + p(x)H'((P_1, x) + +q(x)(P_1, x)]C_1 \\ & + [H''(P_2, x) + p(x)H'((P_2, x) + +q(x)(P_2, x)]C_2 \\ & + [H''(P_3, x) + p(x)H'((P_3, x) + +q(x)(P_3, x)]C_3 \\ & + \ldots \ + [H''(P_{10}, x) + p(x)H'((P_{10}, x) + +q(x)(P_{10}, x)]C_{10} = f(x) \end{aligned} \tag{12}$$

To simplify Eq. (12), consider the definition of the function, $R_i(x)$, $i = 1, 2, 3, \ldots, 10$ be expressed as

$$\begin{aligned} R_i(x) &= H''(P_i, x) + p(x)H'(P_i, x) + q(x)H(P_i, x), \\ i &= 1, 2, 3, \ldots, 10 \end{aligned} \tag{13}$$

By substituting Eq. (13) into Eq. (12), it is possible to obtain the approximation equation of Hosoya as

$$\begin{aligned} &R_1(x)C_1 + R_2(x)C_2 + R_3(x)C_3 + R_4(x)C_4 \\ &\quad + R_5(x)C_5 + R_6(x)C_6 + R_7(x)C_7 \\ &\quad + R_8(x)C_8 + R_9(x)C_9 + R_{10}(x)C_{10} = f(x). \end{aligned} \tag{14}$$

If $x = x_i$ is replaced into Eq. (14), the result can be shown

$$\begin{aligned} &R_1(x_i)C_1 + R_2(x_i)C_2 + R_3(x_i)C_3 + R_4(x_i)C_4 \\ &\quad + R_5(x_i)C_5 + R_6(x_i)C_6 + R_7(x_i)C_7 + R_8(x_i)C_8 \\ &\quad + R_9(x_i)C_9 + R_{10}(x_i)C_{10} = f(x_i) \end{aligned} \tag{15}$$

By substituting $R_j(x_i) = R_{ji}$ for $j = 1, 2, \ldots, 10$ and $f(x_i) = f_i$, then the approximation equation in Eq. (13) can be rewritten in the following form:

$$\begin{aligned} &R_{1,i}C_1 + R_{2,i}C_2 + R_{3,i}C_3 + R_{4,i}C_4 + R_{5,i}C_5 + R_{6,i}C_6 \\ &\quad + R_{7,i}C_7 + R_{8,i}C_8 + R_{9,i}C_9 + R_{10,i}C_{10} = f_i \end{aligned} \tag{16}$$

After that, we substitute the value of i = 1,2,..., 10 into Eq. (16) and then the following equations can be obtained:

$$\begin{aligned} i = 1 : \; &R_{1,1}C_1 + R_{2,1}C_2 + R_{3,1}C_3 + R_{4,1}C_4 + R_{5,1}C_5 \\ &+ R_{6,1}C_6 + \ldots + R_{9,1}C_9 + R_{10,1}C_{10} = f_1 \end{aligned} \tag{17}$$

$$\begin{aligned} i = 2 : \; &R_{1,2}C_1 + R_{2,2}C_2 + R_{3,2}C_3 + R_{4,2}C_4 + R_{5,2}C_5 \\ &+ R_{6,2}C_6 + \ldots + R_{9,2}C_9 + R_{10,2}C_{10} = f_2 \end{aligned} \tag{18}$$

$$\begin{aligned} i = 3 : \; &R_{1,3}C_1 + R_{2,3}C_2 + R_{3,3}C_3 + R_{4,3}C_4 + R_{5,3}C_5 \\ &+ R_{6,3}C_6 + \ldots + R_{9,3}C_9 + R_{10,3}C_{10} = f_3 \end{aligned} \tag{19}$$

$$\vdots$$

$$i = 10 : R_{1,10}C_1 + R_{2,10}C_2 + R_{3,10}C_3 + R_{4,10}C_4 + R_{5,10}C_5$$

$$+ R_{6,10}C_6 + \ldots + R_{9,10}C_9 + R_{10,10}C_{10} = f_{10} \tag{20}$$

Then reconsider the linear system in Eqs. (17)–(20) in the form of a matrix be generated as

$$R\bar{C} = \bar{F} \tag{21}$$

where,

$$R = \begin{bmatrix} R_{1,1} & R_{2,1} & R_{3,1} & \ldots & R_{9,1} & R_{10,1} \\ R_{1,2} & R_{2,2} & R_{3,2} & \ldots & R_{9,2} & R_{10,2} \\ R_{1,3} & R_{2,3} & R_{3,3} & \ldots & R_{9,3} & R_{10,3} \\ \vdots & \vdots & \vdots & & \vdots & \vdots \\ R_{1,9} & R_{2,9} & R_{3,9} & \ldots & R_{9,9} & R_{10,9} \\ R_{1,10} & R_{2,10} & R_{3,10} & \ldots & R_{9,10} & R_{10,10} \end{bmatrix},$$

$$\bar{C} = \begin{bmatrix} C_1 \; C_2 \; C_3 \ldots C_9 \; C_{10} \end{bmatrix},$$

$$\bar{F} = \begin{bmatrix} f_1 \; f_2 \; f_3 \cdots f_9 \; f_{10} \end{bmatrix},$$

3.2 Formulation of the Direct Method

Consider the system of linear equations in Eq. (21) and try to solve them by using the direct method. To facilitate further discussion, the formulation of the direct method should be stated in the form of a matrix. In the implementation of the iteration method, solving the system of linear Eqs. (21) can manipulate its coefficient matrix R by expressing as

$$R = L + D + U \tag{22}$$

where L, U and D are the strictly lower triangular matrix, the strictly upper triangular matrix and the diagonal matrix. Thus, the system of linear Eqs. (21) can be reproduced by using the decomposition of the matrix in Eq. (23) as

$$(L + D + U)\bar{C} = \bar{F} \tag{23}$$

In this study, however, we only consider the direct method in solving the system of linear equations. Thus, the direct method can continue to be implemented by calculating the inverse of the matrix R in (23) and subsequently the approximate solution of Hosoya can be expressed as

$$\bar{C} = R^{-1}\bar{F}. \tag{24}$$

4 Performance Analysis

Three examples of problems have been considered in this study. This study was conducted to test the efficiency of problem solving of linear two-point boundary values. To solve the linear system in Eq. (21) numerically, the full experimental results recorded are obtained through Mathcad 15.0 software.

First Problem: [26]

$$u'' - u' = -e^{(x-1)} - 1, \quad x \in [0, 1] \tag{25}$$

With its boundary conditions:

$$u(0) = u(1) = 0.$$

While the exact solution of the problem (25) is

$$u(x) = x\left(1 - e^{(x-1)}\right). \tag{26}$$

Second Problem: [16]

$$u'' - u = \mathrm{kos}(x), \quad x \in [0, 1] \tag{27}$$

with its boundary values as follows:

$$u(0) = u(1) = 1,$$

While the exact solution of problem (27) is

$$u(x) = c_1 e^x + c_2 e^{-x} - \frac{\mathrm{kos}(x)}{2}. \tag{28}$$

where,

$$c_1 = \frac{-3\mathrm{kosh}(1) + 3\sinh(1) + \mathrm{kos}(1) + 2}{4\sinh(1)},$$

$$c_2 = \frac{3\mathrm{kosh}(1) + 3\sinh(1) - \mathrm{kos}(1) - 2}{4\sinh(1)}.$$

Third Problem: [27]

$$u'' + (x + 1)u' - 2u = \left(1 - x^2\right)e^{-x}, x \in [0, 1] \tag{29}$$

with its boundary values as follows:

$$u(0) = -1, \quad u(1) = 0$$

and its exact solution is

$$u(x) = (x - 1)e^{-x}. \tag{30}$$

Based on the problems (25), (27), and (29), the results for numerical experiments at sixth, eighth, tenth and twelfth orders obtained have already been recorded in Tables 1, 2, 3, and 4. After analyzing the numerical results recorded in Tables 1, 2, 3 and 4, it is clear that the approximate solution for the eighth order Hosoya approximation equation has the smallest absolute error value compared to the other level. This comparative study is analyzed on the maximum absolute error for the first, second, and third problems at these various orders described in Table 5. Therefore, all problems solved using the direct method at the eighth order are more efficient in terms of maximum absolute error as compared to sixth, tenth and twelfth orders.

5 Conclusion

In this study, the problems in Eq. (3) have been discretized through the Hosoya polynomial approach and can form Hosoya approximation equations, as shown in Eq. (12). Through the approximation equation of Hosoya in Eq. (12), a system of linear equations can be generated. Next, this linear equation system is solved in formulating direct methods at various orders, namely sixth, eighth, tenth and twelfth orders. The study results obtained from numerical experiments on the three problems as shown in Tables 1, 2, 3, 4 and 5 have pointed out that the approximate solution at the eighth order Hosoya approximation equation through the direct method is superior in terms of its accuracy as compared to the other orders. Therefore, further research should be continued in studying the use of different problems, namely the problem of singular double-point boundary value and the problem of non-linear two-point boundary value. Because of that the Hosoya approximation equation is potentially to be applied for solving singular and nonlinear two-point boundary value problems.

Table 1 Comparison of the numerical results for first, second and third problems at the sixth order Hosoya approximation function

	First Problem			Second Problem			Third Problem		
x_i	Exact solution	Hosoya solution	Error	Exact solution	Hosoya solution	Error	Exact solution	Hosoya solution	Error
0	0.0000000000	−0.0000003975	3.9751e–07	1.0000000000	0.9999999479	5.2055e–08	−1.0000000000	−1.0000008386	8.3857e–07
0.1	0.0593430340	0.0593203251	2.2709e–05	0.9209801403	0.9209885084	8.3681e–06	−0.8143536762	−0.8143717270	1.8051e–05
0.2	0.1101342072	0.1101179955	1.6212e–05	0.8611278039	0.8611331381	5.3342e–06	−0.6549846025	−0.6549942838	9.6813e–06
0.3	0.1510244089	0.1510134712	1.0938e–05	0.8196945924	0.8196976660	3.0736e–06	−0.5185727545	−0.5185766886	3.9341e–06
0.4	0.1804753456	0.1804622331	1.3112e–05	0.7960185275	0.7960223417	3.8142e–06	−0.4021920276	−0.4021973542	5.3266e–06
0.5	0.1967346701	0.1967169138	1.7756e–05	0.7895198960	0.7895253571	5.4611e–06	−0.3032653299	−0.3032741106	8.7807e–06
0.6	0.1978079724	0.1977898254	1.8147e–05	0.7996988731	0.7997043679	5.4948e–06	−0.2195246544	−0.2195333886	8.7342e–06
0.7	0.1814272455	0.1814154883	1.1757e–05	0.8261348637	0.8261380153	3.1516e–06	−0.1489755911	−0.1489794040	3.8129e–06
0.8	0.1450153975	0.1450131588	2.2387e–06	0.8684875137	0.8684874475	6.6174e–08	−0.0898657928	−0.0898633413	2.4516e–06
0.9	0.0856463238	0.0856493576	3.0338e–06	0.9264993477	0.9264978417	1.5060e–06	−0.0406569660	−0.0406525371	4.4289e–06
1.0	0.0000000000	0.0000003975	3.9752e–07	1.0000000000	0.9999999252	7.4833e–08	0.0000000000	0.0000003354	3.3543e–07

Table 2 Comparison of the numerical results for first, second and third problems at the eighth order Hosoya approximation function

	First Problem			Second Problem			Third Problem		
x_i	Exact solution	Hosoya solution	Error	Exact solution	Hosoya solution	Error	Exact solution	Hosoya solution	Error
0	0.0000000000	0.0000000034	3.4415e–09	1.0000000000	0.9999999987	1.3060e–09	−1.0000000000	−1.0000000036	3.6453e–09
0.1	0.0593430340	0.0593429628	7.1253e–08	0.9209801403	0.9209801493	8.9778e–09	−0.8143536762	−0.8143537388	6.2574e–08
0.2	0.1101342072	0.1101341522	5.4977e–08	0.8611278039	0.8611278098	5.9383e–09	−0.6549846025	−0.6549846436	4.1117e–08
0.3	0.1510244089	0.1510243533	5.5545e–08	0.8196945924	0.8196945979	5.5508e–09	−0.5185727545	−0.5185727924	3.7968e–08
0.4	0.1804753456	0.1804752950	5.0554e–08	0.7960185275	0.7960185319	4.4304e–09	−0.4021920276	−0.4021920591	3.1486e–08
0.5	0.1967346701	0.1967346350	3.5095e–08	0.7895198960	0.7895198980	1.9620e–09	−0.3032653299	−0.3032653484	1.8508e–08
0.6	0.1978079724	0.1978079475	2.4904e–08	0.7996988731	0.7996988735	4.5032e–10	−0.2195246544	−0.2195246674	1.2923e–08
0.7	0.1814272455	0.1814272221	2.3465e–08	0.8261348637	0.8261348640	3.1177e–10	−0.1489755911	−0.1489756074	1.6310e–08
0.8	0.1450153975	0.1450153823	1.5239e–08	0.8684875137	0.8684875130	6.9030e–10	−0.0898657928	−0.0898658079	1.5091e–08
0.9	0.0856463238	0.0856463283	4.5215e–09	0.9264993477	0.9264993446	3.1127e–09	−0.0406569660	−0.0406569728	6.8688e–09
1.0	0.0000000000	0.0000000108	1.0834e–08	1.0000000000	0.9999999966	3.4015e–09	−1.0000000000	−0.0000000123	1.2311e–08

Table 3 Comparison of the numerical results for first, second and third problems at the tenth order Hosoya approximation function

	First problem			Second problem			Third problem		
x_i	Exact solution	Hosoya solution	Error	Exact solution	Hosoya solution	Error	Exact solution	Hosoya solution	Error
0	0.0000000000	0.0000089426	8.9426e-06	1.0000000000	0.9999807002	1.9300e-05	-1.0000000000	-1.0000008628	8.6275e-07
0.1	0.0593430340	0.0593528637	9.8296e-06	0.9209801403	0.9209590411	2.1099e-05	-0.8143536762	-0.8143545967	9.2044e-07
0.2	0.1101342072	0.1101451153	1.0908e-05	0.8611278039	0.8611045524	2.3251e-05	-0.6549846025	-0.6549855845	9.8199e-07
0.3	0.1510244089	0.1510366508	1.2242e-05	0.8196945924	0.8196687324	2.5860e-05	-0.5185727545	-0.5185737994	1.0449e-06
0.4	0.1804753456	0.1804892662	1.3921e-05	0.7960185275	0.7959894637	2.9064e-05	-0.4021920276	-0.4021931320	1.1044e-06
0.5	0.1967346701	0.1967507370	1.6067e-05	0.7895198960	0.7894868481	3.3048e-05	-0.3032653299	-0.3032664816	1.1517e-06
0.6	0.1978079724	0.1978268200	1.8848e-05	0.7996988731	0.7996608156	3.8057e-05	-0.2195246544	-0.2195258271	1.1727e-06
0.7	0.1814272455	0.1814497326	2.2487e-05	0.8261348637	0.8260904514	4.4412e-05	-0.1489755911	-0.1489767360	1.1449e-06
0.8	0.1450153975	0.1450426807	2.7283e-05	0.8684875137	0.8684349893	5.2524e-05	-0.0898657928	-0.0898668267	1.0339e-06
0.9	0.0856463238	0.0856799511	3.3627e-05	0.9264993477	0.9264364300	6.2918e-05	-0.0406569660	-0.0406577541	7.8811e-07
1.0	0.0000000000	0.0000420306	4.2031e-05	1.0000000000	0.9999237517	7.6248e-05	0.0000000000	-0.0000003319	3.319e-07

Table 4 Comparison of the numerical results for first, second and third problems at the twelfth order Hosoya approximation function

	First Problem			Second Problem			Third Problem		
x_i	Exact solution	Hosoya solution	Error	Exact solution	Hosoya solution	Error	Exact solution	Hosoya solution	Error
0	0.0000000000	0.0233606696	2.3361e-02	1.0000000000	1.0315685570	3.1569e-02	-1.0000000000	-0.9983500224	1.6500e-03
0.1	0.0593430340	0.0847877958	2.5445e-02	0.9209801403	0.9556318171	3.4652e-02	-0.8143536762	-0.8125437959	1.8099e-03
0.2	0.1101342072	0.1380405760	2.7906e-02	0.8611278039	0.8995138527	3.8386e-02	-0.6549846025	-0.6529814449	2.0032e-03
0.3	0.1510244089	0.1818640520	3.0840e-02	0.8196945924	0.8626855054	4.2991e-02	-0.5185727545	-0.5163318759	2.2409e-03
0.4	0.1804753456	0.2148351238	3.4360e-02	0.7960185275	0.8448013085	4.8783e-02	-0.4021920276	-0.3996529712	2.5391e-03
0.5	0.1967346701	0.2353353975	3.8601e-02	0.7895198960	0.8457414266	5.6222e-02	-0.3032653299	-0.3003440363	2.9213e-03
0.6	0.1978079724	0.2415130511	4.3705e-02	0.7996988731	0.8656733715	6.5974e-02	-0.2195246544	-0.2161015792	3.4231e-03
0.7	0.1814272455	0.2312269786	4.9800e-02	0.8261348637	0.9051367303	7.9002e-02	-0.1489755911	-0.1448767015	4.0989e-03
0.8	0.1450153975	0.2019634442	5.6948e-02	0.8684875137	0.9651527029	9.6665e-02	-0.0898657928	-0.0848316404	5.0342e-03
0.9	0.0856463238	0.1507118801	6.5066e-02	0.9264993477	1.0473577035	1.2086e-01	-0.0406569660	-0.0342918354	6.3651e-03
1.0	0.0000000000	0.0737823844	7.3782e-02	1.0000000000	1.1541562378	1.5416e-01	0.0000000000	0.0083118100	8.3118e-03

Table 5 Comparison of maximum absolute errors for first, second and third problem at various orders

	First Problem				Second Problem				Third Problem			
x_i	6	8	10	12	6	8	10	12	6	8	10	12
0	3.9751e-07	3.4415e-09	8.9426e-06	2.3361e-02	5.2055e-08	1.3060e-09	1.9300e-05	3.1569e-02	8.3857e-07	3.6453e-09	8.6275e-07	1.6500e-03
0.1	2.2709e-05	7.1253e-08	9.8296e-06	2.5445e-02	8.3681e-06	8.9778e-09	2.1099e-05	3.4652e-02	1.8051e-05	6.2574e-08	9.2044e-07	1.8099e-03
0.2	1.6212e-05	5.4977e-08	1.0908e-05	2.7906e-02	5.3342e-06	5.9383e-09	2.3251e-05	3.8386e-02	9.6813e-06	4.1117e-08	9.8199e-07	2.0032e-03
0.3	1.0938e-05	5.5545e-08	1.2242e-05	3.0840e-02	3.0736e-06	5.5508e-09	2.5860e-05	4.2991e-02	3.9341e-06	3.7968e-08	1.0449e-06	2.2409e-03
0.4	1.3112e-05	5.0554e-08	1.3921e-05	3.4360e-02	3.8142e-06	4.4304e-09	2.9064e-05	4.8783e-02	5.3266e-06	3.1486e-08	1.1044e-06	2.5391e-03
0.5	1.7756e-05	3.5095e-08	1.6067e-05	3.8601e-02	5.4611e-06	1.9620e-09	3.3048e-05	5.6222e-02	8.7807e-06	1.8508e-08	1.1517e-06	2.9213e-03
0.6	1.8147e-05	2.4904e-08	1.8848e-05	4.3705e-02	5.4948e-06	4.5032e-10	3.8057e-05	6.5974e-02	8.7342e-06	1.2923e-08	1.1727e-06	3.4231e-03
0.7	1.1757e-05	2.3465e-08	2.2487e-05	4.9800e-02	3.1516e-06	3.1177e-10	4.4412e-05	7.9002e-02	3.8129e-06	1.6310e-08	1.1449e-06	4.0989e-03
0.8	2.2387e-06	1.5239e-08	2.7283e-05	5.6948e-02	6.6174e-08	6.9030e-10	5.2524e-05	9.6665e-02	2.4516e-06	1.5091e-08	1.0339e-06	5.0342e-03
0.9	3.0338e-06	4.5215e-09	3.3627e-05	6.5066e-02	1.5060e-06	3.1127e-09	6.2918e-05	1.2086e-01	4.4289e-06	6.8688e-09	7.8811e-07	6.3651e-03
1.0	3.9752e-07	1.0834e-08	4.2031e-05	7.3782e-02	7.4833e-08	3.4015e-09	7.6248e-05	1.5416e-01	3.3543e-07	1.2311e-08	3.319e-07	8.3118e-03

References

1. Hosoya, H.: On some counting polynomials in chemistry. Discrete Appl. Math. **19**, 239–257 (1988)
2. Deutsch, E., Velázquez, J.A.R.: The Hosoya polynomial of distance-regular graphs. Discrete Appl. Math. **178**, 153–156 (2014)
3. Ramane, H.S., Shiralashetti, S.C., Mundewadi, R.A., Jummannaver, R.B.: Numerical solution of Fredholm integral equations using Hosoya polynomial of path graphs. Am. J. Numer. Anal. **5**, 11–15 (2017)
4. Aarao, J., Bradshaw-Hajek, B.H., Miklavcic, S.J., Ward, D.A.: The extended domain eigenfunction method for solving elliptic boundary value problems with annular domains. J. Phys. A: Math. Theor. **43**, 185–202 (2010)
5. Robertson, T.N.: The linear two-point boundary value problem on an infinite interval. Math. Comput. **25**(115), 475–481 (1971)
6. Wang, Y.M., Guo, B.Y.: Fourth-order compact finite difference method for fourth-order nonlinear elliptic boundary value problems. J. Comput. Appl. Math. **221**(1), 76–97 (2008)
7. Gupta, Y.: A numerical algorithm for solution of boundary value problems with applications. Int. J. Comput. Appl. **40**(8), 48–51 (2012)
8. Nazan, C., Hikmet, C.: B-spline methods for solving linear system of second order boundary value problems. Comput. Math. Appl. **57**(5), 757–762 (2008)
9. Lin, B.H., Huan, W.H., Yanpeng, C.: Polynomial spline approach for solving second-order boundary-value problems with neumann conditions. Appl. Math. Comput. **32**, 6872–6882 (2011)
10. Hamid, N.N.A., Majid, A.A., Ismail, A.I.M.: Extended cubic B-spline method for linear two-point boundary value problems. Sains Malaysiana **40**(11), 1285–1290 (2011)
11. Rashidinia, J., Sharifi, S.: B-spline method for two-point boundary value problems. Int. J. Math. Model. Comput. **5**(2), 111–125 (2015)
12. Suardi, M.N., Radzuan, N.Z.F.M., Sulaiman, J.: Cubic b-spline solution for two-point boundary value problem with aor iterative method. J. Phys.: Conf. Series. **890**, 012015 (2017)
13. El-Gamel, M.: Comparision of the solution obtained by Adomian decomposition and wavelet-galerkin methods of boundary-value problems. Appl. Math. Comput. **186**(1), 652–664 (2007)
14. Mohsen, A., Gamel, M.E.: On the galerkin and collocation methods for two point boundary value problems using sinc bases. Comput. Math. Appl. **56**, 930–941 (2008)
15. Rahman, M.M., Hossen, M.A., Islam, M.N., Ali, M.S.: Numerical solutions of second order boundary value problems by Galerkin method with Hermite polynomials. Ann. Pure Appl. Math. **1**(2), 138–148 (2012)
16. Jang, B.: Two-point boundary value problems by extended adomian decomposition method. Comput. Appl. Math. **219**(1), 253–262 (2007)
17. Roberts, S.M., Shipman, J.S.: Continuation in shooting methods for two-point boundary value problems. J. Math. Anal. Appl. **18**(1), 45–58 (1967)
18. Price, C.F.: An offset vector iteration method for solving two-point boundary-value problems. The Comput. J. **11**(2), 220–228 (1968)
19. Agarwal, R.P.: On the method of complementary functions for nonlinear boundary-value problems. J. Optim. Theory Appl. **36**(1), 139–144 (1982)
20. Fang, Q., Tsuchiya, T., Yamamoto, T.: Finite difference, finite element and finite volume methods applied to two-point boundary value problems. J. Comput. Appl. Math. **139**(1), 9–19 (2002)
21. Chawla, M.M., Katti, C.P.: Finite difference methods for two point boundary value problems involving high order differential equations. BIT Numer. Math. **19**(1), 27–33 (1979)
22. Pandey, P.K.: Rational finite difference approximation of high order accuracy for nonlinear two point boundary value problems. Sains Malaysiana **43**(7), 1105–1108 (2014)
23. Elbarbary, E.M., El-Kady, M.: Chebyshev finite difference approximation for the boundary value problems. Appl. Math. Comput. **139**(2–3), 513–523 (2003)

24. Pandey, P.K.: Solving two point boundary value problems for ordinary differential equations using exponential finite difference method. Boletim da Sociedade Paranaense de Matemática **34**(1), 45–52 (2016)
25. Suardi, M.N., Sulaiman, J.: Solution of one-dimensional boundary value problem by using Redlich-Kister Polynomial. In: Computational Science and Technology: 7th ICCST 2020, Pattaya, Thailand, 29–30 August, 2020, vol. 724, p. 487 (2021)
26. Caglar, H.N., Caglar, S.H., Elfaituri, K.: B-spline interpolation compared with finite difference, finite element and finite volume method which applied to two point boundary value problems. Appl. Math. Comput. **175**(1), 72–79 (2006)
27. Tat, C.K., Majid, Z.A., Suleiman, M., Senu, N.: Solving linear two-point boundary value. Appl. Math. Sci. **6**(6), 4921–4929 (2012)

The Quaternion Fourier Transform of Finite Measure and Its Properties

Mawardi Bahri, Amran Rahim, Muh. Nur, and Amir Kamal Amir

Abstract The main focus of the present article is to propose the quaternion Fourier transform of finite measure, which is a slight generalization of the Fourier transform of finite measure in real and complex domains. Further, we investigate its essential properties, which including translation, Parseval's formula, positive definite, partial derivative, and measure convergence. It is shown that according to the non-commutative nature of quaternion multiplications some properties of the Fourier transform of finite measure are not valid in the quaternion Fourier transform of finite measure.

Keywords Quaternion fourier transform · Finite measure · Bochner's theorem · Parseval's formula

1 Introduction

It is well known that the quaternion Fourier transform [1, 2] is non-trivial extension of the classical Fourier transform in the framework of quaternion algebra. The quaternion Fourier transform is important mathematical tools in areas like applied mathematics and quaternion-valued signal analysis. Basically, the quaternion Fourier transform sends a suitable function on $\mathbb{R}^2$ into a quaternion-valued function. The quaternion Fourier transform of finite measure is intimately related to the quaternion Fourier transform and can be considered as a slight generalization of the quater-

M. Bahri (✉) · A. Rahim · Muh. Nur · A. K. Amir
Department of Mathematics, Hasanuddin University, Makassar 90245, Indonesia
e-mail: mawardibahri@gmail.com

A. Rahim
e-mail: amran@science.unhas.ac.id

Muh. Nur
e-mail: muhammadnur@unhas.ac.id

A. K. Amir
e-mail: amirkamalamir@yahoo.com

© Institute of Technology PETRONAS Sdn Bhd 2022

S. A. Abdul Karim and A. Shafie (eds.), *Towards Intelligent Systems Modeling and Simulation*, Studies in Systems, Decision and Control 383,
https://doi.org/10.1007/978-3-030-79606-8_6

nion Fourier transform. It transforms a finite measure on $\mathbb{R}^2$ into a quaternion finite measure. The notation of finite measure in the framework of the quaternion Fourier transform and Clifford Fourier transform was introduced by the authors in [3, 4, 15], respectively. They in detail have investigated its interesting properties including the Bochner-Minlos theorem as the fundamental results related to this transformation.

In current article, we aim to continue the work of ones done in [3, 4]. We further investigate useful properties, which were missed such as translation property, Parseval's formula, convergence of measure, partial derivative, and so on. Our first step is to recall the quaternion Fourier transform and its inversion formula, which is needed to define the quaternion Fourier transform of finite measures.

The paper is arranged as follows. In Sect. 2 we shortly remind basic knowledge about quaternion algebra and basic properties that will be needed in later section. Section 3 is concerned with presenting the definition of the quaternion Fourier transform and its inverse. We also discuss σ-bandlimited quaternion function in the quaternion Fourier transform sense. Some of its basic properties are demonstrated in some detail. Section 4 introduce the definition of the quaternion Fourier transform of finite measure. Several properties for the quaternion Fourier transform of finite measure are also discussed in this section. In Sect. 5 we conclude this work.

2 Quaternion Algebra

In this part, we remind the quaternion algebra $\mathbb{H}$ over the real field $\mathbb{R}$, which its elements have the following form

$$\mathbb{H} = \{r = r_0 + \mathbf{i}r_1 + \mathbf{j}r_2 + \mathbf{k}r_3 \mid r_0, r_1, r_2, r_3 \in \mathbb{R}\}, \tag{1}$$

where the multiplication law is defined by relation

$$\mathbf{ij} = -\mathbf{ji} = \mathbf{k}, \quad \mathbf{jk} = -\mathbf{kj} = \mathbf{i}, \quad \mathbf{ki} = -\mathbf{ik} = \mathbf{j}, \quad \mathbf{i}^2 = \mathbf{j}^2 = \mathbf{k}^2 = \mathbf{ijk} = -1. \tag{2}$$

It is easy to see that from relation (2) the quaternion multiplication is not commutative. The element $r = r_0 + \mathbf{i}\,r_1 + \mathbf{j}\,r_2 + \mathbf{k}\,r_3 \in \mathbb{H}$, r_0 is so-called the *scalar* part of r denoted by $S(r)$ and $\mathbf{i}\,r_1 + \mathbf{j}\,r_2 + \mathbf{k}\,r_3$ is called the *vector* (or *pure*) part of r and denoted by $\mathbf{r}$.

We recall the conjugate $\bar{r}$ of the quaternion r as

$$\bar{r} = r_0 - \mathbf{i}\,r_1 - \mathbf{j}\,r_2 - \mathbf{k}\,r_3. \tag{3}$$

and

$$\bar{r} + r = 2S(r). \tag{4}$$

Applying (3) will lead to the norm or modulus of $r \in \mathbb{H}$ expressed as

$$|r| = \sqrt{r\bar{r}} = \sqrt{r_0^2 + r_1^2 + r_2^2 + r_3^2}. \tag{5}$$

Recall that for any $r, p \in \mathbb{H}$. We have

$$\overline{rp} = \bar{p}\bar{r}, \quad |r + p| \leq |r| + |p|. \tag{6}$$

It is straightforward to observe that

$$S(r) \leq |r| \quad \text{and} \quad |\mathbf{r}| \leq |r|. \tag{7}$$

Based on (3) and (5), we can define the inverse of $r \in \mathbb{H} \setminus \{0\}$ as

$$r^{-1} = \frac{\bar{r}}{r\bar{r}}.$$

3 Quaternion Fourier Transform and Bandlimited Functions

Let us start by recalling the definition of the quaternion Fourier transform and its inverse, which will have use of later. A detailed exposition of its properties and application, the reader may consult to [7–14, 16]. We also study σ-bandlimited quaternion function related to the quaternion Fourier transform and its basic properties.

Definition 1 For every $f \in L^2(\mathbb{R}^2; \mathbb{H})$, the quaternion Fourier transform is defined by

$$\mathcal{F}_H\{f\}u, v) = \int_{\mathbb{R}^2} f(x, y) e^{-\mathbf{i}2\pi ux} e^{-\mathbf{j}2\pi vy}\, dxdy. \tag{8}$$

The quaternion Fourier transform is invertible, which means that the original function $f(x, y)$ can be expressed in terms $\mathcal{F}_H\{f\}(u, v)$ as shown in the following theorem:

Theorem 1 *The inverse quaternion Fourier transform is given by*

$$f(x, y) = \int_{\mathbb{R}^2} \mathcal{F}_H\{f\}(u, v) e^{\mathbf{j}2\pi vy} e^{\mathbf{i}2\pi ux}\, du\, dv, \tag{9}$$

provided the integral exists and it is finite.

A useful property of the the quaternion Fourier transform is Plancherel's formula given by

$$\int_{\mathbb{R}^2} |f(x, y)|^2\, dx\, dy = \int_{\mathbb{R}^2} |\mathcal{F}_H\{f\}(u, v)|^2\, du\, dv. \tag{10}$$

Definition 2 We say that a quaternion function $f(x, y)$ is σ-bandlimited if its quaternion Fourier transform can be expressed as

$$\mathcal{F}_H\{f\}(u, v) = 0 \quad for \ u > \sigma, \quad v > \sigma, \tag{11}$$

and the quaternion function $f(x, y)$ is τ-timelimited if it satisfies (compare to [17, 18])

$$f(x, y) = 0 \quad for \ x > \tau, \quad y > \tau. \tag{12}$$

It follows from the inversion theorem for the quaternion Fourier transform (9) that relation (11) may be expressed as

$$f(x, y) = \int_{-\sigma}^{\sigma}\int_{-\sigma}^{\sigma} \mathcal{F}_H\{f\}(u, v) e^{\mathbf{j}2\pi vy} e^{\mathbf{i}2\pi ux} \, du \, dv. \tag{13}$$

Applying Eq. (10) we obtain energy E as

$$E = \int_{\mathbb{R}^2} |f(x, y)|^2 \, dx \, dy = \int_{-\sigma}^{\sigma}\int_{-\sigma}^{\sigma} |\mathcal{F}_H\{f\}(u, v)|^2 \, du \, dv < \infty. \tag{14}$$

The above definition will lead to the following important results.

Theorem 2 *The quaternion Fourier transform of σ-bandlimited function is absolutely integral*

$$\int_{-\sigma}^{\sigma}\int_{-\sigma}^{\sigma} |\mathcal{F}_H\{f\}(u, v)| \, du \, dv < \infty. \tag{15}$$

Proof Applying Schwarz'inequality and Eq. (14) results in

$$\begin{aligned} \left| \int_{-\sigma}^{\sigma}\int_{-\sigma}^{\sigma} \left|\mathcal{F}_H\{f\}(u, v)\right| du \, dv \right|^2 &\leq 4\sigma \int_{-\sigma}^{\sigma}\int_{-\sigma}^{\sigma} \left|\mathcal{F}_H\{f\}(u, v)\right|^2 du \, dv \\ &= 4\sigma E, \end{aligned} \tag{16}$$

and the proof is complete because $E < \infty$.

Theorem 3 *For all $x, y \in \mathbb{R}$, one has*

$$|f(x, y)| \leq 2e^{2\pi\sigma|x|} e^{2\pi\sigma|y|} \sqrt{\sigma E}. \tag{17}$$

Proof If $|u|, |v| < \sigma$, then $\left|e^{\mathbf{i}2\pi ux}\right| < e^{2\pi\sigma|x|}$, $\left|e^{\mathbf{j}2\pi vy}\right| < e^{2\pi\sigma|y|}$. Furthermore,

$$
\begin{aligned}
|f(x,y)| &= \left| \int_{-\sigma}^{\sigma}\int_{-\sigma}^{\sigma} \mathcal{F}_H\{f\}(u,v) e^{\mathbf{j}2\pi vy} e^{\mathbf{i}2\pi ux}\, du\, dv \right| \\
&\leq \int_{-\sigma}^{\sigma}\int_{-\sigma}^{\sigma} \left| \mathcal{F}_H\{f\}(u,v) e^{\mathbf{j}2\pi vy} e^{\mathbf{i}2\pi ux} \right| du\, dv \\
&\leq e^{2\pi\sigma|x|} e^{2\pi\sigma|y|} \int_{-\sigma}^{\sigma}\int_{-\sigma}^{\sigma} \left| \mathcal{F}_H\{f\}(u,v) \right| du\, dv \\
&\leq 2e^{2\pi\sigma|x|} e^{2\pi\sigma|y|} \sqrt{\sigma E},
\end{aligned}
$$

where in the last equality we have applied relation (16). This finishes the proof of the theorem.

4 Quaternion Fourier Transform of Finite Measure

Before we get into the details of definition of the quaternion Fourier transform of finite measure and its essential properties. We first denote the positive measures on $\mathbb{R}^2$.

$$
M^+(\mathbb{R}^2) = \left\{ \mu | \mu \text{ is a positive measure on } \mathbb{R}^2;\ \mu(\mathbb{R}^2) < \infty \right\}. \tag{18}
$$

This will lead to the following definition.

Definition 3 Given μ in $M^+(\mathbb{R}^2)$. The quaternion Fourier transform of μ is the quaternion function $\mathcal{F}_{\mathbf{i},\mathbf{j}}(\mu) : \mathbb{R}^2 \longrightarrow \mathbb{H}$ given by the expression

$$
\begin{aligned}
\mathcal{F}_{\mathbf{i},\mathbf{j}}(\mu)(u,v) &= \int_{\mathbb{R}^2} e^{-\mathbf{i}2\pi ux} e^{-\mathbf{j}2\pi vy}\, d\mu(x,y) \\
&= \int_{\mathbb{R}^2} e^{-\mathbf{i}2\pi ux} e^{-\mathbf{j}2\pi vy} \mu(x,y)\, dx\, dy.
\end{aligned} \tag{19}
$$

Or

$$
\begin{aligned}
\mathcal{F}_{\mathbf{j},\mathbf{i}}(\mu)(u,v) &= \int_{\mathbb{R}^2} e^{-\mathbf{j}2\pi vy} e^{-\mathbf{i}2\pi ux}\, d\mu(x,y) \\
&= \int_{\mathbb{R}^2} e^{-\mathbf{j}2\pi vy} e^{-\mathbf{i}2\pi ux} \mu(x,y)\, dx\, dy.
\end{aligned} \tag{20}
$$

From the above definition, it is straightforward to verify that

$$\mathcal{F}_{\mathbf{i},\mathbf{j}}(\mu)(0,0) = \mathcal{F}_{\mathbf{j},\mathbf{i}}(\mu)(0,0) = \mu(\mathbb{R}^2). \tag{21}$$

Further, μ is called a probability measure if and only if $\mathcal{F}_{\mathbf{i},\mathbf{j}}(\mu)(0,0) = 1$.

Example 1 The quaternion Fourier transform of the uniform measure μ on the rectangle $[a,b] \times [c,d]$ is computed as

$$\begin{aligned}\mathcal{F}_{\mathbf{i},\mathbf{j}}(\mu)(u,v) &= \int_a^b \int_c^d e^{-\mathbf{i}2\pi ux} e^{-\mathbf{j}2\pi vy}\, dx\, dy \\ &= \frac{e^{-\mathbf{i}2\pi ub} - e^{-\mathbf{i}2\pi ua}}{\mathbf{i}2\pi u} \frac{e^{-\mathbf{j}2\pi vd} - e^{-\mathbf{j}2\pi vc}}{\mathbf{j}2\pi v}, \quad u \neq 0, v \neq 0. \end{aligned}\tag{22}$$

In what follows, we show that $\mathcal{F}_{\mathbf{i},\mathbf{j}}(\mu)$ possesses the following properties.

Theorem 4 *The quaternion Fourier transform of $\mu \in M^+(\mathbb{R}^2)$ is bounded and uniformly continuous.*

Proof From the definition, we have

$$\begin{aligned}\left|\mathcal{F}_{\mathbf{i},\mathbf{j}}(\mu)(u,v)\right| &\leq \int_{\mathbb{R}^2} \left|e^{-\mathbf{i}2\pi ux} e^{-\mathbf{j}2\pi vy}\right| d\mu(x,y) \\ &= \int_{\mathbb{R}^2} d\mu(x,y) \\ &= \mu(\mathbb{R}^2) < \infty, \end{aligned}\tag{23}$$

where the last term of (23) does not depend on u and v and is finite. Now with the aid of (6) we obtain for all $(h_1, h_1) \in \mathbb{R} \times \mathbb{R} = \mathbb{R}^2$

$$\begin{aligned}&\left|\mathcal{F}_{\mathbf{i},\mathbf{j}}(\mu)(u+h_1, v+h_2) - \mathcal{F}_{\mathbf{i},\mathbf{j}}(\mu)(u,v)\right| \\ &\leq \int_{\mathbb{R}^2} \left|e^{-\mathbf{i}2\pi(u+h_1)x} e^{-\mathbf{j}2\pi(v+h_2)y} - e^{-\mathbf{i}2\pi ux} e^{-\mathbf{j}2\pi vy}\right| d\mu(x,y) \\ &= \int_{\mathbb{R}^2} \left|e^{-\mathbf{i}2\pi ux}\left(e^{-\mathbf{i}2\pi uh_1} e^{-\mathbf{i}2\pi vh_2} - 1\right) e^{-\mathbf{j}2\pi vy}\right| d\mu(x,y) \\ &= \int_{\mathbb{R}^2} \left|\left(e^{-\mathbf{i}2\pi uh_1} e^{-\mathbf{i}2\pi vh_2} - 1\right)\right| d\mu(x,y) \\ &\leq \int_{\mathbb{R}^2} \left|e^{-\mathbf{i}2\pi uh_1} e^{-\mathbf{i}2\pi vh_2}\right| + |-1|\, d\mu(x,y) \\ &= 2\int_{\mathbb{R}^2} d\mu(x,y). \end{aligned}\tag{24}$$

Notice that the last term of (24) is the independent of u and v. It also tends to 0 as $h_1 \to 0$ and $h_2 \to 0$. According to the dominated convergence, this proves that $\mathcal{F}_{\mathrm{i,j}}(\mu)$ is uniformly continuous on $\mathbb{R}^2$.

Theorem 5 *Let $\mu \in M^+(\mathbb{R}^2)$. For every $f \in L^2(\mathbb{R}^2; \mathbb{H})$ we have*

$$\int_{\mathbb{R}^2} \mathcal{F}_H\{f\}(u,v)\, d\mu(x,y) = \int_{\mathbb{R}^2} f(x,y)\mathcal{F}_{\mathrm{i,j}}(\mu)(u,v)\, dxdy. \tag{25}$$

Proof We see from (8) and the Fubini theorem that

$$\begin{aligned}\int_{\mathbb{R}^2} \mathcal{F}_H\{f\}(u,v)\, d\mu(x,y) &= \int_{\mathbb{R}^2}\int_{\mathbb{R}^2} f(x,y)\mathrm{e}^{-\mathrm{i}2\pi ux}\mathrm{e}^{-\mathrm{j}2\pi vy}\, dxdy\, d\mu(x,y)\\ &= \int_{\mathbb{R}^2} f(x,y)\int_{\mathbb{R}^2} \mathrm{e}^{-\mathrm{i}2\pi ux}\mathrm{e}^{-\mathrm{j}2\pi vy}\, d\mu(x,y)\, dxdy\\ &= \int_{\mathbb{R}^2} f(x,y)\mathcal{F}_{\mathrm{i,j}}(\mu)(u,v)\, dxdy, \end{aligned} \tag{26}$$

and the proof is complete.

Below we obtain another version of Parseval's formula in the context of the quaternion Fourier transform of μ in the following sense:

Theorem 6 *Let $\mu \in M^+(\mathbb{R}^2)$. Suppose that both f and $\mathcal{F}_H\{f\}$ belong to $L^2(\mathbb{R}^2; \mathbb{H})$. Then we have*

$$\int_{\mathbb{R}^2} f(x,y)\, d\mu(x,y) = \int_{\mathbb{R}^2} \mathcal{F}_H\{f\}(u,v)\mathcal{F}_{\mathrm{j,i}}(\mu)(-u,-v))\, dudv. \tag{27}$$

Proof By (9) and the Fubini theorem we have

$$\begin{aligned}\int_{\mathbb{R}^2} f(x,y)\, d\mu(x,y) &= \int_{\mathbb{R}^2}\int_{\mathbb{R}^2} \mathcal{F}_H\{f\}(u,v)\mathrm{e}^{\mathrm{j}2\pi vy}\mathrm{e}^{\mathrm{i}2\pi ux}\, dudv\, d\mu(x,y)\\ &= \int_{\mathbb{R}^2} \mathcal{F}_H\{f\}(u,v)\, dudv \int_{\mathbb{R}^2} \mathrm{e}^{\mathrm{j}2\pi vy}\mathrm{e}^{\mathrm{i}2\pi ux}\, d\mu(x,y)\\ &= \int_{\mathbb{R}^2} \mathcal{F}_H\{f\}(u,v)\mathcal{F}_{\mathrm{j,i}}(\mu)(-u,-v))\, dudv. \end{aligned} \tag{28}$$

This proves the theorem.

Theorem 7 *Let $\mu \in M^+(\mathbb{R}^2)$. For every $b_i \in \mathbb{R}$ with $i = 1, 2$ the following satisfies:*

$$\mathcal{F}_{\mathrm{i,j}}\big(\mu(x - b_1, y - b_2)\big)(u, v) = e^{-\mathrm{i}2\pi u b_1} \mathcal{F}_{\mathrm{i,j}}\big(\mu(x, y)\big)(u, v) e^{-\mathrm{i}2\pi v b_2}. \tag{29}$$

Proof A straightforward computation shows that

$$\begin{aligned}
\mathcal{F}_{\mathrm{i,j}}\big(\mu(x - b_1, y - b_2)\big)(u, v) &= \int_{\mathbb{R}^2} \mathrm{e}^{-\mathrm{i}2\pi u x} \mathrm{e}^{-\mathrm{j}2\pi v y}\, d\mu(x - b_1, y - b_2) \\
&= \int_{\mathbb{R}^2} \mathrm{e}^{-\mathrm{i}2\pi u (b_1 + z_1)} \mathrm{e}^{-\mathrm{j}2\pi v (b_2 + z_2)}\, d\mu(z_1, z_2) \\
&= \mathrm{e}^{-\mathrm{i}2\pi u b_1} \int_{\mathbb{R}^2} \mathrm{e}^{-\mathrm{i}2\pi u z_1} \mathrm{e}^{-\mathrm{j}2\pi v z_2}\, d\mu(z_1, z_2) \mathrm{e}^{-\mathrm{i}2\pi v b_2} \\
&= \mathrm{e}^{-\mathrm{i}2\pi u b_1} \mathcal{F}_{\mathrm{i,j}}\big(\mu(x, y)\big)(u, v), \mathrm{e}^{-\mathrm{i}2\pi v b_2},
\end{aligned} \tag{30}$$

which was to be proved.

Let us introduce the notion of a positive definite function in the context of quaternion algebra.

Definition 4 A continuous function $f : \mathbb{R}^2 \longrightarrow \mathbb{H}$ is said to be positive definite if the inequality

$$\sum_{j,k=1}^{n} b_j \bar{b}_k f(w_j - w_k) \geq 0, \tag{31}$$

is satisfied for any $b_1, b_2, \ldots, b_n \in \mathbb{H}$ and any $w_1, w_2, \ldots, w_n \in \mathbb{R}^2$ with $w_i = (u_i, v_i)$.

It is easily seen that the left-hand side of (31) is a real-valued function. As an immediate consequence of the above definition, we obtain the basic properties of positive definite functions:

Lemma 1 *Let $f : \mathbb{R}^2 \rightarrow \mathbb{H}$ be positive definite. Then*

1. $f(0) \geq 0$
2. $f(-u, -v) = \overline{f(u, v)}$ *for all* $(u, v) \in \mathbb{R}^2$.

Proof Taking $n = 1$ and $w_i = 0$ we obtain

$$|b|^2 f(0, 0) \geq 0 \text{ for all } b \in \mathbb{H}. \tag{32}$$

This implies that

$$f(0, 0) \geq 0, \tag{33}$$

which proves item 1. Now taking $n = 2$ and $(u_2, v_2) = (0, 0)$ we get

$$b_1\bar{b}_2 f(u_1, v_1) + b_2\bar{b}_1 f(-u_1, -v_1) \geq 0, \tag{34}$$

which is valid if $f(u_1, v_1) = \overline{f(-u_1, -v_1)}$. This proves item 2.

The next result shows that if μ is a finite positive measure on $\mathbb{R}^2$, then $\mathcal{F}_{\mathbf{i},\mathbf{j}}(\mu)$ is positive definite.

Theorem 8 *Suppose that $\mu \in M^+(\mathbb{R}^2)$. We have*

$$\sum_{j,k=1}^{n} b_j\bar{b}_k \mathcal{F}_{\mathbf{i},\mathbf{j}}(\mu)(w_j - w_k) \geq 0. \tag{35}$$

Proof By Theorem 4 we have $\mathcal{F}_{\mathbf{i},\mathbf{j}}(\mu)$ is continuous function. So, we see from the basic properties of quaternion (5) that

$$\begin{aligned}
\sum_{j,k=1}^{n} b_j\bar{b}_k \mathcal{F}_{\mathbf{i},\mathbf{j}}(\mu)(w_j - w_k) &= \int_{\mathbb{R}^2} \sum_{j,k=1}^{n} b_j\bar{b}_k e^{-\mathbf{i}2\pi(u_j-u_k)x} e^{-\mathbf{j}2\pi(v_j-v_k)y}\, d\mu(x, y) \\
&= \int_{\mathbb{R}^2} \sum_{j,k=1}^{n} b_j\bar{b}_k e^{-\mathbf{i}2\pi u_j x} e^{\mathbf{i}2\pi u_k x} e^{-\mathbf{j}2\pi v_j y} e^{\mathbf{j}2\pi v_k y}\, d\mu(x, y) \\
&= \int_{\mathbb{R}^2} \sum_{j,k=1}^{n} b_j\bar{b}_k \overline{e^{\mathbf{i}2\pi u_j x}} e^{\mathbf{i}2\pi u_k x} \overline{e^{\mathbf{j}2\pi v_j y}} e^{\mathbf{j}2\pi v_k y}\, d\mu(x, y) \\
&= \int_{\mathbb{R}^2} \left| \sum_{j=k}^{n} b_j e^{\mathbf{i}2\pi u_j x} e^{\mathbf{j}2\pi v_j y} \right|^2 d\mu(x, y) \\
&\geq 0,
\end{aligned} \tag{36}$$

as desired.

The following definition introduces weak convergence of measures.

Definition 5 Let $\mu, \mu_{i,j} \in M^+(\mathbb{R}^2)$. For every $f \in C_0(\mathbb{R}^2; \mathbb{H})$ we say that $\mu_{i,j}$ converges weakly to μ if

$$\lim_{i,j\to\infty} \int_{\mathbb{R}^2} f(x, y)\, d\mu_{i,j}(x, y) = \int_{\mathbb{R}^2} f(x, y)\, d\mu(x, y), \tag{37}$$

where $C_0(\mathbb{R}^2; \mathbb{H})$ is the space of bounded quaternion continuous functions.

The next result gives a sufficient condition of weak convergence in terms of the quaternion Fourier transform of finite measure.

Theorem 9 *Let μ and $\mu_{i,j}$ in $M^+(\mathbb{R}^2)$. If for each $u, v \in \mathbb{R}^2$,*

$$\lim_{i,j\to\infty} \mathcal{F}_{\mathbf{i},\mathbf{j}}(\mu_{i,j})(u,v) = \lim_{i,j\to\infty} \mathcal{F}_{\mathbf{i},\mathbf{j}}(\mu)(u,v) \tag{38}$$

Then

$$\lim_{i,j\to\infty} \int_{\mathbb{R}^2} f(x,y)\, d\mu_{i,j}(x,y) = \int_{\mathbb{R}^2} f(x,y)\, d\mu(x,y), \tag{39}$$

for all $f \in L^2(\mathbb{R}^2; \mathbb{H})$.

Proof In fact, from hypothesis we have

$$\mu_{i,j}(\mathbb{R}^2) = \mathcal{F}_{\mathbf{i},\mathbf{j}}(\mu_{i,j})(0,0) \to \mathcal{F}_{\mathbf{i},\mathbf{j}}(\mu)(0,0). \tag{40}$$

It means that the sequence $\mu_{i,j}(\mathbb{R}^2)$ is bounded. Thus

$$\sup_{i,j} \left|\mathcal{F}_{\mathbf{i},\mathbf{j}}(\mu_{i,j})(u,v)\right| = \sup_{i,j} \mu_{i,j}(\mathbb{R}^2) \leq C < \infty, \tag{41}$$

for some positive constant C. Further, for every $f \in L^2(\mathbb{R}^2; \mathbb{H})$ we have

$$\begin{aligned} \left|f(u,v)\mathcal{F}_{\mathbf{i},\mathbf{j}}(\mu_{i,j})(u,v)\right| &\leq \sup\nolimits_{i,j} \left|f(u,v)\mathcal{F}_{\mathbf{i},\mathbf{j}}(\mu_{i,j})(u,v)\right| \\ &\leq C|f(u,v)|. \end{aligned} \tag{42}$$

Applying the dominated convergence theorem yields

$$\lim_{i,j\to\infty} \int_{\mathbb{R}^2} f(u,v)\mathcal{F}_{\mathbf{i},\mathbf{j}}(\mu_{i,j})(u,v)\, du\, dv = \int_{\mathbb{R}^2} f(u,v)\mathcal{F}_{\mathbf{i},\mathbf{j}}(\mu)(u,v)\, du\, dv. \tag{43}$$

Equation (43) can be rewritten as

$$\begin{aligned} &\lim_{i,j\to\infty} \int_{\mathbb{R}^2}\int_{\mathbb{R}^2} f(u,v)e^{-\mathbf{i}2\pi ux}e^{-\mathbf{j}2\pi vy}\, du\, dv\, d\mu_{i,j}(x,y) \\ &\qquad = \int_{\mathbb{R}^2} f(u,v)e^{-\mathbf{i}2\pi ux}e^{-\mathbf{j}2\pi vy}\, du\, dv\, d\mu(x,y). \end{aligned} \tag{44}$$

Or, equivalently,

$$\lim_{i,j\to\infty} \int_{\mathbb{R}^2} \mathcal{F}_H\{f\}(x,y)\, d\mu_{i,j}(x,y) = \int_{\mathbb{R}^2} \mathcal{F}_H\{f\}(x,y)\, d\mu(x,y). \tag{45}$$

This is the desired result.

Theorem 10 *Let $\mu \in M^+(\mathbb{R}^2)$. If we assume that*

$$\int_{\mathbb{R}^2} |x|\, d\mu(x,y) < \infty, \quad \int_{\mathbb{R}^2} |y|\, d\mu(x,y) < \infty. \tag{46}$$

One has

$$\frac{\partial \mathcal{F}_{\mathbf{i},\mathbf{j}}(\mu)(u,v)}{\partial u} = (-\mathbf{i}2\pi) \int_{\mathbb{R}^2} x\, e^{-\mathbf{i}2\pi ux} e^{-\mathbf{j}2\pi vy}\, d\mu(x,y). \tag{47}$$

and

$$\frac{\partial \mathcal{F}_{\mathbf{i},\mathbf{j}}(\mu)(u,v)}{\partial v} = \int_{\mathbb{R}^2} y\, e^{-\mathbf{i}2\pi ux} e^{-\mathbf{j}2\pi vy}\, d\mu(x,y)(-\mathbf{j}2\pi). \tag{48}$$

Proof We only prove the first statement. In fact, we have

$$\begin{aligned} &\frac{\mathcal{F}_{\mathbf{i},\mathbf{j}}(\mu)(u+h_1,v) - \mathcal{F}_{\mathbf{i},\mathbf{j}}(\mu)(u,v)}{h_1} \\ &= \int_{\mathbb{R}^2} \frac{1}{h_1}\left(\mathrm{e}^{-\mathbf{i}2\pi(u+h_1)x} \mathrm{e}^{-\mathbf{j}2\pi vy} - \mathrm{e}^{-\mathbf{i}2\pi ux} \mathrm{e}^{-\mathbf{j}2\pi vy}\right) d\mu(x,y) \\ &= \int_{\mathbb{R}^2} \mathrm{e}^{-\mathbf{i}2\pi ux} \frac{\left(\mathrm{e}^{-\mathbf{i}2\pi h_1 x} - 1\right)}{h_1} \mathrm{e}^{-\mathbf{j}2\pi vy}\, d\mu(x,y). \end{aligned} \tag{49}$$

It follows that

$$\begin{aligned} \left| \mathrm{e}^{-\mathbf{i}2\pi ux} \frac{\left(\mathrm{e}^{-\mathbf{i}2\pi h_1 x} - 1\right)}{h_1} \mathrm{e}^{-\mathbf{j}2\pi vy} \right| &= \left| \frac{\mathrm{e}^{-\mathbf{i}2\pi h_1 x} - 1}{h_1} \right| \\ &= \left| \mathrm{e}^{-\mathbf{i}\pi h_1 x} \frac{\left(\mathrm{e}^{-\mathbf{i}\pi h_1 x} - \mathrm{e}^{\mathbf{i}\pi h_1 x}\right)}{h_1} \right| \\ &= 2\left| \frac{\sin \pi h_1 x}{h_1} \right| \\ &\le 2\pi |x|. \end{aligned} \tag{50}$$

Applying the dominated convergence theorem we see that

$$h_1 \to 0, \quad \frac{\mathrm{e}^{-\mathbf{i}2\pi h_1 x} - 1}{h_1} \longrightarrow -\mathbf{i}2\pi x, \tag{51}$$

which yields the assertion.

Theorem 11 *If g is a quaternion continuous function defined on $\mathbb{R}^2$ then it is the quaternion Fourier transform of a positive measure if and only if, there exists a constant C such that*

$$\left|\int_{\mathbb{R}^2} \mathcal{F}_H\{f\}(u,v)g(-u,-v))\,dudv\right| \leq C \sup_{(x,y)\in\mathbb{R}^2} |f(x,y)|, \tag{52}$$

for every continuous function $f \in L^1(\mathbb{R}^2; \mathbb{H})$.

Proof Taking $g = \mathcal{F}_{\mathbf{j},\mathbf{i}}(\mu)$ applying Theorem (6) we see that

$$\begin{aligned}\left|\int_{\mathbb{R}^2} \mathcal{F}_H\{f\}(u,v)g(-u,-v))\,dudv\right. &= \left|\int_{\mathbb{R}^2} \mathcal{F}_H\{f\}(u,v)\mathcal{F}_{\mathbf{j},\mathbf{i}}(\mu)(-u,-v))\,dudv\right| \\ &= \left|\int_{\mathbb{R}^2} f(x,y)\,d\mu(x,y)\right| \\ &\leq \sup_{(x,y)} \int_{\mathbb{R}^2} \left|f(x,y)\right| d\mu(x,y). \end{aligned} \tag{53}$$

On the other hand, suppose the inequality is valid. The the linear functional g which map f to is a bounded, continuous linear functional on the set of continuous function f. By the Riesz representation theorem, g maps f. Using Parseval's theorem yields

$$\int_{\mathbb{R}^2} \mathcal{F}_H\{f\}(u,v)g(-u,-v))\,dudv = \int_{\mathbb{R}^2} \mathcal{F}_H\{f\}(u,v)\mathcal{F}_{\mathbf{j},\mathbf{i}}(\mu)(-u,-v))\,dudv. \tag{54}$$

This must hold for all f. So, $g = \mathcal{F}_{\mathbf{j},\mathbf{i}}(\mu)$.

5 Conclusion

The Fourier transform of finite measure has been extended to the quaternion plane so-called the quaternion Fourier transform of finite measure. Several essential properties including translation, Parseval's formula, positive definite, partial derivative, and measure convergence are studied. These properties are generalizations of corresponding properties of the Fourier transform of finite measure in real and complex regions. We have shown that some properties of the Fourier transform of finite measure does not hold in the setting of the quaternion Fourier transform of finite measure.

Acknowledgements The authors would like to acknowledge the financial support of Grant from Ministry of Research, Technology and Higher Education, Indonesia under WCR scheme.

References

1. Ell, T.A.: Quaternion-Fourier transforms for analysis of two-dimensional linear time-invariant partial differential systems. In: Proceeding of the 32nd Conference on Decision and Control, San Antonio, Tx, pp. 1830–1841 (1993)
2. Ell, T.A., Sangwine, S.J.: Hypercomplex Fourier transform of color images. IEEE Trans. Sig. Process. **16**(1), 22–5 (2007)
3. Georgiev, S., Morais, J.: Bochner's theorems in the framework of quaternion analysis. In: Hitzer, E., Sangwine, S.J. (eds) Quaternion and Clifford Fourier Transforms and Wavelets. Trends in Mathematics. Birkhäuser, Basel (2013)
4. Georgiev, S., Morais, J., Kou, K.I., Sprößig, W.: Bochner-Minlos Theorem and Quaternion Fourier Transform. In: Hitzer, E., Sangwine, S.J. (eds) Quaternion and Clifford Fourier Transforms and Wavelets. Trends in Mathematics. Birkhäuser, Basel (2013)
5. Bülow, T.: Hypercomplex Spectral Signal Representations for the Processing and Analysis of Images. Ph.D. thesis, University of Kiel, Germany (1999)
6. Morais, J., Georgiev, S., Sprößig, W.: Real Quaternionic Calculus Handbook. Birkhäuser, New York (2014)
7. Haoui, Y.E., Fahlaoui, S.: Miyachi's Theorem for the quaternion Fourier transform. Circuits Syst. Sig. Process. **39**(4), 2193–2206 (2020)
8. Bahri, M.: Quaternion algebra-valued wavelet transform. Appl. Math. Sci. **5**(71) 3531–3540 (2011)
9. Bahri, M., Lawi, A., Aris, N., Saleh, A.F., Nur, M.: Relationships between convolution and correlation for Fourier transform and quaternion Fourier transform. Int. J. Math. Anal. **7**(43), 2101–2109 (2013)
10. Chen, L.P., Kou, K.I., Liu, M.S.: Pitt's inequality and the uncertainty principle associated with the quaternion Fourier transform. J. Math. Anal. Appl. **423**(1), 681–700 (2015)
11. Hitzer, E.: Quaternion Fourier transform on quaternion fields and generalizations. Adv. Appl. Clifford Algebr. **17**(3), 497–517 (2007)
12. Hitzer, E.: Directional uncertainty principle for quaternion Fourier transform. Adv. Appl. Clifford Algebr. **20**(2), 271–284 (2010)
13. Abouelaz, A., Achak, A., Daher, R., Safouane, N.: Donoho-Stark's uncertainty principle for the quaternion Fourier transform. Bol. Soc. Mat. Mex. (2019). https://doi.org/10.1007/s40590-019-00251-5
14. Lian, P.: Sharp Hausdorff-Young Inequalities for the quaternion Fourier transforms. Proc. Amer. Math. Soc. **148**, 697–703 (2020)
15. Georgiev, S., Jday, R.: Bochner-Minlos theorem in the frame of real Clifford algebras. Rend. Circ. Mat. Palermo, II. Ser (2020). https://doi.org/10.1007/s12215-020-00487-5
16. Loualid, E.M., Elgargati, A., Daher, R.: Quaternion Fourier transform and generalized Lipschitz Classes. Adv. Appl. Clifford Algebr. **31**(14), 271–284 (2021). https://doi.org/10.1007/s00006-020-01098-0
17. Cheng, D., Kou, K.I.: Generalized Sampling Expansions Associated with Quaternion Fourier Transform. https://arxiv.org/abs/1607.06116
18. Papoulis, A.: Signal Analysis. McGraw-Hill Inc., New York (1977)

On Relationship Between Ambiguity Function and Linear Canonical Ambiguity Function

Mawardi Bahri and Samsul Ariffin Abdul Karim

Abstract The linear canonical ambiguity function is a time frequency analysis method which combines the classical ambiguity function with the linear canonical transform. In the present work, we explore a direct relation between the linear canonical ambiguity function and the classical ambiguity function. We then utilize this interesting relation to obtain its fundamental properties such as the inversion theorem, general Moyal's formula and conjugate function. Finally, the linear canonical ambiguity function method is applied to evaluate the linear FM signal.

Keywords Linear canonical ambiguity function · Inversion formula · Moyal's formula · Linear FM signal

1 Introduction

As we know, the linear canonical transform (lCT) is a linear integral transform with three free parameters. It should be noticed that the lCT is a generalized form of the classical Fourier transform and the fractional Fourier transform [1] because they are particular cases of the lCT. Nowadays, the lCT has been broadly studied and applied in different areas of science and technology, including signal processing, optics and others [2, 3]. Many essential properties of this transformation such as the linearity, shifting, scaling, modulation, and inequalities have therefore been investigated in the literature (see, e.g., [4–9]). As proposed in [10–15], it is shown that the lCT plays a central role in constructing the different types of generalized transformations like

M. Bahri (✉)
Department of Mathematics, Hasanuddin University, Makassar 90245, Indonesia
e-mail: mawardibahri@gmail.com

S. A. Abdul Karim
Fundamental and Applied Sciences Department, Centre for Systems Engineering (CSE), Institute of Autonomous System, Universiti Teknologi PETRONAS, Bandar Seri Iskandar, 32610 Seri Iskandar, Perak Darul Ridzuan, Malaysia
e-mail: samsul_ariffin@utp.edu.my

© Institute of Technology PETRONAS Sdn Bhd 2022

S. A. Abdul Karim and A. Shafie (eds.), *Towards Intelligent Systems Modeling and Simulation*, Studies in Systems, Decision and Control 383,
https://doi.org/10.1007/978-3-030-79606-8_7

the Wigner-Ville distribution related to lCT, the ambiguity function associated with lCT, the windowed linear canonical transform and so on.

The main goal of the present research is to provide the different proof of useful properties of the linear canonical ambiguity function such as the inversion theorem, Moyal's formula and conjugate function. We also propose the application of the linear canonical ambiguity function to the linear FM signals. To achieve this, we recall the definition of the classical ambiguity function and its useful properties. Based on the properties we explore link between the linear canonical ambiguity function and classical ambiguity function. We develop this useful relation to find the main properties of the linear canonical ambiguity function.

The outline of the current article is as follows. In Sect. 2 we shortly review the basic notations, definition of the lCT and its useful properties which will be used throughout the following section. Section 3 reminds the definition and well-known properties of the classical ambiguity function. In Sect. 4 we focus our attention to the derivation of essential properties of the linear canonical ambiguity function using the relation between linear canonical ambiguity function and ambiguity Function. The application of the linear canonical ambiguity function to the linear FM signal is presented in Sect. 5. Some conclusions are drawn in Sect. §6.

2 Linear Canonical Transform

In this segment, we start with a brief of the definition and useful properties related to the linear canonical transform (lCT). These properties are very useful in the next section. For some recent results associated with the lCT, we refer the interested reader to the papers [16, 17].

Definition 1 Let $B = (a, b, c, d) = \begin{bmatrix} a & b \\ c & d \end{bmatrix} \in \mathbb{R}^{2\times 2}$ be a matrix parameter satisfying $det(B) = ad - bc = 1$. The lCT of a function $f \in L^1(\mathbb{R})$ is defined as

$$L_B\{f\}(u) = \int_{-\infty}^{\infty} f(x) C_B(u, x)\, dx, \quad b \neq 0 \tag{1}$$

and

$$L_B\{f\}(u) = \sqrt{d} e^{i\frac{cd}{2}u^2} f(du),\ b = 0, \tag{2}$$

where $C_B(u, x)$ is so-called kernel of the lCT described by

$$C_B(u, x) = \frac{1}{\sqrt{2\pi b}} e^{\frac{i}{2}(\frac{a}{b}x^2 - \frac{2}{b}xu + \frac{d}{b}u^2 - \frac{\pi}{2})}. \tag{3}$$

It is obvious that the kernel function of the lCT has the following properties:

$$C_{B^{-1}}(x,u)=\overline{C_B(u,x)}=\frac{1}{\sqrt{2\pi b}}\mathrm{e}^{-\frac{i}{2}(\frac{a}{b}x^2-\frac{2}{b}xu+\frac{d}{b}u^2-\frac{\pi}{2})}, \tag{4}$$

and

$$\int_{-\infty}^{\infty} C_B(x,u)C_{B^{-1}}(u,t)\,du=\delta(x-t),$$

where δ is the Dirac delta function. As can easily seen from Eq. (1) that for $b=0$ the lCT of a signal is essentially a chirp multiplication. In this respect throughout this paper we will consider $b\neq 0$.

It is worthing noting that for $B=\begin{bmatrix} 0 & 1 \\ -1 & 0\end{bmatrix}$, the lCT above coincides with the Fourier transform definition (7) times the factor $\frac{\mathrm{e}^{-i\frac{\pi}{4}}}{\sqrt{2\pi b}}$. The inversion formula of the lCT is expressed as

$$\begin{aligned} f(x) &= \int_{-\infty}^{\infty} L_B\{f\}(u)\,\overline{C_B(u,x)}\,du \\ &= \frac{1}{\sqrt{2\pi b}}\int_{-\infty}^{\infty} L_B\{f\}(u)\,\mathrm{e}^{-\frac{i}{2}(\frac{a}{b}x^2-\frac{2}{b}xu+\frac{d}{b}u^2-\frac{\pi}{2})}\,du. \end{aligned} \tag{5}$$

A useful fact is the natural link between the lCT and the Fourier transform described by

$$L_B\{f\}(u)=\frac{\mathrm{e}^{-i\frac{\pi}{4}}}{\sqrt{2\pi b}}\mathrm{e}^{\frac{id}{2b}u^2}F\{\mathrm{e}^{\frac{ia}{2b}x^2}f\}\left(\frac{u}{b}\right). \tag{6}$$

Here $F\{f\}(u)$ describes the Fourier transform of the complex function $f\in L^1(\mathbb{R})$ given by (see [18])

$$F\{f\}(u)=\int_{-\infty}^{\infty} f(x)\mathrm{e}^{-ixu}\,dx. \tag{7}$$

Some famous properties of the linear canonical transform are collected in the following lemmas.

Lemma 1 *For every $f\in L^2(\mathbb{R})$, the lCT of a shift by real number k is given by*

$$L_B\{T_kf\}(u)=L_B\{f(x-k)\}(u)e^{-\frac{iack^2}{2}+icku}L_B\{f\}(u-ak). \tag{8}$$

Lemma 2 *Given $f \in L^2(\mathbb{R})$ and $u_0 \in \mathbb{R}$, we have*

$$L_B\{M_{u_0} f\}(u) = L_B\{e^{ixu_0} f(x)\}(u) e^{-\frac{ibdu_0^2}{2}+idu_0u} L_B\{f\}(u - bu_0). \tag{9}$$

Lemma 3 *Given $f \in L^2(\mathbb{R})$, we have*

$$L_B\{M_{u_0} T_k f\}(u) = e^{-i(ack^2+bdu_0^2)/2+i(ck+du_0)u-ibcku_0} L_B\{f\}(u - ak - bu_0). \tag{10}$$

Definition 2 For $1 \leq p < \infty$, the Lebesgue space $L^p(\mathbb{R})$ is defined as the Banach space of measurable functions on $\mathbb{R}$ with the norm

$$\|f\|_{L^p(\mathbb{R})} = \left(\int_{-\infty}^{\infty} |f(x)|^p \, dx \right)^{1/p} < \infty. \tag{11}$$

In the space $L^2(\mathbb{R})$, we introduce the inner product as

$$\langle f, g \rangle_{L^2(\mathbb{R})} = \int_{-\infty}^{\infty} f(x)\overline{g(x)} \, dx,$$

and its norm $\|f\|_{L^2(\mathbb{R})} = \sqrt{\langle f, f \rangle_{L^2(\mathbb{R})}}$.

Lemma 4 (lCT Parseval) *Let $f, g \in L^1(\mathbb{R}) \cap L^2(\mathbb{R})$. The direct connection between the functions f, g and their lCT is described by*

$$\langle f, g \rangle_{L^2(\mathbb{R})} = \langle L_B\{f\}, L_B\{g\} \rangle_{L^2(\mathbb{R})}, \tag{12}$$

and

$$\|f\|^2_{L^2(\mathbb{R})} = \|L_B\{f\}\|^2_{L^2(\mathbb{R})}. \tag{13}$$

3 Ambiguity Function

In this section we introduce the definition of the ambiguity function and collect well-known properties. For more details we refer the interested reader to [19, 20] and the references therein.

Definition 3 Let $f, g \in L^2(\mathbb{R})$. the classical ambiguity function of f and g is defined by

$$\mathcal{A}_{f,g}(t, u) = \int_{-\infty}^{\infty} f\left(x + \frac{t}{2}\right)\overline{g\left(x - \frac{t}{2}\right)} \mathrm{e}^{-iux} \, dx. \tag{14}$$

Some essential properties of the classical ambiguity function are summarized as follows.

1. Conjugate function

$$\overline{\mathcal{A}_{f,g}(t,u)} = \mathcal{A}_{g,f}(-t,-u) \tag{15}$$

2. Shifting

$$\mathcal{A}_{T_k f, T_k g}(t,u) = \mathrm{e}^{-iuk}\mathcal{A}_{f,g}(t,u) \tag{16}$$

3. Modulation

$$\mathcal{A}_{M_{u_0} f, M_{u_0} g}(t,u) = \mathrm{e}^{-iu_0 k}\mathcal{A}_{f,g}(t,u-u_0) \tag{17}$$

4. Moyal's formula

$$\int_{-\infty}^{\infty}\int_{-\infty}^{\infty} \mathcal{A}_{f_1,g_1}(t,u)\mathcal{A}_{f_2,g_2}(t,u)\,dtdu = (f_1,f_2)(g_1,g_2) \tag{18}$$

5. Inversion theorem

$$f(t) = \frac{1}{2\pi\overline{g(0)}}\int_{-\infty}^{\infty} \mathcal{A}_{f,g}(\frac{t}{2},u)\,\mathrm{e}^{iut}\,du, \tag{19}$$

provided $\overline{g(0)} \neq 0$.

In [15] the author defined the classical ambiguity function in the frequency domain as follows.

Definition 4 Let $F\{f\}, F\{g\} \subset L^2(\mathbb{R})$. The ambiguity function related to the Fourier transform of f and g is defined by

$$\mathcal{A}_{f,g}(t,u) = \int_{\infty}^{\infty} F\{f\}(w+\frac{u}{2})\overline{F\{g\}(w-\frac{u}{2})}\mathrm{e}^{iwt}\,dw. \tag{20}$$

4 Linear Canonical Ambiguity Function (lCAF)

In this part, we first provide the definition of the linear canonical ambiguity function (see [11, 12, 21]). We introduce an interesting relation between the linear canonical ambiguity function and the classical ambiguity function. We develop this relation to establish the fundamental properties of the linear canonical ambiguity function, which their proofs are different from the literature. We start with a formal definition of the linear canonical ambiguity function.

Definition 5 Suppose that two complex functions f, g belong to $L^2(\mathbb{R})$. Define the linear canonical ambiguity function of f and g associated with the matrix parameter B by

$$\begin{aligned}\mathcal{A}^B_{f,g}(t,u) &= \int_{-\infty}^{\infty} f\big(x+\frac{t}{2}\big)\overline{g\big(x-\frac{t}{2}\big)}C_B(u,x)\,dx \\ &= \int_{-\infty}^{\infty} f\big(x+\frac{t}{2}\big)\overline{g\big(x-\frac{t}{2}\big)}\frac{1}{\sqrt{2\pi b}}\mathrm{e}^{\frac{i}{2}(\frac{a}{b}x^2-\frac{2}{b}xu+\frac{d}{b}u^2-\frac{\pi}{2})}\,dx. \end{aligned} \tag{21}$$

Observe that when $B = \begin{bmatrix} 0 & 1 \\ -1 & 0 \end{bmatrix}$ we obtain

$$\mathcal{A}^B_{f,g}(t,u) = \frac{1}{\sqrt{2\pi}}\mathrm{e}^{-i\frac{\pi}{4}}\mathcal{A}_{f,g}(t,u), \tag{22}$$

which shows that the ambiguity function is a special case of the linear canonical ambiguity function. Now plugging (3) into the right-hand side of (21) gives

$$\mathcal{A}^B_{f,g}(t,u) = \int_{-\infty}^{\infty} R_{f,g}(x,t)C_B(u,x)\,dx, \tag{23}$$

where

$$R_{f,g}(x,t) = f\big(x+\frac{t}{2}\big)\overline{g\big(x-\frac{t}{2}\big)}. \tag{24}$$

Suppose that the lCAF is the lCT of the function $R_{f,g}(x,t)$ with respect to x. Then we have

$$\mathcal{A}^B_{f,g}(t,u) = L_B\{R_{f,g}(x,t)\}(u). \tag{25}$$

Theorem 1 *The linear ambiguity function of $f \in L^2(\mathbb{R})$ is related to the classical ambiguity function via*

$$e^{-i\frac{d}{2b}u^2}\mathcal{A}^B_{f,g}(t,u) = \mathcal{A}_{\check{f},g}(t,\frac{u}{b}), \tag{26}$$

where

$$\check{f}\big(x+\frac{t}{2}\big) = \frac{e^{-i\frac{\pi}{4}}}{\sqrt{2\pi b}}e^{i\frac{a}{2b}x^2}f\big(x+\frac{t}{2}\big). \tag{27}$$

Proof A straightforward computation shows that

$$
\begin{aligned}
\mathcal{A}^B_{f,g}(t,u) &= \int_{-\infty}^{\infty} f(x+\frac{t}{2})\overline{g(x-\frac{t}{2})}\frac{1}{\sqrt{2\pi b}}e^{\frac{i}{2}(\frac{a}{b}x^2-\frac{2}{b}xu+\frac{d}{b}u^2-\frac{\pi}{2})}\,dx \\
&= e^{i\frac{d}{2b}u^2}\int_{-\infty}^{\infty}\frac{e^{-i\frac{\pi}{4}}}{\sqrt{2\pi b}}e^{i\frac{a}{2b}x^2}f(x+\frac{t}{2})\overline{g(x-\frac{t}{2})}e^{-i\frac{xu}{b}}\,dx \\
&= e^{i\frac{d}{2b}u^2}\int_{-\infty}^{\infty}\frac{e^{-i\frac{\pi}{4}}}{\sqrt{2\pi b}}e^{i\frac{a}{2b}x^2}f(x+\frac{t}{2})\overline{g(x-\frac{t}{2})}e^{-i\frac{xu}{b}}\,dx \\
&= e^{i\frac{d}{2b}u^2}\int_{-\infty}^{\infty}\check{f}(x+\frac{t}{2})\overline{g(x-\frac{t}{2})}e^{-i\frac{xu}{b}}\,dx \\
&= e^{i\frac{d}{2b}u^2}A_{\check{f},g}(t,\frac{u}{b}),
\end{aligned}
\tag{28}
$$

where in the fourth equality we have applied relation (14). This yields the desired result.

Now taking $x = \frac{t}{2}$, Eq. (27) can be rewritten as

$$
\check{f}(t) = \frac{e^{-i\frac{\pi}{4}}}{\sqrt{2\pi b}}e^{i\frac{a}{2b}(\frac{t}{2})^2}f(t). \tag{29}
$$

Let us now apply Theorem 1 to derive the fundamental properties of the linear canonical ambiguity function.

Theorem 2 *For any function $f, g \in L^2(\mathbb{R})$, we have*

$$
f(t) = \frac{1}{\overline{g(0)}}\int_{-\infty}^{\infty}\mathcal{A}^B_{f,g}(t,u)\mathcal{C}_{B^{-1}}(u,\frac{t}{2})\,du. \tag{30}
$$

Proof Replacing f by $\check{f}$ defined by (29) both sides of the inversion theorem for the classical ambiguity function described by (19) we see that

$$
\begin{aligned}
\check{f} &= \frac{1}{2\pi\overline{g(0)}}\int_{-\infty}^{\infty}A_{\check{f},g}(t,u)\,e^{i\frac{1}{2}ut}\,du \\
&= \frac{1}{2b\pi\overline{g(0)}}\int_{-\infty}^{\infty}A_{\check{f},g}(t,\frac{u}{b})\,e^{i\frac{1}{2b}ut}\,du
\end{aligned}
\tag{31}
$$

According (26) and (29) we have

$$\frac{e^{-i\frac{\pi}{4}}}{\sqrt{2\pi b}} e^{i\frac{a}{2b}(\frac{t}{2})^2} f(t) = \frac{1}{2b\pi \overline{g(0)}} \int_{-\infty}^{\infty} e^{-i\frac{d}{2b}u^2} \mathcal{A}_{f,g}^B(t,u) e^{i\frac{t}{2b}ut}\, du \tag{32}$$

Simplifying it and using (21) we arrive at

$$\begin{aligned} f(t) &= \frac{1}{\overline{g(0)}} \int_{-\infty}^{\infty} \mathcal{A}_{f,g}^B(t,u) \frac{1}{\sqrt{2\pi b}} e^{-\frac{i}{2}(\frac{a}{b}(\frac{t}{2})^2 - \frac{1}{b}tu + \frac{d}{b}u^2 - \frac{\pi}{2})}\, du \\ &= \frac{1}{\overline{g(0)}} \int_{-\infty}^{\infty} \mathcal{A}_{f,g}^B(t,u) C_{B^{-1}}(u, \frac{t}{2})\, du, \end{aligned} \tag{33}$$

and the prof is complete.

Theorem 3 *For all functions $f_1, f_2, g_1, g_2 \in L^2(\mathbb{R})$, then the following result holds*

$$\int_{-\infty}^{\infty}\int_{-\infty}^{\infty} \mathcal{A}_{f_1,g_1}^B(t,u) \overline{\mathcal{A}_{f_2,g_2}^B(t,u)}\, dt du = \langle f_1, f_2\rangle_{L^2(\mathbb{R})} \langle g_1, g_2\rangle_{L^2(\mathbb{R})} \tag{34}$$

Especially, we have

$$\int_{-\infty}^{\infty}\int_{-\infty}^{\infty} \left|\mathcal{A}_{f,g}^B(t,u)\right|^2 dt du = \| f\|^2_{L^2(\mathbb{R})} \|g\|^2_{L^2(\mathbb{R})} \tag{35}$$

and

$$\int_{-\infty}^{\infty}\int_{-\infty}^{\infty} \mathcal{A}_{f,f}^B(t,u) \overline{\mathcal{A}_{g,g}^B(t,u)}\, dt du = \left|\langle f, g\rangle_{L^2(\mathbb{R})}\right|^2. \tag{36}$$

Proof We only prove the first assertion. According to Moyal's formula for the classical ambiguity function defined by (18) we obtain

$$\frac{1}{2\pi} \int_{-\infty}^{\infty}\int_{-\infty}^{\infty} \mathcal{A}_{\check{f}_1,g_1}(t,u) \overline{\mathcal{A}_{\check{f}_2,g_2}(t,u)}\, dt du = \langle \check{f}_1, \check{f}_2\rangle_{L^2(\mathbb{R})} \langle g_1, g_2\rangle_{L^2(\mathbb{R})}. \tag{37}$$

Setting $u = \frac{u}{b}$, we can write the above identity in the form

$$\frac{1}{2\pi b}\int_{-\infty}^{\infty}\int_{-\infty}^{\infty} \mathcal{A}_{\check{f}_1,g_1}(t,\frac{u}{b})\overline{\mathcal{A}_{\check{f}_2,g_2}(t,\frac{u}{b})}\,dtdu = \langle \check{f}_1, \check{f}_2\rangle_{L^2(\mathbb{R})}\langle g_1, g_2\rangle_{L^2(\mathbb{R})}. \tag{38}$$

Here,

$$\check{f}_1(t) = \frac{e^{-i\frac{\pi}{4}}}{\sqrt{2\pi b}} e^{i\frac{a}{2b}(\frac{t}{2})^2} f_1(t), \tag{39}$$

and

$$\check{f}_2(t) = \frac{e^{-i\frac{\pi}{4}}}{\sqrt{2\pi b}} e^{i\frac{a}{2b}(\frac{t}{2})^2} f_2(t). \tag{40}$$

Observe first that

$$\begin{aligned}\langle \check{f}_1, \check{f}_2\rangle_{L^2(\mathbb{R})} &= \int_{-\infty}^{\infty} \frac{e^{-i\frac{\pi}{4}}}{\sqrt{2\pi b}} e^{i\frac{a}{2b}(\frac{y}{2})^2} f_1(y)\overline{\frac{e^{-i\frac{\pi}{4}}}{\sqrt{2\pi b}} e^{i\frac{a}{2b}(\frac{y}{2})^2} f_2(y)}\,dy \\ &= \frac{1}{2\pi b}\int_{-\infty}^{\infty} f_1(y)\overline{f_2(y)}\,dy \\ &= \frac{1}{2\pi b}\langle f_1, f_2\rangle_{L^2(\mathbb{R})}.\end{aligned} \tag{41}$$

Thus

$$\begin{aligned}\frac{1}{2\pi b}\int_{-\infty}^{\infty}\int_{-\infty}^{\infty} e^{-i\frac{d}{2b}u^2}\mathcal{A}^B_{f_1,g_1}(t,u)\overline{e^{-i\frac{d}{2b}u^2}\mathcal{A}^B_{f_2,g_2}(t,u)}\,dtdu \\ = \tfrac{1}{2\pi b}\langle f_1, f_2\rangle_{L^2(\mathbb{R})}\langle g_1, g_2\rangle_{L^2(\mathbb{R})}.\end{aligned} \tag{42}$$

This implies that

$$\int_{-\infty}^{\infty}\int_{-\infty}^{\infty} \mathcal{A}^B_{f_1,g_1}(t,u)\overline{\mathcal{A}^B_{f_2,g_2}(t,u)}\,dtdu = \langle f_1, f_2\rangle_{L^2(\mathbb{R})}\langle g_1, g_2\rangle_{L^2(\mathbb{R})}, \tag{43}$$

and the prof is complete.

We now provide the different proof of the conjugate function property of the the linear canonical ambiguity function by applying Theorem 1.

Theorem 4 *Let f, g be in $\in L^2(\mathbb{R})$. Then we have*

$$\overline{\mathcal{A}^B_{f,g}(t,u)} = \mathcal{A}^{B^{-1}}_{g,f}(-t,u). \tag{44}$$

Proof Including $\check{f}$ into the conjugate function property of the classical ambiguity function defined by (15) we easily obtain

$$\overline{A_{\check{f},g}(t,\frac{u}{b})} = A_{g,\check{f}}(-t,-\frac{u}{b}). \tag{45}$$

The right-hand side of (45) can be rewritten as

$$\begin{aligned} A_{g,\check{f}}(-t,-\frac{u}{b}) &= \int_{-\infty}^{\infty} g(x-\frac{t}{2})\overline{\check{f}(x+\frac{t}{2})}e^{\frac{iux}{b}}\,dx \\ &= \int_{-\infty}^{\infty} g(x-\frac{t}{2})\overline{\frac{e^{-i\frac{\pi}{4}}}{\sqrt{2\pi b}}e^{i\frac{a}{2b}x^2}f(x+\frac{t}{2})}e^{\frac{iux}{b}}\,dx \\ &= \int_{-\infty}^{\infty} g(x-\frac{t}{2})\overline{f(x+\frac{t}{2})}e^{-\frac{i}{2}(\frac{a}{b}x^2-\frac{2}{b}xu-\frac{\pi}{2})}\,dx. \end{aligned} \tag{46}$$

Applying (26) into the left-hand side of (45) yields

$$\begin{aligned} \overline{\mathcal{A}^{B}_{f,g}(t,u)} &= \int_{-\infty}^{\infty} g(x-\frac{t}{2})\overline{f(x+\frac{t}{2})}e^{-\frac{i}{2}(\frac{a}{b}x^2-\frac{2}{b}xu+\frac{d}{b}u^2-\frac{\pi}{2})}\,dx \\ &= \mathcal{A}^{B^{-1}}_{g,f}(-t,u), \end{aligned} \tag{47}$$

which was to be proved.

The next result presents another form of relation (35).

Theorem 5 *Suppose that $f, g \in L^2(\mathbb{R})$. For any u and t, one has*

$$\frac{1}{2}\int_{-\infty}^{\infty}\int_{-\infty}^{\infty} \left|\mathcal{A}^{B}_{f,g}(t,u)\right|^2 du\,dt = \|f\|^2_{L^2(\mathbb{R})}\|g\|^2_{L^2(\mathbb{R})}. \tag{48}$$

Proof Taking into account Plancherel's formula for the lCT (13) we see that

$$\begin{aligned} \int_{-\infty}^{\infty} \left|L_B\{R_{f,g}(x,t)\}(u)\right|^2 du &= \int_{-\infty}^{\infty} \left|R_{f,g}(x,t)\right|^2 dx \\ &= \int_{-\infty}^{\infty} \left|f(x+\frac{t}{2})\overline{g(x-\frac{t}{2})}\right|^2 dx. \end{aligned} \tag{49}$$

We further obtain

$$
\begin{aligned}
&\int_{-\infty}^{\infty}\int_{-\infty}^{\infty} \left|L_B\{R_{f,g}(x,t)\}(u)\right|^2 du\,dt \\
&= \int_{-\infty}^{\infty}\int_{-\infty}^{\infty} \left| f(x+\frac{t}{2})\overline{g(x-\frac{t}{2})}\right|^2 dx\,dt \\
&= \int_{-\infty}^{\infty} \left| f(x+\frac{t}{2})\right|^2 dt \int_{-\infty}^{\infty} \left| g(x-\frac{t}{2})\right|^2 dx \\
&= 2\int_{-\infty}^{\infty} |f(y)|^2 dy \int_{-\infty}^{\infty} |g(z)|^2 dz,
\end{aligned}
\tag{50}
$$

and (48) result.

Based on Definition 4 we propose the new definition of the linear canonical ambiguity function in terms of the lCT $L_B\{f\}$ and $L_B\{g\}$ as follows.

Definition 6 Suppose that $L_B\{f\}$, $L_B\{g\}$ belong to $L^2(\mathbb{R})$. Define the linear canonical ambiguity function of the lCT of the functions f and g associated with the matrix parameter B by

$$
\mathcal{A}^B_{f,g}(t,u) = \int_{-\infty}^{\infty} L_B\{f\}(w+\frac{u}{2})\overline{L_B\{g\}(w-\frac{u}{2})}\frac{1}{\sqrt{2\pi b}}e^{\frac{i}{2}(\frac{a}{b}w^2-\frac{2}{b}wt+\frac{d}{b}t^2-\frac{\pi}{2})}\,dw.
\tag{51}
$$

5 Application

In the following, the simple application of the linear canonical ambiguity function is described in some detail.

Theorem 6 *If the linear FM signal is expressed as*

$$
h(t) = f(t)e^{i\frac{a}{2b}t^2}.
\tag{52}
$$

Then we have

$$e^{i\frac{d}{2b}\alpha^2t^2}e^{-i\frac{d}{b}u\alpha t}\mathcal{A}_{h,h}^{B}(t,u)=\mathcal{A}_{f,f}^{B}(t,u-\alpha t). \tag{53}$$

Proof It is easy to get

$$R_{h,h}(x,t)=f(x+\frac{t}{2})\overline{f(x-\frac{t}{2})e^{i\frac{\alpha}{b}xt}}. \tag{54}$$

This implies that

$$\mathcal{A}_{h,h}^{B}(t,u)=\int_{-\infty}^{\infty}f(x+\frac{t}{2})\overline{f(x-\frac{t}{2})}e^{i\frac{\alpha}{b}xt}\frac{1}{\sqrt{2\pi b}}e^{\frac{i}{2}(\frac{a}{b}x^2-\frac{2}{b}xu+\frac{d}{b}u^2-\frac{\pi}{2})}\,dx. \tag{55}$$

The above expression may be rewritten in the form

$$\begin{aligned}
&\mathcal{A}_{h,h}^{B}(t,u)\\
&=\int_{-\infty}^{\infty}f(x+\frac{t}{2})\overline{f(x-\frac{t}{2})}\frac{1}{\sqrt{2\pi b}}e^{\frac{i}{2}(\frac{a}{b}x^2-\frac{2}{b}(u-\alpha t)x+\frac{d}{b}(u-\alpha t)^2-\frac{\pi}{2})}\,dx\\
&=\int_{-\infty}^{\infty}e^{-i\frac{d}{2b}\alpha^2t^2}e^{i\frac{d}{b}u\alpha t}f(x+\frac{t}{2})\overline{f(x-\frac{t}{2})}\frac{1}{\sqrt{2\pi b}}e^{\frac{i}{2}(\frac{a}{b}x^2-\frac{2}{b}(u-\alpha t)x+\frac{d}{b}(u-\alpha t)^2-\frac{\pi}{2})}\,dx\\
&=e^{-i\frac{d}{2b}\alpha^2t^2}e^{i\frac{d}{b}u\alpha t}\mathcal{A}_{f,f}^{B}(t,u-\alpha t).
\end{aligned}$$

This is required result.

It is straightforward to verify that for $B=\begin{bmatrix}0 & 1\\ -1 & 0\end{bmatrix}$ we get

$$\mathcal{A}_{h,h}^{B}(t,u)=\mathcal{A}_{f,f}(t,u-\alpha t). \tag{56}$$

In what follows, we evaluate the linear canonical ambiguity function of the Gaussian function, which will be useful later.

Example 1 For the Gaussian function $f(t)=e^{-\alpha t^2}$, we have

$$\begin{aligned}
\mathcal{A}^B_{f,f}(t,u) &= \int_{-\infty}^{\infty} e^{-\alpha(x+\frac{t}{2})^2} e^{-\alpha(x-\frac{t}{2})^2} \frac{1}{\sqrt{2\pi b}} e^{\frac{i}{2}(-\frac{2}{b}xu+\frac{d}{b}u^2-\frac{\pi}{2})}\, dx \\
&= \int_{-\infty}^{\infty} e^{-\alpha(x^2+\frac{t^2}{2})} \frac{1}{\sqrt{2\pi b}} e^{\frac{i}{2}(\frac{a}{b}x^2-\frac{2}{b}xu+\frac{d}{b}u^2-\frac{\pi}{2})}\, dx \\
&= \int_{-\infty}^{\infty} e^{-\alpha(x^2+\frac{t^2}{2})} \frac{1}{\sqrt{2\pi b}} e^{\frac{i}{2}(-\frac{2}{b}xu+\frac{d}{b}u^2-\frac{\pi}{2})}\, dx \\
&= \frac{1}{\sqrt{2\pi b}} e^{-\frac{\alpha t^2}{2}} e^{\frac{i}{2}(\frac{d}{b}u^2-\frac{\pi}{2})} \int_{-\infty}^{\infty} e^{-\alpha x^2} e^{-i\frac{xu}{b}}\, dx \\
&= \frac{1}{\sqrt{2\pi b}} e^{-\frac{\alpha t^2}{2}} e^{\frac{i}{2}(\frac{d}{b}u^2-\frac{\pi}{2})} \sqrt{\frac{\pi}{\alpha}} e^{-\frac{u^2}{4\alpha b}} \\
&= \frac{1}{\sqrt{2\alpha b i}} e^{-\frac{\alpha t^2}{2}} e^{\frac{-1+2\alpha d i}{4\alpha b}u^2}. \qquad (57)
\end{aligned}$$

Now we consider theLFM signal of the form

$$h(t) = e^{-\alpha t^2} e^{i\frac{a}{2b}t^2}. \qquad (58)$$

Applying Theorem 6 and Example 1 we immediately obtain

$$\begin{aligned}
\mathcal{A}^B_{h,h}(t,u) &= e^{i\frac{d}{2b}\alpha^2 t^2} e^{-i\frac{d}{b}u\alpha t} \mathcal{A}^B_{f,f}(t, u-\alpha t) \\
&= e^{i\frac{d}{2b}\alpha^2 t^2} e^{-i\frac{d}{b}u\alpha t} \frac{1}{\sqrt{2\alpha b i}} e^{-\frac{\alpha t^2}{2}} e^{\frac{-1+2\alpha d i}{4\alpha b}(u-\alpha t)^2}. \qquad (59)
\end{aligned}$$

6 Conclusion

The linear canonical ambiguity function has been introduced. It is an extension of the classical ambiguity function in the lCT domains. Some of its fundamental properties were obtained by applying the relation between the classical ambiguity function and the linear canonical ambiguity function. We also have introduced the new definition of the linear canonical ambiguity function, which is a generalized form of the definition of the ambiguity function in the frequency domain. The simple application of the linear canonical ambiguity function method is studied to evaluate the linear FM signal.

Acknowledgements The first author would like to acknowledge the financial support of Grant from Ministry of Research, Technology and Higher Education, Indonesia under PDUPT scheme.

References

1. Zayed, Z.I.: On the relationship between the Fourier and fractional Fourier transforms. IEEE Sig. Process. Lett. **3**(12), 310–311 (1996)
2. Shi, J., Liu, Q., He, L.: Sampling and reconstruction in arbitrary measurement and approximation spaces associated with the linear canonical transform. IEEE Trans. Sig. Process. **64**(24), 6379–6391 (2016)
3. Gou, Y., Li, B.Z.: Blind image watermarking on linear canonical wavelet transform and QR decomposition. IET Sig. Process. **10**(10), 773–786 (2016)
4. Pei, S.C., Ding, J.J.: Eigen functions of linear canonical transform. IEEE Trans. Sig. Process. **50**, 11–26 (2002)
5. Zhao, J., Tao, R., Li, Y.L., Wang, Y.: Uncertainty principles for linear canonical transforms. IEEE Trans. Sig. Process. **57**(7), 2856–2858 (2009)
6. Wei, D., Ran, Q., Li, Y.: A convolution and correlation theorem for the linear canonical transform and its application. Circuits Syst. Sig. Process. **31**(1), 301–312 (2012)
7. Wei, D., Ran, Q., Li, Y.: New convolution theorem for the linear canonical transform and its translation invariance property. Optik-Int. J. Light Elecctron. Opt. **123**(16), 1478–1481 (2012)
8. Shi, J., Liu, X., Zhang, N.: On uncertainty principle for signal concentrations with fractional Fourier transform. Sig. Process. **92**(12), 2830–2836 (2012)
9. Feng, Q., Li, B.Z.: Convolution and correlation theorems for the two-dimensional linear canonical transform and its applications. IET Sig. Process. **10**(2), 125–132 (2016)
10. Bahri, M., Ashino, R.: Convolution and correlation theorem for Wigner-Ville distribution associated with linear canonical transform. In: Proceeding of 2015 12th International Conference on Information Technology-New Generation, Las Vegas, Nevada, pp. 341–346 (2015)
11. Che, T.W., Li, B.Z., Xu, T.Z.: Ambiguity function associated with the linear canonical transform. EURASIP J. Adv. Sig. Process. **138**, 1–14 (2012)
12. Bahri, M., Haddade, A., Toaha, S.: Some useful properties of ambiguity function associated with linear canonical transform. Far East J. Electron. Commun. **17**(2), 455–473 (2017)
13. Bahri, M., Ashino, R.: Convolution and Correlation theorems for Wigner-Ville distribution associated with linear canonical transform. In: 12th International Conference on Information Technology-New Generations, Las Vegas (2015)
14. Bahri, M., Ashino, R.: Some properties of windowed linear canonical transform and its logarithmic uncertainty principle. Int. J. wavelet Multiresolut Inf. Process. **14**(3) (2016)
15. Zhao, H., Ran, Q.W., Maa, J., Ta, L.Y.: Linear canonical ambiguity function and linear canonical transform moments. Optik-Int. J. Light Elecctron. Opt. **122**(6), 540–543 (2011)
16. Moshinsky, M., Quesnee, C.: Linear canonical transform and their unitary representations. J. Math. Phys. **12**, 1772–1783 (1971)
17. Collins, S.A.: Lens-system diffraction integral written in term of matrix optics. J. Opt. Soc. Am. **60**(9), 1168–1177 (1970)
18. Bracewell, R.: The Fourier Transform and its Applications. McGraw Hill, Boston (2000)
19. Debnath, L., Shah, F.A.: Wavelet Transforms and their Applications. Springer Nature, Switzerland (2020)
20. Gröchenig, K.: Foundation of Time-Frequency Analysis, Birkhäuser, Boston (2001)
21. Zhang, Z.C.: Novel Wigner distribution and ambiguity function associated with the linear canonical transform moments. Optik-Int. J. Light Elecctron. Opt. **127**, 4995–5012 (2016)

Numerical Solution of Oil Spill Weathering with Two Points Block Backward Differentiation Formula

Nooraini Zainuddin, Hira Soomro, Noraini Jamaludin, Siti Ainor Mohd Yatim, and Zarina Bibi Ibrahim

Abstract In this paper, the oil-weathering model, which consists of four first order ordinary differential equations (ODEs), were solved by using two-point block backward differentiation formula (2BDF). The 2BDF method gives two approximations for each integration step with a fixed ratio of changing step sizes. These four ODEs provide solutions for the oil slick area, fraction of evaporation, fraction of emulsification, and fraction of dissolution for the oil spill. Subsequently, these solutions were used to calculate the changes in the oil spill density and viscosity. The result was generated for crude oil at the temperatures of 273K and 283K with the water content at a maximum of 80%. Numerical solutions gained from 2BDF were plotted against the solutions from ode45. Error of computations between these two solvers suggests that the 2BDF can be an alternative method for the simulation of oil-weathering models.

Keywords Oil weathering model · Ordinary differential equations · Block BDF

N. Zainuddin (✉) · H. Soomro
Fundamental and Applied Sciences Department, Universiti Teknologi PETRONAS, 32610 Seri Iskandar, Bandar Seri Iskandar, Perak Darul Ridzuan, Malaysia
e-mail: aini_zainuddin@utp.edu.my

N. Jamaludin
Centre for Foundation Studies, Universiti Teknologi PETRONAS, 32610 Seri Iskandar, Bandar Seri Iskandar, Perak Darul Ridzuan, Malaysia
e-mail: noraini.jamaludin@utp.edu.my

S. A. M. Yatim
School of Distance Education, Universiti Sains Malaysia, USM, 11800 Penang, Malaysia
e-mail: ainor@usm.my

Z. B. Ibrahim
Department of Mathematics, Faculty of Science, Universiti Putra Malaysia (UPM), 43400 Selangor, Malaysia
e-mail: zarinabb@upm.edu.my

© Institute of Technology PETRONAS Sdn Bhd 2022
S. A. Abdul Karim and A. Shafie (eds.), *Towards Intelligent Systems Modeling and Simulation*, Studies in Systems, Decision and Control 383,
https://doi.org/10.1007/978-3-030-79606-8_8

In this study, the oil-weathering model, which consists of four first order ordinary differential equations (ODEs), were solved by using two-point block backward differentiation formula (2BDF). The 2BDF method gives two approximations for each integration step with a fixed ratio of changing step sizes. These four ODEs provide solutions for the oil slick area, fraction of evaporation, fraction of emulsification, and fraction of dissolution for the oil spill. Subsequently, these solutions were used to calculate the changes in the oil spill density and viscosity. The result was generated for crude oil at the temperatures of 273 and 283 K with the water content at a maximum of 80%. Numerical solutions gained from 2BDF were plotted against the solutions from ode45. Error of computations between these two solvers suggests that the 2BDF can be an alternative method for the simulation of oil-weathering models.

1 Introduction

Marine oil spill incident is defined as an accidental release or discharge of petroleum hydrocarbons at seas caused by human errors or natural disaster [1, 2]. Rafferty [3] listed nine of the biggest oil spills in the history. Some are listed here. In August 1983, the Castillo de Bellver oil tanker caught fire. The tanker was about 80 miles away from Cape Town, South Africa, in the South Atlantic Ocean. When the fire broke out, the tanker kept cruising and split into two sections. The stern, filled with nearly 110,000 tonnes of crude oil, cruised twenty-four miles away from the coast before it disappeared. The front part of the tanker was separated from the coast, where it was sunk by engineers. A part of the oil slick, however, was ignited during the eruption, causing much of the crude oil produced on the surface to be trapped in the Benguela current and flowed into the sea until they dispersed.

Another incident was caused by a rusted pipeline in the Russian Arctic, which led to the Kolva Rivers oil spill in 1994. Eight months of oil pooled around the ruptured spot, containing a dike which once replaced, caused nearly 84 million gallons of oil to leak into the Kolva Channel. About 186 km of tundra and wetlands were polluted with oil. Oil spills in the area released gallons of petroleum into the countryside, and the Kolva River.

The infamous Deepwater Horizon incident on 20 April 2010 occurred when a natural gas surge blew up a concrete wall, which had recently been constructed to screen a well drilled by the deep-water Horizon oil platform. The explosion caused a massive incidental oil leak in the antiquity of the Gulf of Mexico. The gas went up the platform's riser and ignited it, killing 11 staff and injuring 17 others. Two days later, the oil rig collapsed and fell. Some months later, by September 17, approximately 134 MGs of oil had spilled before the well was capped (as the United States District Court found), and almost 2100 km of areas of the United States Gulf Coast from Texas to Florida were oil coated.

Once the oil spill occurs, the spillage oil begins to change its chemical and physical properties over time. These changes are called oil-weathering process, which includes, among others, dispersion, evaporation, and spreading. The rate of these changes varies depending on the properties of the environment and oil spill. The environmental properties include wind speed, temperature of seawater, and density of seawater. The oil spillage properties refer to the oil viscosity and density, water content of the spillage oil, and the initial volume of oil spill.

Oil spills will definitely affect the marine life and environment, particularly in the immediate surroundings. These incidents come with a wide range of consequences and are often depicted in the reports as long-term environmental accidents. Such beliefs are appropriate, given the distressing pictures of oiled birds and other animals that frequently accompany them. Additionally, oil spills have both long- and short-term impacts on animals and their ecosystems.

Oil spills can have an effect on the atmosphere mostly in the following manners:

- Physical smothering of organisms: Smothering will hinder the body's ability to regulate essential functions like ventilation, feeding, and thermoregulation. High-viscosity oils worsen the issue.
- Chemical toxicity: This is typical of lighter chemical compounds that are more bio-available, i.e., ingested into muscles, tissues, and cells, and may have harmful effects that are sub-lethal or lethal.
- Ecological changes: The lack of main species with a special role in an ecological system causes this. Different organisms may perform similar roles in their place, and the consequences for the ecosystem as a whole will not be as serious. More harmful, though, is the replacement of a community's niche with species that conduct entirely different roles, altering ecosystem dynamics.
- Indirect effect: Oiling or cleanup activities result in the loss of shelter or habitat.

The potential impact of oils on open seas is mostly on the top of the column, but there can be a drowning of some types of oils and seabed impacts. The animals that are often affected by an open sea spill are planktons, whales, seabirds, marine mammals, and reptiles. The vulnerability of aquatic life due to elevated amounts of biologically or chemically released oil is the most common source of oil impacts in nearshore waters. Because of the large number of habitat types and species that live in this aquatic zone, the effects of oil spills in deeper inshore waters and shorelines are difficult to predict.

Therefore, study on oil weathering is vital as this awareness is essential for creating a spill response model and contingency plans [4]. Abascal et al. [5] had developed a statistical oil spill respond model which later is verified by using the real data from Prestige accident. The model is based on a Lagrangian transportation model simulation of a hypothetical oil spill scenario database. A re-analysis database was used for the simulation, which consisted of a 44-year wind and wave aftermath. By comparing the real data with the numerical results, the model's reliability was assessed. The agreement between the real and numerical data indicates that the established statistical oil spill model is a useful tool for spill response planning.

Giwa and Jimoh [6] researched the spreading capacity of various crude oil types by developing model equations based on the power law for each of them. The experiments were conducted with seven different forms of crude oil such as Brass Light Mbede and Heavy H crude oils. It was revealed that Brass Light Mbede had the highest spreading rate, while Heavy H crude oil type possessed the lowest. The statistical evidence indicates that the established model equation is considerable.

On the Louisiana continental shelf, the oil/water/dispersant mixture from the 2010 Deepwater Horizon oil spill was compared to the yearly production of oxygen-depleted bottom waters. It was unclear whether the spilled oil would exacerbate the size or magnitude of the seasonal hypoxic area formation in 2010. Instead, higher river discharge, higher nitrate-N load, an easterly (westward) wind, and decreased wind speed were found to increase the hypoxic area in all the years studied. Despite the fact that the oil spill had no effect on the magnitude of the 2010 hypoxic region, there was proof that the 2010 hypoxia season was similar to the long-term reversal [7].

To predict oil spill weathering processes, a number of new oil spill fate and trajectory models were recently created. These models, however, foresee the outcome of spilled oil and changes in oil properties separately. The most important oil-weathering procedures, such as distribution, vaporisation, scattering into the water bulk, emulsification, and variations in viscosity and density, are thus used to create a system dynamic model. To explain these key oil-weathering processes, which fluctuate and contain diverse effects on one another, differential equations were used.

In this study, an attempt was made to solve the ordinary differential equations in the oil-weathering model by means of the block method. Frequently, these ODEs were solved by Runge Kutta based method [8–12]. The proposed block method is derived from the form of backward differentiation formulas, which has specifically been found to be very efficient in solving stiff ODEs in terms of reducing the computational time and improving the accuracy of the approximate solution. Because of their high stability, backward differentiation formulas (BDFs) are commonly used.

This chapter is structured as follows; in Sect. 2, a detailed introduction about 2BDF with its derived formulas is given. Section 3 elaborates on the oil-weathering model that predicts the oil spill volume, spill area, and other oil spill properties. Results and discussion about the proposed model are presented in Sect. 4, and followed by the conclusion in Sect. 5.

2 Two Points Block BDF (2BDF)

The method of 2-point block backward differentiation formula (2BDF) was proposed by Ibrahim et al. [13] for solving the problems of ODEs. The method gives two approximation solutions at each integration step. Later, Yatim et al. [14] introduced the variable order and variable step version of 2BDF known as VSVO-BBDF. The method in Yatim et al. [14] is of order 3, 4, and 5 with a fixed variation in changing the step size (h). Step size is only allowed to differ by the ratio (r), of 1, 2 and 10/19.

In this chapter, it is proposed to apply only the method of third order 2BDF with variable step size strategy for solving the ODEs in oil-weathering models. Similar variation of step size as in Yatim et al. [14] is used. The third order 2BDF method for $r = 1, 2$ and 10/19 are given by,

$$r = 1; \quad \begin{cases} R_{n+1} = -\frac{1}{3}R_{n-1} + 2R_n - \frac{2}{3}R_{n+2} + 2hR'_{n+1}, \\ R_{n+2} = \frac{2}{11}R_{n-1} - \frac{9}{11}R_n + \frac{18}{11}R_{n+1} + \frac{6}{11}hR'_{n+2}, \end{cases}$$

$$r = 2; \quad \begin{cases} R_{n+1} = -\frac{1}{8}R_{n-1} + \frac{9}{4}R_n - \frac{9}{8}R_{n+2} + 3hR'_{n+1}, \\ R_{n+2} = \frac{1}{21}R_{n-1} - \frac{4}{7}R_n + \frac{32}{21}R_{n+1} + \frac{4}{7}hR'_{n+2}, \end{cases}$$

$$r = \frac{10}{19}; \quad \begin{cases} R_{n+1} = -\frac{361}{480}R_{n-1} + \frac{841}{380}R_n - \frac{841}{1824}R_{n+2} + \frac{29}{19}hR'_{n+1}, \\ R_{n+2} = \frac{6859}{13195}R_{n-1} - \frac{576}{455}R_n + \frac{4608}{2639}R_{n+1} + \frac{48}{91}hR'_{n+2}, \end{cases} \tag{1}$$

where R is the parameter involved in the oil weathering models and R' is their response rate. During the implementation, in the case of a successful step, the step size is allowed to be fixed ($r = 1$) or to increase by a factor of 1.9 ($r = 10/19$). However, when a failure step occurs, the current step is reiterated by halving the step size ($r = 2$).

3 Oil-Weathering Model

Oil weathering refers to the series of changes in the chemical and physical properties of oil when oil spill occurs. The initial oil properties, the spill amount, hydrodynamic, and weather conditions affect the oil-weathering process. It is reported that many oil-weathering models have been developed based on empirical and semi-empirical approaches. For the sake of applying the 2BDF method, which is for the first order ODEs, this paper utilised the oil-weathering models used in Zhong and You [15]. The models predict the oil spill volume, spill area, and other oil spill properties.

The spreading of oil spill includes three phases: gravity-inertial spreading, gravity-viscous spreading, and tension-viscous spreading. Most models consider mainly the second phase, which is gravity-viscous spreading as it covers the planning horizon in most cases. The rate of change for the second phase [10, 16–18] is modeled as:

$$\frac{dA}{dt} = \frac{K_1 V^{4/3}}{A} \tag{2}$$

A is the spreading area of oil spill, V is the volume of oil spill and K_1 is constant with a value of 150 s^{-1}.

The rate of spillage oil evaporating from the sea surface is modelled by Sebastiao and Soares [10],

$$\frac{dF_E}{dt} = \frac{K_{ev} A}{V} \exp\left(A_{ev} - \frac{B_{ev}}{T_K}(T_O + T_G F_E) \right) \quad (3)$$

From Buchanan and Hurford [19], the evaporation mass transfer coefficients K_{ev} is calculated as $K_{ev} = 2.5 \times 10^{-3} WIND^{0.78}$. T_O is the initial boiling point (K) and T_G is the modified gradient temperature (K). T_O and T_G are calculated by $T_O = 532.98 - 3.125API$ and $T_G = 985.62 - 13.597API$ respectively [20]. A_{ev} and B_{ev} are the constant for oil-weathering process with the values of 6.3 and 10.3 respectively.

The rate of emulsification that incorporate water into oil is given by Mackay et al. [16],

$$\frac{dF_W}{dt} = K_{em} \times (WIND + 1)^2 \times \left(1 - \frac{F_W}{C_3}\right) \quad (4)$$

At the beginning, the fractional water content, F_W is zero, $F_{W(t=0)} = 0$. K_{em} is an empirical constant equivalent to 2×10^{-6}, *WIND* is the wind speed and C_3 is the maximum water content in oil (%). In this paper, the corrected value for K_{em} is used which considers the effect of emulsion formation [21]. The corrected value for K_{em} is given by

$$K_{em} = K_{ev} \times (1 - F_W) \quad (5)$$

Another oil-weathering process is oil dissolution. Shortly after the oil spillage, the most soluble component of oil will dissolve quickly, which will affect the other oil-weathering process [22]. Typically, only a small amount of crude oil dissolved in seawater [21]. The rate of dissolution is estimated by Shen et al. [23] as,

$$\frac{dF_d}{dt} = K_d A \left(\frac{S_o e^{at}}{1000 \rho_r} \right) \quad (6)$$

F_d is the volume of oil dissolved in seawater (m^3), ρ_r is the reference oil density (km/m^3), S_o is the initial solubility of oil in water ($30\ g/m^3$) and a is a constant with the value of -0.1.

The oil density and oil viscosity are influenced by the evaporation and emulsification process. These two properties of oil after the spillage are given by Lehr [24] as,

$$\rho_t = F_w \rho_w + \rho_r (1 - F_w)(1 + 0.18 F_E) \quad (7)$$

$$\mu_t = \mu_0 \exp(C_1 F_E) \exp\left(\frac{2.5 Y_w}{1 - 0.65 Y_w} \right). \quad (8)$$

ρ_t is the oil density at time t (km/m^3), ρ_w is the density of seawater, μ_t is the oil viscosity at time t (cP), μ_0 is the initial oil viscosity, and C_1 is a constant with value 10 for crude oil.

4 Results and Discussion

The set of ODEs involved in the oil-weathering process were frequently solved by using the Runge Kutta method [10, 15, 21, 25]. In this research, we attempt to solve ODEs given by Eqs. (2)–(6) by using the 2BDF method. The results were later applied to Eqs. (7) and (8) for the calculation of the oil density and oil viscosity, respectively. The numerical computations were then compared with the ode45 in MATLAB, which is a combination of the fourth and fifth order of the Runge Kutta methods.

The following values were made constant throughout the simulation of oil weathering models (Table 1).

The simulation was conducted for ten days, assuming that the emulsification process only began after the fifth day. At the beginning of the simulation, none of the oil evaporated and dissolved. The oil temperature was assumed to be equivalent to the sea surface temperature. Two crude oils with different properties were tested (Table 2).

(i) Oil Spreading Area

After ten days, with 1000 m^3 of oil spill, the spreading area was approximately 1,600,000 m^2 with oil thickness reaching 0.06 cm. The plotting between ode45 and 2BDF showed the same agreement in the simulation result as demonstrated by the small boxes in Figs. 1, 2, 3, 4, 5, 6, 7, 8, 9, 10, 11, and 12.

(ii) Evaporation

Table 1 Constant for the models' parameter

Parameter	Value
Volume of oil spill (m^3)	1000
Wind speed (m/s)	5
Density of sea water (kg/m^3)	1024

Table 2 Properties of the crude oil

Temperature (K)	Reference API	Density (kg/m^3)	Viscosity (cP)	Maximum water content in oil (%)
283	26.8	898	206.1	80
273	26.8	906	393.4	80

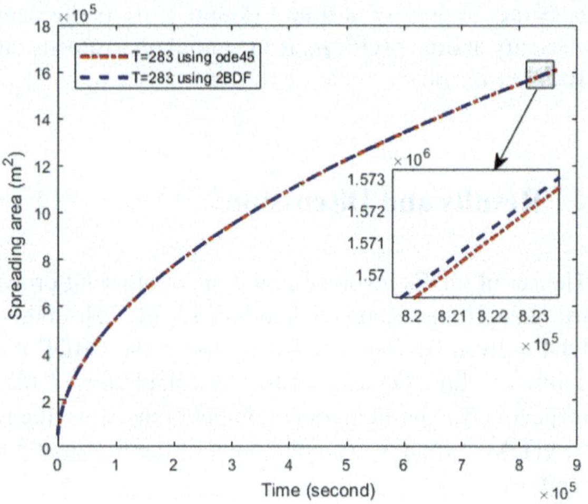

Fig. 1 Spreading area at temperature 283 K

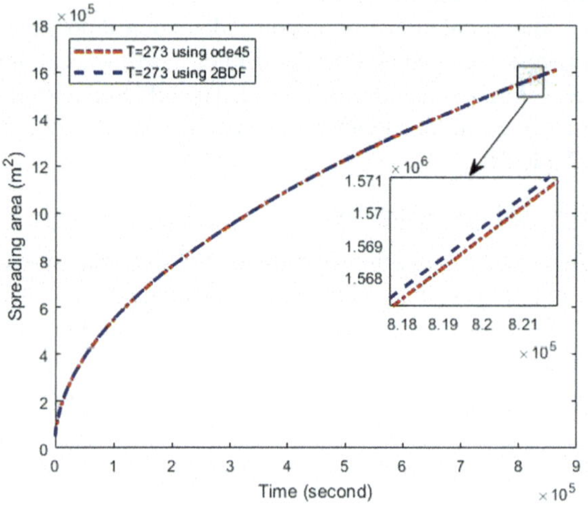

Fig. 2 Spreading area at temperature 273 K

After ten days, the oil spillage with the temperature 283 K achieved 40% of oil evaporation, while the oil spillage at temperature 273 K achieved 35% of oil evaporation.

(iii) Emulsification

Emulsification process only began after the fifth day of the oil spill incident. From the simulation results, the water content in the oil spill achieved its maximum of 80% only one day after the process started.

(iv) Dissolution

At both temperatures of 283 and 273 K, the fraction of oil dissolution was approximately only 0.017%. This process happened immediately after the oil spill incident.

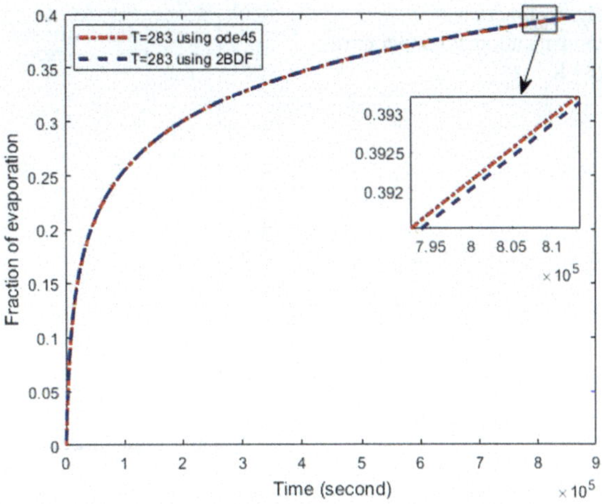

Fig. 3 Fraction of evaporation at temperature 283 K

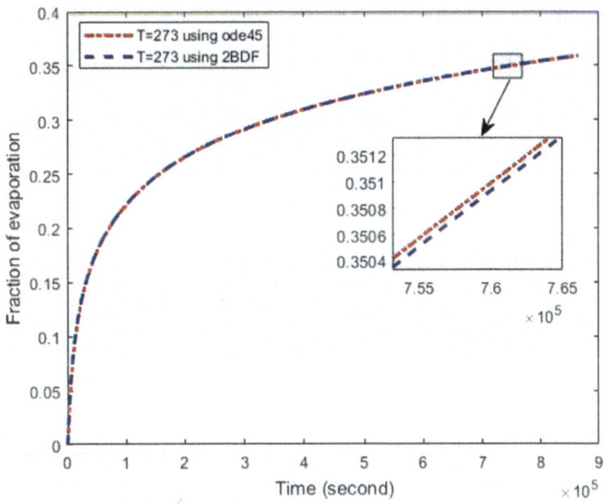

Fig. 4 Fraction of evaporation at temperature 273 K

Although only a small amount of oil dissolved into the seawater, the toxic component of the dissolved oil can still cause catastrophic injury to the aquatic organisms [26].

(v) Density and Viscosity

Both the density and viscosity of oil changed rapidly once the emulsification process started. The oil density remained stagnant at approximately 1010 km/m^3 at the end of the simulation, which is 98% of seawater density. This difference in the oil and seawater density at the end of the simulation can be attributed to the 80% of water content in the oil. Once the emulsification started, the viscosity of oil spill increased with time. By the end of the simulation, the viscosity of oil spillage at temperature 273 K is higher than the oil spillage at temperature 283 K.

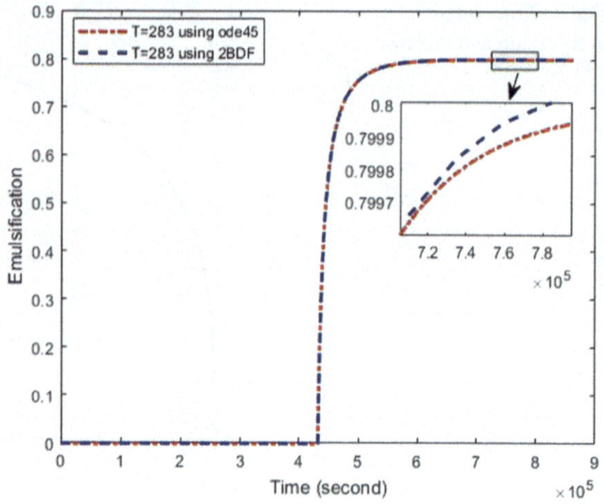

Fig. 5 Fraction of emulsification at temperature 283 K

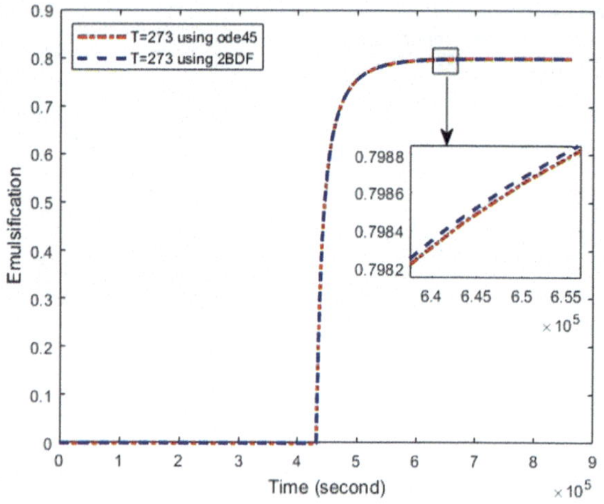

Fig. 6 Fraction of emulsification at temperature 273 K

The percentage of computational error at the end of simulation time between the 2BDF and ode45 are tabulated in Table 3. These percentage of errors were calculated by $|y_{ref} - y_{app}|/y_{ref} \times 100$. y_{ref} is the solution from ode45 with the RelTol = AbsTol = 5e–14, and y_{app} is the solution from 2BDF.

Error of computation showed that the solution given by 2BDF converged to the solution by ode45 as the tolerance (TOL) decreased. The percentage of relative error was less than 0.8% for all the parameters in the model of oil weathering.

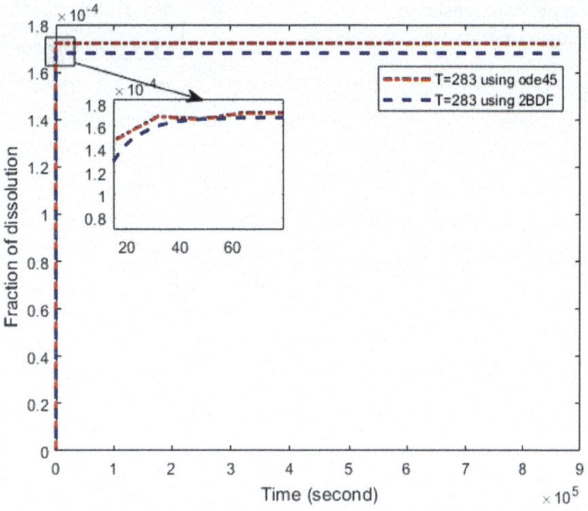

Fig. 7 Fraction of dissolution at temperature 283 K

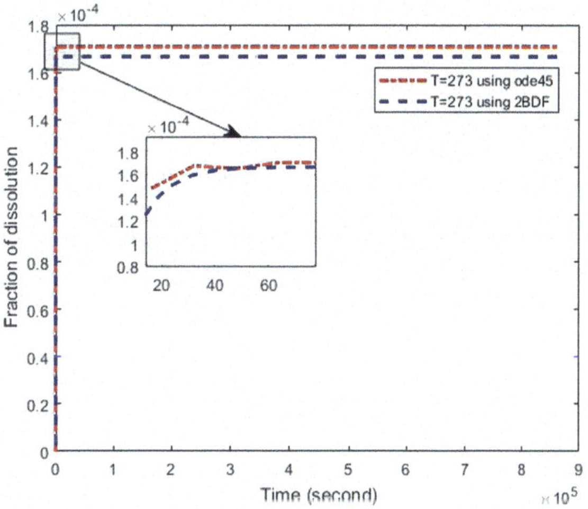

Fig. 8 Fraction of dissolution at temperature 273 K

5 Conclusion

Oil spills lead to serious environmental disaster affecting the sensitive aquatic system. Thus, the forecasting of oil spill weathering process is important for the planning of effective response operation. To forecast the oil-weathering process, its related models were extensively studied and were eventually solved with the existing solver.

In this paper, the method of 2BDF is proposed to solve the oil-weathering models. The numerical computations were compared with ode45, which is the frequent solver for the ODEs in the oil-weathering models. The numerical results justified

Fig. 9 Oil density at temperature 283 K

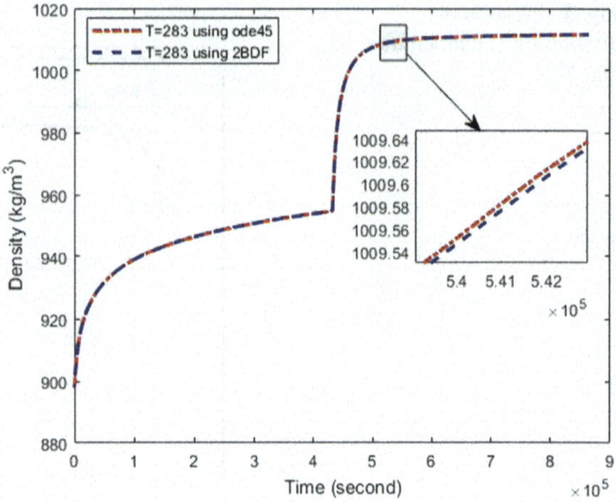

Fig. 10 Oil density at temperature 273 K

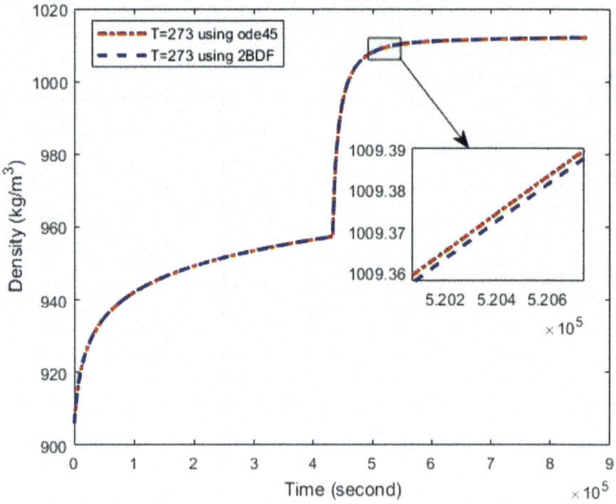

that the proposed 2BDF can be an alternative solver for solving the ODEs in the oil-weathering process.

Fig. 11 Oil viscosity at temperature 283 K

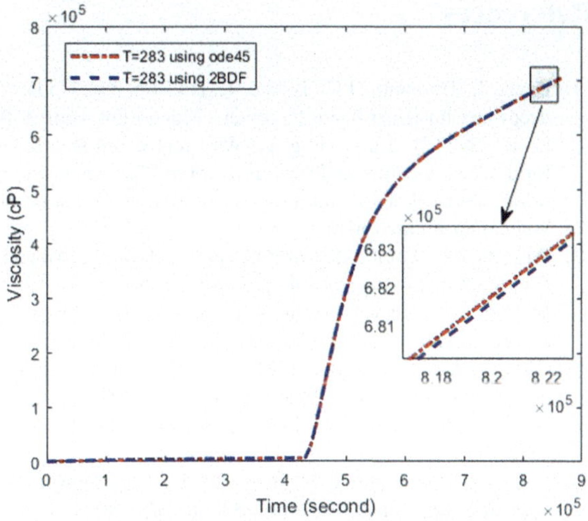

Fig. 12 Oil viscosity at temperature 273 K

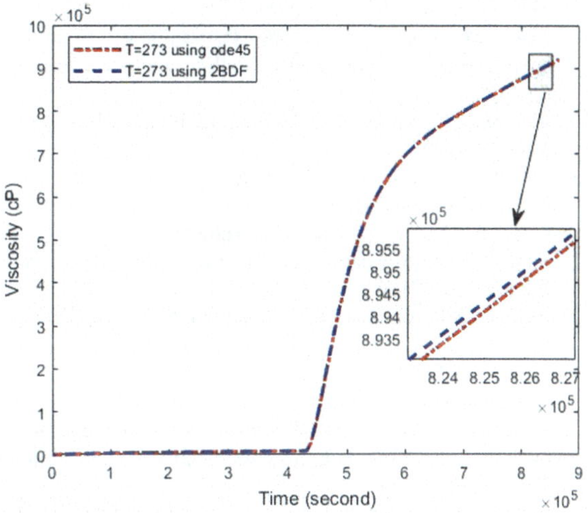

Table 3 Percentage of computational error between ode45 and 2BDF

Temperature (K)	TOL	A (%)	F_E (%)	F_w (%)	F_d (%)	ρ (%)	μ (%)
283	10^{-4}	0.547897	0.617005	0.278164	0.147482	0.005848	0.029209
	10^{-6}	0.014433	0.024326	0.006006	0.042206	0.000013	0.044489
	10^{-8}	0.000286	0.000404	0.000026	0.000905	0.000002	0.001376
273	10^{-4}	0.547897	0.458912	0.278164	0.147439	0.007812	0.776417
	10^{-6}	0.013950	0.014019	0.006006	0.041945	0.000120	0.001772
	10^{-8}	0.000286	0.000209	0.000026	0.000937	0.000001	0.000526

References

1. Beyer, J., Trannum, H.C., Bakke, T., Hodson, P.V., Collier, T.K.: Environmental effects of the deepwater horizon oil spill: a review. Mar. Pollut. Bull. **110**, 28–51 (2016)
2. Li, P., Chen, B., Li, Z., Jing, L.: An agent-based simulation-optimization coupling approach for device allocation and operation control in response to offshore oil spills. In: Proceedings of the International Conference on Marine and Freshwater Environments (iMFE 2014)-Our Water, Our Future (2014)
3. Rafferty, J.P.: 9 of the Biggest Oil Spills in History, Britannica (2020). https://www.britannica.com/list/9-of-the-biggest-oil-spills-in-history
4. McCay, D.F.: Development and application of damage assessment modeling: example assessment for the North Cape oil spill. Mar. Pollut. Bull. **47**(9–12), 341–359 (2003)
5. Abascal, A.J., Castanedo, S., Medina, R., Liste, M.: Analysis of the reliability of a statistical oil spill response model. Mar. Poll. Bull. **60**, 2099–2110 (2010)
6. Giwa, A., Jimoh, A.: Development of models for the spreading of crude oil. AUJ.T. **14**, 1–6 (2010)
7. Rabalais, N.N., Smith, L.M., Turner, R.E.: The deepwater horizon oil spill and Gulf of Mexico shelf hypoxia. Cont. Shelf Res. **152**, 98–107 (2018)
8. Ashino, R., Nagase, M., Vaillancourt, R.: Behind and beyond the MATLAB ODE suite. Comput. Math. Appl. **40**, 491–512 (2000)
9. Nazir, M., Khan, F., Amyotte, P., Sadiq, R.: Multimedia fate of oil spills in a marine environment—an integrated modelling approach. Process Saf. Environ. Prot. **86**, 141–148 (2008)
10. Sebastiao, P., Soares, C.G.: Modeling the fate of oil spills at sea. Spill Sci. Technol. Bull. **2**, 121–131 (1995)
11. Sublette, K.L., Thoma, G., Duncan, K.: Risk reduction and soil ecosystem restoration in an active oil producing area in an ecologically sensitive setting (Report No. DE-FC26-01BC15332). United States Government (2006)
12. Tkalich, P.: Numerical simulation of oil spills and oil combating techniques. WIT Trans. Ecol. Environ. **44** (2006)
13. Ibrahim, Z.B., Othman, K.I., Suleiman, M.B.: Implicit r-point block backward differentiation formula for solving first-order stiff ODEs. Appl. Math. Comput. **186**, 558–565 (2007)
14. Yatim, S.A.M., Ibrahim, Z.B., Othman, K.I., Suleiman, M.B.: A numerical algorithm for solving stiff ordinary differential equations. Math. Probl. Eng. Article ID 989381 (2013)
15. Zhong, Z.X., You, F.Q.: Oil spill response planning with consideration of physicochemical evolution of the oil slick: a multiobjective optimization approach. Comput. Chem. Eng. **35**, 1614–1630 (2011)
16. Mackay, D., Buistt, I.A., Mascarenhas, R., Paterson, S.: Oil spill processes and model. Environment Canada Manuscript Report No. EE-8. Ottawa, Ontario (1980)
17. Reed, M., Johansen, I., Brandvik, P.J., Daling, P., Lewis, A., Fiocco, R., et al.: Oil spill modeling towards the close of the 20th century: overview of the state of the art. Spill Sci. Technol. Bull. **5**, 3–16 (1999)
18. Spaulding, M.L.: A state-of-the-art review of oil spill trajectory and fate modeling. Oil Chem. Pollut. **4**, 39–55 (1988)
19. Buchanan, L., Hurford, N.: Methods for predicting the physical changes in oil spilt at sea. Oil Chem. Pollut. **4**(4), 311–328 (1988)
20. NOAA, ADIOS2 (automated data inquiry for oil spills), National Oceanic and Atmospheric Administration, Seattle, WA (2000)
21. Mishra, A.K., Kumar, G.S.: Weathering of oil spill: modeling and analysis. Aquat. Procedia **4**, 435–442 (2015)
22. Spaulding, M.L.: State of the art review and future directions in oil spill modeling. Mar. Pollut. Bull. **115**, 7–19 (2017)

23. Shen, H.T., Yapa, D.P., Wang, D.S., Yang, X.Q.: A mathematical model for oil slick transport and mixing in rivers. Special Report. Cold Regions Research & Engineering Laboratory, Hanover, NH, German (1993)
24. Lehr, W.J.: Review of modeling procedures for oil spill weathering behavior. Adv. Ecol. Sci. **9**, 51–90 (2001)
25. Guilherme, D.B.D.O.: Estimation of ecological risk of oil spills in ice-covered waters: a surface slick model coupled with a food-web bioaccumulation model. Masters Thesis, Memorial University of Newfoundland (2019)
26. Guo, W.J., Wang, Y.X.: A numerical oil spill model based on a hybrid method. Mar. Pollut. Bull. **58**, 726–734 (2009)

Visualization of Electric Field Response for a Semi Circle Transmitter in Seabed Logging

Elsayed Dahy and Afza Shafie

Abstract Sea bed logging (SBL) is a technique that makes use of remote application of the control source electromagnetic method (CSEM) to detect presence of an oil reservoir. In this technique, electromagnetic (EM) waves transmitted from a horizontal electric dipole (HED) are used to detect the difference in resistivity between the different mediums. A curved electric dipole antenna (CED) is proposed as a new source of EM waves. This chapter presents a 3D visualization of electric field response of the CED. The results show that the CED has potential to be used for hydrocarbon exploration.

Keywords Electromagnetics · Curved electric dipole · Visualization and hydrocarbon exploration

1 Introduction

One of the techniques used to detect the presence of hydrocarbon is the seismic technique. Seismic method uses acoustic waves to detect the resistive layer; however, this technique has the drawback of being unable to distinguish between hydrocarbon and salt water trapped in the well. Control source electromagnetic (CSEM) is one of the techniques used to detect the presence of hydrocarbon layer beneath the sea bed. This method uses a generated electric and magnetic field [1–4] by a transmitter to excite the earth under the ocean's surface. The transmitter is approximately being towed 30–40 m above the seafloor and an array of receivers placed on the sea bed is used to detect subsurface resistivity contrasts. In this method, a horizontal electric

E. Dahy (✉)
Department of Mathematics, Faculty of Science, University Al-Azhar, Assuit 71524, Egypt
e-mail: dahy@azhar.edu.eg

A. Shafie
Fundamental and Applied Sciences Department, Universiti Teknologi PETRONAS, 32610, Bandar Seri Iskandar, Perak Darul Ridzuan, Malaysia
e-mail: afza@utp.edu.my

© Institute of Technology PETRONAS Sdn Bhd 2022

S. A. Abdul Karim and A. Shafie (eds.), *Towards Intelligent Systems Modeling and Simulation*, Studies in Systems, Decision and Control 383,
https://doi.org/10.1007/978-3-030-79606-8_9

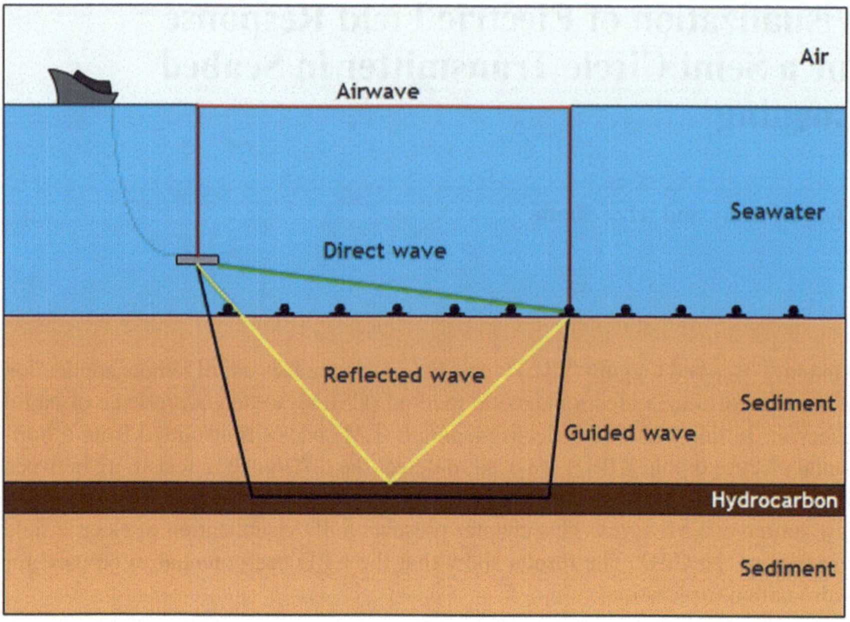

Fig. 1 Marine controlled source electromagnetic environment showing an on the seabed, a network of receivers and a ship towing an HED transmitter

dipole (HED) antenna is used to produce ultralow frequency (~0.1–5 Hz) electromagnetic waves as shown in Fig. 1. The EM response consists of airwaves, direct waves, reflected waves and the guided waves.

The underwater magnetotelluric process has been used in the analysis of resistivity contrast under the seafloor. As one of the early pioneers in marine CSEM, Coggon and Harrison [5], proposed a high frequency vertical magnetic dipole source to estimate shallow seafloor structures. The vertical dipole is least affected by the air waves in shallow waters and propagates EM waves to lesser depths compared with the horizontal dipole. It is therefore best suited for shallow targets in shallow waters. Cox et al. [6] introduced the marine CSEM process, which has been used to study the lithosphere underneath oceanic systems with great results. Cox [7] later suggested the use of a HED antenna sounding to study the seafloor layers. Cox [8] expanded on the theory of marine CSEM and described a deep-sea experimentation of this technique. Due to the smaller effect of airwaves, the HED was very successful in deep waters for either shallow targets or relatively deep targets. The use of magnetometric resistivity in the sea field was later expanded by Edwards et al. [9], Wolfgram et al. [10], Constable et al. [11] and Hoversten et al. [12].

Since then, dipole antenna has been adopted as the source transmitter for CSEM. A low frequency signal (in the range of 0.1–10 Hz) is emitted by this EM source into surrounding. The signal is then recorded by many stationery receivers on the sea bed having the magnetic and electric sensors as illustrated in Fig. 1.

The presence of the air waves will affect the detection of HC reservoir and make the gathered CSEM information from a survey difficult especially at water column less than 500 m. Many studies have been carried out on resolution of the airwave problem. The signals, which consist of energy refracted at the interface of air-water, traverse the air layer, and constantly transfer energy through the water, are heavily influenced by the airwave as studied by Wait et al. [13], Bannister et al. [14] and Amundsen et al. [15]. Thus, the airwave effect depends on its vertical propagation path through the water as observed by Um and Alumbaugh [16].

The SBL technique has been successful in deep water (>500 m) environment and detection of hydrocarbon reservoirs up to 1500 m deep as shown by Gelius et al. [17], Sodal [18], Sodal [19], Eidesmo et al. [1], Dell'Aversana and Vivier [20], Wang et al. [21] and Mittet and Morten [22]. This technique has remained a challenge in shallow water (<500 m). This is because, while doing the SBL survey, especially in shallow water, the presence of the airwaves reduces the signal to noise ratio recorded by the receiver, hence masking the detection of possible presence of hydrocarbon reservoir as shown by Gelius [23]. Furthermore, the current HED used in SBL is only able to detect hydrocarbon depth of about 2000 m due to the limited strength of the transmitted EM waves. A possible solution to mitigate this problem is a new design of the antenna developed by Yahya et al. [24] and Dahy et al. [25]. It has been shown that this design is able to disperse the airwaves hence the potential to be used in SBL.

With many implementations of the marine CSEM process, there has recently been an increase in interest in preliminary research using computational modeling of EM responses in order to come up with the best survey variables and learn more about the targeted responses [26–30].

This chapter presents the behaviour of the electric field of semi-circle antenna in 3D visualization. An antenna whose height (h) equals radius (r) is considered. Section 2 presents the general equation for the electric field of the antenna. The equations are based on the Maxwell equations for EM. Section 3 will discuss on the results while Sect. 4 is the conclusion and recommendation.

2 General Equation of the Electric Field Response

Consider an electric dipole curved antenna with r = h, here h denotes the antenna's height and r denotes the radius, shown in Fig. 2. The length of antenna is 270 m and the antenna is located 30 m above the sea bed. The antenna center is coinciding with z axis and its alignment is in the x-z plane, and y axis is orthogonal to this plane, (refer Fig. 2).

From Dahy et al. [31] the electric field response equations are:

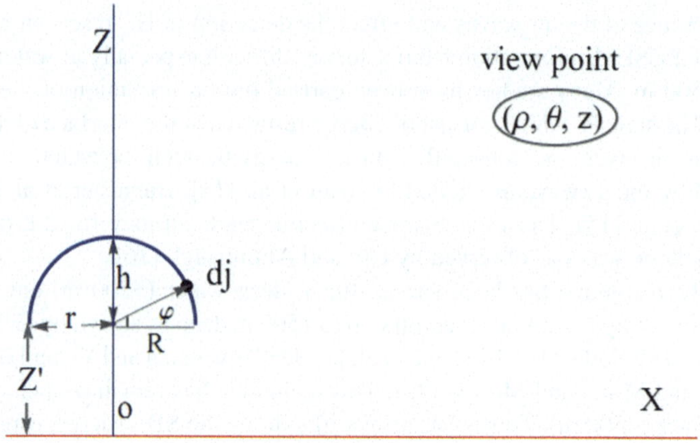

Fig. 2 Diagram showing the geometry of the antenna

$$E_x(\rho, z) = i\omega a \left(\begin{array}{l} \int_0^\infty \left(kJ_0(k\rho)\sin^2\theta + \frac{J_1(k\rho)}{\rho}\cos 2\theta \right) \frac{\partial h_\psi}{\partial z'} \\ + \left(-kJ_0(k\rho)\cos^2\theta + \frac{J_1(k\rho)}{\rho}\cos 2\theta \right) \frac{\partial h_\pi}{\partial z} \end{array} \right) \mathrm{d}k \tag{1}$$

$$E_y(\rho, z) = i\omega a \sin\theta\cos\theta \int_0^\infty kJ_2(k\rho) \left(\frac{\partial h_\psi}{\partial z'} + \frac{\partial h_\pi}{\partial z} \right) \mathrm{d}k \tag{2}$$

$$E_z(\rho, z) = i\omega a \cos\theta \int_0^\infty k^2 J_1(k\rho) G_\pi \mathrm{d}k \tag{3}$$

where $a = \frac{\mu I R^2}{4}$, $b_0 = -i\omega\mu\sigma_0$, θ is the azimuth and J_1, J_2 are first-kind Bessel functions of order 1, 2 respectively.

The inline axis in the form of the x axis was used in this analysis (see Fig. 2), and $h_\psi = G_\psi - G_\psi^0$ and $\frac{\partial h_\pi}{\partial z'} = \frac{\partial G_\pi}{\partial z'} - \frac{\partial G_\pi^0}{\partial z'}$ where the zero in the superscript corresponds to the Greens function with k = 0.

$$G_\psi\left(z, z'\right) = -\frac{e^{-\beta_0\left|z-z'\right|} + R_L^{PM} e^{-2\beta_0 H} e^{-\beta_0\left(z+z'\right)} + R_U^{PM} e^{\beta_0\left(z+z'\right)} + R_L^{PM} R_U^{PM} e^{-2\beta_0 H} e^{\beta_0\left|z-z'\right|}}{2\beta_0\left(1 - R_L^{PM} R_U^{PM} e^{-2\beta_0 H}\right)} \tag{4}$$

$$G_\pi\left(z, z'\right) = -\frac{e^{-\beta_0\left|z-z'\right|} + R_L^{TM} e^{-2\beta_0 H} e^{-\beta_0\left(z+z'\right)} + R_U^{TM} e^{\beta_0\left(z+z'\right)} + R_L^{TM} R_U^{TM} e^{-2\beta_0 H} e^{\beta_0\left|z-z'\right|}}{2\beta_0\left(1 - R_L^{TM} R_U^{TM} e^{-2\beta_0 H}\right)} \tag{5}$$

where $\beta_0^2 = k^2 - i\omega\mu\sigma_0$ (not real) the parameter of ocean diffusion, σ_0 is the water layer's electric conductivity, $k^2 = k_x^2 + k_y^2$ is the horizontal wavenumber and R_A^B is the interaction coefficients.

Consider the earth as a stack of N conductivity-varying layers σ_i and breadth h_i terminated by a half-space, which may be an insulator, conductor, or ideal conductor, Chave [32] give stable recursive expressions in the form for the lower diffusion interaction coefficient:

$$R_i^{TM} = \frac{\chi_i + e^{-2\beta_{i+1}h_{i+1}} R_{i+1}^{TM}}{1 + \chi_i e^{-2\beta_{i+1}h_{i+1}} R_{i+1}^{TM}} \quad (6)$$

$$R_i^{PM} = \frac{\alpha_i + e^{-2\beta_{i+1}h_{i+1}} R_{i+1}^{PM}}{1 + \alpha_i e^{-2\beta_{i+1}h_{i+1}} R_{i+1}^{PM}} \quad (7)$$

Interface interaction coefficients are as follows:

$$\chi_i = \frac{\beta_i\sigma_{i+1} - \beta_{i+1}\sigma_i}{\beta_i\sigma_{i+1} + \beta_{i+1}\sigma_i} \quad (8)$$

$$\alpha_i = \frac{\beta_i - \beta_{i+1}}{\beta_i + \beta_{i+1}} \quad (9)$$

3 Simulation of the Electric Field in Frequency Domain Results and Discussion

The behavior of the electric field response is numerically solved using the Gauss–Kronrod quadrature technique. MATLAB® software is then used to generate the numerical solutions. Table 1 lists the conditions that were taken into consideration for the solution.

The source is 30 m high. The receivers are placed on seabed. The inline orientation is X, while the broadside direction is any one that has an angle with the x axis as seen in Fig. 2. The parameters of the source are r = h = R = 85.9 m with arc length of antenna is 270 m. Figure 3 defines the model of geology considered for the simulation.

4 Results and Discussion

The results of the simulation are shown in Figs. 4, 5, 6 and 7. In these figures, the blue outline represents the response in the presence of a hydrocarbon layer while the red one represents the absence of a hydrocarbon layer.

Table 1 Simulation parameters

Parameter	Value
Air conductivity	10^{-11} s/m
Sea water conductivity	4 s/m
Soil wet conductivity (overburden)	1.5 s/m
Conductivity of under burden	3/2 s/m
Conductivity for oil	10^{-3} s/m
permeability for all layered	$4\pi \times 10^{-7}$ NA^{-2}
Sea water permeability	$0.999991 \times 4\pi \times 10^{-7}$ NA^{-2}
Thickness of hydrocarbon layer	100 m
Thickness of overburden layer	1000 m
Water depth	500 m
Distance of offset	10000 m

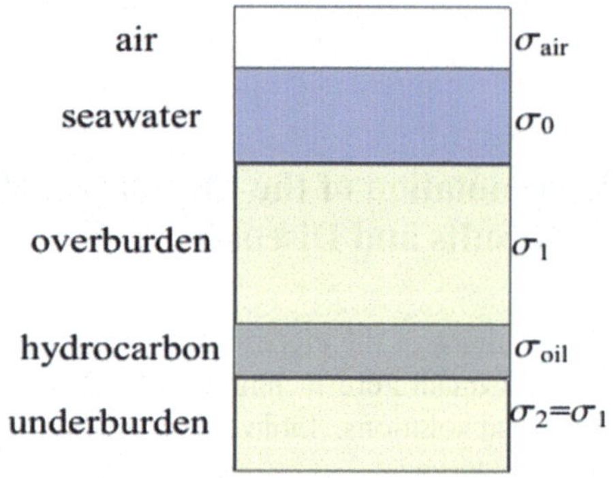

Fig. 3 The geological design for model simulation

In general, at full azimuth ($\theta = 0 \rightarrow 2\pi$), the electric field data response is recorded at all directions along the offset, where θ represent the azimuth angle of the proposed electric field wave equations. From trigonometric function behavior, it will have a maximum and minimum values that could reach to zero. For Ez and Ey components, the maximum or minimum values are very clear, but for Ex component it has different case, as shown in Eqs. (1), (2) and (3).

From Fig. 4, the electric field for X component has response at all azimuth, but it is high at 0, $\pi/2$, π and $3\pi/2$ while it is weak at another azimuth ($\pi/4$, $3\pi/4$, $5\pi/4$ and $7\pi/4$). The electric field for Y component has different behavior in which it has very weak respond (may reach to zero) at azimuths (0°, $\pi/2$, π and $3\pi/2$), while having good response for other azimuths, but it will be high at azimuths $\pi/4$, $3\pi/4$, $5\pi/4$ and $7\pi/4$ (see Fig. 5). The electric field for Z component has very weak response

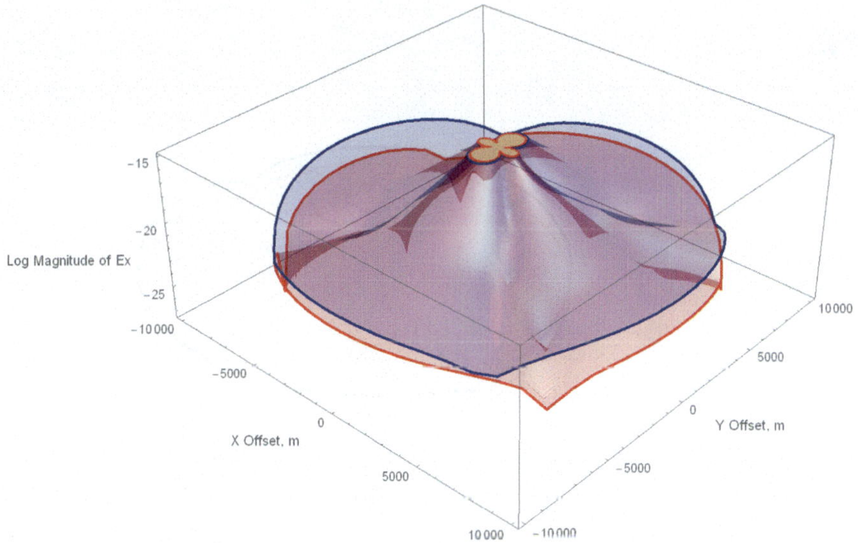

Fig. 4 Electric field of X component full azimuth, with and without hydrocarbon

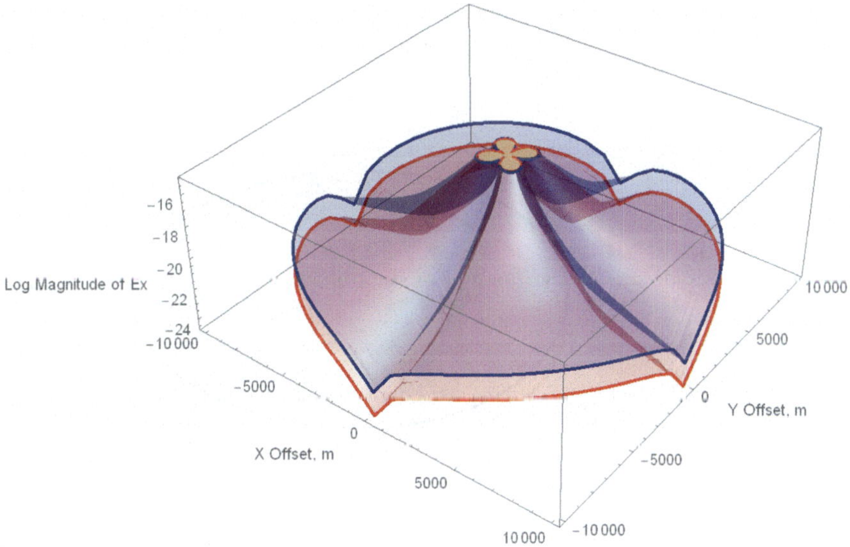

Fig. 5 Electric field of Y component full azimuth, with and without hydrocarbon

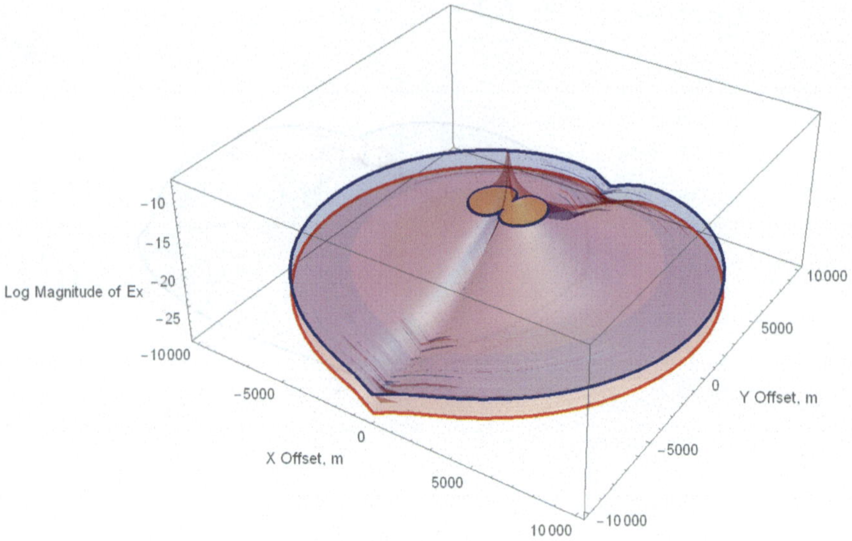

Fig. 6 Electric field of Z component full azimuth, with and without hydrocarbon

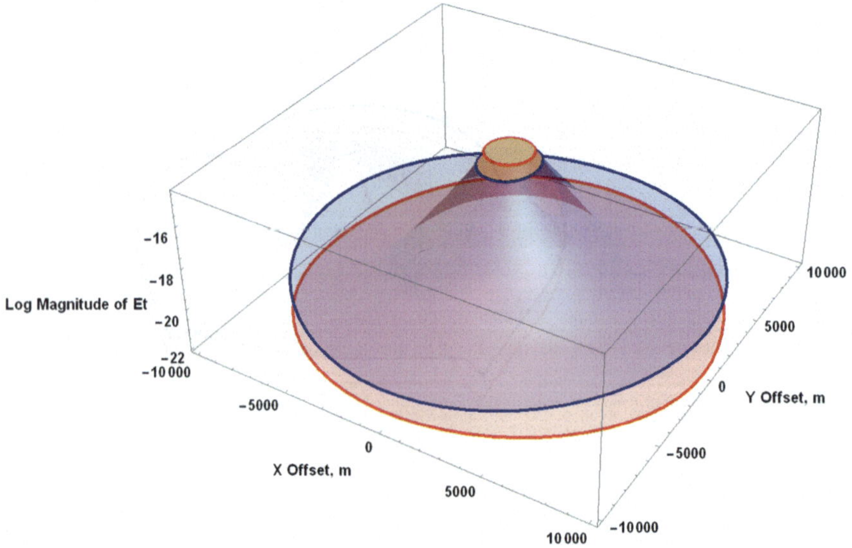

Fig. 7 Total of electric field full azimuth, with and without hydrocarbon

(may reach to zero) at azimuths ($\pi/2$, and $3\pi/2$) as shown in Fig. 6 and, it will be high at azimuths 0 and π. These responses are due to the design of the antenna. For all the three components, the electric field is symmetry at inline direction and at $\pi/2$

direction. But for total E field, the response at all azimuth is almost similar and high as shown in Fig. 7.

The 3D plot is able to show good response of electric field for detecting the presence of hydrocarbon. In addition, it provide more information about which azimuth is best for which component to collect the response data.

The difference between with and without hydrocarbon in term of percentage (%) can be calculated using Eq. (10).

$$\text{percentage}(\%) = \frac{|with\ HC - without\ HC|}{with\ HC} \times 100 \tag{10}$$

From Figs. 8 and 9, the percentage difference of electric field between with and without hydrocarbon presence is clearly demonstrated. For electric field for X component the difference is noticeable starting at end of near offset (2000 m) and continue to increase proportionally with offset. At azimuths ($-\pi/4$ to $\pi/4$ and $3\pi/4$ to $5\pi/4$),

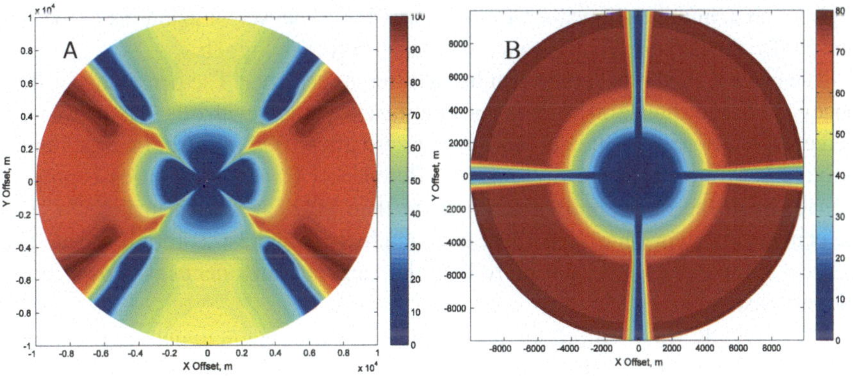

Fig. 8 Percentage difference of Ex (**a**) and Ey (**b**)

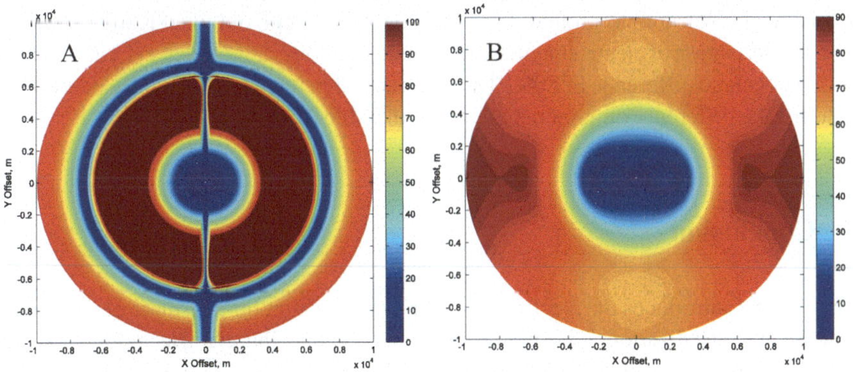

Fig. 9 Percentage differences of Ez (**a**) and Etotal (**b**)

Table 2 Percentage difference of electric field between components

Component	Percentage difference	
Ex	**>75% for azimuth between $-\pi/4$ to $\pi/4$, $3\pi/4$–$5\pi/4$**	**>65% others**
Ey	>65% all azimuth	0 around 0, $\pi/2$, π and $3\pi/2$
Ez	>80% all azimuth	0 around $\pi/2$ and $3\pi/2$
Etotal	>80% for azimuth between $-\pi/4$ to $\pi/4$, $3\pi/4$–$5\pi/4$	>65% other

the percentage difference reach 75% and above as shown in Fig. 8a. The percentage difference of Ex is weak between Azimuths ($\pi/4$–$3\pi/4$ and $5\pi/4$–$\pi/4$), with value below 65%. At azimuth ($\pi/4$, $3\pi/4$, $5\pi/4$ and $7\pi/4$), the percentage difference reach zero around two lines that has slope (1, −1) as shown in Fig. 8a.

For the Y component, difference in the magnitude of the E field starts at 1500 m offset and continue to increase with increasing offset. The percentage difference is high for all azimuths with value of more than 65%. However, at azimuths (0, $\pi/2$, π and $3\pi/2$), the percentage difference will reach zero, see Fig. 8b.

The electric field of z component has a different behavior. The percentage difference can be seen after 1500 m offset and continue to increase with offset. However, the percentage difference will decrease to near zero after 7000 m offset in all azimuth. Around 8000 m offset, the percentage difference of Ez starts to increase proportionally with the offset until it reach more than 80%, as shown in Fig. 9a. The percentage difference has zero value at azimuths ($\pi/2$ and $3\pi/2$) for all offset. The vertical component is affected more than the tangent components as shown in the 2D response.

For complete wave (E_{total}), the percentage difference starts to appear at azimuth (around $\pi/2$ and $3\pi/2$) before appearing again at azimuths (around 0 and π). The percentage difference of E_{total} for azimuths between $-\pi/4$ to $\pi/4$ and $3\pi/4$ to $5\pi/4$ has a value of more than 80%, and has a value of less than 65% at others azimuths as shown in Fig. 9b. By looking at all the figures, it is observed that, there is a symmetrical behavior of electric field around azimuth (0 and $\pi/2$). The obtained results showed that, the derivation of electromagnetic response equation is able to detect hydrocarbon with percentage difference of 80%. The summary of percentage difference of electric field for the different components are given in Table 2.

5 Conclusion

This chapter introduces the electric field response equation for a semi-circle antenna, and its 3D visualization. The results have shown that the electric field response has different values for different azimuths. The values of some component can be high at some azimuths; however other component is small at the same azimuth.

Furthermore, the 3D visualization is able to provide more information on determining the best azimuth for which component to collect the data. In addition, the derived electromagnetic wave equation for the antenna is able to detect hydrocarbon with percentage difference of up to 80% between with and without hydrocarbon. These results show the potential for a semi-circle antenna utilisation in CSEM for hydrocarbon exploration.

References

1. Eidesmo, T., Ellingsrud, S., MacGregor, L., Constable, S., Sinha, M., Johansen, S., et al.: Sea bed logging (SBL), a new method for remote and direct identification of hydrocarbon filled layers in deepwater areas. First Break **20** (2002)
2. Constable, S., Srnka, L.J.: An introduction to marine controlled-source electromagnetic methods for hydrocarbon exploration. Geophysics **72**, WA3–WA12 (2007)
3. Stefatos, A., Boulaenko, M., Hesthammer, J.: Marine CSEM technology performance in hydrocarbon exploration-limitations or opportunities? First Break **27** (2009)
4. Løseth, L.O., Pedersen, H.M., Schaug-Pettersen, T., Ellingsrud, S., Eidesmo, T.: A scaled experiment for the verification of the seabed logging method. J. Appl. Geophys. **64**, 47–55 (2008)
5. Coggon, J., Morrison, H.: Electromagnetic investigation of the sea floor. Geophysics **35**, 476–489 (1970)
6. Cox, C., Filloux, J., Larsen, J.: Electromagnetic studies of ocean currents and electrical conductivity below the ocean floor. The Sea **4**, 637–693 (1971)
7. Cox, C.: Electromagnetic induction in the oceans and inferences on the constitution of the earth. Geophys. Surv. **4**, 137–156 (1980)
8. Young, P.D., Cox, C.S.: Electromagnetic active source sounding near the East Pacific Rise. Geophys. Res. Lett. **8**, 1043–1046 (1981)
9. Edwards, R., Law, L., Wolfgram, P., Nobes, D., Bone, M., Trigg, D., et al.: First results of the MOSES experiment: sea sediment conductivity and thickness determination, Bute Inlet, British Columbia, by magnetometric offshore electrical sounding. Geophysics **50**, 153–160 (1985)
10. Wolfgram, P., Edwards, R., Law, L., Bone, M.: Polymetallic sulfide exploration on the deep sea floor: the feasibility of the MINI-MOSES experiment. Geophysics **51**, 1808–1818 (1986)
11. Constable, S.C., Orange, A.S., Hoversten, G.M., Morrison, H.F.: Marine magnetotellurics for petroleum exploration Part I: a sea-floor equipment system. Geophysics **63**, 816–825 (1998)
12. Hoversten, G.M., Constable, S.C., Morrison, H.F.: Marine magnetotellurics for base-of-salt mapping: Gulf of Mexico field test at the Gemini structure. Geophysics **65**, 1476–1488 (2000)
13. Wait, J.R.: Propagation of electromagnetic pulses in a homogeneous conducting earth. Appl. Sci. Res. Section B **8**, 213–253 (1960)
14. Bannister, P.R.: New simplified formulas for ELF subsurface-to-subsurface propagation. IEEE J. Oceanic Eng. **9**, 154–163 (1984)
15. Amundsen, L., Løseth, L., Mittet, R., Ellingsrud, S., Ursin, B.: Decomposition of electromagnetic fields into upgoing and downgoing components. Geophysics **71**, G211–G223 (2006)
16. Um, E.S., Alumbaugh, D.L.: On the physics of the marine controlled-source electromagnetic method. Geophysics **72**, WA13–WA26 (2007)
17. Gelius, L.-J.: Multi-component processing of sea bed Logging data. Department of Geoscience, University of Oslo, Norway. Piers Online **2** (2006)
18. Sodal, A.R.: Transmitter antenna. United States Patent, 2006
19. Sodal, A. R.: (NO): Transmitter antenna. United States Patent, 2006

20. Dell'Aversana, P., Vivier, M.: Expanding the frequency spectrum in marine CSEM applications. Geophys. Prospect. **57**, 573–590 (2009)
21. Wang, Y.-X., Jin, R.-H., Liang, X.-L., Geng, J.-P.: Propagation of SLF/ELF electromagnetic waves excited by an underground HED in the lower ionosphere. IEEE Trans. Antennas Propagat. **60**, 5412–5418 (2012)
22. Mittet, R., Morten, J.P.: The marine controlled-source electromagnetic method in shallow water. Geophysics **78**, E67–E77 (2013)
23. Gelius, L.-J.: Multi-component processing of sea bed logging data. Department of Geoscience, University of Oslo, Norway. Piers Online **2**, 200 (2006)
24. Yahya, N., Nasir, N., Akhtar, M.N., Kashif, M., Zaid, H.M., Shafie, A.: Modeling of antenna for deep target hydrocarbon exploration (2012)
25. Dahy, E., Shafie, A., Yahya, N.: Electric field behavior in time domain for semicircle antenna in homogeneous multi layered media. In: Abdul Karim, S.A. (ed.) Theoretical, Modelling and Numerical Simulations Toward Industry 4.0, pp. 35–51. Springer Singapore, Singapore (2021)
26. Zhdanov, M.S.: Electromagnetic geophysics: notes from the past and the road ahead. Geophysics **75**, 75A49–75A66 (2010)
27. Frenkel, M.A., Davydycheva, S.: To CSEM or not to CSEM? Feasibility of 3D marine CSEM for detecting small targets. Lead. Edge **31**, 435–446 (2012)
28. Key, K.: Marine electromagnetic studies of seafloor resources and tectonics. Surv. Geophys. **33**, 135–167 (2012)
29. Strack, K.: Future directions of electromagnetic methods for hydrocarbon applications. Surv. Geophys. **35**, 157–177 (2014)
30. Jang, H., Jang, H., Kim, H.J.: Three-dimensional electromagnetic responses of disk-shaped hydrocarbon reservoir in marine sediments. Geosci. J. **19**, 305–312 (2015)
31. Dahy, E., Shafie, A., Yahya, N.: A study on the behavior of electromagnetic waves for a curved antenna. Appl. Math. **10**, 2101–2108 (2016)
32. Chave, A.D., Luther, D.S.: Low-frequency, motionally induced electromagnetic fields in the ocean: 1. Theory. J. Geophys. Res. Oceans **95**, 7185–7200 (1990)

Component Analysis of Electromagnetic Wave Propagation in a Layered Model for CSEM Hydrocarbon Surveying

Amir Rostami, Afza Shafie, Hassan Soleimani, and Surajudeen Olalekan Sikiru

Abstract Controlled source electromagnetic (CSEM) refers to a technique to investigate and explore possible geophysical resistive anomalies surrounded by conductive layers. This method employs an extra low frequency electromagnetic (EM) shooting and, as a result, recorded signal interpretations determine a resistivity contrast image of the strata. Therefore, CSEM is a strongly efficient technique to explore deep-water marine hydrocarbon reservoirs. However, the shallow-water unsolved limitations of EM exploration methods still exist. The main complexity of shallow-water exploration is due to airwave. Airwave is produced by the upward propagated EM signal, inside the seawater. The up-going wave travels at the interface of seawater and air, due to their large contrast of the resistivity. The airwave is then transmitted back to the seawater and will be detected by the array of receivers. In shallow-waters, air wave is the most significant recorded signal. Thus, it can mask the wanted data (called guided wave) which is due to presence of a resistive anomaly. Studying EM propagation in a layered model helps to have a clearer understanding of the physics of CSEM. Therefore, here we present a finite element based result-driven model to investigate two main EM polarized waves (i.e. transverse electric (TE) and transverse magnetic (TM) modes) in a plane layered model, consist of air, seawater, overburden, hydrocarbon and under-burden layers. TE mode refers to a polarization that electric field is perpendicular to the propagation's direction and in TM mode magnetic field is perpendicular to the direction of propagation. For a tangible comparison between

A. Rostami (✉)
Shale Gas Research Group, Universiti Teknologi PETRONAS, 32610 Seri Iskandar, Bandar Seri Iskandar, Perak Darul Ridzuan, Malaysia
e-mail: amir.rostami@utp.edu.my

A. Shafie · H. Soleimani · S. O. Sikiru
Fundamental and Applied Sciences Department, Universiti Teknologi PETRONAS, 32610 Seri Iskandar, Bandar Seri Iskandar, Perak Darul Ridzuan, Malaysia
e-mail: afza@utp.edu.my

H. Soleimani
e-mail: Hassan.soleimani@utp.edu.my

S. O. Sikiru
e-mail: surajudeen_18000257@utp.edu.my

© Institute of Technology PETRONAS Sdn Bhd 2022

S. A. Abdul Karim and A. Shafie (eds.), *Towards Intelligent Systems Modeling and Simulation*, Studies in Systems, Decision and Control 383,
https://doi.org/10.1007/978-3-030-79606-8_10

TM and TE modes, a shallow-water model is assumed. The results illustrate that airwave contributes in TE mode, whereas it is quite absent in TM mode. However, the contribution of TM mode is due to presence of the thin hydrocarbon layer, as a resistive anomaly. EM component analysis and propagation behaviour study improve the interpretation techniques of CSEM hydrocarbon surveying.

Keywords Controlled source electromagnetic · Airwave · Transverse electric · Transverse magnetic

1 Introduction

Marine controlled source electromagnetic (MCSEM) or seabed logging (SBL), is a marine hydrocarbon (HC) reservoir exploration method to complement seismic technique [1, 2]. In MCSEM, a massive horizontal electric dipole (HED) is deployed over the seafloor to produce extra-low frequency (0.1–10 Hz) EM signals [2, 3]. Low frequency (LF) enables the EM wave to penetrate deep into the geological strata [3–6]. This provides a resistivity image of the strata based on the amplitude and phase of the received signal which is collected by arrays of receivers located on the seafloor [7]. Considering the significant resistivity of an oil reservoir that is at least one figure larger than its surrounding environment, the contrast of SBL data is superior to that of the seismic generated density-based image [2]. Thus, the uncertainty of liquid reservoirs delineation (i.e. distinguishing oil from water) in seismic can be verified by a SBL survey.

SBL consists of a stratified environment which are classified based on the resistivity of each layer [8]. It consists of 1- air as the most resistive layer (10^{10} Ωm), 2- seawater which is the most conductive layer due to its salinity, 3- overburden, that is sedimentary layer above a HC reservoir, 4- HC reservoir, if exists, with a significant resistivity, and 5- under-burden which is the sedimentary formation beneath the HC reservoir. Depending on rock and fluid properties, resistivity of seawater and sediment can be from 0.1 to 5 Ωm. Whereas the HC layer is highly resistive. Its resistivity varies between 10 and 500 Ωm, depending on the oil saturation.

Analysing how the EM wave emits in a multi-layered geophysical model is essential to enhance our perception of the physics of MCSEM [9]. Therefore, the aim of this work is a detailed component analysis of a standard MCSEM model. In order to do so, we address the following objectives.

(a) To compute each component of EM response in a standard MCSEM model.
(b) To validate the contribution of EM components to the airwave and guided wave.

This chapter is coordinated as follows. Section 1 is a concise presentation of the context the present work relies in, and an overview of the structure of this chapter. Section 2 is a brief introduction to the CSEM components and EM transverse modes. It elucidates the contribution of main EM components to the CSEM data. Section 3 elaborates on the methodology. Section 4 outlines the simulation results followed by a comprehensive discussion in Sect. 5. The final section presents the summary of this chapter.

2 CSEM Components: Transverse Electric and Transverse Magnetic Modes

EM-generated data is dependent on the resistivity of geological layers [10]. The electrically resistive layers are mainly either shale or HC reservoir. Shale, as a consolidated dense sedimentary rock with an inconsiderable fluid permeability, produces a clear seismic image after data processing of the formation, unlike the liquid reservoirs that have densities of a same order [11]. The highly resistive HC reservoir is surrounded by low resistive (1–5 Ωm) formations including rocks and sediment saturated by conductive seawater.

2.1 CSEM Components

Figure 1 represents the EM wave components in an MCSEM standard model. These components are categorized into three main elements:

1. Direct waves: The signal that is directly transmitted from the source to the sensors [12, 13]. Due to the exponential attenuation of the direct wave, the impact of this component is negligible at mid-offsets.
2. Airwave: Which is generated by the upward propagated EM propagation inside the seawater [14–18]. This up-going will be trapped at the interface of seawater and air, due to their huge resistivity contrast, and travels at the sea surface. The produced guided wave at the air-seawater interface is then transmitted back into the seawater and will be recorded by the receivers located on the seabed. The attenuation of wave guided by the interface of the two media is as slow as r^{-3} [18, 19]. Therefore, airwave affects the long offsets, even up to several tens of kilometres.
3. Guided wave: This component is produced by down-going wave that indicates presence of a resistive anomaly. The down-going EM signal is then guided by the interface of water saturated rocks and HC reservoir, and then reflects to the seafloor to be captured by the EM sensors [20, 21]. Presence of highly resistive reservoir affects the amplitude and phase of the recorded data [18]. This can be interpreted as delineation of the HC reservoir.

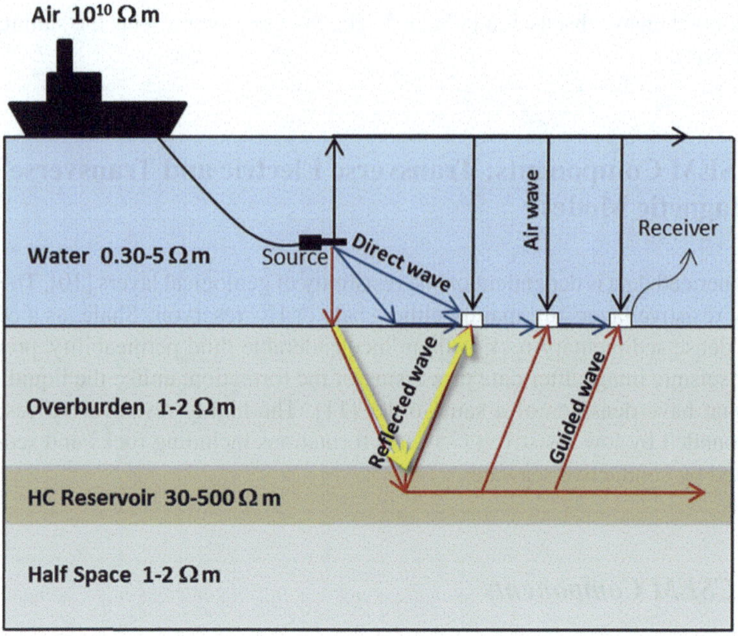

Fig. 1 EM waves in CSEM. Black arrows: refracted transmission via the air-water surface (airwave), blue arrows: direct transmission, and refracted transmission along the seabed (direct wave), red arrows: guided wave, and yellow arrows: reflected wave via high resistive anomaly [20]

2.2 *Transverse Electric and Transverse Magnetic Modes*

Decomposition of EM field in CSEM is a useful method to better understanding of the nature of CSEM. In MCSEM induction in a stratified formation, EM field is decomposed into transverse electric (TE) and transverse magnetic (TM) modes [12, 19, 22]. TE mode contributes to the tangential electric polarization, where the vertical electric component is missing. Whereas TM mode is the tangential magnetic polarized plane wave, where the vertical component of magnetic field is zero [19].

Mathematical derivation of the transverse modes makes an explicit comprehension of EM wave propagation. We start from the EM wave equation when it travels through a dielectric medium.

$$\begin{bmatrix} \boldsymbol{E}(x, y, z) \\ \boldsymbol{H}(x, y, z) \end{bmatrix} = \begin{bmatrix} \boldsymbol{E}^0(x, y) \\ \boldsymbol{H}^0(x, y) \end{bmatrix} e^{-\gamma z}. \tag{1}$$

where $\boldsymbol{E}$ and $\boldsymbol{H}$ denote the travelled electric and magnetic waves, respectively, $\boldsymbol{E}^0$ and $\boldsymbol{H}^0$ are the amplitudes of electric and magnetic waves, γ is attenuation factor, and z is direction of propagation. Equations below denote Ampere's and Faraday's laws for a source-free environment.

$$\nabla \times \begin{bmatrix} \boldsymbol{E} \\ \boldsymbol{H} \end{bmatrix} = \mathrm{j}\omega \begin{bmatrix} -\mu \boldsymbol{H} \\ \varepsilon \boldsymbol{E} \end{bmatrix} \tag{2}$$

where j is complex number, ω is the angular frequency, μ is magnetic permeability and ε is electric permittivity. Applying Eq. (2) to Eq. (1), we obtain partial differential equations (PDEs) as follows.

$$\partial_y E_z + \gamma E_y = -\mathrm{j}\omega\mu H_x, \tag{3}$$

$$\partial_x E_z + \gamma E_x = \mathrm{j}\omega\mu H_y, \tag{4}$$

$$\partial_x E_y - \partial_y E_x = -\mathrm{j}\omega\mu H_z, \tag{5}$$

$$\partial_y H_z + \gamma H_y = \mathrm{j}\omega\varepsilon E_x, \tag{6}$$

$$\partial_x H_z + \gamma H_x = -\mathrm{j}\omega\varepsilon E_y, \tag{7}$$

$$\partial_x H_y - \partial_y H_x = \mathrm{j}\omega\varepsilon E_z. \tag{8}$$

We can rearrange (3) and (7), and combine (4) and (6) to obtain E_x and H_x. Similarly, we can combine (5) and (8) to generate H_y and E_y.

$$H_x = -\gamma h^{-2} \partial_x H_z + \mathrm{j}\omega\varepsilon h^{-2} \partial_y E_z, \tag{9}$$

$$H_y = -\gamma h^{-2} \partial_x H_z - \mathrm{j}\omega\varepsilon h^{-2} \partial_x E_z, \tag{10}$$

$$E_x = -\gamma h^{-2} \partial_x E_z - \mathrm{j}\omega\varepsilon h^{-2} \partial_y H_z, \tag{11}$$

$$E_y = -\gamma h^{-2} \partial_x E_z + \mathrm{j}\omega\varepsilon h^{-2} \partial_x H_z, \tag{12}$$

where $h^2 = \gamma^2 + \beta^2$ and, $\beta = \omega\sqrt{\mu\varepsilon}$, is the wave number. In Eqs. (9)–(12) the expressions of the transverse components, E_x, E_y, H_x, and H_y are in terms of the vertical components of E and H. Now for the TE- and TM- modes scenarios we have:

TE mode: $E_z = 0$ and $H_z \neq 0$;
TM mode: $E_z \neq 0$ and $H_z = 0$.

To solve PDEs for TE- and TM- modes we recall Helmholtz equation:

$$\nabla^2 \mathrm{A} + k^2 \mathrm{A} = 0. \tag{13}$$

For TE mode, A is our H_z and k is equivalent to β. We obtain

$$\nabla^2 H_z + \beta^2 H_z = 0. \tag{14}$$

Expand:

$$\partial_x^2 H_z + \partial_y^2 H_z + \partial_z^2 H_z + \beta^2 H_z = 0, \tag{15}$$

Substituting $\partial_z^2 H_z = -\gamma^2 H_z^0(x, y)\exp(-\gamma z)$, and a simple mathematical simplification, we obtain

$$\partial_x^2 H_z^0 + \partial_y^2 H_z^0 + h^2 H_z = 0. \tag{16}$$

For TM mode, we can repeat this steps, where we need to substitute E_z in (13).

$$\nabla^2 E_z + \beta^2 E_z = 0, \tag{17}$$

and thus

$$\partial_x^2 E_z^0 + \partial_y^2 E_z^0 + h^2 E_z = 0. \tag{18}$$

In MCSEM, seawater surface behaves as a waveguide due to the huge resistivity difference between air and seawater [19]. TE and TM modes introduce two independent sets of parameters [i.e. (E_y, H_x, H_z) and (H_y, E_x, E_z), respectively]. This is due to the primary Maxwell's discovered principle, i.e. electric and magnetic components of EM wave are perpendicular to each other, considering that the guided wave by a waveguide propagates along a single direction. Where in the TE mode the magnetic field is perpendicular to the direction of propagation. Thus, electric field is the only guided wave propagated along the waveguide path, and vice versa for the TM mode.

TE and TM modes propagate independently within a stratified structure and are connected only by the controlled source that in general excites both. In the TE mode an inductive EM coupling occurs between neighbouring layers. Thus, a thin resistive anomaly has almost no effect on the propagation of TE mode. In the TM-mode EM coupling between neighbouring layers is galvanic. Thus, even a very thin resistive anomaly behaves as a strong obstacle protecting the underneath conductive layers from the TM wave penetration. Therefore, the response from the resistive layer possesses mere TM mode characteristic. In the next sections, we discuss this phenomenon based on electric field response in a layered MCSEM model.

3 Method

CST EM Studio was used to investigate the relationships and characteristics of changes in parametric values to EM signals in the MCSEM model. This study is mainly on the bases of the Low Frequency Domain Solver (LFDS). The LFDS is an ideal tool to solve periodic eddy current PDEs such as SBL. A multi-layer geophysical model including air, seawater, overburden sedimentary rock, oil reservoir, and under-burden sedimentary formation is considered to obtain resistivity dependant electric field response.

The models are designed to investigate the behaviour of airwave and guided wave in the electric field components, for an efficient use of MCSEM data. For airwave analysis, A region of 25 × 25 km with varying seawater level between 500 m and 2500 m, at 500 m intervals, is modelled (Fig. 2). Although the water level of 2500 m is considered as deep water in MCSEM, airwave cannot be neglected. Therefore, another additional model was designed, where the air layer is eliminated (Fig. 3). Let call this "half-space" model. Removal of air layer prevents airwave generation and enables us to compare the contribution of airwave to each electric field component. Every specific seawater level is modelled for both with and without a 100 m tick resistive layer located 1000 m below the seabed, representing the oil reservoir.

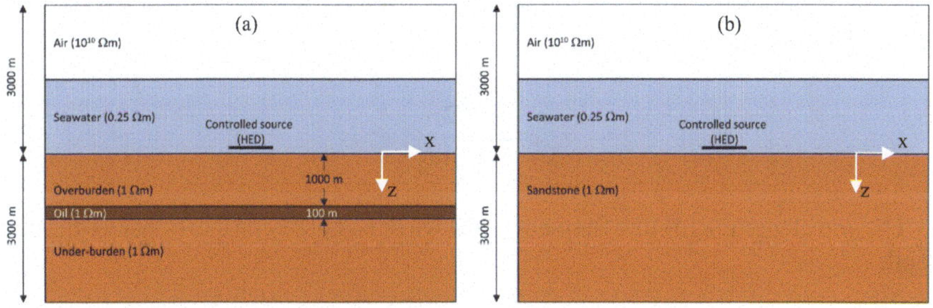

Fig. 2 Multi-layered standard model **a** with oil reservoir and **b** without oil layer

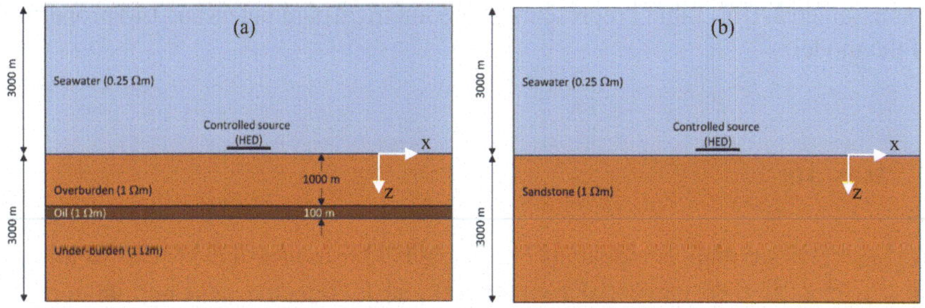

Fig. 3 Half-space model **a** with oil reservoir and **b** without oil layer

Table 1 Parametric values of simulation models

Parameters	Air	Seawater	Sediment	Oil
ρ (Ωm)	1.0×10^{10}	0.25	1.0	50
Relative ε	1.0	74	13	2.33
Relative μ	1.0	1.0	1.0	1.0
Thickness (m)	0–2500	500–3000	3000	100
Surface area	25 km × 25 km			

Three main EM parameters describe the EM properties of the model: electric permittivity ε, magnetic permeability μ, and electric resistivity ρ. Having the angular frequency, ω, wave number is defined to describe the propagation:

$$\beta = \omega\sqrt{\mu\varepsilon} + \mathrm{j}\mu/\rho\omega = \omega/c + \mathrm{j}/\delta, \tag{19}$$

where, c is the wave velocity and δ is the skin depth. The skin depth is the distance that EM wave can travel with less than 63% of attenuation. In Eq. (19) the first term is an indication of displacement current, from Maxwell's equations, and the second term describes the conduction current. Therefore, the equation above depicts that the displacement current is negligible in low frequency approach, where $\omega\varepsilon << 1/\rho$. In a quasi-static approach, when the frequency is extremely low, the skin depth [23, 24] can be obtained by

$$\delta = \sqrt{\frac{2\rho}{\mu\omega}}. \tag{20}$$

The wavelength, λ, of low frequency EM wave is directly proportional to the skin depth.

$$\lambda = 2\pi\delta. \tag{21}$$

The applied frequency was taken 0.125 Hz and the length of the controlled source was assumed 270 m. Table 1 represents the geometrical parameters and EM properties of the model.

4 Results

We analyzed two main components of electric field response in the MCSEM model, i.e. E_x (electric field tangential component) and E_z (electric field normal component). Plots of E_x and E_z magnitudes are generated versus offset. Where offset is the electric source and receiver horizontal displacement. In EM surveys, Magnitude versus offset (MVO) is a data interpretation method to investigate the relation between magnitude/normalized magnitude of electric and/or magnetic field response with respect to the offset.

In this work MVO graphs of the layered model are obtained, where the seawater level varied from 500 to 3000 m. There is no air layer in the model with 3000 m water level. At each seawater level two models of without and with HC layer are computed.

Figure 4 illustrates the magnitude of E_x at different water levels. In this figure MVO graphs represent a comparison of x component response for with and without oil layer.

Figure 5 displays variation of E_x response versus seawater change, for with and without HC reservoir models.

Figure 6 represents the capability of vertical electric field component in HC detection. MVO responses of this figure of E_z for with and without oil models at each seawater lever are plotted.

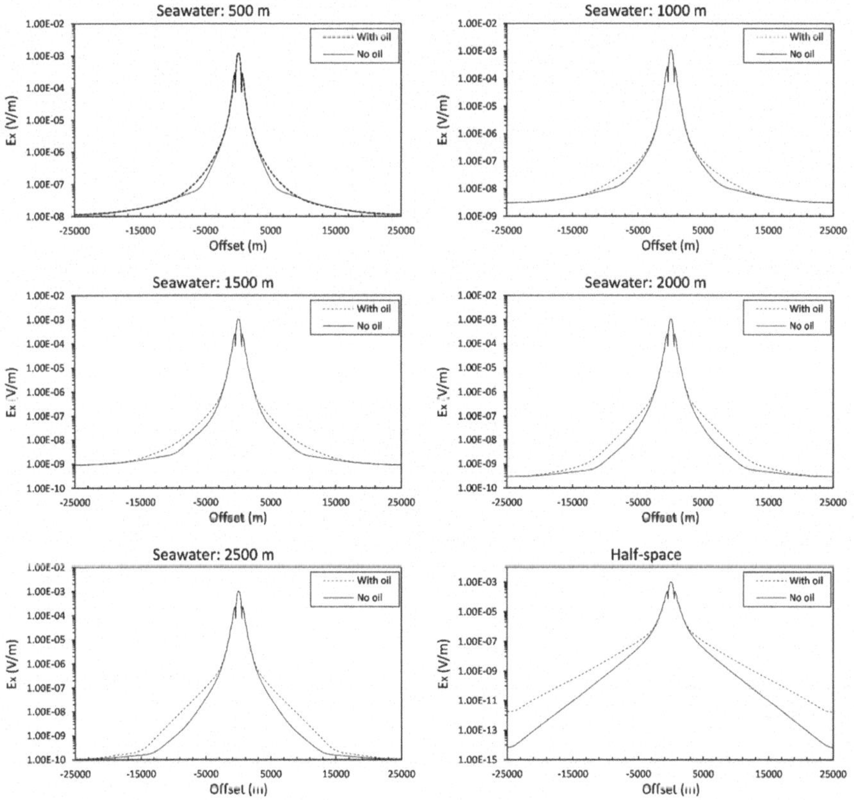

Fig. 4 Comparison of the response of E_x for with and without oil layer models at different water levels and half-space

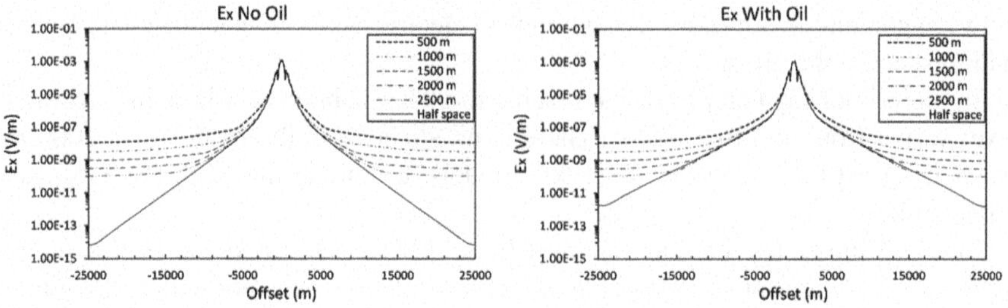

Fig. 5 E_x versus offset at different water levels for with and without oil layer models

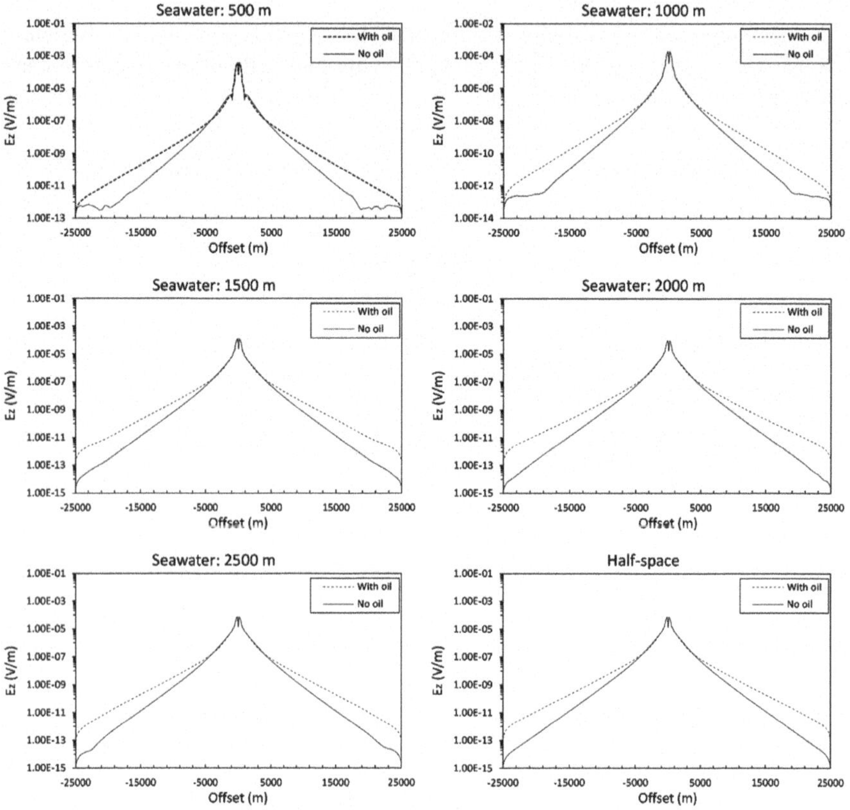

Fig. 6 Comparison of the response of E_z for with and without oil layer models at different water levels and half-space

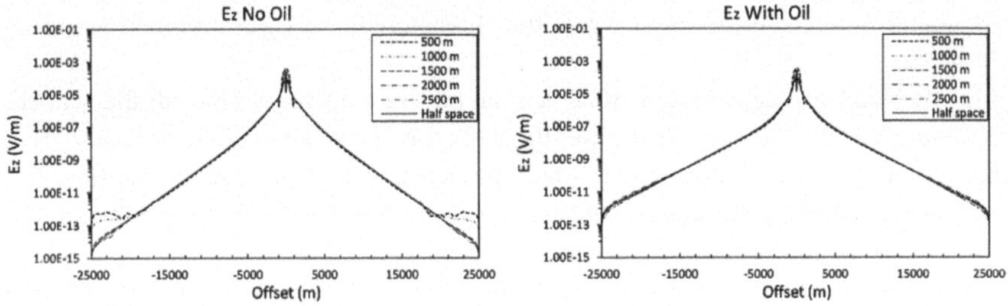

Fig. 7 E_Z versus offset at different water levels for with and without oil layer models

Figure 7 shows the behavior of E_y while changing the seawater, for both with and without HC models.

5 Component Analysis of Electromagnetic Shooting Response

We described the nature of TE and TM modes earlier in Sect. 2.2. In this section we present a discussion on the contribution of the TE and TM modes to MVO of x and z components of electric field, obtained from simulation results.

Taking a closer look at Fig. 4, which shows E_x amplitude vs offset at different water levels, reveals that the effect of a resistive anomaly in MVO response is more distinguished. This is due to the fact that in shallow water environments airwave is the dominating wave in MCSEM data. This can be seen more clearly in Fig. 5, where the E_x response is given for different water levels. By increasing the water level, the amplitude of E_x decreases, dramatically. For half-space model, where the air layer is eliminated to ensure removal of airwave, E_x shows a continuous exponential decay. Since EM wave attenuates exponentially in a conductive medium, the exponential fall is a characteristic of the direct wave. The huge gap between the E_x amplitude in standard and half-space models confirms that airwave is the dominant component of the E_x response.

Knowing that airwave is a pure TE mode phenomenon, we can conclude that the most significant EM wave mode in E_x is TE mode. Although the existence of an oil layer can be observed in E_x versus offset graphs, E_x component is not capable of delineating a HC layer in shallow waters due to dominance of the inductive TE mode [22].

In terms of HC exploration, the scenario becomes interesting by studying the vertical electric field component, E_Z. The graphs in Fig. 6 show an exponential decay for without oil models. However, by introducing an oil layer we figure out a sharp shift in E_Z response. MVO graphs in Fig. 7 verify that there is fairly no effect of airwave in vertical electric component, since there is no gap between E_Z response of standard models at different water levels and that of half-space model. Thus, neglecting the exponentially attenuating direct wave, the characteristic of E_z component is purely TM mode.

In MCSEM survey, where a HC layer can be located deeply beneath the seabed, attenuation of the EM wave can be significant before approaching to the resistive HC. Thus, to increase the certainty of HC exploration removal of airwave is essential [25]; since airwave dominates even at long-range offsets (see Fig. 5 where the attenuation of E_x is very slow after 5 km offset, in presence of air layer).

A useful concept of EM wave propagation in MCSEM survey is decomposition of the wave to TE- and TM- modes. Both transverse modes are excited by the electric controlled source and propagate independently through a layered formation. The EM coupling between the neighbouring structures in TE mode is inductive. Thus, a thin resistive anomaly, representing an oil reservoir, does not affect the TE mode. However, neighbouring formations in TM mode are galvanically coupled [19]. Therefore, characteristic of guided wave, which is due to presence of a resistive HC layer, is absolutely TM mode. From the results it can be depicted that TE mode is dominating in E_x response. Therefore, distinguishing the minorly contributed TM mode, which carries the guided wave information, has a very low certainty. Whereas, TE mode is merely absent in E_z response. Thus, E_z component consists of guided wave, as a pure TM mode characteristic, and the exponentially attenuating direct wave.

6 Summary

We presented an electromagnetic component analysis for MSCEM. Standard multi-layered SBL models with varying water level for two scenarios of with and without HC layer were studied. To investigate HC exploration, E_x and E_z amplitudes were analysed. Presented results showed a major contribution of TE mode to the E_x response versus a minor contribution of TM mode. Whereas, E_z carries the guided wave, which is purely TM mode. The absence of TE mode in E_z response, enables it to clearly distinguish a very thin resistive anomaly.

References

1. Eidesmo, T., Ellingsrud, S., MacGregor, L., Constable, S., Sinha, M., Johansen, S., Kong, F., Westerdahl, H.: Sea bed logging (SBL), a new method for remote and direct identification of hydrocarbon filled layers in deepwater areas. First Break **20**(3) (2002)
2. Ellingsrud, S., Eidesmo, T., Johansen, S., Sinha, M., MacGregor, L., Constable, S.: Remote sensing of hydrocarbon layers by seabed logging (SBL): results from a cruise offshore Angola. Lead. Edge **21**(10), 972–982 (2002)
3. Løseth, L.O., Pedersen, H.M., Ursin, B., Amundsen, L., Ellingsrud, S.: Low-frequency electromagnetic fields in applied geophysics: waves or diffusion? Geophysics **71**(4), W29–W40 (2006)

4. Chave, A.D., Constable, S.C., Edwards, R.N.: Electrical exploration methods for the seafloor. In: Electromagnetic Methods in Applied Geophysics: Volume 2, Application, Parts A and B: Society of Exploration Geophysicists, pp. 931–966 (1991)
5. Chave, A.D.: Seafloor electromagnetic exploration methods. Gorda Ridge, pp. 191–200. Springer (1990)
6. Constable, S., Cox, C., Chave, A.: Offshore electromagnetic surveying techniques.In: SEG Technical Program Expanded Abstracts 1986: Society of Exploration Geophysicists, pp. 81–82 (1986)
7. Constable, S.: Marine electromagnetic methods—a new tool for offshore exploration. Lead. Edge **25**(4), 438–444 (2006)
8. Løseth, L., Ursin, B.: Electromagnetic fields in planarly layered anisotropic media. Geophys. J. Int. **170**(1), 44–80 (2007)
9. Ansari, A., Shafie, A.B., Said, A.B.M.: Relationship of resistivity contrast and thickness depth of hydrocarbon for seabed logging application. Int. J. Comput. Sci. Issues (IJCSI) **9**(3), 214 (2012)
10. Constable, S.: Ten years of marine CSEM for hydrocarbon exploration. Geophysics **75**(5), 75A67–75A81 (2010)
11. Alvarez, P., Alvarez, A., MacGregor, L., Bolivar, F., Keirstead, R., Martin, T.: Reservoir properties prediction integrating controlled-source electromagnetic, prestack seismic, and well-log data using a rock-physics framework: Case study in the Hoop Area, Barents Sea, Norway. Interpretation **5**(2), SE43–SE60 (2017)
12. Løseth, L., Amundsen, L.: On the signal propagation in marine CSEM. In: 69th EAGE Conference and Exhibition incorporating SPE EUROPEC 2007, 2007: European Association of Geoscientists & Engineers, pp. cp-27-00148
13. Nyamasvisva, T.E., Yahya, N., Hasbullah, H.B., Rauf, M., Rostami, A., Dahy, E.: Estimating direct waves with respect to receiver offset, antenna current and frequency in MCSEM. In: 2016 International Conference on Computer and Communication Engineering (ICCCE), pp. 480–484. IEEE (2016)
14. Nyamasvisva, T.E., Hasbullah, H.B., Yahya, N., Nayan, Y.B., Rostami, A., Rauf, M.: Computer algorithm for airwave prediction in marine controlled source electromagnetics data. In: 2016 3rd International Conference on Computer and Information Sciences (ICCOINS), pp. 386–390. IEEE (2016)
15. Rostami, A., Soleimani, H., Yahya, N., Nyamasvisva, T.E., Rauf, M.: The impact of airwave on tangential and normal components of electric field in seabed logging data. In: AIP Conference Proceedings, vol. 1787, no. 1, p. 060006. AIP Publishing LLC (2016)
16. Gelius, L.-J.: Multi-component processing of sea bed logging data. Piers Online **2**(6), 589–593 (2006)
17. Løseth, L.O., Amundsen, L., Jenssen, A.J.: A solution to the airwave-removal problem in shallow-water marine EM. Geophysics **75**(5), A37–A42 (2010)
18. Kong, F., Johnstad, S., Park, J.: Wavenumber of the guided wave supported by a thin resistive layer in marine controlled-source electromagnetics. Geophys. Prospect. **58**(4), pp. 711–723 (2010)
19. Weidelt, P.: Guided waves in marine CSEM. Geophys. J. Int. **171**(1), 153–176 (2007)
20. Rostami, A., Soleimani, H., Yahya, N., Nyamasvisva, T.E., Rauf, M.: A study of electric field components in shallow water and water half-space models in seabed logging. In: AIP Conference Proceedings, vol. 1787, no. 1, p. 060007. AIP Publishing LLC (2016)
21. Yahya, N., Akhtar, M.N., Nasir, N., Kashif, M., Shafie, A., Zaid, H.M.: Guided and direct wave evaluation of controlled source electromagnetic survey using finite element method (2012)
22. Weidelt, P.: Guided waves in marine CSEM and the adjustment distance in MT: a synopsis. In: Electromagnetic Colloquium, Czech Republic (2007)
23. Løseth, L.O., Pedersen, H., Schaug-Pettersen, T., Ellingsrud, S., Eidesmo, T.: A scaled experiment for the verification of the SeaBed Logging method. J. Appl. Geophys. **64**(3–4), 47–55 (2008)

24. Rauf, M., Khan, A.M., Ansari, A., Jilani, M.T., Shahzeb, T.: Skin depth verification of the electromagnetic waves for hydrocarbon detection. Int. J. Appl. Electromagnet. Mech. **60**(3), 313–326 (2019)
25. Abdulkarim, M., Shafie, A., Ahmad, W.F.W., Razali, R.: A simulation study on factors affecting airwaves using factorial design. In: International Visual Informatics Conference, pp. 206–217. Springer (2013)

A Review on the Important Key Properties of Mathematical Models Describing Photovoltaic/Thermal (PV/T) Solar Collectors System

Saiful Azmi Husain, Mahmod Othman, and Noran Nur Wahida Khalili

Abstract Solar energy is one of the most environmentally friendly renewable energy. There are several solar energy systems that have been studied and one of the studies made is to combine solar energy systems, which aims to optimize the energy efficiency. Photovoltaic/thermal solar collector; also known as hybrid solar collector, combines photovoltaic solar cells which convert sunlight into electricity with a solar thermal collector which transfer the wasted heat from the photovoltaic module to a heat transfer fluid. The hybrid technology has shown higher overall energy efficiency than solar photovoltaic or solar thermal alone. The target of this book is to review and present the use of important key properties of mathematical models that describe the PV/T solar collector system. A review on the current development of new mathematical models to achieve higher energy efficiency as well as challenges and limitations of photovoltaic/thermal solar collector system will also be presented.

Keywords Solar energy systems · Key properties · Mathematical models · Energy efficiency · Photovoltaic/thermal (PV/T) solar collector system

1 Introduction

The studies of collector design are among the most important modules of Photo voltaic/Thermal (PV/T) system, which represent an idea that generated electrical and thermal energy simultaneously. The first concept of hybrid PV/T system were

S. A. Husain
Mathematical Sciences Department, Faculty of Science, Universiti Brunei Darussalam, Tungku Link Road, Gadong, BE 1410, Brunei Darussalam
e-mail: Saiful.husain@ubd.edu.bn

M. Othman (✉) · N. N. W. Khalili
Fundamental and Applied Sciences Department, Universiti Teknologi PETRONAS, 32610 Seri Iskandar, Bandar Seri Iskandar, Perak Darul Ridzuan, Malaysia
e-mail: mahmod.othman@utp.edu.my

© Institute of Technology PETRONAS Sdn Bhd 2022

S. A. Abdul Karim and A. Shafie (eds.), *Towards Intelligent Systems Modeling and Simulation*, Studies in Systems, Decision and Control 383,
https://doi.org/10.1007/978-3-030-79606-8_11

presented since the mid-1970s by Wolf [15] and Florschuetz [5]. An amount of experimental and theoretical studies have been discussed & presented on PV/T systems. New designs have been developed for PV/T system collector and compared, in terms of heat absorption and lower production costs [14]. A review of the recent literature on Photovoltaic/Thermal (PV/T) technology and application had been presented by various researchers and scientists. Various types of PV/T collectors such as air, solar, water/air and water PV/T collectors have been described in terms of performance, design, fabrication, simulation and experimental evaluation [13]. There are also quite a number of important mathematical models that described the PV/T system that come with the new design. Specifically, there are many researches describing the use of mathematical models that described the PV/T solar collector system. For instance, Khelifa et al. [8] used mathematical model, based on the equations of the energy balances written for the various nodes of the system, and the coupled differential equations obtained are solved by using the finite difference method. Thus, the temperatures of the various layers of PV/T solar collector and the coolant temperature can be predicted. Abu Bakar et al. [1] also utilized 2D steady state energy balance equations, for the design of bi-fluid (air and water) PV/T solar collector configuration to achieve higher energy efficiency.

The main objectives of the present studies are:

(a) To present the important key properties of mathematical models that described the PV/T solar collector system
(b) To present the most recent development of mathematical models that described the PV/T solar collector to achieve higher efficiency
(c) To present the challenges & limitations of the system.

This chapter is organized as follows. In Sect. 1, we give some basic introduction as well as related literature review. Section 2 discusses the presentation of important key properties of mathematical models that are normally used to describe the PV/T solar collector system. Section 3 presents one of the recent numerical examples of PV/T solar collector, to illustrate the important key properties of mathematical models being used. Section 4 presents the most recent development of mathematical models that described the PV/T solar collector to achieve higher efficiency as well as the challenges & limitations of these systems. Summary will be given in the final section.

2 Important Key Properties of Mathematical Models

There are many researches describing the use of mathematical models that described the PV/T solar collector system. The fundamental key components in mathematical models to describe the PV/T solar collector system are the following:-

(i) Assumptions of the mathematical models: these are important components to guarantee that the mathematical models are applicable in that particular design system and these assumptions are used to simplify the analysis. For example, in order to model the finned double-pass PV/T solar air heater [10], assumptions have been made to simplify the steady-state one-dimensional analysis in closed-form solutions. In this particular model, the assumption is such that the steady state of energy transfer has been achieved, whereas, capacity effects of the glass cover, solar cells and back plate have been neglected. Also, the temperatures of the glass cover, solar cells and plates vary only in the direction of working fluid flow and the side losses from the system are negligible.

(ii) Conservation of Energy & Energy Balance Equations: Another important key component in modelling the PV/T solar collector system is a fundamental concept for thermal analysis of any thermal system, which is the conservation of energy. Conservation of energy can be analyzed through energy balance calculation under steady state conditions. In steady state, the useful energy output of the collector is the difference between the absorbed solar radiation and the total thermal losses from the collector, i.e.

Useful energy = Absorbed solar energy − Thermal losses.

For example, the energy balance is expressed as

$$Q_u = A_c\left[S - U_L\left(T_{plate} - T_{ambient}\right)\right], \tag{1}$$

where

S represent the absorbed solar radiation;
U_L represent the total losses;
T_{plate} represent the temperature of the absorbing plate;
$T_{ambient}$ represent the temperature of the air;
A_c represent the area of the collector surface;
Q_u represent the useful energy.

For instance, with reference to Eq. 1, the general heat balance in the study of flat plate solar collector [4] can be written as

$$c_i V_i \rho_i \frac{dT_i}{dt} = \frac{k_i}{\delta_i}(T_{ab} - T_i) + h_{i,am}(T_{am} - T_i),$$

where

c_i represents the specific heat capacity of the insulation;
ρ_i represents the density of the insulator;
V_i represents the volume of the insulator;
$\frac{dT_i}{dt}$ represents the rate of change of temperature of the insulation;
k_i represents the thermal conductivity of the insulator;

δ_i represents the thickness of the insulator
T_{ab} represents the temperature of the absorber;
T_{am} represents the temperature of the air;
T_i represents the temperature of the insulator;
$h_{i,am}$ represents the heat transfer coefficients.

(iii) Heat Transfer Coefficients: Another important key component in modelling the PV/T solar collector is the heat transfer coefficients, due to the fact that there is a heat transfer occurring from the photovoltaic module to the fluid. There is a need to calculate the heat transfer coefficients in order to check how much the wasted heat had been transferred, so that better modification of the design solar collector system can be made. For instance, the heat transfer coefficients are calculated using the formula [4] below, for the radiation between the absorber (ab) and the glass cover (g):-

$$h_{r,j} = \frac{\sigma\left(T_{ab,j}^2 + T_{g,j}^2\right)\left(T_{ab,j} + T_{g,j}\right)}{\left(\frac{1}{\varepsilon_{ab}}\right) + \left(\frac{1}{\varepsilon_g}\right) - 1} \tag{2}$$

where

$h_{r,j}$ represents the heat transfer coefficients;
σ represents the Stefan-Boltzmann constant;
ε represents the emissivity;
j represents the node number in the flow direction (z);
T represents the temperature.

(iv) Thermal Energy Efficiency: Another important key component in modelling the solar collector is to find out how much the thermal efficiency for that specific particular design system. The higher the useful energy output from a particular design, the higher the expected efficiency. Thermal efficiency of the collector is an important parameter to consider in this kind of analysis as it creates the basis for comparison of different materials and modifications of collector systems. The thermal efficiency η is defined as

$$\eta = \frac{Q_u}{A_c G_T}$$

where Q_u is the useful energy output from a collector, G_T is the incident solar radiation flux (irradiance), and A_c is the area of the collector. The denominator here is the total energy input for the collector, whereas G_T is the parameter characterizing the external conditions, and it is usually known from practical measurements (with a pyranometer) or assumptions for a specific location.

3 Numerical Example

In this section, a numerical example of a one-dimensional energy balance equation representing the heat transfer of a single passage air PV/T solar collector is discussed. The model has been developed based on the properties of the solar collector system. The system consists of a 50 W monocrystalline PV panel, a back-plate attached with a set of fins, and a set of fans to create an airflow for heat removal.

To illustrate the important key properties of mathematical model being used, assumptions have been made to simplify the model development. It is assumed that the energy transfer of the system has achieve steady state. The heat capacity effects of the back-plate and solar cells, and the side losses are neglected. It is also assumed that the temperature of the layers in the collector vary only in the airflow direction.

The energy balance equations for each node of temperature are written as follows: Eq. 3 shows the heat transfer equation for the nodes of the solar cells in the PV panel, Eq. 4 is for the air temperature nodes while Eq. 5 is for the nodes of the surface of the back-plate with fins [7, 16].

$$\overbrace{\tau_g\alpha_p(1-PF)G}^{1}+\overbrace{\tau_g\alpha_{pv}(PF)(1-\eta_{pv})G}^{2}=\overbrace{h_{cpf}(T_p-T_f)}^{3}$$
$$+\overbrace{h_{rpbp}\frac{A_c}{A_{ab}}(T_p-T_{bp})}^{4}+\overbrace{h_{rps}(T_p-T_s)}^{5}+\overbrace{h_{cpw}(T_p-T_w)}^{6} \tag{3}$$

$$\frac{\dot{m}C_f}{W}\frac{dT_f}{dx}=\overbrace{h_{cpf}(T_p-T_f)}^{3}+\overbrace{h_{cbpf}\frac{A_{ab}}{A_c}\eta_p(T_{bp}-T_f)}^{7} \tag{4}$$

where fin effectiveness, $\eta_p=1-\frac{A_{fin}}{A_{ab}}(1-\eta_{fin})$; fin efficiency, $\eta_{fin}=\frac{\tanh mh_{fin}}{mh_{fin}}$; and

$$m=\left(\frac{2h_{cbpf}}{k_{fin}w_{fin}}\right)^{\frac{1}{2}}$$

$$\overbrace{U_{bp}(T_{bp}-T_a)}^{8}+\overbrace{h_{cbpf}\frac{A_{ab}}{A_c}\eta_p(T_{bp}-T_f)}^{7}+\overbrace{h_{rpbp}\frac{A_c}{A_{ab}}(T_p-T_{bp})}^{4}=0 \tag{5}$$

The numbered heat transfer terms are: (1) The rate of the solar energy absorbed by the absorber plate of the PV module per unit area. (2) The rate of the solar energy received by solar cells of the PV module after transmission and the rate of electrical energy available per unit area. (3) The rate of heat transfer to the air flow per unit area. (4) The rate of energy radiated to the surface of the back plate and fins per unit area. (5) The rate of energy radiated to the sky per unit area. (6) The rate of heat transfer to the wind per unit area. (7) The rate of heat transfer to the air flow per unit area. (8) The rate of heat lost to ambient through back plate per unit area.

The performance of the solar collector is measured using thermal and electrical efficiency as in Eqs. 6 [16] and 7 [9].

$$\eta_{th} = \frac{\dot{m}_f C_f (T_0 - T_i)}{A_c G} \tag{6}$$

$$\eta_{ele} = \eta_{ref}\left(1 - \beta_{ref}\left(T_{pv} - T_{ref}\right)\right) \tag{7}$$

The model has been simulated in MATLAB to observe the performance of the solar collector under the change of the air mass flowrate created by the fan. Figure 1 shows the (a) thermal efficiency and (b) electrical efficiencies of the solar collector with changes of air mass flowrate, under the ambient temperature set to be 306 K and various solar irradiance. From the results in Fig. 1, it can be seen that the efficiency of the solar collector increases when the air mass flowrate increases. This is aligned with the fact that solar cells are highly sensitive towards temperature. Any increase in temperature could reduce its efficiency, hence the additional airflow created help to reduce the operating temperature and maintain the collector's performance.

4 Recent Development of Mathematical Models: Challenges and Limitations

It was found out that from analytical and numerical models that PV/T efficiencies could range from over 70% for a perfect collector and to less than 60% for a low quality collector [2]. But a perfect collector is so costly due to the materials chosen and expensive design concepts. The steady state mathematical models have been used to study the performance of single and double pass PV/T air solar collectors, and the results showed that double pass have superior efficiencies (40–45% for combined (thermal and electrical)-Sopian et al. [11, 12]. Furthermore, an explicit dynamic model had been developed, based on control volume finite difference approach for a single glazed PV/T collector [3]. The model can generate results for hourly performance analysis, including instantaneous thermal/electrical gains and efficiencies. The main challenges is that one need to find the suitable mathematical models which better described the new design types. Also, new or modified models need to be developed to predict certain parameters in the system process. For instance, new or modified models needs to be developed to predict the optimum mass flow rate, that was reported to be 0.015 kg/s m^2 with only using experimental study [6].

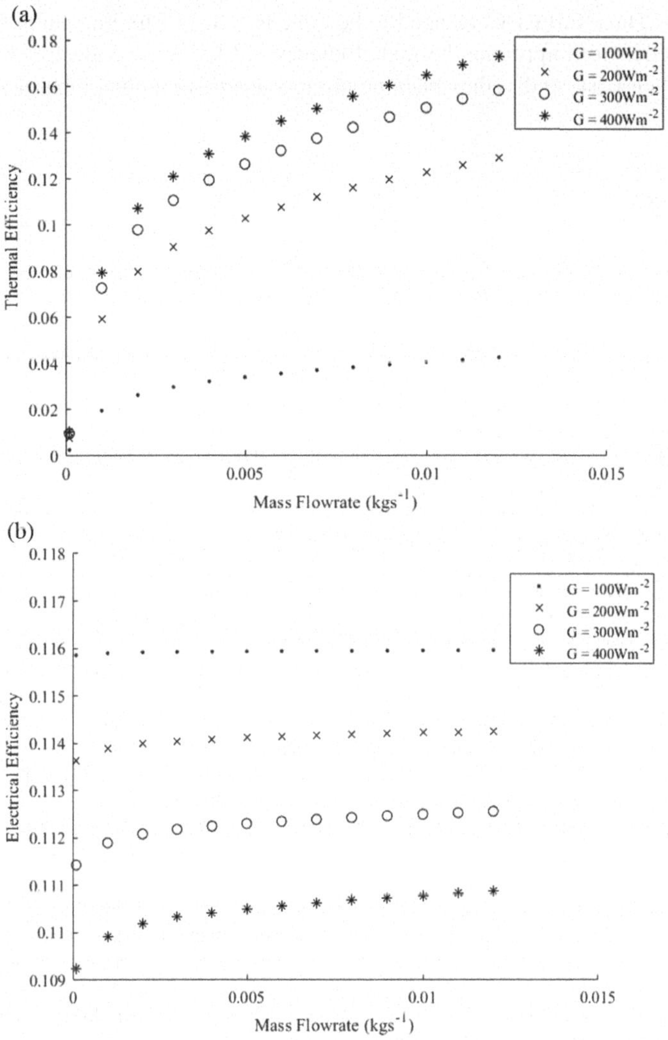

Fig. 1 Thermal and electrical efficiencies of the solar collector under ambient temperature 306 K and various solar irradiance, *G*

5 Summary

The hybrid technology has shown higher overall energy efficiency than solar photovoltaic or solar thermal alone. Important key properties in mathematical models have been presented as well as discussion of challenges and limitations in terms of finding suitable mathematical models that better described the new design system and predicting certain parameters, which currently have been found by experimental study only. Thus, further work need to be done in terms of finding suitable mathematical models for improving thermal efficiency of PV/T solar collector system and reducing their cost, so that they become more competitive in the global market.

References

1. Abu Bakar, M.N., Othman, M., Din, M.H., Manaf, N.A., Jarimi, H.: Design concept and mathematical model of a bi-fluid photovoltaic/thermal (PV/T) solar collector. Renew. Energy **67**, 153–164 (2014)
2. Charalambous, P.G., Maidment, G.G., Kalogirou, S.A., Yiakoumetti, K.: Photovoltaic thermal (PV/T) collectors: a review. Appl. Therm. Eng. **27**(2–3), 275–286 (2007)
3. Chow, T.T.: Performance analysis of photovoltaic-thermal collector by explicit dynamic model. Sol. Energy **75**, 143–152 (2003)
4. Duffie, J., Beckmann, W.: Solar engineering of thermal processes, 3rd edn. Wiley Interscience, New York (2006)
5. Florschuetz, L.W.: On heat rejection from terrestrial solar cell arrays with sunlight concentration. In: Proceedings of the 11th Photovoltaic Specialist Conference, pp 318–26. IEEE (1975)
6. Garg, H.P., Agarwal, R.K.: Some aspects of a PV/T collector/forced circulation flat plate solar water heater with solar cells. Energy Convers. Manage **36**(2), 87–99 (1995)
7. Jarimi, H., Bakar, M.N.A., Othman, M., Din, M.H.: Bi-fluid photovoltaic/thermal (PV/T) solar collector: experimental validation of a 2-D theoretical model. Renew. Energy **85**, 1052–1067 (2016)
8. Khelifa, A., Touafek, K., Ben Moussa, H., Tabet, I.: Modeling & detailed study of hybrid photovoltaic thermal (PV/T) solar collector. Sol. Energy **135**, 169–176 (2016)
9. Michael, J.J., Iniyan, S., Goic, R.: Flat plate solar photovoltaic–thermal (PV/T) systems: a reference guide. Renew. Sustain. Energy Rev. **51**, 62–88 (2015)
10. Othman, M.Y., Yatim, B., Sopian, K., Abu Bakar, M.N.: Performance studies on a finned double-pass photovoltaic-thermal (PV/T) solar collector. Desalination **209**, 43–49 (2007)
11. Sopian, K.S., Yigit, H.T., Liu, H.T., Kakac, S., Veziroglu, T.N.: Performance analysis of photovoltaic/thermal air heaters. Energy Convers. Manage **37**(11), 1657–1670 (1996)
12. Sopian, K.S., Liu, H.T., Kakac, S., Veziroglu, T.N.: Performance of a double pass photovoltaic thermal solar collector suitable for solar drying systems. Energy Convers. Manage **41**, 353–365 (2000)
13. Sultan, S.M., Ervina Efzan, M.N.: Review on recent Photovoltaic/Thermal (PV/T) technology advances and applications. Sol. Energy **173**, 939–954 (2018)
14. Touafek, K., Khelifa, A., Adouane, M.: Theoretical and experimental study of sheet and tubes hybrid PVT collector. Energy Convers. Manage. **80**, 71–77 (2014)
15. Wolf, M.: Performance analyses of combined heating and photovoltaic power systems for residences. Energy Convers. **16**, 79–90 (1976)
16. Zondag, H.A., De Vries, D.W., Van Helden, W.G.J., Van Zolingen, R.J.C., Van Steenhoven, A.A.: The yield of different combined PV-thermal collector designs. Sol. Energy **74**(3), 253–269 (2003)

Trisection and Pentasection Method: A Modification of the Bisection Method for Solving Non-linear Algebraic Equations

Saratha Sathasivam and Shehab Abdulhabib Saeed Alzaeemi

Abstract The bisection method is one of the most used methods of solving non-linear equations, it is based on existence of a solution (root) in a given interval on a real number line, the interval is divided into two equal parts, it can be obviously seen that the root will be contained by one of the intervals, the interval that contained the root is later divided into another two sub-intervals while the interval that does not contain the root is dropped and never used again. The processes continued until the desired stopping criteria is reached, the interval that contain the roots is reduced to set of points, the root and its neighborhood. The root and its neighborhood may approximately serve as the desired root. In this research a new method, a modification of bisection method namely the trisection method and pentasection method are introduced, this method will be tested using several selected functions, the obtained results are then analyzed based on the number of iterations, the CPU time. The result from this research showed that the higher the number of the divisions of the interval, the higher the CPU time, while the number of iterations that yielded the result (root) reduces drastically. Moreover, the trisection method works faster than the Pentasection method.

Keywords Bisection method · Trisection method · Pentasection method · Non-linear equations

Among the widely implemented methods, the bisection approach can be used to solve the linear equations, as well as the non-linear equations. This method draws upon the existence of a solution (root) in a given interval on a real number line, the interval is divided into two equal parts, it can be obviously seen that the root will be contained by one of the intervals, the interval that contained the root is later divided into another two sub-intervals while the interval that does not contain the root is dropped and never used again. The processes continued until the desired stopping criteria are reached, the interval that contains the roots is reduced to a set of points, the

S. Sathasivam (✉) · S. A. S. Alzaeemi
School of Mathematical Sciences, Universiti Sains Malaysia, 11800 USM, Penang, Malaysia
e-mail: saratha@usm.my

© Institute of Technology PETRONAS Sdn Bhd 2022

S. A. Abdul Karim and A. Shafie (eds.), *Towards Intelligent Systems Modeling and Simulation*, Studies in Systems, Decision and Control 383,
https://doi.org/10.1007/978-3-030-79606-8_12

root, and its neighborhood. The root and its neighborhood may approximately serve as the desired root. In this research a new method, a modification of bisection method namely the trisection method and pentasection method are introduced, this method will be tested using the selected function, the obtained results are then analyzed based on the iterations' number and CPU time. The results of this study showed that the higher the number of the divisions of the interval, the higher the CPU time, while the number of iterations that yielded the result (root) reduces drastically. Also, this trisection method functions faster than the Pentasection method.

1 Introduction

Numerical analysis is a field of mathematics, as well as computer science, which produces, analyses, and uses algorithms to solve numerical problems related to continuous mathematics. There are many equations whose roots cannot be evaluated analytically by any methods. The approximate values of the roots of such equations can be found either by a graphical approach, or the number of iterative methods, or by a combination of both processes. In numerical methods of solving linear and non-linear equations or root finding, the most popular methods are the Bisection method [1], Newton's method, and Secant method. Several studies have shown that many researchers are interested in the development and the application of the Bisection method such as the trisection method [2] and pentasection method.

2 Bisection Method

The bisection approach is the easiest method using an interval that is divided into two or half. The root is known to be lying in either one of these intervals. The division of this interval into two continues until the root is found, up to a certain degree of accuracy. Doron [3] assume that this method has opposite signs at both edges of the intervals where $(a)f(b) < 0$. It is acknowledged that (x) possesses one root at least in an interval, $[a, b]$. This method can proceed and continue the iteration until converging to a specific point in a tolerance range; it can find the x value like $f(x) = 0$ or roughly 0 [4]. A better approximation is obtained if a point $(c, 0)$ is found where the secant line L joining the points $(b, f(b))$ and $(a, f(a))$ crosses the x-axis [5]. This method is mainly advantageous because of its behavior, which is convergent always since the method brackets a specific root quite swiftly compared to the method of the incremental search. The algorithm of bisection represents the most primitive. However, it is robust in locating the real root as it can always converge to the specific root [6]. Hence, the Bisection method is easy to use and apply; it enjoys a wide array of different applications in various developments. However, the division of the interval into two sections leads to slow convergence of the Bisection Method. Hence, a new method is proposed named the trisection method and pentasection method. These methods are tested based on number iterations, and CPU times.

3 The Trisection Method

The trisection method is a modification of the classical Bisection method which is a three-section method. The method of bisection, which can divide an interval in two sections, leads to slow convergence. This new scheme divided the interval into three sections. The root is then identified either in the first, second, third interval. This will lead to faster convergence of the root and provide a faster calculation of the roots. The algorithm of the trisection method is represented in the following proceedings:

I. The input: xL ('Entering the interval's left endpoint'); xR ('Entering the interval's right endpoint'), $f(x)$, ε.

II. Take the interNu = 0, m = 0.

III. interNu = interNu + 1.

IV. When $f(xL) = 0$, thus root = xL and go to Step 17.

V. When $f(xR) = 0$, thus root = xR and go to Step 17.

VI. $xC = \frac{xR+xL}{2}$.

VII. When $f(xC) = 0$, thus root = xC and go to Step 17.

VIII. m = m + 1.

IX. Calculating points: $x_m = xL + (xR - xL)\frac{l^k_{n-m+1}}{l^k_{n-m+3}}$ and $x'_m = xL + (xR - xL)\frac{l^k_{n-m+2}}{l^k_{n-m+3}}$.

X. When $f(x_m) = 0$, thus root = x_m and go to Step 17.

XI. When $f(x'_m) = 0$, thus root = x'_m and go to Step 17.

XII. First Case: $f(xL)f(x_m) < 0$, thus $xR = x_m$.
Second Case: $f(x_m)f(xC) < 0$, thus $xL = x_m$ and $xR = xC$.
Third Case: $f(xC)f(x'_m) < 0$, thus $xL = xC$ and $xR = x'_m$.
Fourth Case: $f(xR)f(x'_m) < 0$, thus $xL = x'_m$.

XIII. When $(xR\text{-} xL) > \varepsilon$ proceed to the third step.

XIV. The root = xL.

XV. Locate the points, whereby the given function can take the maximum, as well as the minimum values.

XVI. Check out if the selected points are the function's roots or not.

XVII. Finally, come to an end.

4 The Pentasection Method

The pentasection method is a modification of the classical Bisection method which is the fifth section method. The bisection method which divides the interval into two sections leads to slow convergence. This new scheme divided the interval into five sections. The root is then identified either in the first, second, third, fourth, fifth interval. This will lead to faster convergence of the root and provide a faster calculation of the roots. The algorithm of the pentasection method is represented in the following steps:

I. The input: xL ('Enter the interval's left endpoint'); xR ('Enter the interval's right endpoint'), $f(x)$, ε.
II. Take the interNu = 0, m = 0.
III. interNu = interNu + 1.
IV. When $f(xL) = 0$, thus root $= xL$ and go to Step 19.
V. When $f(xR) = 0$, thus root $= xR$ and go to Step 19.
VI. $xC = \frac{xR+xL}{2}$.
VII. When $f(xC) = 0$, thus root $= xC$ and proceed to Step No. 19.
VIII. m = m + 1.
IX. Calculate the given points: $x_m = xL+(xR-xL)\frac{l^k_{n-m+1}}{l^k_{n-m+5}}$, $x'_m = xL+(xR-xL)\frac{l^k_{n-m+2}}{l^k_{n-m+5}}$, $x''_m = xL + (xR - xL)\frac{l^k_{n-m+3}}{l^k_{n-m+5}}$, $x'''_m = xL + (xR - xL)\frac{l^k_{n-m+4}}{l^k_{n-m+5}}$.
X. If $f(x_m) = 0$ then root $= x_m$ and go to Step 19.
XI. If $f(x'_m) = 0$ then root $= x'_m$ and go to Step 19.
XII. If $f(x''_m) = 0$ then root $= x''_m$ and go to Step 19.
XIII. If $f(x'''_m) = 0$ then root $= x'''_m$ and go to Step 19.
XIV Case I. $f(xL)f(x_m) < 0$ then $xR = x_m$.
Case II. $f(x_m)f(xC) < 0$ then $xL = x_m$ and $xR = xC$.
Case III. $f(xC)f(x'_m) < 0$ then $xL = xC$ and $xR = x'_m$.
Case IV. $f(x'_m)f(x''_m) < 0$ then $xL = x'_m$ and $xR = x''_m$.
Case V. $f(x''_m)f(x'''_m) < 0$ then $xL = x''_m$ and $xR = x'''_m$.
Case VI. $f(xR)f(x'''_m) < 0$ then $xL = x'''_m$.
XV. If $(xR\text{-}\ xL) > \varepsilon$ go to Step 3.
XVI. root $= xL$.
XVII. Find the points where the function takes its maximum and minimum values.
XVIII. Check whether these points are the roots of the function or not.
XIX End.

5 Results and Discussion

The results are presented for the following function. This function has been tested through Matlab [7] to check for the number of iterations and CPU times based on the trisection method and pentasection method. The numerical result based on the test run of both methods using the standard test function is listed below. A companion is held for analyzing and comparing the trisection method and the pentasection method by two examples. First example: $f(x) = x^2 + x - 6$ on an interval $[-2, 10]$. The root of the $f(x)$ is x = 2 on an interval $[-2, 10]$. Based on Table 2, if the function f(x) root is near the interval's endpoints, the trisection method can function faster compared to the pentasection method.

Table 1 The number of the iteration and CPU time for the trisection method and pentasection method for $f(x) = x^2 + x - 6$

k = 5	Number of the iteration for trisection	Number of the iteration for pentasection	CPU time for trisection (S)	CPU time for pentasection (S)
[1, 5]	2	28	10	30
[1, 10]	27	34	15	42
[1, 20]	28	40	21	54
[−2, 4]	30	44	29	61
[−2, 5]	38	48	36	72
[−2, 10]	44	58	47	79

Table 1 is the numerical results for the number of iterations and the CPU times for first equation in the first example. Based on Table 1 shows that the number of iteration is increasing when the number of intervals increased. Therefore, the pentasection method possesses a higher number of iteration compared to the trisection method. However, in terms of CPU time, the CPU time increased significantly when the interval increased. Therefore, the pentasection method possessed higher CPU time compared to trisection. The significant increase of CPU time is due to the division of the interval and determination of $f(a)f(b) < 0$ in each interval.

Second example: $f(x) = e^x + x - 1$ on an interval $[-6, 6]$. The root of the $f(x)$ is $x = 0$ on an interval $[-6, 6]$. Based on Table 2, if the function $f(x)$ root is near the interval's endpoints, the trisection method can function faster compared to the pentasection method.

Table 2 is the numerical results for the number of iterations and the CPU times for second equation in the second example. Based on Table 2 shows that the number of iteration is increasing when the number of intervals increased. Therefore, the pentasection method possesses a higher number of iteration compared to the trisection method. However, in terms of CPU time, the CPU time increased significantly when the interval increased. Therefore, the pentasection method possessed higher CPU time compared to trisection. The significant increase of CPU time is due to the division of the interval and determination of $f(a)f(b) < 0$ in each interval.

Table 2 The number of the iteration and CPU time for the trisection method and pentasection method for $f(x) = e^x + x - 1$

k = 5	Number of the iteration for trisection	Number of the iteration for pentasection	CPU time for trisection (S)	CPU time for pentasection (S)
[−6, 6]	6	24	9	13
[−4, 9]	24	28	13	22
[−9, 1]	28	30	19	34

6 Conclusion

In this study, a new modification of Bisection has been proposed. This modification is based on an odd number of intervals. The new method is named as the trisection method and pentasection method. Based on the results of this work, if researchers are interested in the method, which possesses a lower number of iteration and considers the CPU time, it is recommended that they use the trisection method as it converges faster than the pentasection method with a lower CPU time to solve linear and non-linear algebraic equations.

Acknowledgements This research was funded by the Fundamental Research Grant Scheme (FRGS) (203/PMATHS/6711689) by the Ministry of Higher Education Malaysia and Universiti Sains Malaysia.

References

1. Yakoubsohn, J.: Numerical analysis of a bisection-exclusion method to find zeros of univariate analytic functions. J. Complex. **21**(5), 652–690 (2005)
2. Demir, A.: Trisection method by k-Lucas numbers. Appl. Math. Comput. **198**(1), 339–345 (2008)
3. Doron, L.: Introduction to numerical analysis. Departments of Mathematics and Center of Scientific Computation and Mathematical Modeling (CSCAMM), University of Maryland. Retrieved from http://www2.math.umd.edu/~dlevy/books/na.pdf (2010)
4. Solanki, C., Thapliyal, P., Tomar, K.: Role of bisection method. Int. J. Comput. Appl. Technol. Res. **3**(8), 533–535 (2014)
5. Mathews, J.H., & Fink, K.D.: Numerical methods using MATLAB, 4th edn. Pearson. Print, Upper Saddle River, N.J. (2004)
6. Burden, R.L., Faires, J.D.: Numerical analysis, 5th edn. PWS Kent Pub. Co. Print, Boston (1993)
7. https://matlab.mathworks.com/

Predicting and Assessing Road Accidents Using Autoregressive Model and Value at Risk Approach

Teh Raihana Nazirah Roslan, Chee Keong Ch'ng, Masnita Misiran, and Nattakorn Phewchean

Abstract Road accidents have claimed many lives with approximately 1.35 million deaths worldwide and deemed as a critical issue in most countries. Understanding the common factors that contribute to road accidents is not enough, and it is essential to assess the risk involved to prepare for precautionary actions. In this study, we utilize the autoregressive model and value at risk approach in predicting Malaysian road accidents occurrences and assessing the respective risk. We construct a current risk analysis theoretical framework pertain to the vehicle's condition, forecast and investigate the relationship between the involved variables, and obtain the value at risk for road accidents. From our findings, the road accidents will increase 1.12% higher than the year 2019, along with a 2.77% increase in the number of transportations in 2020. In addition, there is a 95% confidence that in year 2020, the number of road accidents will reduce not more than 3.81%. Coherent from the analysis, there is a potential to adopt these approaches more extensively for this issue as the quantitative

T. R. N. Roslan (✉)
Othman Yeop Abdullah Graduate School of Business, Universiti Utara Malaysia, 50300 Kuala Lumpur, Malaysia
e-mail: raihana@uum.edu.my

C. K. Ch'ng · M. Misiran
School of Quantitative Sciences, Universiti Utara Malaysia, 06010 Sintok, Kedah, Malaysia
e-mail: chee@uum.edu.my

M. Misiran
e-mail: masnita@uum.edu.my

T. R. N. Roslan · M. Misiran
Institute of Strategic Industrial Decision Science Modeling, School of Quantitative Sciences, Universiti Utara Malaysia, 06010 Sintok, Kedah, Malaysia

N. Phewchean
Department of Mathematics, Faculty of Science, Mahidol University, Rama 6 rd, Bangkok 10400, Thailand

Centre of Excellence in Mathematics, Perdo, Thailand

N. Phewchean
e-mail: nattakorn.phe@mahidol.ac.th

© Institute of Technology PETRONAS Sdn Bhd 2022

S. A. Abdul Karim and A. Shafie (eds.), *Towards Intelligent Systems Modeling and Simulation*, Studies in Systems, Decision and Control 383,
https://doi.org/10.1007/978-3-030-79606-8_13

analysis are feasible. In addition, a strong positive relationship is found between the likelihood of road accidents and the number of transportations. Thus, prediction of road accidents and identification of its value at risk on yearly basis are beneficial to project the best course of action to deal with road accidents occurrences in the country.

Keywords Value at risk · Road accidents · Quantitative risk analysis · Autoregressive · Correlation

1 Introduction

Road accidents have claimed many lives daily and deemed as a critical problem in most countries [1]. World Health Organization (WHO) [2] reported that approximately 1.35 million people died worldwide due to road accidents. In fact, road transportation is recognized as the most dangerous transport compared to all modes of transport in the world. Some common road accidents are sleep deprivation, using phone while driving, the effects of alcohol, speeding, reckless driving, as well as road and weather conditions. Kamarudin et al. [3] provided alarming statistics of death caused by road accidents currently averaging at 19 people daily, which showed an increase of 70% from year 1990 to 2011 in Malaysia. In addition, this situation is worsened during festive seasons. By considering Shah Alam which is one of the concentrated cities in Selangor in their study, the authors reported that 70% rise of fatal accidents had been recorded there. The dense population in that area, along with limited monitoring and enforcement of road regulation makes it difficult to prevent road accidents.

However, rural areas recorded a higher number of road traffic fatalities (deaths) at 66% in contrast to the accidents occurring in the urban areas. Inferior quality of road designs, defects, and inefficient response at accident sites are among the identified factors that might have caused this problem. Meanwhile, expressways attained the highest rate of road traffic deaths per kilometer. It was notable to mention that most deaths occur on straight roads and the Y/T junctions [4]. This may be due to limitation of road users' visibility in junctions, thus restricting their viewing angle. It is also reported that young drivers were more susceptible to get involved with road accidents. This may be due to their less experience in handling the vehicles [5, 6]. Heavy vehicles also posed greater risk [7, 8], while behavioral changes due to stress after working also contributed to this issue [9, 10].

Although series of safety programs and some initiatives have been continuously introduced, there is not much significant improvement on the number of accidents. The number was still considered high in most states in Malaysia [11]. Darma et al. [4] emphasized that even though the government has outlined initiatives in regard to road traffic deaths, well thought plan of action and implementation detailed in the Malaysian Road Safety Plan 2006 needs to be enhanced and be more specified. Kamarudin et al. [3] were also in agreement with Darma et al. [4]. They highlighted the combination of factors that contributed to this tragedy namely drivers' behavior, volume of traffic, road infrastructure and quality, as well as the environment. Drivers' behavior includes poor driving habits or attitudes such as not adhering to the rules of traffic, and frequent use of mobile phones while driving. In Malaysia, most current efforts are focusing more on the improvement of road condition and vehicles. Selamat and Surienty [12] also believed that road and environment conditions are among the important factors that contributed to this issue. In Malaysia, bad road conditions especially in dangerous locations can increase the risk of accident.

Understanding the common factors that contribute to road accidents is not enough to improve the current condition of this problem. These factors are widely subjective to be solved in a short term and require massive adjustment and funding which involved not only limited resources in terms of monetary, but also time factor and policy changes. This arises the need to predict the number of transportations and road accidents, as well as assess the value at risk of road accidents on a yearly basis to prepare for the precautionary actions. Since the value at risk approach had been widely used in financial area [13–15], this approach can be extended to assess this road accident risk whilst provide us with some quantitative value for easier plan of action.

2 Risk Analysis Theoretical Framework

In this study, we first constructed a risk analysis theoretical framework that originally focused on IR 4.0 evolution in vehicle's condition to fit with our problem in hand. The uncertainty assessment was constructed based on the Fourth Industrial Revolution [16, 17] and the types of future technology in vehicles [18–20]. The illustration is presented in Fig. 1. We later conducted the following steps to predict road accidents and assess its value at risk. First, the data of transportations for year 2020 was forecasted by analyzing the number of transportations and the road accidents using the data from year 2009 to 2019. Then, the correlation analysis between the number of transportations and the road accidents was performed to investigate the strength of relationship between these variables. Finally, the Value at Risk (VaR) was computed to quantify the level of risk on road accidents for the upcoming year based on the historical data available.

We used secondary data obtained from Malaysia Road Safety Department [21] for the number of transportations, together with the number of road accidents from year 2009 until 2019. The variance method was adopted for value at risk, assisted prior by autoregressive (AR) model approach. In the AR model, the variable's current value is defined as a function of its previous values plus error term. The y_t was taken as the function of the time-tagged value itself as dependent variable. The model can be written as

$$y_t = \beta_0 + \beta_1 y_{t-1} + \beta_2 y_{t-2} + \cdots + \beta_k y_{t-k} + \varepsilon_t \qquad (1)$$

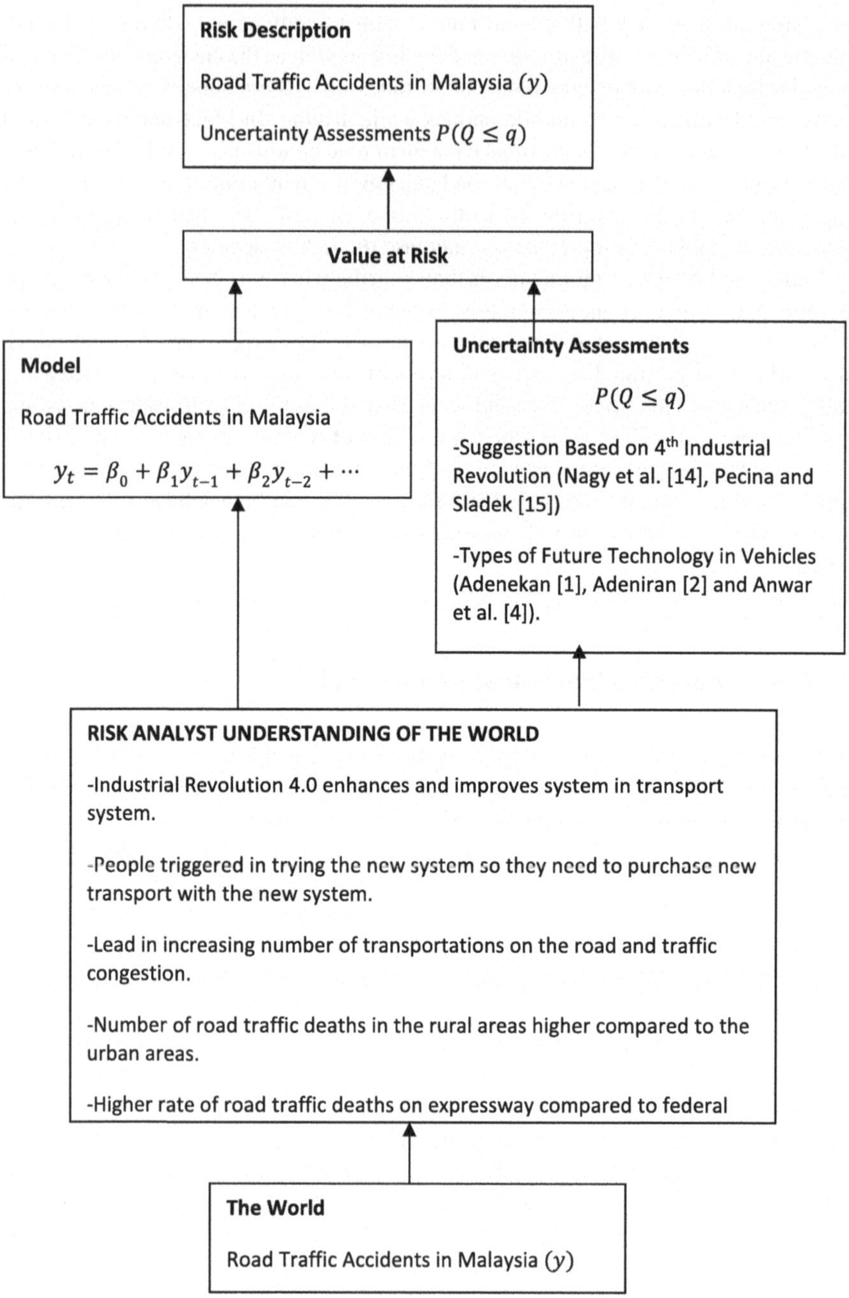

Fig. 1 Risk analysis theoretical framework

where y_t represents the dependent variable, β_0 is the intercept, β_k is the coefficient value, y_{t-k} is the kth order of the lagged dependent or current value, and ε_t is the error term which is assumed independent and identically distributed with mean zero.

In selecting the best model,

- First, the third-order autoregressive model was fitted to the average of dependent variable and the significance of the third-order autoregressive parameter was tested (Use $\alpha = 0.05$)

AR (3)

$$y_t = \beta_0 + \beta_1 y_{t-1} + \beta_2 y_{t-2} + \beta_3 y_{t-3} + \varepsilon_t \tag{2}$$

- If necessary, the second-order autoregressive model would be fitted to the average of dependent variable and the significance of the second-order autoregressive parameter would be tested (Use $\alpha = 0.05$)

AR (2)

$$y_t = \beta_0 + \beta_1 y_{t-1} + \beta_2 y_{t-2} + \varepsilon_t \tag{3}$$

- Then, if necessary, the first-order autoregressive model would be fitted to the average of dependent variable and the significance of the first-order autoregressive parameter would be tested (Use $\alpha = 0.05$)

AR (1)

$$y_t = \beta_0 + \beta_1 y_{t-1} + \varepsilon_t \tag{4}$$

- Data were forecasted using the best model under consideration.

3 Correlation Analysis

This analysis was conducted to measure the degree of linear relationship between the number of transportations on the road versus the occurrence of road accidents. The formula is given by

$$r_{xy} = \frac{\sum(x_i - \bar{x}) - (\sum y_i - \bar{y})}{\sqrt{\sum (x_i - \bar{x})^2 \sum (y_i - \bar{y})^2}} \tag{5}$$

where

r_{xy} is the correlation coefficient of the linear relationship between the variables x and y,

Table 1 Coefficient of correlation

Coefficient of correlation	Description
$r_{xy} > 0.70$	Very strong positive linear relationship
$0.40 < r_{xy} < 0.69$	Strong positive linear relationship
$0.30 < r_{xy} < 0.39$	Moderate positive linear relationship
$0.20 < r_{xy} < 0.29$	Weak positive linear relationship
$0.01 < r_{xy} < 0.19$	No or negligible linear relationship
$r_{xy} = 0$	No linear relationship
$-0.01 < r_{xy} < -0.19$	No or negligible linear relationship
$-0.20 < r_{xy} < -0.29$	Weak negative linear relationship
$-0.30 < r_{xy} < -0.39$	Moderate negative linear relationship
$-0.40 < r_{xy} < -0.69$	Strong negative linear relationship
$r_{xy} > -0.70$	Very strong negative linear relationship

x_i is the values of the x-variable in a sample,
$\bar{x}$ is the mean of the values of the x-variable,
y_i is the values of the y-variable in a sample,
$\bar{y}$ is the mean of the values of the y-variable (Table 1).

4 Value at Risk

Using the formulation of value at risk and historical data, the value at risk of road accidents that occurred was identified by considering a 95% confidence interval. Here, the historical simulation approach was implemented in Microsoft Excel using these steps:

- The changes in the number of the road accidents were calculated using

$$\text{Changes} = \frac{(\text{Current Year's Number of Road Accidents} - \text{Previous Year's Number of Road Accidents})}{\text{Previous Year's Number of Road Accidents}}$$

- All the changes obtained were sorted in increasing order.
- The total count of the changes was computed using 'count' function.

The sorted return corresponding to the 5% of the total count was taken as Var (95) representing the 95% confidence level.

5 Results and Discussion

Forecasting Number of Transportations on the Road.

Table 2 displays the results of all AR models considered. The third-order Autoregressive and the second-order autoregressive B_3 and B_2 showed that there were no significances since both p-values were greater than 0.05. Thus, we need to eliminate these models. Observe that the first-order AR model is significant towards the B_1 since the p-value was less than 0.05.

Based on Table 3, we used the coefficients obtained to be incorporated in our model as follows

$$y_t = 2301361 + 0.951035y_{t-1}$$

Next, we present the data regarding the number of transportations on the road from 2009 until 2019. Thus, the forecast for year 2020 can now be predicted as

$$y_{2020} = 2301361 + 0.951035y_{2019}$$

Table 2 Regression analysis of autoregressive models to forecast number of transportations on the road

		p-value	Conclusion
Third-order autoregressive model AR(3)	Intercept	0.789594	Not significant
	B_1	0.60339	
	B_2	0.615604	
	B_3	0.939892	
Second-order autoregressive model AR(2)	Intercept	0.577016	Not significant
	B_1	0.236094	
	B_2	0.522182	
First-order autoregressive model AR(1)	Intercept	0.00860646	Significant
	B_1	3.36E-09	

Table 3 Coefficient table

	Coefficients	Standard error	t stat	p-value
Intercept	2,301,361	637,223.4	3.611545	0.00860646
B_1	0.951035	0.026467	35.93317	3.36E–09

$$y_{2020} = 2301361 + 0.951035(30002960)$$

$$y_{2020} = 30835226$$

Based on this projection, we can see that in year 2020, the number of transportations on the road would increase up to 30,835,226 units, which is higher compared to the year 2019. This indicates a 2.77% increase in 2020 as compared to the previous year (Table 4).

Forecasting Road Accidents

In Table 5, the third order and second order AR models B_3 and B_2 showed that there were no significances since the p-values were greater than 0.05. These implies that

Table 4 Numbers of transportations on the road from 2009 until 2019

Number of transportations on the road	
Year	Number
2009	18,933,237
2010	20,188,565
2011	21,401,269
2012	22,702,221
2013	23,819,256
2014	25,101,192
2015	26,301,952
2016	27,613,125
2017	28,738,194
2018	29,127,844
2019	30,002,960
2020 (predicted year)	30,835,226

Table 5 Regression analysis of autoregressive models to forecast road accidents

		p-value	Conclusion
Third-order Autoregressive Model AR (3)	Intercept	0.8644	Not significant
	B_1	0.41132	
	B_2	0.675278	
	B_3	0.53674	
Second-order Autoregressive Model AR (2)	Intercept	0.607371	Not significant
	B_1	0.241951	
	B_2	0.440898	
First-order Autoregressive Model AR (1)	Intercept	0.438394	Significant
	B_1	0.001895	

Table 6 Coefficient table

	Coefficients	Standard error	*t* stat	*p*-value
Intercept	70,065.73	85,280.81	0.821588	0.438394
B_1	0.885097	0.183165	4.832245	0.001895

these models could not be used in forecasting. However, the first order AR model of B_1 was significant as the p-value was greater than 0.05. Thus, the first-order AR model may be used in forecasting our data.

We continue with Table 6 which displays the coefficient values for the first order regression.

As a result, the forecasting model becomes

$$y_t = 70065.73 + 0.885097 y_{t-1}.$$

By following this, we focus on the data for road accidents from 2009 until 2019 as presented in Table 7. The forecasted data for year 2020 is as follows

$$y_{2020} = 70065.73 + 0.885097 y_{2019}$$

$$y_{2020} = 70065.73 + 0.885097(555628)$$

$$y_{2020} = 561,850.$$

Table 7 Data of road accidents

Road accidents	
Year	Number
2009	397,330
2010	414,421
2011	449,040
2012	462,423
2013	427,204
2014	476,196
2015	489,606
2016	521,466
2017	533,875
2018	548,598
2019	555,628
2020 (predicted year)	561,850

Table 8 The correlation table

		Number of transportations on the road
Road accidents	Pearson correlation	0.9542
	Sig. (2-Tailed)	<0.001
	N	10

The result above implies that for year 2020, the road accidents would increase to 561,850 occurrences, which is 1.12% higher than the year 2019.

Table 8 concludes that the correlation between road accidents and number of transportations on the road was significant, since the p-values was less than 0.05.

The correlation value between road accidents and the number of transportations on the road was 0.9542. This indicates that there is a strong positive linear relationship between road accidents and the number of transportations on the road, suggesting that the number of transportations on the road could significantly affect the occurrence of road accidents.

Value at Risk

In this subsection, we implement the steps described in the previous section to calculate the respective value at risk. Table 9 exhibits the results executed from Microsoft Excel.

At 95% confident interval, the value at risk for the number of road accidents was obtained by multiplying the number of "count" which is 10 with 0.05, which gives us the value of 0.5. By referring to the column "number", the value 0.5 is associated with the respective sorted changes of −3.808093456. This implies that there is a 95%

Table 9 Value at risk calculation for road accidents

Year	Number of road accidents	Changes	Sorted changes	Number	Count
2009	397,330				10
2010	414,421	4.301462261	−7.616186911	1	
2011	449,040	8.353582468	1.281448347	2	
2012	462,423	2.980358097	2.379637407	3	
2013	427,204	−7.616186911	2.757761648	4	
2014	476,196	11.46805742	2.816067334	5	
2015	489,606	2.816067334	2.980358097	6	
2016	521,466	6.507273195	4.301462261	7	
2017	533,875	2.379637407	6.507273195	8	
2018	548,598	2.757761648	8.353582468	9	
2019	555,628	1.281448347	11.46805742	10	

confidence that in year 2020, the number of road accidents will reduce not more than 3.81%.

In year 2020, the number of transportations on the road would increase to 30,835,226 (2.77% increase from year 2019) vehicles. Meanwhile, the road accidents would also increase to 561,850 (1.12% higher in year 2020 compared to 2019) occurrences. There was a strong positive linear relationship between the number of road accidents and the number of transportations on the road. It can be concluded that as the number of transportations on the road increase, the road accidents would also increase. Since the increasing number of vehicles is difficult to be controlled, thus assessing its risk should be conducted differently. By using some statistical measures and 95% confidence interval, we found out that there is a 95% confidence that in year 2020, the number of road accidents will reduce not more than 3.81%. This percentage of difference is quite small since the maximum number of road accidents that can be reduced is only 21,169. Thus, identifying the value at risk on yearly basis is able to project the best course of action to better deal with the risk of road accidents in the country.

6 Summary

In this chapter, we utilize the autoregressive model and value at risk approach that is commonly used in financial sector into assessing the risk posed in Malaysian road accidents occurrences. The findings suggest that there is potential to adopt these approaches more extensively for this issue as the quantitative analysis are feasible. In addition, strong correlation is found between the number of vehicles with the likelihood of road accidents to occur. As the number of vehicles increase, so is the accidents, which is agreeable with common perception. Rohayu et al. [22] for example, has predicted escalate of road fatalities to 10,716 cases in the year 2020. Such grim outlook needs to be urgently addressed. These untimely deaths will cause many unnecessary wastes in country's resources, with the direct impact of immediate loss in the trained and future human capital of the country. Government will lose its initial investment in training and subsidizing the wellbeing of its citizen from the day they were born. By meeting avoidable fatalities from road accidents, the expected output from these citizens in terms of productive contribution to the country will be stopped prematurely. Besides, medical facilities are also not in their optimum, as expensive medical care (medical specialist, bed occupancy, and other medical apparatus) needs to be allocated into avoidable medical emergency such as accident-based emergency. Note also that public medical facilities in Malaysia is heavily subsidized by the government's budget allocation. Currently, Malaysia has yet to impose medical insurance initiatives such as *Medicare* for all citizen, thus most medical expenses on accident cases will be borne by this allocation. These wastes of resources hinder progress into focusing more on pressing epidemic issues, economic developments, and overall improvement for citizen in any functioning country.

As we see severe impacts posed by road accidents, it is crucial to see this threat similar as how we give emphasis on economic and financial development in the country. Thus, rigorous assessment that is normally applied to financial sector should also be applied while solving this issue. Regulatory bodies overseeing this issue should be properly functioning, similarly as how the central bank and securities commission regulate and monitor the progress of financial services. Thus, the valuation of risk on frequent terms should also be conducted to aggressively apply the preventive measures, instead of clean-the-mess method of operative. Prevention is better than cure. In this case, prevention is the only way to reduce surge number of wasted resources.

References

1. Roehler, D.R., Sann, S., Kim, P., Bachani, A.M., Campostrini, S., Florian, M., Ballesteros, M.F.: Motorcycle helmet attitudes, behaviours and beliefs among Cambodians. Int. J. Inj. Contr. Saf. Promot. **20**(2), 179–183 (2013)
2. World Health Organization Road Traffic Injuries. https://www.who.int/news-room/fact-sheets/detail/road-traffic-injuries (2018)
3. Kamarudin, M.K.A., Wahab, N.A., Umar, R., Saudi, A.S.M., Saad, M.H.M., Rosdi, N.R.N., Ridzuan, A.M.: Road traffic accident in Malaysia: trends, selected underlying, determinants and status intervention. Int. J. Eng. Technol. **7**(4.34), 112–117 (2018)
4. Darma, Y., Mohamed, R.K., Sulaiman, A.: An analysis of Malaysia road traffic death distribution by road environment. Sadhana **42**(9), 1605–1615 (2017)
5. Masuri, M.G., Isa, K.A.M., Tahir, M.P.M.: Children, youth and road environment: road traffic accident. Procedia Soc. Behav. Sci. **38**, 213–218 (2012)
6. Mohamad, F.F., Mohamad, J., Abdullah, A.S.: Do young Malaysian drivers involve in speeding behavior: An investigation at selected accident hotspots in peninsular Malaysia. J. Adv. Res. Soc. Behav. Sci. **9**(1), 75–83 (2017)
7. Pylkkönen, M., Sihvola, M., Hyvärinen, H.K., Puttonen, S., Hublin, C., Sallinen, M.: Sleepiness, sleep, and use of sleepiness countermeasures in shift-working long-haul truck drivers. Accid. Anal. Prev. **80**, 201–210 (2015)
8. Useche, S.A., Ortiz, V.G., Cendales, B.E.: Stress-related psychosocial factors at work, fatigue, and risky driving behavior in bus rapid transport (BRT) drivers. Accid. Anal. Prev. **104**, 106–114 (2017)
9. Butlewski, M., Hankiewicz, K.: Psychomotor performance monitoring system in the context of fatigue and accident prevention. Procedia Manuf. **3**, 4860–4867 (2015)
10. Fida, R., Paciello, M., Tramontano, C., Fontaine, R.G., Barbaranelli, C., Farnese, M.L.: An integrative approach to understanding counterproductive work behavior: the roles of stressors, negative emotions, and moral disengagement. J. Bus. Ethics **130**(1), 131–144 (2015)
11. Road Safety Plan of Malaysia 2014–2020. http://www.mot.gov.my › Darat › Road_Safety_Plan_2014–2020_booklet-EN (2014)
12. Selamat, M.N., Surienty, L.: An examination of commuting accident in Malaysia. In: 3rd Scientific Conference on Occupational Safety and Health: Sci-Cosh 2014 (2015)
13. Ahmadi-Javid, A., Fallah-Tafti, M.: Portfolio optimization with entropic value-at-risk. Eur. J. Oper. Res. **279**(1), 225–241 (2019)
14. Bernard, C., Rüschendorf, L., Vanduffel, S., Yao, J.: How robust is the value-at-risk of credit risk portfolios? Eur. J. Finan. **23**(6), 507–534 (2017)
15. Filippi, C., Guastaroba, G., Speranza, M.G.: Applications of conditional value-at-risk beyond finance: a literature review. In: Technical Report 2. University of Brescia, Department of Economics and Management (2017)

16. Nagy, J., Oláh, J., Erdei, E., Máté, D., Popp, J.: The role and impact of Industry 4.0 and the internet of things on the business strategy of the value chain—the case of Hungary. Sustainability **10**(10), 3491 (2018)
17. Pecina, P., Sladek, P.: Fourth industrial revolution and technical education. https://www.researchgate.net/publication/315352615_Fourth_Industrial_Revolution_andTechnical_Education (2017)
18. Adenekan, S.: 5 ways IoT can help to reduce automatic vehicle accidents. https://readwrite.com/2019/05/01/5-ways-iot-can-help-to-reduce-automatic-vehicle-accidents/ (2019)
19. Adeniran, A.O.: Idea of smart development in the fourth industrial revolution emphasis on smart road. SF J. Telecommunic, **1**(1) (2017)
20. Anwar, A.K.K., Hanis, H.H., Amirudin, M.R.M., Rabihah, I., Wong, S.V.: Advancement in vehicle safety in Malaysia from planning to implementation. Asian Transp. Stud. **4**(4), 704–714 (2017)
21. Malaysia Road Safety Department. http://www.jkjr.gov.my/ms/
22. Rohayu, S., Sharifah Allyana, S.M.R., Jamilah, M.M., Wong, S.V.: Predicting Malaysian road fatalities for year 2020, MRR 06/2012, Malaysian Institute of Road Safety Research, Kuala Lumpur. https://www.researchgate.net/publication/283299518_Predicting_Malaysian_Road_Fatalities_for_Year_2020 (2012)

Global Stability Analysis of Dengue Model with Awareness, Vector Control and Time Delays

Jay Michael R. Macalalag, Timothy Robin Y. Teng, and Elvira P. de Lara-Tuprio

Abstract In this paper, a mathematical model for a single-strain dengue virus transmission, incorporating vector control, disease awareness among susceptible humans, and both the latent delays for human and mosquitoes, is proposed and studied. The global stability properties of disease-free equilibrium and endemic equilibrium are completely established through Lyapunov functionals and LaSalle's invariance principle. The global dynamics of the equilibrium points are characterized by the value of basic reproductive number R_0. If $R_0 < 1$, then the disease-free equilbrium is globally asymptotically stable, and is unstable if $R_0 > 1$. Furthermore, the endemic equilibrium is globally asymptotically stable if $R_0 > 1$. Numerical simulations are presented to illustrate the theoretical results.

Keywords Dengue fever · Delay differential equations · Global stability · Lyapunov functional · Hopf bifurcation

1 Introduction

Dengue is a mosquito-borne viral infection caused by the dengue virus (DENV) that has four serotypes (DEN1, DEN2, DEN3 and DEN4) and which belong to the genus Flavivirus, family Flaviviridae. Dengue causes a severe flu-like illness which may sometimes lead to a more complicated condition called severe dengue. Severe

J. M. R. Macalalag
College of Mathematics and Natural Sciences, Caraga State University, Butuan City 8600, Agusan Del Norte, Philippines
e-mail: jrmacalalag@carsu.edu.ph

T. R. Y. Teng · E. P. de Lara-Tuprio (✉)
Department of Mathematics, School of Science and Engineering, Ateneo de Manila University, Loyola Heights, Quezon City 1108, Metro Manila, Philippines
e-mail: edelara-tuprio@ateneo.edu

T. R. Y. Teng
e-mail: tteng@ateneo.edu

© Institute of Technology PETRONAS Sdn Bhd 2022

S. A. Abdul Karim and A. Shafie (eds.), *Towards Intelligent Systems Modeling and Simulation*, Studies in Systems, Decision and Control 383,
https://doi.org/10.1007/978-3-030-79606-8_14

dengue (formerly known as dengue hemorrhagic fever) was first recognized in the 1950s during dengue epidemics in the Philippines and Thailand. This has become a leading cause of hospitalization and deaths among children and adults in the Asian and Latin American countries [15].

Dengue virus is transmitted to humans through the bite of infective females *Aedes aegypti* and *Aedes albopictus* mosquitoes. Once humans are infected with the virus, they can transmit the infection to susceptible mosquitoes after the first symptoms appear, following an incubation period of 4–10 days [15]. When the virus enters the mosquito's blood, the virus will require an additional 8-12 days incubation before it can then be transmitted to another human. The mosquito remains infected for the remainder of its life, which might be several days or a few weeks [2]. A human who recovers from a dengue infection of one serotype becomes immune to that particular serotype. However, subsequent infection from the other serotypes increases the risk of developing severe dengue [15].

There are already some vaccines under clinical trials and one licensed, the Dengvaxia, which has been recommended for use in countries with high number of dengue cases [15]. However, until the safety and reliability of this vaccine is fully established, the most effective way to control the spread of dengue in the community is by taking preventive measures. These can be done through reduction of mosquito population and personal protection awareness against mosquito bites [2]. Mosquito reduction includes the elimination of mosquito breeding sites and the use of insecticides to kill mosquitoes. On the other hand, personal protection includes avoiding mosquito bites (by using mosquito net when sleeping, window and door screens, long-sleeved clothes and long pants, mosquito repellent, etc.).

Mathematical models provide a useful tool in understanding the spread of dengue virus in a community. Several compartmental models for dengue have been introduced and studied (see [5, 7, 8]). In 1998, Esteva and Vargas [5] introduced a model for dengue virus transmission by using ordinary differential equations. This model assumed that when a susceptible mosquito consumes a blood meal from an infectious human, that mosquito immediately becomes infectious. They further assumed that when an infectious mosquito bites a susceptible human, that bitten human becomes infectious instantaneously. In reality, it takes some time before these infected humans and mosquitoes become infectious [15]. These time periods are called latent periods. In 2008, Wei et al. [14] considered an SIR-SI vector-borne disease model incorporating a latent delay for the infected vectors. In 2012, Lashari et al. [11] studied how the latent delay for infected humans affects the dynamics of the SIR-SI vector-borne disease model.

In 2017, Guan et al. [8] introduced a model for dengue fever transmission with latent time delays for both humans and mosquitoes, and found out that these delays resulted to the occurrence of Hopf bifurcation (see [10, p. 58]). They used the following system of differential equations

$$S'_H(t) = \mu_H N_H(t) - \mu_H S_H(t) - \lambda_H I_V(t) S_H(t) \quad (1)$$
$$I'_H(t) = \lambda_H I_V(t - \tau_1) S_H(t) - (\mu_H + \delta) I_H(t) \quad (2)$$
$$R'_H(t) = \delta I_H(t) - \mu_H R_H(t) \quad (3)$$
$$S'_V(t) = A_2 - \lambda_V I_H(t) S_V(t) - \mu_V S_V(t) \quad (4)$$
$$I'_V(t) = \lambda_V I_H(t - \tau_2) S_V(t) - \mu_V I_V(t), \quad (5)$$

where S_H, I_H, R_H, S_V and I_V represent the number of susceptible humans, infected humans, recovered humans, susceptible mosquitoes and infected mosquitoes, respectively. The parameters μ_H, μ_V and δ are all constants which represent the death rate of humans, death rate of mosquitoes and recovery rate of humans from the disease. Here, $N_H(t)$ represents the total human population at time t. It is also assumed that humans and mosquitoes in the infected compartments at time t are not all capable of transmitting the virus. Hence, the latent delays τ_1 and τ_2 for mosquitoes and humans, respectively, are included in the model. However, in (1), the term $-\lambda_H I_V(t) S_H(t)$, where λ_H is the effective contact rate sufficient to transmit the virus from an infected mosquito to a susceptible human, implies that all infected mosquitos at time t are infectious. Similarly, in (4), the term $-\lambda_V I_H(t) S_V(t)$, where λ_V is the effective contact rate sufficient to transmit the virus from an infected human to a susceptible mosquito, implies that all infected humans at time t are all infectious. These contradict the assumption that infected mosquitoes and humans at time t are not all immediately capable of transmitting the virus.

In this paper, we make improvements to the dengue model in [8] so that the above problem is addressed. Moreover, human protection awareness and vector control are introduced in the model to study their effects on the spread of dengue virus in the community.

The rest of the paper is organized as follows. The mathematical model is presented in Sect. 2. In Sect. 3, we determine the equilibrium points and the basic reproductive number R_0. We then establish conditions for the global stability of the equilibrium points in Sect. 4. In Sect. 5, we illustrate the theoretical results using some numerical simulations. Finally, we conclude in Sect. 6 with some suggestions for future work.

2 The Mathematical Model

The human population, with size $N_H(t)$ at time t, is partitioned into three compartments of individuals who are susceptible $S_H(t)$, infected $I_H(t)$ and removed $R_H(t)$. On the other hand, the total mosquito population size at time t, denoted by $N_V(t)$, is partitioned into only two compartments of mosquitoes that are susceptible $S_V(t)$ and infected $I_V(t)$.

Assume that all recruitment of human population and of mosquito population goes only in the susceptible compartments at constant rates A_1 and A_2, respectively. Denote the natural death rates per unit time of humans and mosquitoes by μ_H and μ_V, respectively. Assume also that an infected human dies at a rate ε due to the severity and

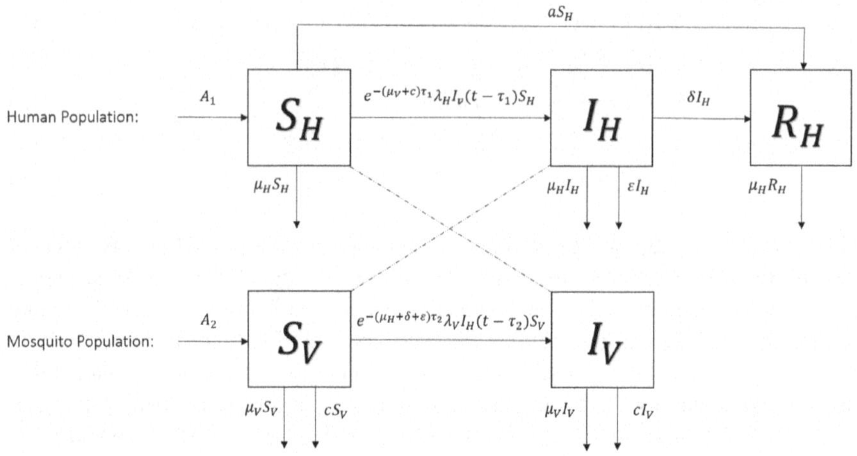

Fig. 1 Progression of dengue infection for the human and mosquito population

complications of the disease. Furthermore, susceptible and infected mosquitoes may die due to vector control at a rate c. Susceptible humans who practice self-protection (awareness) against the mosquito bites move to the removed compartment at a rate a. Moreover, an infected human recovers from the disease at a rate δ and goes to the removed compartment. Only one strain of dengue virus is considered in this study, hence, the recovered individuals will acquire permanent immunity from that particular serotype. On the other hand, once a mosquito becomes infectious, such mosquito is capable of transmitting the virus for the rest of its life.

We now introduce the time delays τ_1 and τ_2 in the model describing the incubation periods of the dengue virus in mosquitoes and humans, respectively. Observe that at time t, the infectious mosquitoes are those who were infected by the virus τ_1 units of time ago, provided that they are still alive after that τ_1 period. Hence, the incidence term for infected humans is given by $e^{-(\mu_V+c)\tau_1}\lambda_H I_V(t-\tau_1)S_H(t)$, where $e^{-(\mu_V+c)\tau_1}$ represents the probability that an infected mosquito remains alive after a period of τ_1 units. Similarly, at time t, the infectious humans are those who were infected by the virus τ_2 units of time ago, provided that they are still alive after that τ_2 period. This suggests that the incidence term of infected vectors is $e^{-(\mu_H+\delta+\epsilon)\tau_2}\lambda_V I_H(t-\tau_2)S_V(t)$, where $e^{-(\mu_H+\delta+\epsilon)\tau_2}$ represents the probability that an infected human remains alive and infected after a period of τ_2 units.

The mathematical model for the dengue virus transmission as described schematically in Fig. 1 is given by the following system of differential equations:

$$S'_H(t) = A_1 - e^{-(\mu_V+c)\tau_1}\lambda_H I_V(t-\tau_1)S_H(t) - (\mu_H + a)S_H(t) \tag{6}$$
$$I'_H(t) = e^{-(\mu_V+c)\tau_1}\lambda_H I_V(t-\tau_1)S_H(t) - (\mu_H + \epsilon + \delta)I_H(t) \tag{7}$$
$$R'_H(t) = aS_H(t) + \delta I_H(t) - \mu_H R_H(t) \tag{8}$$
$$S'_V(t) = A_2 - e^{-(\mu_H+\delta+\epsilon)\tau_2}\lambda_V I_H(t-\tau_2)S_V(t) - (\mu_V + c)S_V(t) \tag{9}$$
$$I'_V(t) = e^{-(\mu_H+\delta+\epsilon)\tau_2}\lambda_V I_H(t-\tau_2)S_V(t) - \mu_V I_V(t) - cI_V(t) \tag{10}$$

where $N_H(t) = S_H(t) + I_H(t) + R_H(t)$, $N_V(t) = S_V(t) + I_V(t)$, S_H, I_H, R_H, S_V and I_V are differentiable real-valued functions on $\mathbb{R}$. The parameters are all positive constants. The system of Eqs. (6)–(10) will be referred to as system (M).

For $\tau = \max\{\tau_1, \tau_2\} > 0$, let $\mathcal{C} := C\left([-\tau, 0], \mathbb{R}^5\right)$ be the Banach space of continuous functions from $[-\tau, 0]$ into $\mathbb{R}^5$ with norm $||\phi||_* = \sup_{-\tau\le\theta\le 0} ||\phi(\theta)||$ for each $\phi \in \mathcal{C}$.

For any differentiable function $X = (S_H, I_H, R_H, S_V, I_V) : [-\tau, \zeta) \to \mathbb{R}^5$ with $\zeta > 0$, define $X_t \in \mathcal{C}$ for each $t \ge 0$ by

$$X_t(\theta) = X(t+\theta) = (S_H(t+\theta), I_H(t+\theta), R_H(t+\theta), S_V(t+\theta), I_V(t+\theta))$$

for all $\theta \in [-\tau, 0]$.

Denote by $\mathbb{R}_{0+}$ the set of all nonnegative real numbers. We only consider solutions $X = (S_H, I_H, R_H, S_V, I_V)$ on $[-\tau, +\infty)$ of system (M) with initial condition $\phi = (\phi_1, \phi_2, \phi_3, \phi_4, \phi_5)$ from the set

$$\Phi = \left\{\phi \in C([-\tau, 0], \mathbb{R}^5_{0+}) : \phi_1(0) > 0, \phi_4(0) > 0\right\}.$$

That is, for all $\theta \in [-\tau, 0]$, we have

$$S_H(\theta) = \phi_1(\theta), \quad I_H(\theta) = \phi_2(\theta), \quad R_H(\theta) = \phi_3(\theta),$$
$$S_V(\theta) = \phi_4(\theta) \quad I_V(\theta) = \phi_5(\theta).$$

The following theorems can be established by using the properties of retarded functional differential equations (see [4, 10]).

Theorem 1 *For each $\phi = (\phi_1, \phi_2, \phi_3, \phi_4, \phi_5) \in \Phi$, the solution to system (M) with initial condition ϕ exists and is unique.*

Theorem 2 *Every solution to system (M) with initial condition $\phi \in \Phi$ is nonnegative and bounded on $[0, +\infty)$.*

Hereafter, we shall denote by Ω the space of all solutions to system (M) with initial condition $\phi \in \Phi$.

3 The Basic Reproductive Number and Equilibrium Points

System (M) always has the disease-free equilibrium

$$E^0 = \left(S_{H_0}, I_{H_0}, R_{H_0}, S_{V_0}, I_{V_0}\right) = \left(\frac{A_1}{\mu_H + a}, 0, \frac{aA_1}{(\mu_H + a)\mu_H}, \frac{A_2}{\mu_V + c}, 0\right)$$

in Ω. By using the Next Generation Matrix approach, we obtained the **basic reproductive number**

$$R_0 = \frac{A_1 A_2 e^{-(\mu_V + c)\tau_1} e^{-(\mu_H + \delta + \epsilon)\tau_2} \lambda_H \lambda_V}{(\mu_H + a)(\mu_V + c)^2 (\mu_H + \epsilon + \delta)},$$

which represents the average number of secondary infections that a single infectious human can produce in a totally susceptible population of humans and mosquitoes.

It is important to note that the dynamics of system (M) is characterized by R_0. In particular, the value of R_0 determines that number of equilibrium points, as well as the stability of these points which will be shown in the next section.

When $R_0 > 1$, the system has another equilibrium point, the endemic equilibrium $E^* = (S_{H*}, I_{H*}, R_{H*}, S_{V*}, I_{V*})$, where

$$S_{H*} = \frac{(\mu_V + c)\left[A_1 e^{-(\mu_H + \delta + \epsilon)\tau_2} \lambda_V + (\mu_V + c)(\mu_H + \epsilon + \delta)\right]}{\lambda_V e^{-(\mu_H + \delta + \epsilon)\tau_2} \left(A_2 e^{-(\mu_V + c)\tau_1} \lambda_H + (\mu_H + a)(\mu_V + c)\right)}, \tag{11}$$

$$I_{H*} = \frac{[\mu_H + a][\mu_V + c]^2 [R_0 - 1]}{e^{-(\mu_H + \delta + \epsilon)\tau_2} \lambda_V \left[(\mu_H + a)(\mu_V + c) + A_2 \lambda_H e^{-(\mu_V + c)\tau_1}\right]}, \tag{12}$$

$$R_{H*} = \frac{\delta I_{H*} + a S_{H*}}{\mu_H}, \tag{13}$$

$$S_{V*} = \frac{(\mu_H + \epsilon + \delta)\left[A_2 e^{-(\mu_V + c)\tau_1} \lambda_H + (\mu_H + a)(\mu_V + c)\right]}{e^{-(\mu_V + c)\tau_1} \lambda_H \left[A_1 e^{-(\mu_H + \delta + \epsilon)\tau_2} \lambda_V + (\mu_H + \epsilon + \delta)(\mu_V + c)\right]}, \tag{14}$$

$$I_{V*} = \frac{(\mu_H + a)(\mu_V + c)(\mu_H + \epsilon + \delta)[R_0 - 1]}{e^{-(\mu_V + c)\tau_1} \lambda_H \left[A_1 e^{-(\mu_H + \delta + \epsilon)\tau_2} \lambda_V + (\mu_V + c)(\mu_H + \epsilon + \delta)\right]}. \tag{15}$$

Ideally, R_0 must decrease to a value below 1 in order to control the spread of the disease. Because R_0 is dependent on the values of the model parameters, disease control and intervention strategies that can influence the values of certain parameters so as to achieve $R_0 < 1$ may be introduced.

In the following remarks, we confirm mathematically that the level of human awareness a and the level of effective vector control c affect the value of R_0.

Remark 1 For fixed parameters A_1, A_2, μ_V, μ_H, δ, ϵ, λ_H, λ_V, τ_1 and τ_2, R_0 is a decreasing function of a.

This is because

$$\frac{\partial R_0}{\partial a} = -\frac{A_1\, A_2\, e^{-(\mu_V+c)\tau_1} e^{-(\mu_H+\epsilon+\delta)\tau_2} \lambda_H\, \lambda_V}{(\mu_H+a)^2\, (\mu_V+c)^2\, (\mu_H+\epsilon+\delta)} < 0.$$

This simply implies that increasing one's knowledge of the methods of preventing mosquito bites and practicing these help decrease the rate of spread of dengue virus in the community.

Remark 2 For fixed parameters A_1, A_2, μ_V, μ_H, δ, ϵ, λ_H, λ_V, τ_1 and τ_2, R_0 is a decreasing function of c.

Again, observe that

$$\begin{aligned}\frac{\partial R_0}{\partial c} = &-\frac{A_1\, A_2\, \tau_1\, e^{-(\mu_V+c)\tau_1} e^{-(\mu_H+\epsilon+\delta)\tau_2} \lambda_H\, \lambda_V}{(\mu_H+a)\, (\mu_V+c)^2\, (\mu_H+\epsilon+\delta)} \\ &- 2\,\frac{A_1\, A_2\, e^{-(\mu_V+c)\tau_1} e^{-(\mu_H+\epsilon+\delta)\tau_2} \lambda_H\, \lambda_V}{(\mu_H+a)\, (\mu_V+c)^3\, (\mu_H+\epsilon+\delta)} < 0\end{aligned}$$

which justifies the above remark. Hence, intensifying vector control measures is essential in slowing down the spread of dengue virus in the community.

4 Global Stability Analysis

In this section, we will analyze the global stability of the equilibrium points using Lyapunov functionals and the LaSalle's Invariance Principle [10]. Because the variable R_H is not explicit in Eqs. (6), (7), (9), and (10), we may first consider the following subsystem (M_R) of system (M):

$$S_H'(t) = A_1 - e^{-(\mu_V+c)\tau_1} \lambda_H I_V(t-\tau_1) S_H(t) - (\mu_H+a) S_H(t) \tag{16}$$
$$I_H'(t) = e^{-(\mu_V+c)\tau_1} \lambda_H I_V(t-\tau_1) S_H(t) - (\mu_H+\epsilon+\delta) I_H(t) \tag{17}$$
$$S_V'(t) = A_2 - e^{-(\mu_H+\epsilon+\delta)\tau_2} \lambda_V I_H(t-\tau_2) S_V(t) - (\mu_V+c) S_V(t) \tag{18}$$
$$I_V'(t) = e^{-(\mu_H+\epsilon+\delta)\tau_2} \lambda_V I_H(t-\tau_2) S_V(t) - (\mu_V+c) I_V(t) \tag{19}$$

Certainly, system (M_R) always has the disease-free equilibrium $E_0 = \left(S_{H_0}, 0, S_{V_0}, 0\right)$, where $S_{H_0} = \dfrac{A_1}{\mu_H+a}$ and $S_{V_0} = \dfrac{A_2}{\mu_V+c}$. Also, if $R_0 > 1$, then system (M_R) has a unique endemic equilibrium $E_* = (S_{H*}, I_{H*}, S_{V*}, I_{V*})$, where S_{H*}, I_{H*}, S_{V*} and I_{V*} are given in (11), (12), (14) and (15), respectively.

We will only consider solutions $\bar{X} = (S_H, I_H, S_V, I_V)$ on $[-\tau, +\infty)$ of system (M_R) with initial condition $\phi = (\phi_1, \phi_2, \phi_4, \phi_5)$ from the set

$$\Phi_R = \left\{\phi = (\phi_1, \phi_2, \phi_4, \phi_5) \in C\left([-\tau, 0], \mathbb{R}^4_{0+}\right) : \phi_1(\theta) > 0, \phi_4(\theta) > 0\right\}$$

Denote by Ω_R the space of all solutions to system (M_R) with initial condition $\phi \in \Phi_R$.

Observe also that for any $\phi = (\phi_1, \phi_2, \phi_4, \phi_5) \in \Phi_R$, there exists a unique solution to system (M_R) with initial condition ϕ, which is nonnegative and bounded on $[0, +\infty)$.

Theorem 3 *If $R_0 < 1$, then the disease-free equilibrium $\left(S_{H_0}, 0, S_{V_0}, 0\right)$ of system (M_R) is globally asymptotically stable in Ω_R.*

Proof Define the functional $V : \Phi_R \to \mathbb{R}$ by

$$V(\gamma) = V_1(\gamma) + V_2(\gamma) + V_3(\gamma) + V_4(\gamma) + V_5(\gamma) + V_6(\gamma), \tag{20}$$

where

$$V_1(\gamma) = \frac{e^{-(\mu_H+\epsilon+\delta)\tau_2}\lambda_V A_2}{(\mu_H + \epsilon + \delta)(\mu_V + c)}\left[\gamma_1(0) - S_{H_0} - S_{H_0}\ln\frac{\gamma_1(0)}{S_{H_0}}\right],$$

$$V_2(\gamma) = \left[\gamma_4(0) - S_{V_0} - S_{V_0}\ln\frac{\gamma_4(0)}{S_{V_0}}\right],$$

$$V_3(\gamma) = \frac{e^{-(\mu_H+\epsilon+\delta)\tau_2}\lambda_V A_2}{(\mu_H + \epsilon + \delta)(\mu_V + c)}\gamma_2(0) + \gamma_5(0),$$

$$V_4(\gamma) = e^{-(\mu_H+\epsilon+\delta)\tau_2}\lambda_V\frac{A_2}{\mu_V + c}\int_{-\tau_2}^{0}\gamma_2(\theta)d\theta,$$

$$V_5(\gamma) = \frac{e^{-(\mu_H+\epsilon+\delta)\tau_2}e^{-(\mu_V+c)\tau_1}\lambda_H\lambda_V A_1 A_2}{(\mu_V + c)(\mu_H + a)(\mu_H + \epsilon + \delta)}\int_{-\tau_1}^{0}\gamma_5(\theta)d\theta,$$

$$V_6(\gamma) = \frac{e^{-(\mu_H+\epsilon+\delta)\tau_2}\lambda_V A_2(\mu_H + a)}{(\mu_H + \epsilon + \delta)(\mu_V + c)}\int_{-\tau}^{0}\frac{(\gamma_1(\theta) - S_{H_0})^2}{\gamma_1(\theta)}d\theta$$

$$+ (\mu_V + c)\int_{-\tau}^{0}\frac{(\gamma_4(\theta) - S_{V_0})^2}{\gamma_4(\theta)}d\theta \;+\; (\mu_V + c)(1 - R_0)\int_{-\tau}^{0}\gamma_5(\theta)d\theta.$$

Certainly, V is continuous and nonnegative on Φ_R if $R_0 < 1$. Furthermore, $V(\gamma) = 0$ if and only if $\gamma = E_{0*}$, where $E_{0*} = \left(S_{H_{0*}}, I_{H_{0*}}, S_{V_{0*}}, I_{V_{0*}}\right) \in \Phi_R$, the function identically equal to $E_0 = (S_{H_0}, 0, S_{V_0}, 0)$.

Let $\bar{X} = (S_H, I_H, S_V, I_V)$ be a solution to system (M_R) with initial condition $\phi = (\phi_1, \phi_2, \phi_4, \phi_5) \in \Phi_R$. Then $\bar{X}_t \in \Phi_R$ for $t \geq 0$, with $\bar{X}_0 = \phi$, and $V(\bar{X}_t) = V_1(\bar{X}_t) + V_2(\bar{X}_t) + V_3(\bar{X}_t) + V_4(\bar{X}_t) + V_5(\bar{X}_t) + V_6(\bar{X}_t)$, where

$$V_1(\bar{X}_t) = \frac{e^{-(\mu_H+\epsilon+\delta)\tau_2}\lambda_V A_2}{(\mu_H+\epsilon+\delta)(\mu_V+c)}\left[S_H(t) - S_{H_0} - S_{H_0}\ln\frac{S_H(t)}{S_{H_0}}\right],$$

$$V_2(\bar{X}_t) = \left[S_V(t) - S_{V_0} - S_{V_0}\ln\frac{S_V(t)}{S_{V_0}}\right]$$

$$V_3(\bar{X}_t) = \frac{e^{-(\mu_H+\epsilon+\delta)\tau_2}\lambda_V A_2}{(\mu_H+\epsilon+\delta)(\mu_V+c)} I_H(t) + I_V(t)$$

$$V_4(\bar{X}_t) = e^{-(\mu_H+\epsilon+\delta)\tau_2}\lambda_V \frac{A_2}{\mu_V+c}\int_{t-\tau_2}^{t} I_H(\theta)d\theta$$

$$V_5(\bar{X}_t) = \frac{e^{-(\mu_H+\epsilon+\delta)\tau_2}e^{-(\mu_V+c)\tau_1}\lambda_H\lambda_V A_1 A_2}{(\mu_V+c)(\mu_H+a)(\mu_H+\epsilon+\delta)}\int_{t-\tau_1}^{t} I_V(\theta)d\theta$$

$$V_6(\bar{X}_t) = \frac{e^{-(\mu_H+\epsilon+\delta)\tau_2}\lambda_V A_2(\mu_H+a)}{(\mu_H+\epsilon+\delta)(\mu_V+c)}\int_{t-\tau}^{t}\frac{(S_H(\theta)-S_{H_0})^2}{S_H(\theta)}d\theta$$

$$+ (\mu_V+c)\int_{t-\tau}^{t}\frac{(S_V(\theta)-S_{V_0})^2}{S_V(\theta)}d\theta \ + \ (\mu_V+c)(1-R_0)\int_{t-\tau}^{t} I_V(\theta)d\theta.$$

It can be shown that the derivative of V along the solutions of system (M_R) with initial condition $\phi \in \Phi_R$ is given by

$$\begin{aligned}\frac{d}{dt}V(\bar{X}_t) = &- \frac{e^{-(\mu_H+\epsilon+\delta)\tau_2}\lambda_V A_2(\mu_H+a)}{(\mu_H+\epsilon+\delta)(\mu_V+c)}\frac{\left(S_H(t-\tau)-S_{H_0}\right)^2}{S_H(t-\tau)}\\ &- (\mu_V+c)\frac{\left(S_V(t-\tau)-S_{V_0}\right)^2}{S_V(t-\tau)}\\ &- (\mu_V+c)(1-R_0)I_V(t-\tau).\end{aligned} \tag{21}$$

Thus, if $R_0 < 1$, then $V'(X_t) = \frac{d}{dt}V(X_t) \leq 0$ for $t \geq 0$.

Note that in (21), $V'(\bar{X}_t) = 0$ if $S_H(t) = S_H(t-\tau) = S_{H_0}$, $S_V(t) = S_V(t-\tau) = S_{V_0}$, $I_H(t-\tau) = 0$ and $I_V(t-\tau) = 0$. For simplicity, let

$$k_1 = \frac{e^{-(\mu_H+\epsilon+\delta)\tau_2}\lambda_V A_2(\mu_H+a)}{(\mu_H+\epsilon+\delta)(\mu_V+c)} \text{ and } k_2 = (\mu_V+c).$$

Now, if $V'(\bar{X}_t) = 0$, then

$$-k_1\frac{\left(S_H(t-\tau)-S_{H_0}\right)^2}{S_H(t-\tau)} - k_2\frac{\left(S_V(t-\tau)-S_{V_0}\right)^2}{S_V(t-\tau)} = k_2(1-R_0)I_V(t-\tau). \tag{22}$$

When $R_0 < 1$, the left-hand side of (22) is always nonpositive while the right-hand side is always nonnegative. Thus, (22) holds only when

$$k_1 \frac{\left(S_H(t-\tau) - S_{H_0}\right)^2}{S_H(t-\tau)} + k_2 \frac{\left(S_V(t-\tau) - S_{V_0}\right)^2}{S_V(t-\tau)} = 0 \quad \text{and} \tag{23}$$

$$k_2(1 - R_0) I_H(t-\tau) = 0. \tag{24}$$

Hence, (23) and (23) imply that $S_H(t-\tau) = S_{H_0}$, $S_V(t-\tau) = S_{V_0}$, $I_H(t-\tau) = 0$, and $I_V(t-\tau) = 0$. Therefore, for all $t \geq 0$, $V'(\bar{X}_t) = 0$ if and only if $S_H(t) = S_H(t-\tau) = S_{H_0}$, $S_V(t) = S_V(t-\tau) = S_{V_0}$, $I_H(t-\tau) = 0$ and $I_V(t-\tau) = 0$.

By LaSalle's Invariance Principle [10], $\bar{X}_t \to (S_{H_{0*}}, I_{H_{0*}}, S_{V_{0*}}, I_{V_{0*}})$ as $t \to +\infty$. Accordingly, $\bar{X}(t)$ converges to $(S_{H_0}, 0, S_{V_0}, 0)$ as $t \to +\infty$, that is, the disease-free equilibrium $(S_{H_0}, 0, S_{V_0}, 0)$ is globally asymptotically stable in Ω_R.

Corollary 1 *If $R_0 < 1$, then the disease-free equilibrium $\left(S_{H_0}, 0, R_{H_0}, S_{V_0}, 0\right)$ of system (M) is globally asymptotically stable in Ω.*

Theorem 4 *If $R_0 > 1$, then the disease-free equilibrium $\left(S_{H_0}, 0, S_{V_0}, 0\right)$ of system (M_R) is unstable.*

Proof By linearizing the system (M_R) at the disease-free equilibrium $\left(S_{H_0}, 0, S_{V_0}, 0\right)$, we obtain the Jacobian matrix $J(\left(S_{H_0}, 0, S_{V_0}, 0\right)) - \lambda I$ with eigenvalue λ as

$$\begin{bmatrix} -\mu_H - a - \lambda & 0 & 0 & -s_2\left(\dfrac{A_1}{\mu_H + a}\right) \\ 0 & -(\mu_H + \epsilon + \delta) - \lambda & 0 & s_2\left(\dfrac{A_1}{\mu_H + a}\right) \\ 0 & -s_1\left(\dfrac{A_2}{\mu_V + c}\right) & -(\mu_V + c) - \lambda & 0 \\ 0 & s_1\left(\dfrac{A_2}{\mu_V + c}\right) & 0 & -(\mu_V + c) - \lambda \end{bmatrix},$$

where $s_1 = e^{-(\mu_H+\delta+\epsilon)\tau_2 - \lambda\tau_2}\lambda_V$ and $s_2 = e^{-(\mu_V+c+\lambda)\tau_1}\lambda_H$.

Accordingly, $-\dfrac{\mu_V + c}{2}$ and $-\dfrac{\mu_H + a}{2}$ are negative eigenvalues of the Jacobian matrix. The other eigenvalues are the roots of the equation

$$\begin{aligned} &4(\mu_V + c)(\mu_H + a)\lambda^2 + 2(\mu_V + c)(\mu_H + a)(c + \delta + \epsilon + \mu_H + \mu_V)\lambda \\ &= A_1 A_2 e^{-\tau_1(\lambda+\mu_V)} e^{-\tau_2(\mu_H+\delta+\epsilon+\lambda)}\lambda_H\lambda_V - (\mu_V + c)^2(\delta + \epsilon + \mu_H)(\mu_H + a) \end{aligned}$$

at least one of which is positive.

Corollary 2 *If $R_0 > 1$, then the disease-free equilibrium $\left(S_{H_0}, 0, R_{H_0}, S_{V_0}, 0\right)$ of system (M) is unstable in Ω.*

Denote by Φ_+ and Φ_{R+} the sets of initial conditions for systems (M) and (M_R), respectively, such that for $\phi \in \Phi$ and $\psi \in \Phi_R$, we have $\phi(\theta) > 0$ and $\psi(\theta) > 0$

for all $\theta \in [-\tau, 0]$. Also, let Ω_+ and Ω_{R+} be the spaces of all positive solutions to systems (M) and (M_R) with initial conditions in Φ_+ and Φ_{R+}, respectively.

Theorem 5 *If $R_0 > 1$, then the endemic equilibrium $(S_{H*}, I_{H*}, S_{V*}, I_{V*})$ of system (M_R) is globally asymptotically stable in Ω_{R+}.*

Proof Following similar arguments as in the proof of Theorem 3 and with Lyapunov functional $V : \Phi_{R+} \to \mathbb{R}$ defined by

$$V(\gamma) = V_1(\gamma) + V_2(\gamma) + V_3(\gamma) + V_4(\gamma) + V_5(\gamma) + V_6(\gamma) + V_7(\gamma) + V_8(\gamma),$$

where

$$V_1(\gamma) = \frac{e^{-(\mu_H+\epsilon+\delta)\tau_2}\lambda_V S_{V*} I_{H*}}{e^{-(\mu_V+c)\tau_1}\lambda_H I_{V*} S_{H*}} \left[\gamma_1(0) - S_{H*} - S_{H*} \ln \frac{\gamma_1(0)}{S_{H*}}\right],$$

$$V_2(\gamma) = \frac{e^{-(\mu_H+\epsilon+\delta)\tau_2}\lambda_V S_{V*} I_{H*}}{e^{-(\mu_V+c)\tau_1}\lambda_H I_{V*} S_{H*}} \left[\gamma_2(0) - I_{H*} - I_{H*} \ln \frac{\gamma_2(0)}{I_{H*}}\right],$$

$$V_3(\gamma) = \gamma_4(0) - S_{V*} - S_{V*} \ln \frac{\gamma_4(0)}{S_{V*}},$$

$$V_4(\gamma) = \gamma_5(0) - I_{V*} - I_{V*} \ln \frac{\gamma_5(0)}{I_{V*}},$$

$$V_5(\gamma) = e^{-(\mu_H+\epsilon+\delta)\tau_2}\lambda_V S_{V*} I_{H*} \int_0^{\tau_1} \left[\frac{\gamma_5(-\theta)}{I_{V*}} - 1 - \ln \frac{\gamma_5(-\theta)}{I_{V*}}\right] d\theta,$$

$$V_6(\gamma) = e^{-(\mu_H+\epsilon+\delta)\tau_2}\lambda_V S_{V*} I_{H*} \int_0^{\tau_2} \left[\frac{\gamma_2(-\theta)}{I_{H*}} - 1 - \ln \frac{\gamma_2(-\theta)}{I_{H*}}\right] d\theta$$

$$V_7(\gamma) = \frac{e^{-(\mu_H+\epsilon+\delta)\tau_2}\lambda_V S_{V*} I_{H*}(\mu_H + a)}{e^{-(\mu_V+c)\tau_1}\lambda_H I_{V*}} \int_{-\tau}^{0} \left[\frac{\gamma_1(\theta)}{S_{H*}} - 2 + \frac{S_{H*}}{\gamma_1(\theta)}\right] d\theta,$$

$$V_8(\gamma) - (\mu_V + c) S_{V*} \int_{-\tau}^{0} \left[\frac{\gamma_4(\theta)}{S_{V*}} - 2 + \frac{S_{V*}}{\gamma_4(\theta)}\right] d0,$$

we conclude that the endemic equilibrium $(S_{H_*}, I_{H_*}, S_{V_*}, I_{V_*})$ is globally asymptotically stable in Ω_{R+}.

Corollary 3 *If $R_0 > 1$, then the endemic equilibrium $(S_{H*}, I_{H*}, R_{H*}, S_{V*}, I_{V*})$ of system (M) is globally asymptotically stable in Ω_+.*

5 Numerical Simulations

The objective of this section is to illustrate the results obtained in Sec. 4 by performing numerical simulations using MATLAB. For simplicity, only constant functions shall be considered as initial conditions. Moreover, all simulations will be based on the parameter values presented in Table 1.

Simulation 1. Suppose $A_1 = 200$, $A_2 = 200$, $a = 0.31$, $c = 0.42$, $\lambda_H = 0.0016$, $\lambda_V = 0.1311$, $\tau_1 = 11$ and $\tau_2 = 13$. Then $R_0 = 0.544$ and the disease-free equilibrium is (645, 0, 407, 0). We also consider the following sets of initial conditions:
(a) $(\phi_1, \phi_2, \phi_4, \phi_5) = (850, 500, 550, 250)$;
(b) $(\phi_1, \phi_2, \phi_4, \phi_5) = (250, 150, 275, 125)$;
(c) $(\phi_1, \phi_2, \phi_4, \phi_5) =(1000, 250, 750, 575)$; and

Table 1 Parameter values of SIR-SI dengue model

Parameter	Description	Value	Unit	Taken from
μ_H	Natural death rate for humans	0.0000397	day^{-1}	Assumed
μ_V	Natural death rate for mosquitoes	0.0714	day^{-1}	[7]
ϵ	Dengue disease death rate	0.00035	day^{-1}	Assumed
δ	Human recovery rate	0.143	day^{-1}	Assumed
λ_H	Effective contact rate (mosquito to human)	(0.0146, 0.2241)	day^{-1}	[12]
λ_V	Effective contact rate (human to mosquito)	(0.1299, 0.6821)	day^{-1}	[12]
A_1	Recruitment rate for humans	Varies	day^{-1}	Assumed
A_2	Recruitment rate for mosquitoes	Varies	day^{-1}	Assumed
a	Degree of protection awareness	[0, 1)	–	Assumed
c	Degree of vector control	[0, 1)	–	Assumed
τ_1	Latency period for infected human	[1, 11]	days	[6]
τ_2	Latency period for infected mosquito	[1, 16]	days	[6]

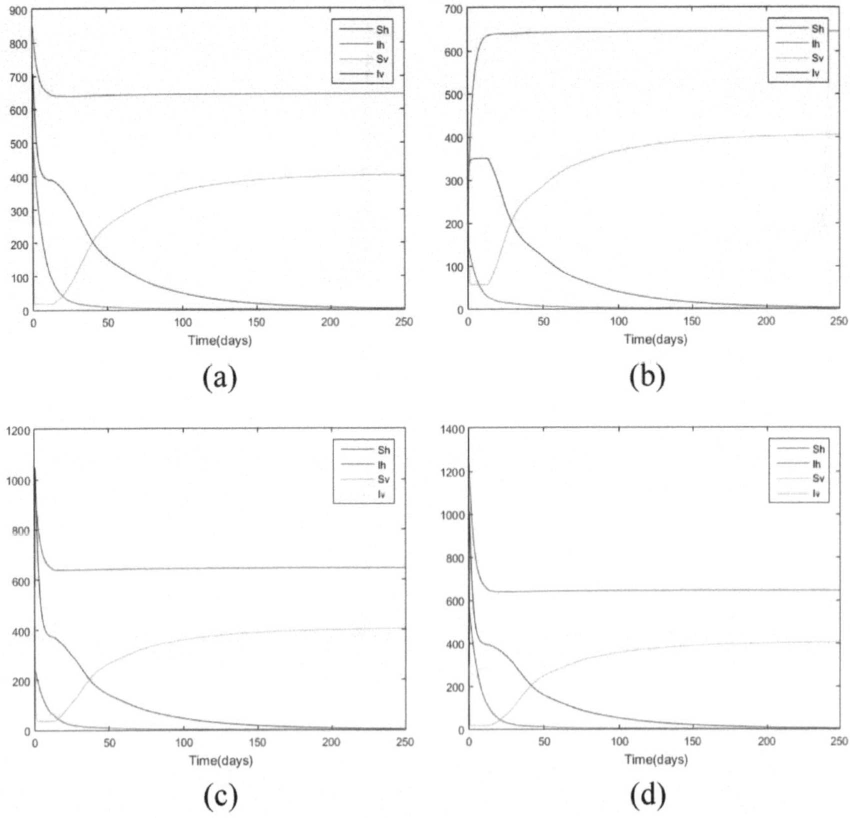

Fig. 2 The solutions considered in Simulation 1 converge to the disease-free equilibrium

(d) $(\phi_1, \phi_2, \phi_4, \phi_5) = (1250, 570, 850, 300)$.

It can be seen in Fig. 2 that the corresponding solution curves all converge to $(645, 0, 407, 0)$. This agrees with the result that the disease-free equilibrium is globally asymptotically stable whenever $R_0 < 1$.

Simulation 2. Suppose $A_1 = A_2 = 100$, $\lambda_H = 0.0016$, $\lambda_V = 0.1311$, $\tau_1 = 9$, $a = 0.27$ and $c = 0.29$. Then $R_0 = 2.487$ and the disease-free equilibrium is $(370, 0, 277, 0)$. By randomly picking 100 different initial conditions $(\phi_1, \phi_2, \phi_4, \phi_5)$, it can be observed in Fig. 3 that the corresponding solution curves do not converge to $(370, 0, 277, 0)$. This agrees with the result that the disease-free equilibrium becomes unstable whenever $R_0 > 1$.

Remark 3 Note that in Simulation 2, the R_0 is 2.487 and the endemic equilibrium now exists, which is found to be $(S_{H*}, I_{H*}, S_{V*}, I_{V*}) = (357.1, 24.8, 115.3, 116.3)$. Furthermore, each of the solution curves of S_H, I_H, S_V, and I_V converges to S_{H*}, I_{H*}, S_{V*}, and I_{V*}, respectively, as illustrated in Fig. 3. This agrees with the result that if $R_0 > 1$, then the endemic equilibrium is globally asymptotically stable.

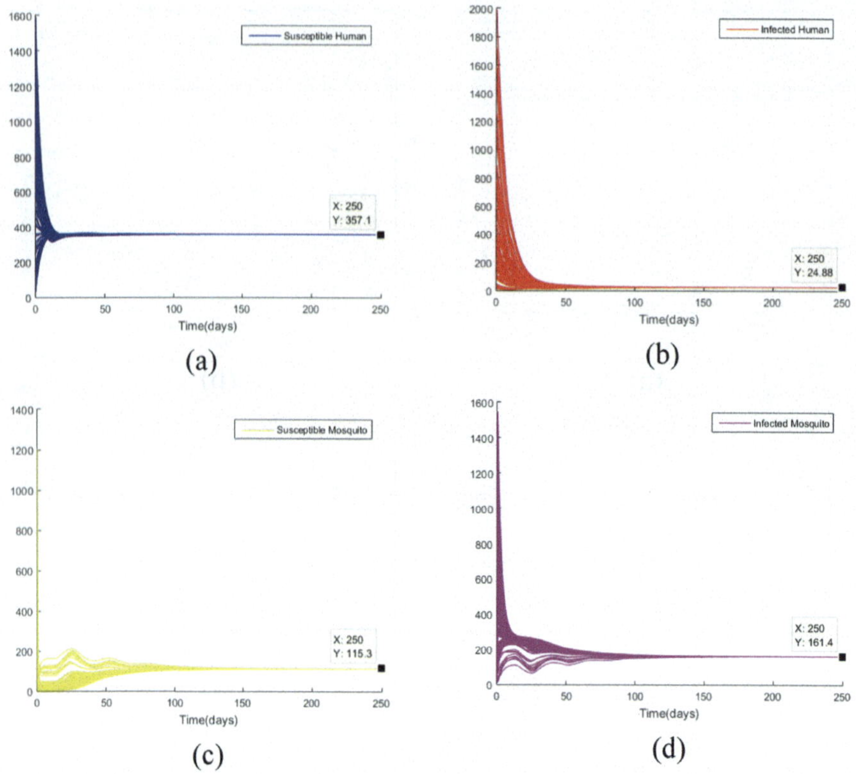

Fig. 3 The solutions in Simulation 2 converge to the endemic equilibrium

Simulation 3. Take $A_1 = A_2 = 100$, $\tau_1 = 9$, $\tau_2 = 5$, $a = 0.15$ and $c = 0.2$. Hence, $R_0 = 9.270661$ and the endemic equilibrium is (500, 174, 9, 360). Consider the following initial conditions:
(a) $(\phi_1, \phi_2, \phi_4, \phi_5) = (350, 186, 300, 190)$;
(b) $(\phi_1, \phi_2, \phi_4, \phi_5) = (250, 150, 275, 125)$;
(c) $(\phi_1, \phi_2, \phi_4, \phi_5) =(1000, 250, 750, 575)$; and
(d) $(\phi_1, \phi_2, \phi_4, \phi_5) = (1250, 570, 850, 300)$.

Figure 4 shows that the corresponding trajectories of the solutions converge to (500, 174, 9, 360). This is again consistent with the result that the endemic equilibrium is globally asymptotically stable if $R_0 > 1$.

Simulation 4. Suppose $A_1 = 200$, $A_2 = 200$, $c = 0.42$, $\lambda_H = 0.0016$, $\lambda_V = 0.1311$, $\tau_1 = 11$, $\tau_2 = 13$, and take the initial condition $(\phi_1, \phi_2, \phi_4, \phi_5) = (850, 500, 550, 250)$. We vary the values of a and obtain the following (see Fig. 5):
(a) When $a = 0.05$, we have $R_0 = 3.37$ and the endemic equilibrium (3842, 54, 126, 281) is globally asymptotically stable;
(b) When $a = 0.15$, we have $R_0 = 1.12$ and the endemic equilibrium (1330, 3, 363, 44) is still globally asymptotically stable;

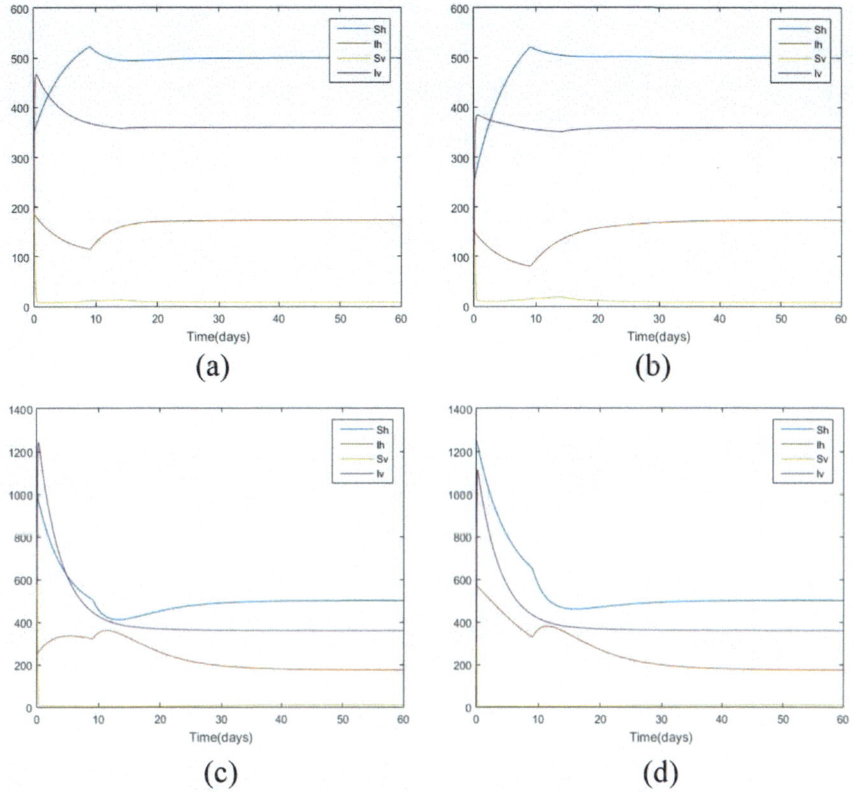

Fig. 4 The solutions in Simulation 3 converge to the endemic equilibrium

(c) When $a = 0.25$, we have $R_0 = 0.67$ and the disease-free equilibrium (800, 0, 407, 0) is now globally asymptotically stable; and
(d) When $a = 0.35$, we have $R_0 = 0.48$ and the disease-free equilibrium (571, 0, 407, 0) is also globally asymptotically stable.

The results in Simulation 4 coincide with Remark 1, that is, the basic reproductive number R_0 is a decreasing function of the level of self-protection awareness a. This result is useful in identifying ways in order to decrease the spread of dengue virus in the community.

We now illustrate how the level of vector control c affects the stability of the equilibrium points.

Simulation 5. Suppose $A_1 = 200, A_2 = 200, a = 0.27, \lambda_H = 0.0016, \lambda_V = 0.1311, \tau_1 = 11, \tau_2 = 13$, and initial condition $(\phi_1, \phi_2, \phi_4, \phi_5) = (850, 500, 550, 250)$. We vary the values of c and obtain the following (see Fig. 6):
(a) When $c = 0.05$, we have $R_0 = 599.63$ and the endemic equilibrium (208, 1002, 10, 1638) is globally asymptotically stable;

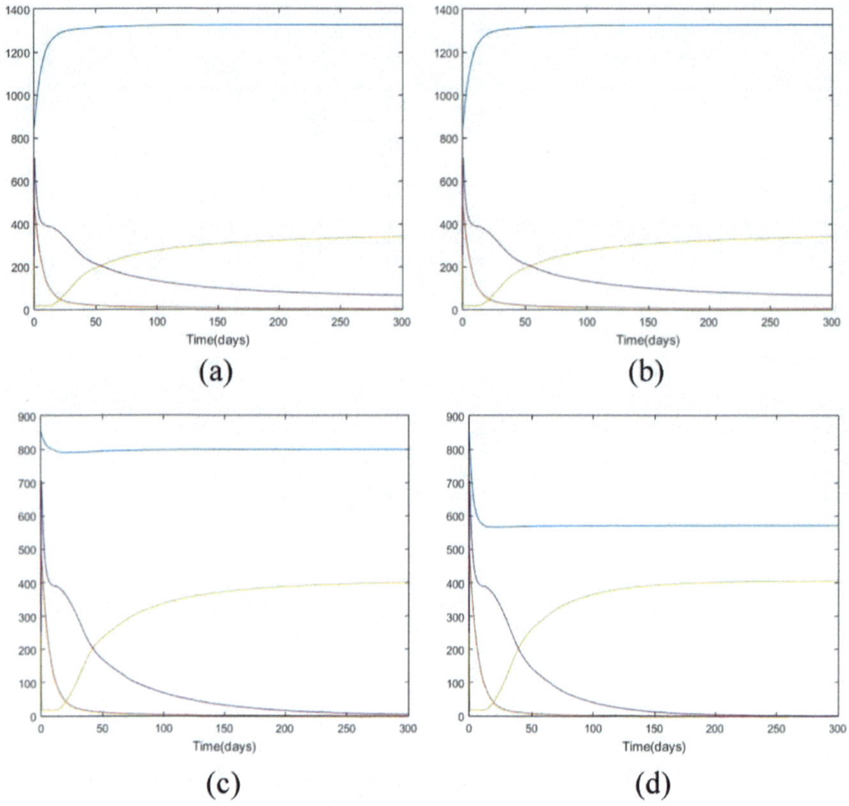

Fig. 5 The effect of varying the values of a to the stability of equilibrium points

(b) When $c = 0.25$, we have $R_0 = 9.47$ and the endemic equilibrium (676, 121, 72, 550) is still globally asymptotically stable;
(c) When $c = 0.40$, we have $R_0 = 0.84$ and the disease-free equilibrium (740, 0, 424, 0) is now globally asymptotically stable; and
(d) When $c = 0.45$, we have $R_0 = 0.39$ and the disease-free equilibrium (740, 0, 384, 0) is also globally asymptotically stable.

Simulation 5 illustrates that R_0 is also a decreasing function of the level of vector control c, as stated in Remark 2. This simply implies that intensifying vector control measures is essential in slowing down the spread of dengue virus in the community.

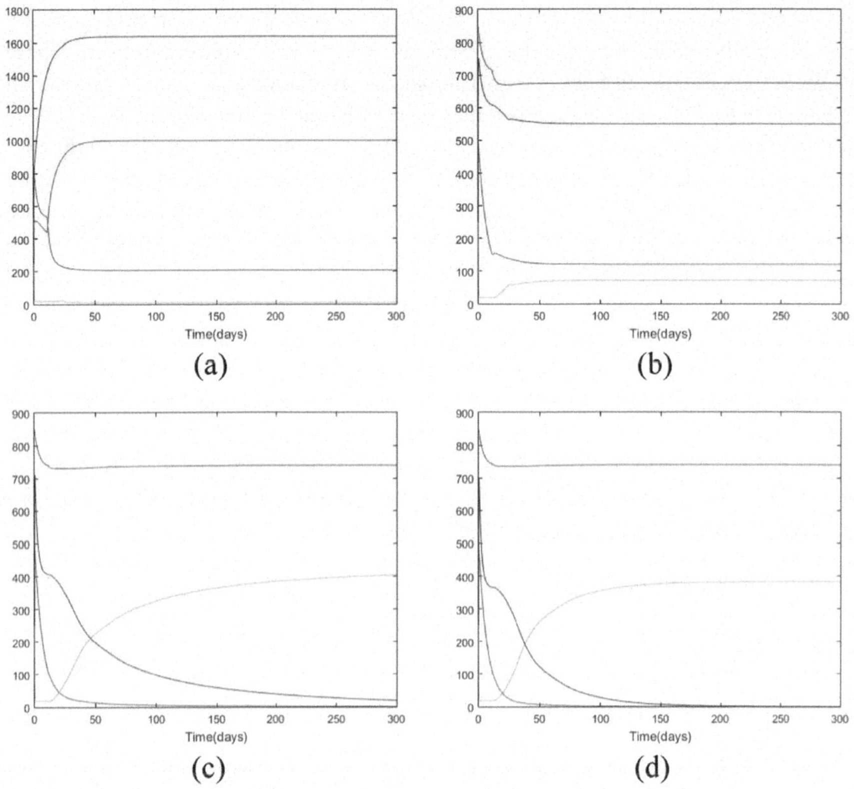

Fig. 6 The effect of varying the values of c to the stability of equilibrium points

6 Concluding Remarks and Future Works

In this paper, we incorporated the incubation times of dengue virus for human and mosquito in an SIR-SI human-mosquito dengue transmission model. The human population is subdivided into compartments of susceptible, infected and recovered individuals, and the mosquito population is subdivided into the compartments of susceptible and infected mosquitoes. We discussed the model and the boundedness and nonnegativity of solutions of the system in Sect. 2. In Sect. 3, we presented the basic reproductive number R_0 and calculated the equilibrium points of the model. We then utilized the Lyapunov functionals and LaSalle's Invariance Principle to establish the global stability of the equilibrium points in Sect. 4. In particular, the solutions converge to the disease-free equilibrium when $R_0 < 1$, indicating that the disease will eventually die out. If $R_0 > 1$, then the disease-free equilibrium is unstable and the endemic equilibrium exists. In this case, the endemic equilibrium is globally asymptotically stable, implying that the disease will remain in the community. These results were validated in Sect. 5 using numerical simulations.

Contrary to the work in [8], this paper shows that the presence of latent delays τ_1 and τ_2 does not lead to the occurrence of Hopf bifurcation. Moreover, this study confirms that increasing the level of human self-protection awareness helps in controlling the spread of the dengue virus in the community. Futhermore, the use of extensive vector control methods can minimize the virus transmission in the community.

In 2019, Chunqing Wu and Patricia J. Y. Wong [16] introduced a similar model for dengue transmission in the paper entitled *Dengue transmission: mathematical model with discrete time delays and estimation of the reproduction number.* Vector controls and human awareness are not considered in the model. Moreover, the mosquito population is assumed to be constant. In contrast to our paper, they have only established the local stability of the equilibrium points. They have additional results on the estimation of reproductive number when the population is *partially susceptible* to the disease. A parallel study on such estimation may be done in our proposed model. We leave this for future work.

References

1. Brauer, F., Chavez, C.C.: Mathematical Models in Population Biology and Epidemiology. Springer (2012)
2. Centers for Disease Control and Prevention. https://www.cdc.gov/dengue/epidemiology/index.html. Accessed 24 Jan 2019
3. Derouich, M., Boutayeb, A., Twizell, E.H.: A model of dengue fever. Biomed. Eng. Online (2003). https://doi.org/10.1186/1475-925X-2-4
4. Driver, R.: Ordinary and Delay Differential Equations. Springer, New York (1977)
5. Esteva, L., Vargas, C.: Analysis of a dengue disease transmission model. Math. Biosci. **150**, 131–151 (1998)
6. Favier, C., Degallier, N., Rosa-Freitas, M.G., Boulanger, J.P., Costa Lima, J.R., Luitgards-Moura, J.F., Menkes, C.E., Mondet, B., Oliveira, C., Weimann, E.T.S., Tsouris, P.: Early determination of the reproductive number for vector-borne diseases: the case of dengue in Brazil. Trop. Med. Int. Health **11**, 332–340 (2006)
7. Feng, Z., Velasco-Hernández, J.X.: Competitive exclusion in a vector-host model for the dengue fever. J. Math. Biol. **35**(5), 523–544 (1997). 10.1007s002850050064
8. Guan, J., Wu, L., Chen, M., Dong, X., Tang, H., Chen, Z.: The stability and Hopf bifurcation of the dengue fever model with delay. Ital. J. Pure Appl. Math. **37**, 139–156 (2017)
9. Hale, J.K., Verduyn Lunel, S.: Introduction to Functional Differential Equations. Springer, New York (1993)
10. Kuang, Y.: Delay Differential Equations with Applications in Population Dynamics. Academic Press Inc., California (1993)
11. Lashari, A., Hattaf, K., Zaman, G.: A delay differential equation model of a vector borne disease with direct transmission. Int. J. Ecol. Econ. Stat. (IJEES) **27**(4) (2012)
12. Pandey, A., Mubayi, A., Medlock, J.: Comparing vector-host and SIR models for dengue transmission. J. Math. Biol. **246**(2), 252–259 (2013)
13. Smith, H.: An Introduction to Delay Differential Equations with Applications to the Life Sciences. Springer (2011)

14. Wei, H.M., Li, X.Z., Martcheva, M.: An epidemic model of a vector-borne disease with direct transmission and time delay. J. Math. Anal. Appl. **342**, 895–908 (2008)
15. World Health Organization. https://www.who.int/denguecontrol/en/
16. Wu, C., Wong, P.J.Y.: Dengue transmission: mathematical model with discrete time delays and estimation of the reproduction number. J. Biol. Dyn. **13**(1), 1–25 (2019). https://doi.org/10.1080/17513758.2018.1562572

A Copula Based-Approach for Image Splicing Detection

Aniq A. Rohmawati and Rimba W. Ciptasari

Abstract Measuring properties in pixels to reveal an image description is of great essential for image processing. One of obstacles in image splicing detection is an unavailability of complete-reference image. In this study, a statistical method is proposed using a cumulative joint distribution of colours pixel. Consider an image consisting of particular pixels with RGB (Red, Green, Blue) components, each colour provides unique behaviour of probability function that implicitly sticking in marginal distributions. Copulas provide a cumulative joint distribution that consist of the univariate marginals and load their dependency structure, since the deficiency of linear dependency under non-linear strictly transformation. Copulas can determine the distinguish the cumulative joint distribution in the edge boundaries of image splicing. According to experimental results, Gaussian and Archimedean copulas provide deeper description to detect the contrast copula parameters of splicing image distribution.

Keywords Archimedean · Blind authentication · Copula parameters · Dependency · Gaussian · Joint distribution · Tampered image · Margin distribution · Pixels · RGB colour

1 Introduction

Nowadays, with the advanced digital technology and software in image processing, a tampered image is easier to create without needed professional skills. Image forgery may cause chaos in public because an important fragment may be erased from an original image. In most events, image forgery phenomenon inherently leads to malicious manipulation to mislead the viewers on purpose. Furthermore, today's information channels are growth rapidly, thus possibly that digital image forgery becomes more widespread as well. Therefore, improving some effective approaches

---

A. A. Rohmawati (✉) · R. W. Ciptasari
School of Computing, Telkom University, Bandung, Indonesia
e-mail: aniqatiqi@telkomuniversity.ac.id

© Institute of Technology PETRONAS Sdn Bhd 2022
S. A. Abdul Karim and A. Shafie (eds.), *Towards Intelligent Systems Modeling and Simulation*, Studies in Systems, Decision and Control 383,
https://doi.org/10.1007/978-3-030-79606-8_15

to identify digital images forgeries is of great significant and become major concern in area of image forensic. Currently technologies for digital image detection maybe roughly classified into two types, the blind (passive) authentication and active based-image reference. The various approaches of watermarking algorithms with some priori knowledge based [13] is beneficial knowledge to authenticate image tempering, also [12] presented the techniques of digital image watermarking based a survey. Later, [7, 8] shown comprehensive methods for digital image forensic referred to blind approaches. More related image detection-works utilized image forgeries over image tempering. In recent years, among all the types of authentication, the blind detection of a tampered image attracts lots of attention. These blind-approaches analyse image tampering problem by looking into different angel. There is no prior information such signatures or embedded watermarks in this approaches. The clue behind the blind detection is the fact that once though primary fragment(s) of image are vanishing, an underlying statistical artefacts of an image would be inevitably transformed by particular image tampering.

We are also noticed that the image tampering has two common issues, authenticated to image splicing and copy-move. The authentication based image splicing leads to detect manipulation which combines fragment(s) of the image into another different image. In comparison, a copy-move detection aims to track the availability of similar fragments in single image. The blind approach for image splicing is a preliminary but valuable to study for recovering people's trust in digital image that has been developed recently. The higher order moment spectra based approach has been tested in as series of paper by Ng et al. [10]. It is shown good-accuracy for identifying image splicing. The Hilbert-Huang transform and wavelet decomposition, proposed by Fu et al. [3], is beneficial for capturing non-linear and non-stationary structure in image splicing with accuracy 80.15%. He et al. [4] suggested Markov features in discrete wavelet domain for detecting digital image splicing. The image model consisting first and second order moment proposed by Dong et al. [1] achieved a promising accuracy with value 91.87%. Recent statistical-based [14] and [4] approach extend the detection the image tampering. The quantifying of statistic features is of great essential for image splicing detection. Image reflect with respect to the spectral distribution of ambient surface reflectance and the separate physical factors, that give rise to the perceived colour of an object. The colours pixel of image represent a random observation that provide unique probabilistic behaviour, besides reveal correlation structure (dependency) among colour variations.

Intuitively, image forgeries definitely embed additional fragment(s) to its original image, it is reasonable bringing a transformation on image characteristics and some statistical features. In the present work, we turn to reveal statistical characteristics on image using cumulative joint distribution based colours pixel. Meanwhile, copulas allow us to construct a flexible joint distribution function with dissimilar marginal distributions. This research resorts to copula model which allows us to observe the discontinuity of statistical features caused by splicing in term of estimated parameter of copula model. The first and second moments of conditional copula constructed by Patton [11]. Moreover, copula may exploit an alternative dependency among colours pixel, because of the limitations of linear correlation. Linear correlation show not invariant over non-linear strictly transformation, it is a major shortage. The use of non-linear dependency derived from copulas have broader advantages [6, 9, 17]. However, the colour-based approaches face difficulties in bias detection of image splicing in the presence of complex in background image and its variations. We propose the comparison result related to empirical joint distribution function and two kind of copula model: Gaussian and Archimedean copulas to handle a wide variety of variations in image splicing. The rest of the work is organized as follows. In Sect. 2, we introduce empirical marginal distribution and construct marginal distribution of RGB pixel of spliced image. In Sect. 3, we develop cumulative joint distribution of RGB pixel marginal distributions and propose an estimated parameter of copula model. In Sect. 4, we observe the result of copula model over various dataset. In Sect. 5, we analyse the copula model performance and compare the result using chi-square test. We conclude the paper and discuss the possibility of future works of blind image splicing detection in Sect. 6.

2 RGB Synthesized Images Distribution

Currently, the technology has been raised related to transfer information, the insecure information is transmitted at large in scale in form of digital images. On the other hand, a description of image property has considered to be rich in revealing a colour pixel and aggregates pixel to exploit image characteristics. In this work, we approach colour image to understand colour variations due to chaotic properties related to image tampering are based on RGB (Red, Green, Blue) components. The spliced images have drawbacks to use RGB components and their dependency in detecting such discontinues trend of pixel behaviours, non-uniform pixels distribution and chaotic parameters in model representation. Tong et al. [16] presented a novel colour image encryption from two one-dimensional chaotic systems with good statistical properties. Later, [18] presented the disadvantages by reducing the encryption of the RGB components and the relationship between components.

Consider a general image $M \times M \times 3$ pixels of spliced image are random variables, where M are the rows also columns, 3 are the RGB, and each of the components has an $M \times M$ with 0 to 255 range values. Let X_1, X_2, X_3 are a double RGB image, respectively, each X_i is $M^2 \times 1$ one-dimensional vector. A description for a random vector has to provide probabilistic description of its behaviour components simultaneously. For example, we have three images, the three different images possibly reveal similar patterns in histogram, but their RGB components particularly show unique probabilistic ranges in Fig. 1. Thus, we possibly to consider probability concept to construct the empirical marginal distributions of each magnitude of RGB pixels completely as follows:

$$\Phi_{X_i}(a) = \mathbf{P}(X_i \leq a) = \frac{\#X_i \leq a}{M^2} \tag{1}$$

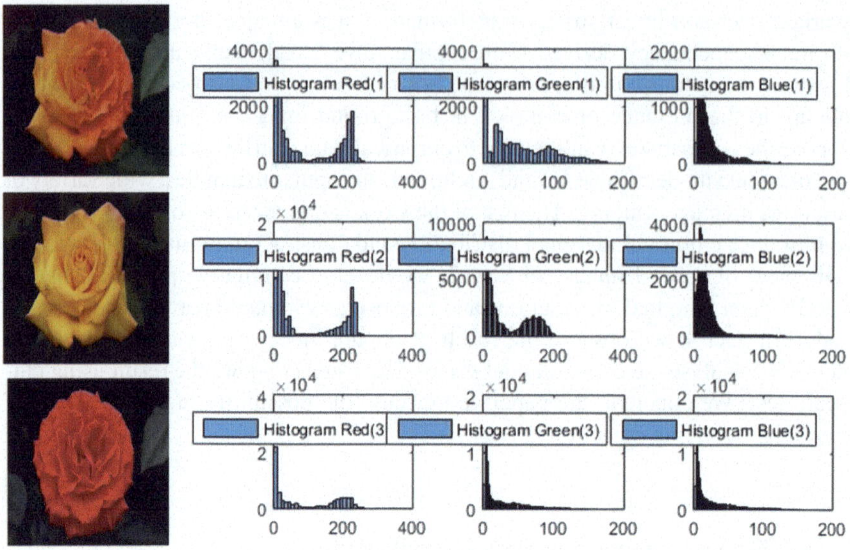

Fig. 1 The histograms reveal that the frequency of RGB components of each image scatter uniquely by the ranges

where a is random sample which $a \in X_i$. The marginal distributions function of X_i written Φ_{X_i}, or we may simply write Φ, is the distribution function of that RGB colours considered individually. Another an important point of view in this spliced image detection is how the RGB marginal distributions of fragment(s) that come from one image. Since every random variables has unique cumulative distribution that can reveal the dissimilar structure of dependency in different ranges of behaviour. Hence, we also may observe similar and dissimilar structures among fragment(s) image, which are part from a spliced image and an original image, such in Fig. 2. The key concept at this work is constructing the representative model of joint distribution based on RGB marginals of fragments in particular image. Thus, if the fragment(s) of image are suspicious, we definitely take comparison of the parameters of copula-a joint distribution function in each of the fragment(s), to observe the alteration in copula parameters.

Sklar's Theorem

Let F denote an d-dimensional joint distribution with marginal distribution functions $F_1, F_2, ..., F_d$. Then, there exists a copula $C : [0, 1]^d \rightarrow [0, 1]$ for all real $x_1, x_2, ..., x_d$ such that:

$$\begin{aligned} F(x_1, ..., x_d) &= \mathbf{P}(X_1 \leq x_1, ..., X_d \leq x_d) \\ &= C\left(\mathbf{P}(X_1 \leq x_1, ..., X_d \leq x_d)\right) \\ &= C\left(F_1(x_1), ..., F_d(x_d)\right) \end{aligned} \tag{2}$$

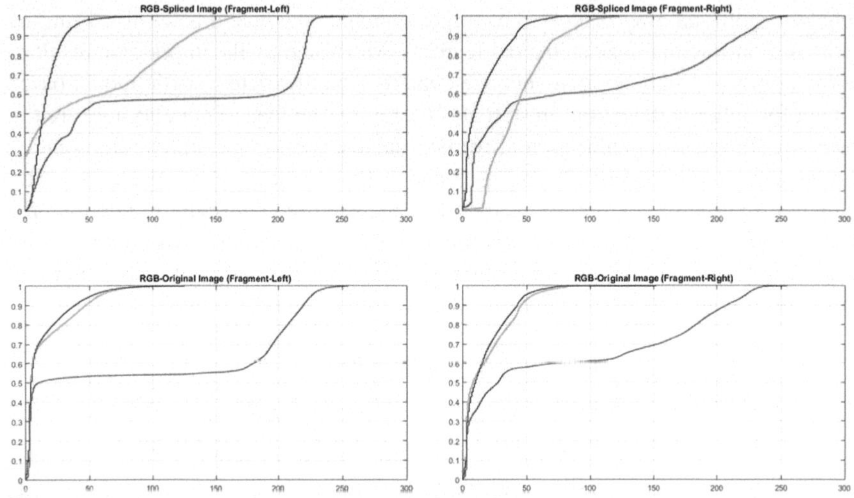

Fig. 2 The dissimilar of probabilistic behaviours over RGB-distributions between fragments of spliced (up) and original image (bottom)

Since a marginal distribution and its dependence structure can express any joint distribution function, if the marginals are continuous which is derived as follows:

$$\begin{aligned} f(x_1, ..., x_d) &= \frac{\partial F(x_1, ..., x_d)}{\partial x_1 ... \partial x_d} \\ &= \frac{\partial C(u_1, ..., u_d)}{\partial u_1 ... \partial u_d} \times \prod_i \frac{\partial F_i(x_i)}{\partial x_i} \\ &= c(\tilde{u}) \times \prod_i f_i(x_i) \end{aligned} \quad (3)$$

Here, $u_i = F_i(x_i)$ for $i = 1, ..., d, \tilde{u} = (u_1, ..., u_d)$, c is a copula density function, and f_i is the density function of F_i. Sklar [15] theorem proposed fundamental properties of a copula describing the dependent function which relates with the univariate marginal distributions and joint distribution function.

3 Parameter Setting of Gaussian and Archimedean Copulas

Due to a comprehensive study shown by Shi et al. [14], a run-length based order of moment reported as a promising method in steganalysis. The objective of steganalysis is to reveal if there is any alteration in the image, and in most occasion it is hard to observe a suspicious image over human eyes. Since image forgery definitely converts additional fragment(s) to its original image, in which it is feasible bringing a transformation on image statistical features. Motivated by the impressive result of run-length based statistic moments, we construct multivariate statistic model based copula for splicing detection. In statistics literature, the idea of copulas appeared in the case of discussion of non-normality marginal distributions in multivariate context. It have to be taken to notice from Eqs. 2 and 3, there is always a unique copula (C) which given by particular marginal distributions and joint distribution function (F). It has clearly to show that copula and the marginals also completely determine joint distribution function. Suppose now that we are given the marginal distributions of RGB are Φ_R, Φ_G, and Φ_B, the corresponding copula can described as:

$$\begin{aligned} C(r, g, b) &= \mathbf{P}(R \leq r, G \leq g, B \leq b) \\ &= \mathbf{P}(\Phi_R(R) \leq r, \Phi_G(G) \leq g, \Phi_B(B) \leq b) \\ &= \mathbf{P}\left(R \leq \Phi_R^{-1}(r), G \leq \Phi_G^{-1}(g), B \leq \Phi_B^{-1}(b)\right) \\ &= H_{R,G,B}\left(\Phi_R^{-1}(r), \Phi_G^{-1}(g), \Phi_B^{-1}(b)\right) \end{aligned} \tag{4}$$

Thus,

$$h_{R,G,B}(r, g, b) = c(r, g, b)\phi_R(r)\phi_G(g)\phi_B(b) \tag{5}$$

where H is joint distribution function based on copula, $c(r, g, b)$ correspond to the particular copula density, and $h_{R,G,B}$ represents joint probability density function which involve each marginals probability density function (ϕ). Sklar's theorem shows that every copula is unique by the ranges of the marginal distributions. An essential features of copula is the marginal distributions possibly can be dissimilar to each other.

In our case, we employ Gaussian copula considering multivariate Gaussian as of the prominent choice based on the Law of Large Numbers (LLN). However, it also has limitation that only linear correlation can be accounted. In most situations, the simultaneous occurrence of pixel-transformation are of particular interest and that would be relevant to provide non-linear correlation. We there consider the Archimedean copulas. To take into account non-linear dependency structure, we adopt Archimedean copulas: Clayton and Gumbel.

3.1 Empirical Copula

$$C(r, g, b) = \frac{\#\{(u, v, w) : \Phi(r) \leq u, \Phi(g) \leq v, \Phi(b) \leq w\}}{n} \tag{6}$$

where N represent number of sample from marginal distributions. The empirical copula is constructed by empirical cumulative distribution of the data, Eq. 1. According to active authentication, as reference image is available, the empirical copula of RGB components of reference image will be valuable significantly to detect distinction structure and reveal the similarity distance over tampered one. However, the hardest working in active authentication for image splicing disclosure is discovering a full-reference image. Indeed, without reference, we suppose refer to parameters as statistical feature in copula model. Thus, we construct copula based empirical copula using Gaussian and Archimedean approaches.

3.2 Gaussian Copula

$$C(r, g, b; \Sigma_3) = \Phi_\rho\left(\Phi^{-1}(r), \Phi^{-1}(b), \Phi^{-1}(b)\right) \tag{7}$$

where Σ_3 presents a sample covariance 3×3 matrix, $\rho \in [-1, 1]$, and Φ^{-1} is the inverse of marginal distribution function of standard multivariate Gaussian. Let $\mathbf{y}' = (r_l, g_l, b_l)$ for $l = 1, 2, ..., L$ be the transpose of vector $\mathbf{y}$, thus, the joint density function of $C(r, g, b; \Sigma_3)$ through multivariate Gaussian density given as follows:

$$c_\Sigma(\mathbf{y}) = \frac{1}{\left|\Sigma_3^{1/2}\right|} \exp\left(\frac{-y'(\Sigma_3^{-1} - \mathbf{I})y}{2}\right) \tag{8}$$

Using Maximum Likelihood of Eq. 8, we determine $\widehat{\Sigma_3}$ as a sample covariance matrix estimation, as follows

$$\widehat{\Sigma_3} = \frac{1}{L}\sum_{l=1}^{L} y_r y_r'$$

3.3 Archimedean Copula

$$C(r, g, b) = \zeta^{-1}\left(\zeta(r) + \zeta(b) + \zeta(b)\right) \tag{9}$$

Then ζ be an Archimedean generator, ζ is 3-monotone on $[0, \infty]$ and $r, g, b \in [0, 1]^3$. For $\theta \in (0, \infty)$, the following Archimedean generator for Clayton and Gumbel, respectively as follows:

$$\text{Clayton}: \zeta(y) = \frac{y^{-\theta} - 1}{\theta}, \theta > 0$$
$$\text{Gumbel}: \zeta(y) = (-\ln y)^{\theta}, \theta \geq 1 \tag{10}$$

The density of a 3-dimensional Archimedean copula Eq. 9 if only if ζ^{-1} is strictly monotonic on $\mathbb{R}_{+}$, as follows:

$$c_{\theta}(\mathbf{y}) = 3\left(\zeta_{\theta}^{-1}\right)'(3\zeta_{\theta})\,\zeta_{\theta}^{-1}(y), \quad y \in [0, 1]$$

Given a random distribution follows a uniform distribution, $\mathbf{y}_i \sim U[0, 1]^3$, for $i \in 1, ..., L$, [5] present diagonal maximum-likelihood estimator to find admissible $\widehat{\theta}$ and obtain

$$\widehat{\theta} = \text{argsup}_{\theta \in \Theta} \sum_{i=1}^{L} \log c_{\theta}(\mathbf{y})$$

It is guaranteed to estimate parameters for multivariate Archimedean copula.

4 Verification Scheme Performance

In this section, we perform simulation considering number of partitions in image, we employ in spliced and original images. First, for each suspicious image, we conduct partition(s) in single image to fragments, with scheme as Fig. 3. Then, for each fragment of partition(s), we extract the RGB marginal distribution functions and construct copulas distribution, conducting Gaussian and Archimedean copula, respectively. Afterwards, we estimate the parameters copula in each fragment, and compare the parameters of copulas to reassure that those fragments have the identical distribution. From flowchart Fig. 4, to integrate the result of copula models and parameters estimation, we evaluate copula distributions under sample dataset using z-test. In our experiments, four benchmarks are considered to construct the copula model, then evaluated the parameters estimation.

4.1 The Spliced Image Configuration

We address in detail three partition schemes on Fig. 3 into two spliced images with different level of tampered: smooth and harsh, respectively.

According to Fig. 7, at a first impression, the copula-joint distribution plots impressively reveal substantial gap, as we see, their size fragments compared is affected to the distance between two copula distribution plots. In first scheme, Fig. 6, we have

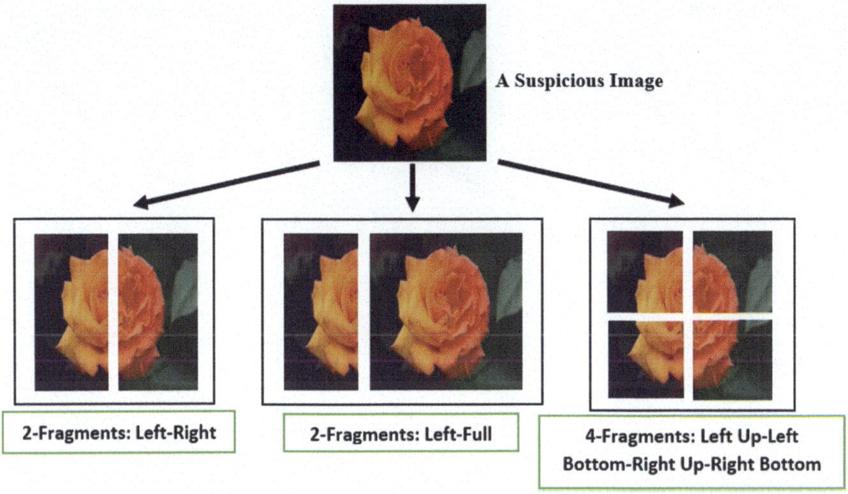

Fig. 3 The partition scheme to observe a RGB-fragment joint distribution

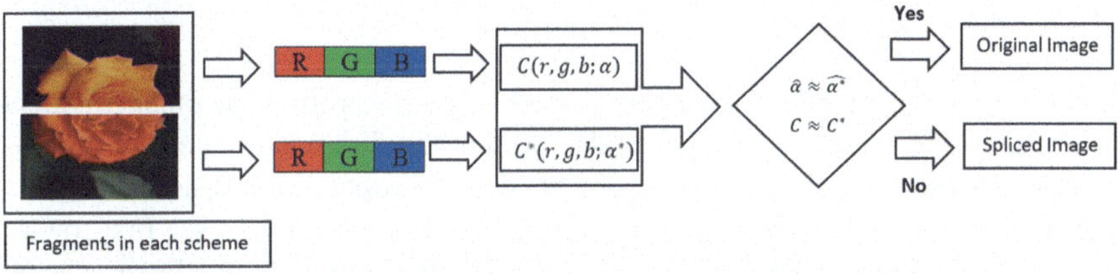

Fig. 4 The verification scheme to reassure the copula-joint distribution of fragments are identical or not

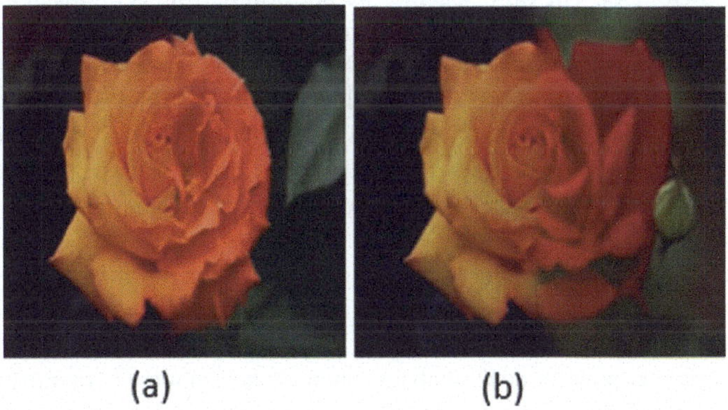

Fig. 5 In a sense, we slightly notice that (**b**) is a spliced image, but it is hard to spot that (**a**) is a tampered ones

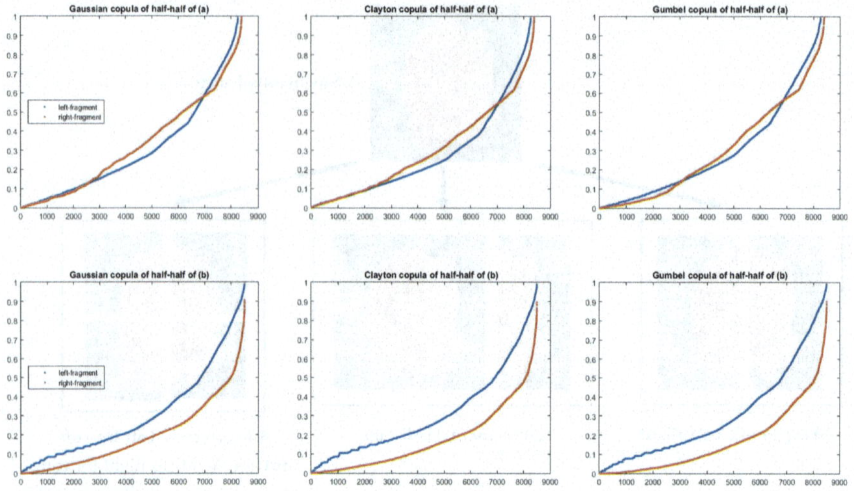

Fig. 6 The first scheme (left-right fragments) copula distribution plots: the columns refer to Gaussian, Clayton and Gumbel, and the rows are pointed into (**a**) and (**b**)

similar size of fragments, whereas the second scheme considers the other fragment is in full size. However, both scheme also present the dissimilar pattern of copula-joint distribution functions. Several concerns are the copula-joint distribution plots between left and right side of (a) are remarkably close each other, see first row of Fig. 6, notably that this image has soft-tampered structure, thus, the latter scheme applied to (a) showing that certainly overlapped plots. Also, copula-joint distributions of each image show identically patterns both of Gaussian, and Archimedean: Clayton and Gumbel. The third scheme, we perform partitions in image to get four fragments with same size. Intuitively, (a) and (b) might consist of two images that merged to be left and right side, respectively. Hence, the left-up and left-bottom (a) and (b) image may lead to have identical distributions, and the right-up and right-bottom as well. As stated by that facts, in Fig. 9, both of Gaussian and Archimedean portrayed the homogeneous copula distribution plots among those similar fragments, also for dissimilar fragments they show quite gaps revealed in initial values of left-up and left bottom fragment.

We added analysis in copula parameters, as characteristic that describe unique structures in copula-joint distribution. The gap (or distance) between copula distribution plots denotes a dependency in terms of concordance. Under pseudo-observation margins, copula conduct a deep analysis whether the compared fragments in each scheme have dependency pattern and behaviour distribution by estimating and evaluating parameters, according to the visualization that was shown in the previous.

According to copula parameters estimation based on (a) in Table 1, at first scheme, the dependencies structure at right and left side indicates heterogeneous values of among Gaussian, Clayton and Gumbel parameters. However, in the latter scheme, magnitude of dependency of Clayton and Gumbel at similar fragment such left-up

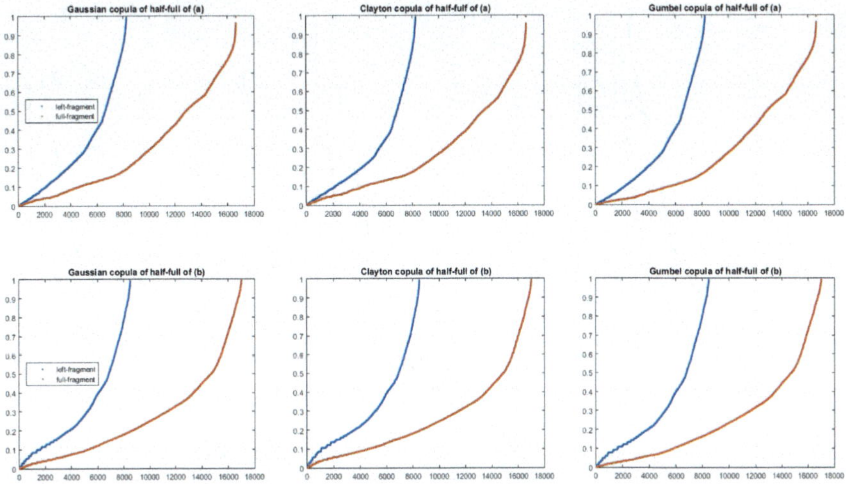

Fig. 7 The second scheme (half-full) joint distribution plots, the wider gap between half and full copula plot is because of dissimilar size of ranges at x-coordinate

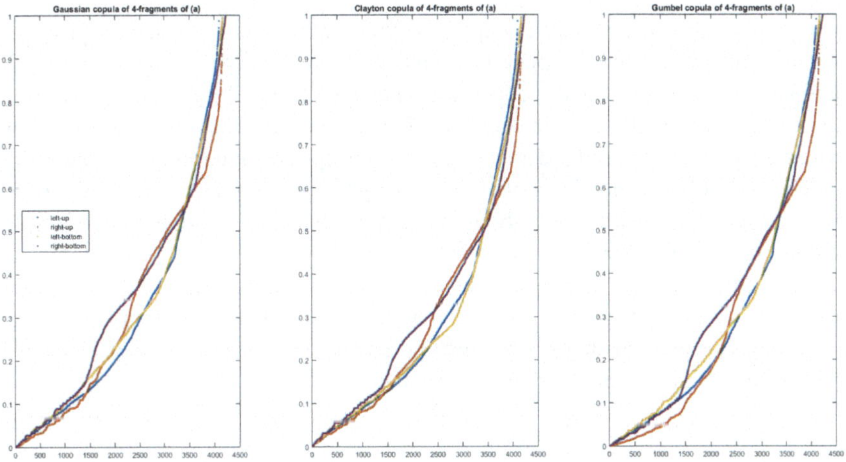

Fig. 8 The latter scheme that applied to **a** shows overlapping among four plots and all fragments disclosure different patterns

and left-bottom reveals slightly different, presumably because of limitation in number of observation, more partitions lead to less pixel to observe. Indeed, Gaussian parameters performs well to identify different dependencies among similar fragments, the latter scheme, it reveals a dependency between left-up and left bottom. Another result, right-up and right-bottom present identical values, and the magnitude become significantly different in suspicious fragments, such left-up and right-up. To another spliced image, Table 2 presents parameters estimation related to frag-

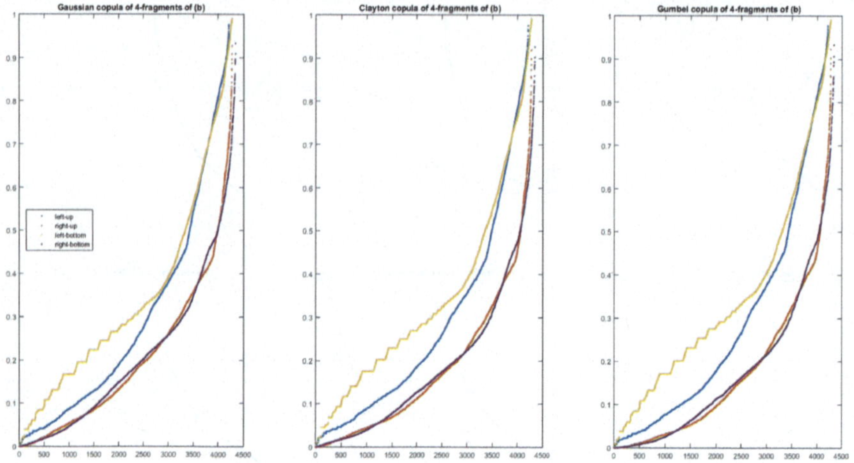

Fig. 9 The latter scheme applied to (**b**) remaining several anomalies gap in left-up and left bottom at initial ranges, but it quickly turned to close proximity

ments extracted from (b), that is expressed that Archimedean constantly provides prominent magnitude over suspicious fragments. Moreover, of the three schemes, all Gaussian remain heterogeneous dependency between similar fragments, such left-up and right-up. But, it also remains that covariance matrix of Gaussian describing low performance to reveal similar dependency of right-up and right bottom fragments of (b). The 'small but sharp' leave that arise in right-bottom side of (b), may turn to noise that set a gap between right-bottom and right-up. Thus, to confirm all of result, the noise presence in image also may be considered during computing. Contemplate from copula parameters over fragments in (a) and (b), these parameter estimation have to conduct proper number of pixel that observed. The key point of parameters estimation, a suspicious image is certainly preferred to perform Archimedean copula to describe details of joint distribution structure.

4.2 The Original Image Configuration

Of the four benchmarks that we observe, two images are original. In a series of schemes, we extended to evaluate the performance of copula model to identify if the image is original.

As can be seen from Fig. 11, copula performs well under original image, both of Gaussian and Archimedean describe identical and precisely structure as on left and right side of (i). Corresponding the previous result Fig. 9, the difference size of fragments in second scheme implies to the distance copula distribution plots become quite larger, but the reverse is two copula distribution plots indeed similar patterns

Table 1 Copula parameters estimation conducting fragments of (**a**) in all scheme

Fragments	$\widehat{\Sigma}$-Gaussian	$\mathbb{E}(\widehat{\theta})$-Clayton	$\mathbb{E}(\widehat{\theta})$-Gumbel
Left	$\begin{bmatrix} 1 & 0.8384 & 0.3810 \\ 0.8384 & 1 & 0.3751 \\ 0.3810 & 0.3751 & 1 \end{bmatrix}$	0.4325	2.5334
Right	$\begin{bmatrix} 1 & 0.7513 & 0.5207 \\ 0.7513 & 1 & 0.7863 \\ 0.5207 & 0.7863 & 1 \end{bmatrix}$	0.9089	1.9547
Full	$\begin{bmatrix} 1 & 0.7862 & 0.4530 \\ 0.7862 & 1 & 0.4447 \\ 0.4530 & 0.4447 & 1 \end{bmatrix}$	0.5941	2.1614
Left-up	$\begin{bmatrix} 1 & 0.8481 & 0.3545 \\ 0.8481 & 1 & 0.3592 \\ 0.3545 & 0.3592 & 1 \end{bmatrix}$	2.2216	2.6182
Right-up	$\begin{bmatrix} 1 & 0.6962 & 0.4184 \\ 0.6962 & 1 & 0.7758 \\ 0.4184 & 0.7758 & 1 \end{bmatrix}$	1.8878	1.6514
Left-bottom	$\begin{bmatrix} 1 & 0.8274 & 0.4133 \\ 0.8274 & 1 & 0.4020 \\ 0.4133 & 0.4020 & 1 \end{bmatrix}$	1.2094	2.4097
Right-bottom	$\begin{bmatrix} 1 & 0.8049 & 0.5986 \\ 0.8049 & 1 & 0.7812 \\ 0.5986 & 0.7812 & 1 \end{bmatrix}$	2.1920	2.4367

(i) (ii)

Fig. 10 These couple original images reveal the different brightness, we can see that (i) more luminous than (ii)

Table 2 Copula parameters estimation conducting fragments of (**b**) in all scheme

Fragments	$\widehat{\Sigma}$-Gaussian	$\mathbb{E}(\widehat{\theta})$-Clayton	$\mathbb{E}(\widehat{\theta})$-Gumbel
Left	$\begin{bmatrix} 1 & 0.9149 & 0.3376 \\ 0.9149 & 1 & 0.3612 \\ 0.3376 & 0.3612 & 1 \end{bmatrix}$	0.5717	3.6863
Right	$\begin{bmatrix} 1 & 0.2291 & 0.1765 \\ 0.2291 & 1 & 0.4821 \\ 0.1765 & 0.4821 & 1 \end{bmatrix}$	0.1654	1.0625
Full	$\begin{bmatrix} 1 & 0.8342 & 0.2257 \\ 0.8342 & 1 & 0.3360 \\ 0.2257 & 0.3360 & 1 \end{bmatrix}$	0.3483	2.5611
Left-up	$\begin{bmatrix} 1 & 0.9100 & 0.3245 \\ 0.9100 & 1 & 0.3154 \\ 0.3245 & 0.3154 & 1 \end{bmatrix}$	0.3663	1.2423
Right-up	$\begin{bmatrix} 1 & 0.2887 & 0.0922 \\ 0.2887 & 1 & 0.5258 \\ 0.0922 & 0.5258 & 1 \end{bmatrix}$	0.1219	1.0477
Left-bottom	$\begin{bmatrix} 1 & 0.9537 & 0.3572 \\ 0.9537 & 1 & 0.3971 \\ 0.3572 & 0.3971 & 1 \end{bmatrix}$	0.5157	1.4077
Right-bottom	$\begin{bmatrix} 1 & 0.2224 & 0.2212 \\ 0.2224 & 1 & 0.5567 \\ 0.2212 & 0.5567 & 1 \end{bmatrix}$	0.2619	1.1301

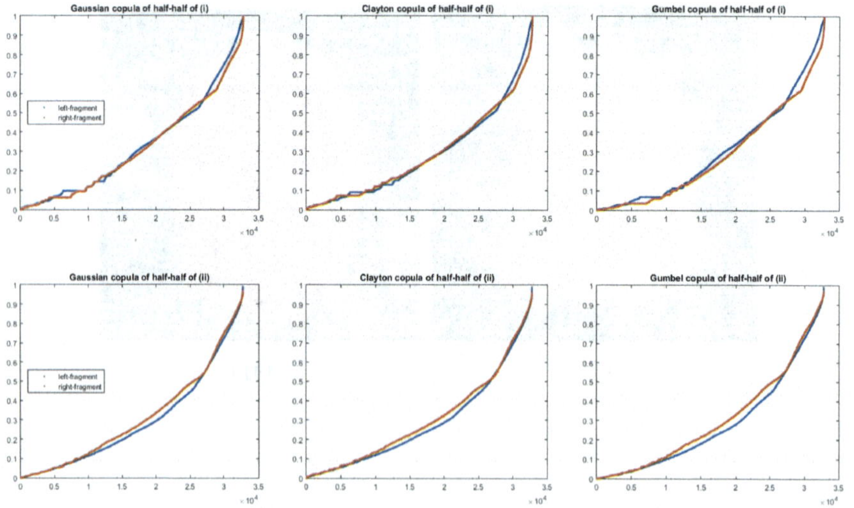

Fig. 11 The first scheme (left-right fragments) copula distribution plots: the columns refer to Gaussian, Clayton and Gumbel, and the rows are pointed into (i) and (ii)

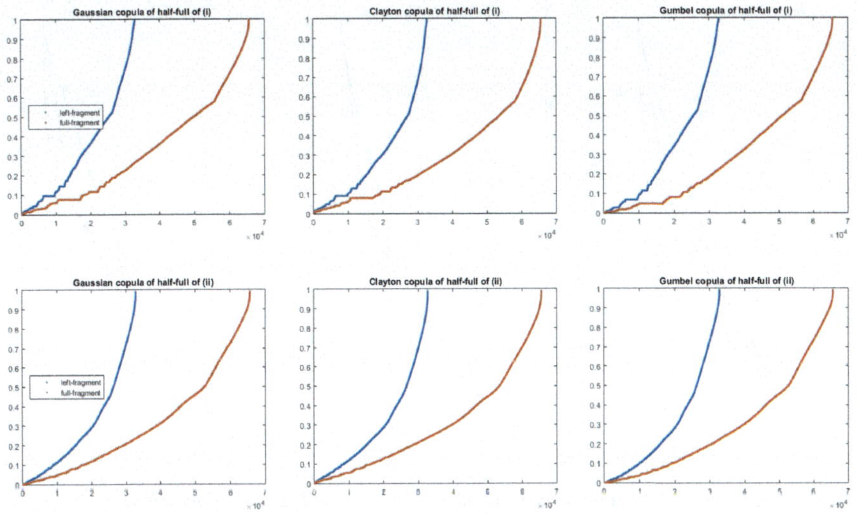

Fig. 12 The second scheme (half-full) joint distribution plots of original image that disclosure homogeneous trends

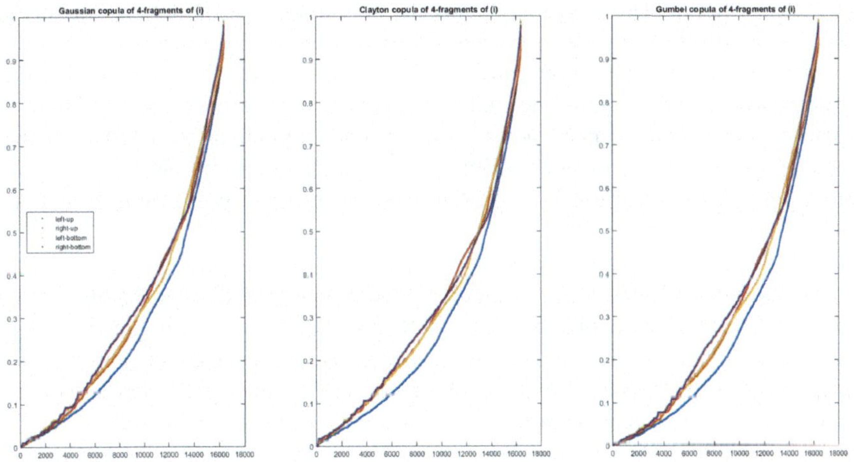

Fig. 13 The latter scheme applied to (i): the copula distributions plots have identical trend, but left-up fragment is slightly far away from corresponding fragments at initial ranges

between left side and full size, see Fig. 12, the similar patterns are expressed clearly in fragments of (i) in initial ranges.

Due to the result of spliced images, in the latter scheme, Fig. 12 shows that copulas disclosure not only identical trends, but also the overlap plots over fragments that come from original image. Note that this numerically challenging to observe unique structure with noise presence, as describe in Fig. 10, it might be happened because

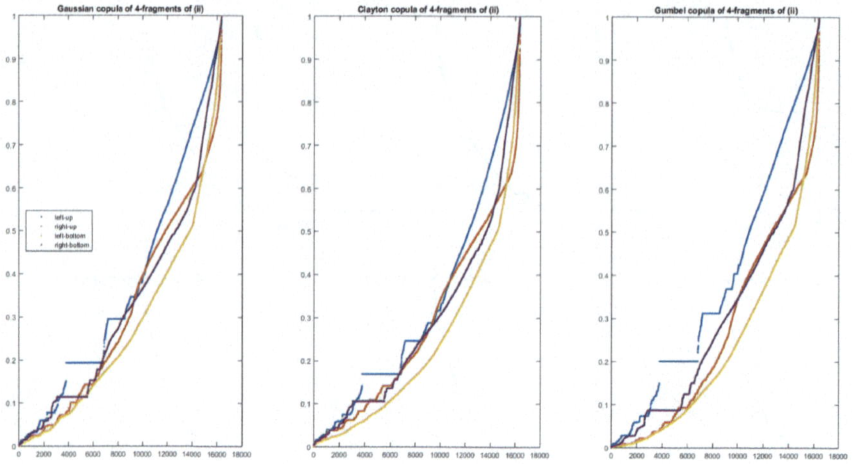

Fig. 14 The latter scheme applied to (ii) remaining unstable patterns, notably left side

number of leaves in (ii) that appear on up and bottom fragment of image have significantly different. Thus, the Fig. 14 reveals discontinues plot in left-up fragment, also both of left-up and right-up contemplating heterogeneous trends, in a sense.

There are several conditions, such as the one addressed about noise presence where an availability of unique dependency structures to observe is required. Numerical issue in computing dependencies estimator arise for multivariate distribution, the maximum likelihood estimator performs well according to reveal identical parameters in fragment(s) of original image. The result is at Fig. 13 point out that the left-up fragment distribution views distinct pattern compared to left-bottom, but from Table 3, therefore, Clayton parameter is markedly similar dependency among that fragments. Table 3 is parameter estimation of each copula displaying that the Gaussian and Archimedean copula performs well according to various fragments of (i), they state similar dependencies among fragments, also the reversed. In Table 4, it has clearly visible, the fragments left-up and left-bottom, that expected to have homogeneous structure, approximate different parameters Gaussian and Clayton may possibly mislead the conclusion. Surprising result, the parameters estimation of Clayton and Gumbel constantly show the identical values over similar fragments of (ii), this already looks promising, in fact that Fig. 14 presents large distance in copula distribution plots. Sometimes, integrating results of modelling may not be able to generalize overall performance, because we deal with a size of sample data. Independently of any context of research, we usually consider meta-analysis under P-value and a sample size, to reassure that the effect of sample size is zero, that what we call as null hypothesis. In the last section, concern in statistics of small samples enhanced [2], subsequently we perform of z-test due to the result of copula models and parameters estimation.

Table 3 Copula parameters estimation conducting fragments of (i) in all scheme

Fragments	$\widehat{\Sigma}$-Gaussian	$\mathbb{E}(\widehat{\theta})$-Clayton	$\mathbb{E}(\widehat{\theta})$-Gumbel
Left	$\begin{bmatrix} 1 & 0.9098 & 0.3384 \\ 0.9098 & 1 & 0.4140 \\ 0.3384 & 0.4140 & 1 \end{bmatrix}$	0.5604	1.2796
Right	$\begin{bmatrix} 1 & 0.9394 & 0.4144 \\ 0.9394 & 1 & 0.4598 \\ 0.4144 & 0.4598 & 1 \end{bmatrix}$	0.6324	1.3689
Full	$\begin{bmatrix} 1 & 0.9394 & 0.4144 \\ 0.9394 & 1 & 0.4598 \\ 0.4144 & 0.4598 & 1 \end{bmatrix}$	0.5102	1.3141
Left-up	$\begin{bmatrix} 1 & 0.8779 & 0.2455 \\ 0.8779 & 1 & 0.3499 \\ 0.2455 & 0.3499 & 1 \end{bmatrix}$	0.5222	1.1947
Right-up	$\begin{bmatrix} 1 & 0.9222 & 0.3748 \\ 0.9222 & 1 & 0.4454 \\ 0.3748 & 0.4454 & 1 \end{bmatrix}$	0.6081	1.3543
Left-bottom	$\begin{bmatrix} 1 & 0.9301 & 0.4228 \\ 0.9301 & 1 & 0.4694 \\ 0.4228 & 0.4694 & 1 \end{bmatrix}$	0.5463	1.3757
Right-bottom	$\begin{bmatrix} 1 & 0.9521 & 0.4381 \\ 0.9521 & 1 & 0.4679 \\ 0.4381 & 0.4679 & 1 \end{bmatrix}$	0.6647	1.3692

5 Evaluation Over Dataset

Research results have long conducted meta-analysis to integrate information considered P-value that present a straightforward approach. The convenient approach in this splicing image is conducting z statistical test. Fisher et al. [2] considered P-value, test to evaluate a model that assumes a particular direction under samples is correct.

Indeed, we evaluate how well a P-value on z-test can describe the similarity of distribution among fragments in a general image, such when we collected diverse fragments from identical image. The yields of z-test under 5% significance level can be reflected to support for a series of results in previous sections may collectively unclear indicate significance. In this work, P-value are counted under from alternative hypothesis that expected the difference distribution between fragments of other related fragment(s). We also assume the test is Gaussian distributed. In order to apply z-test of each scheme, we set mean and variance of particular fragment and test to joint distribution of other fragment, such we consider mean and variance left fragments against joint distribution data on right fragment at first scheme. The decision related to similar/dissimilar distribution of both two fragments in each scheme has

Table 4 Copula parameters estimation conducting fragments of (ii) in all scheme

Fragments	$\widehat{\Sigma}$-Gaussian	$\mathbb{E}(\widehat{\theta})$-Clayton	$\mathbb{E}(\widehat{\theta})$-Gumbel
Left	$\begin{bmatrix} 1 & 0.4981 & 0.5854 \\ 0.4981 & 1 & 0.8667 \\ 0.5854 & 0.8667 & 1 \end{bmatrix}$	1.0507	1.8092
Right	$\begin{bmatrix} 1 & 0.4285 & 0.5054 \\ 0.4285 & 1 & 0.9079 \\ 0.5054 & 0.9079 & 1 \end{bmatrix}$	1.1567	1.6768
Full	$\begin{bmatrix} 1 & 0.4492 & 0.5346 \\ 0.5346 & 1 & 0.8892 \\ 0.5346 & 0.8892 & 1 \end{bmatrix}$	1.0883	1.7182
Left-up	$\begin{bmatrix} 1 & 0.6518 & 0.7112 \\ 0.6518 & 1 & 0.8209 \\ 0.7112 & 0.8209 & 1 \end{bmatrix}$	1.2032	2.5031
Right-up	$\begin{bmatrix} 1 & 0.4148 & 0.4489 \\ 0.4148 & 1 & 0.9399 \\ 0.4489 & 0.9399 & 1 \end{bmatrix}$	1.1658	1.5977
Left-bottom	$\begin{bmatrix} 1 & 0.2781 & 0.3873 \\ 0.2781 & 1 & 0.9090 \\ 0.3873 & 0.9090 & 1 \end{bmatrix}$	0.7583	1.4832
Right-bottom	$\begin{bmatrix} 1 & 0.4664 & 0.5649 \\ 0.4664 & 1 & 0.8635 \\ 0.5649 & 0.8635 & 1 \end{bmatrix}$	1.0593	1.8303

Table 5 P-value among different fragments at first and second schemes

Fragments	Image	Right	Full
Left	(a)	0.0000	0.0121
	(b)	0.0000	0.0000
	(i)	0.6912	0.9637
	(ii)	1	1

present clearly in Tables 5 and 6. Therefore, the P-value turns to bigger than 5% at left-bottom and right-up of (a), that needs appropriate scenario about this yields. We addressed the result of from (b)-spliced image that the parameter estimation right-up and right-bottom are different, that supposed to be identical, corresponding to P-value, we obtain significance outcome that both of fragments are identical. And, z-test also outperforms as well at (ii)-original image between left-up and left-bottom, P-value is above of 5%.

Table 6 P-value among different fragments at latter scheme

Fragments	Image	Right-up/Right-bottom	Left-up/Left-bottom
Left-up	(a)	0.0002/0.0000	-/0.2379
	(b)	0.0000/0.0000	-/1.0000
	(i)	1.0000/1.0000	-/1.0000
	(ii)	0.3200/1.0000	-/0.9998
Left-bottom	(a)	0.9495/0.0000	0.2379/-
	(b)	0.0000/0.0000	1.0000/-
	(i)	0.9969/0.9998	1.0000/-
	(ii)	1.0000/0.9998	0.9998/-
Right-up	(a)	-/0.7329	0.0002/0.9495
	(b)	-/0.9996	0.0000/0.0000
	(i)	-/0.8331	1.0000/0.9969
	(ii)	-/0.9998	0.3200/1.0000

6 Discussion

In this work, we have proposed a blind detection technique based on copula model to identify spliced images. Copula as joint distribution has involved marginal distributions of RGB components. Gaussian and Archimedean copula has shown their superior performance in detecting tampered image. The fragments of spliced reveal irregularity trend in copula distribution and therefore cause changes in the parameters of copula. Combining with z-test report, the dependency of Clayton and Gumbel copula indicated that they can be regarded as promising approaches for presenting inconsistency magnitude among the fragments of spliced image. However, there exists anomaly copula features when identify similar fragments of original image. Therefore, under that situation, finding an appropriate preprocessing image, such handling noise over image, is desirable and should be considered.

References

1. Dong, J., Wang, W., Tan, T., Shi, Y.Q.: Run-length and edge statistics based approach for image splicing detection. In: International Workshop on Digital Watermarking, pp. 76–87. Springer (2008)
2. Fisher, R., Immer, F., Tedin, O.: The genetical interpretation of statistics of the third degree in the study of quantitative inheritance. Genetics **17**, 107 (1932)
3. Fu, D., Shi, Y.Q., Su, W.: Detection of image splicing based on hilbert-huang transform and moments of characteristic functions with wavelet decomposition. In: International Workshop on Digital Watermarking, pp. 177–187. Springer (2006)
4. He, Z., Lu, W., Sun, W., Huang, J.: Digital image splicing detection based on markov features in dct and dwt domain. Pattern Recognit. **45**, 4292–4299 (2012)

5. Hofert, M., Mächler, M., McNeil, A.J.: Likelihood inference for archimedean copulas in high dimensions under known margins. J. Multivariate Anal. **110**, 133–150 (2012)
6. Huang, J.J., Lee, K.J., Liang, H., Lin, W.F.: Estimating value at risk of portfolio by conditional copula-garch method. Insurance Math. Econ. **45**, 315–324 (2009)
7. Luo, W., Qu, Z., Pan, F., Huang, J.: A survey of passive technology for digital image forensics. Front. Comput. Sci. China **1**, 166–179 (2007)
8. Mahdian, B., Saic, S.: A bibliography on blind methods for identifying image forgery. Signal Process. Image Commun. **25**, 389–399 (2010)
9. Nelsen, R.B.: Dependence and order in families of archimedean copulas. J. Multivariate Anal. **60**, 111–122 (1997)
10. Ng, T.T., Chang, S.F., Sun, Q., 2004. Blind detection of photomontage using higher order statistics. In: 2004 IEEE International Symposium on Circuits and Systems (IEEE Cat. No. 04CH37512), pp. V–V. IEEE
11. Patton, A.J.: Modelling time-varying exchange rate dependence using the conditional copula (2001)
12. Potdar, V.M., Han, S., Chang, E., 2005. A survey of digital image watermarking techniques. In: INDIN'05. 2005 3rd IEEE International Conference on Industrial Informatics, pp. 709–716. IEEE (2005)
13. Rey, C., Dugelay, J.L.: A survey of watermarking algorithms for image authentication. EURASIP J. Adv. Signal Process. **I**(2002), 1–9 (2002)
14. Shi, Y.Q., Chen, C., Xuan, G., Su, W.: Steganalysis versus splicing detection. In: International Workshop on Digital Watermarking, pp. 158–172. Springer (2007)
15. Sklar, A.: Random variables, distribution functions, and copulas: a personal look backward and forward. Lecture notes-monograph series **1–14** (1996)
16. Tong, X., Cui, M.: Image encryption with compound chaotic sequence cipher shifting dynamically. Image Vis. Comput. **26**, 843–850 (2008)
17. Vandenhende, F., Lambert, P.: Improved rank-based dependence measures for categorical data. Stat. Probab. Lett. **63**, 157–163 (2003)
18. Wang, L., Song, H., Liu, P.: A novel hybrid color image encryption algorithm using two complex chaotic systems. Opt. Lasers Eng. **77**, 118–125 (2016)

Artificial Intelligence Forecasting for Transmission Line Ampacity

Yaman Hamed, Muhamad Safwan Abd Rahman, Mohd Zainal Abidin Ab Kadir, Miszaina Osman, Azrul Mohd Ariffin, and Nur Fadilah Ab Aziz

Abstract Overhead transmission lines are used to transmit electrical energy at a high voltage over long distances. The capacity of the transmission system is usually capped at a certain level to comply with the safety and reliability conditions of the system. The limit of the current capacity is estimated based on the worst weather conditions. However, during normal weather conditions, the conductor usually has more transition capacity than its limited level. Dynamic line ratings (DLR) provide an estimation of the line current capacity based on the actual weather conditions. Hence, the transmission line can be used to its full capacity when needed. Dynamic line ratings can be modelled physically, statistically, and using machine learning and artificial intelligence. This chapter aims to address the implementation of machine learning and artificial intelligence algorithms in predicting the DLR of transmission systems. Using artificial intelligence to forecast the measurements of DLR enables operators to estimate the current maximum capacity in real-time. The advantages and limitations of each approach are detailed in this chapter as well.

Keywords Artificial intelligence · Machine learning · Statistical modelling · Dynamic line rating · Overhead lines

1 Introduction

The demand for electric energy supply is increasing constantly over the years. The overhead transmission line system (OHL) is one of the methods of transmitting electrical energy at a high voltage from one point to another, especially over long distances. The thermal stress caused by the current passing through the transporting line harms the system components. The deterioration of the mechanical parts of the OHL, the risk caused by excessive elongation of the conductor, conductor ground clearance, the increment in the sag, and damage to connectors hardware are some of the hazards that affect the reliability and security of the transporting system [1].

Y. Hamed (✉) · M. S. Abd Rahman · M. Z. A. Ab Kadir · M. Osman · A. M. Ariffin · N. F. Ab Aziz
Institute of Power Engineering, Universiti Tenaga Nasional (UNITEN), 43000 Kajang, Malaysia

© Institute of Technology PETRONAS Sdn Bhd 2022

S. A. Abdul Karim and A. Shafie (eds.), *Towards Intelligent Systems Modeling and Simulation*, Studies in Systems, Decision and Control 383,
https://doi.org/10.1007/978-3-030-79606-8_16

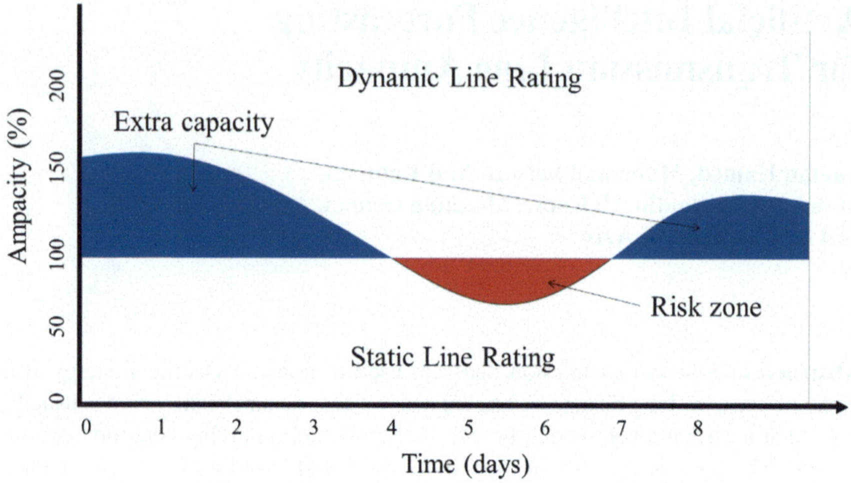

Fig. 1 SLR versus DLR ampacity limits

Therefore, system operators tend to operate the conductors under conservative conditions that assure the safety of the entire power transportation system. These regulations are known as Static Line Ratings (SLR) [2]. The SLR calculates the maximum allowable current capacity, namely ampacity, based on the worst conditions that are involved in the heat exchange process of a certain conductor [3]. However, the worst-case conditions occur rarely during the lifetime of the conductor. Therefore, the SLR constraints usually limit the capabilities of the OHL, which encourages the operators to find a real-time estimate of the condition of the conductor so that it can operate at its full capacity when needed.

Dynamic line rating is a technology that can be used to improve the transmission efficiency and the capacity of the existing power system without changing the system structure or breaking the current technical specification. It is an economical and feasible method to meet the increasing power demands and the need for new energy integration [4]. The difference between the line ampacity limited by SLR and the actual capacity when estimated using DLR is illustrated in Fig. 1.

2 Line Rating Methods

The implementation of the DLR technique requires tracking the conductor temperature in real-time. The estimation of the real-time parameters is classified in two methods:

2.1 Direct Methods

The direct methods in estimating the DLR measures the quantities related directly to the conductor's temperature. Sensors are installed over the conductor to measure mechanical and thermal variables such as line tension, conductor ground clearance, conductor sag, or conductor temperature directly [5].

The direct DLR methods measure the average conductor temperature of an overhead line or span section regardless of the surrounding weather conditions or current intensity. These methods are precise and can achieve high accuracy, however, the costs of implementing such tools can be financially high.

This chapter will focus on the indirect methods in predicting dynamic line ratings for overhead transmission lines.

2.2 Indirect Methods

The indirect methods estimate the DLR based on the heat balance transfer between the conductor and the surrounding environment. The weather parameters at specific locations along the transmission line are used to estimate the DLR in either deterministic or probabilistic approaches. The deterministic approaches investigate the physical or the mathematical relationship between the conductor's temperature and the surrounding meteorological parameters without any random factor. While the probabilistic approaches investigate the related variables as their probability distribution. The ampacity estimated by the deterministic models using certain inputs generates the same output. On the other hand, the probabilistic models may result in more general outputs depending on the input distribution function. In this chapter, deterministic approaches are discussed in detail for DLR prediction. The probabilistic methods are to be discussed in future work.

2.3 Physical Models

There are two available physical standards used to estimate the transmission line thermal rating namely, The IEEE Standard 738 [6] and the CIGRE Technical Brochure 601 [7]. Both models estimate the ampacity and conductor temperature based on the thermal behavior of the conductor using the assumptions of heat balance. The heat balance concept assumes that the conductor heat gain through the passing current is equivalent to the heat loss to the surrounding environment.

The IEEE standard estimates the thermal line rating based on the heat balance equation given in (1):

$$q_c + q_r = q_s + I^2\ R(T_c) \tag{1}$$

where: q_c and q_r are the heat loss by convection and radiation to surrounding air respectively, q_s is the heat gained from solar radiation, $I^2 R(T_c)$ is the heat generated by the current flowing through the conductor. I is the line loading and $R(T_c)$ is the conductor resistance at the conductor core temperature T_c.

The maximum current allowable to cross through the transmission line is determined using (2)

$$I = \sqrt{\frac{q_c + q_r - q_s}{R(T_c)}} \tag{2}$$

The conductor may have different temperatures from span to span as the weather conditions vary along the transmission line. Therefore, the final line rating is estimated based on the minimum allowable current along with the whole transmission system.

The heat balance equation presented in (1) assumes that weather conditions remain constant along the whole line. Thus, it calculates the line ampacity in a steady-state condition. However, the line current, temperature, and environmental conditions continuously vary along the overhead power conductor sections. Therefore, the conductor heat capacity can be included within the heat balance equation to present transient state conditions, as given in (3):

$$q_c + q_r + mC_p \frac{dT_c}{dt} = q_s + I^2\ R(T_c) \tag{3}$$

where the $m\ C_p$ is the heat capacity of the conductor.

The CIGRE standard estimates the thermal line rating based on an expanded heat balance equation compared to IEEE standard as given in (4):

$$q_c + q_r + q_w = q_s + q_m + q_i + I^2\ R(T_c) \tag{4}$$

where q_w is the heat loss to evaporation, q_m is magnetic heat gain, and q_i is the conductor coronal heating.

The additional parameters included in the CIGRE standard are found to have minimal effect on the heat balance equation during normal conditions [8]. However, it is noteworthy that for certain cases of weather parameters (such as a wind speed that is above 5 m/s) the ampacity determined by the IEEE and CIGRE model is significantly different [9]. Therefore, in general conditions, the heat balance equation presented in (1) can be sufficient for thermal rating calculations.

In addition to the physical models, the ampacity of the transmission line can be estimated using the mathematical relationship between the conductor's temperature and the surrounding weather conditions. Such DLR modelling approaches are known as numerical weather prediction (NWP) models [10].

3 Forecasting Line Ampacity Using Weather Parameters

Weather conditions have a great role in predicting line ampacity. The main weather parameters that are used in DLR prediction are ambient temperature, wind speed, wind direction, and solar radiation. When transmission line ampacity is required for the present time, the beforementioned meteorological data can be obtained from national weather stations. The weather services provide the weather parameters in time intervals from short a horizon of one minute to one hour. These measurements are used to estimate future meteorological parameters for a future horizon between 1 and 24 h. Some artificial intelligence methods can predict the weather measurements for even a whole week ahead. This precise knowledge of future transmission line capacity will greatly facilitate the purchase of competitively priced electricity from remote locations [11].

The weather models are classified as stochastic or deterministic. In this chapter, several models from both approaches will be presented. All NWP models are expected to estimate the ampacity of the transmission line based on the weather parameters as depicted in Fig. 2.

The solar radiation parameter is usually modelled with a Fourier series as a function of the position of the sun [10], ambient temperature, and wind speed can be generated using Fourier series as well [11]. On the other hand, there are several methods to predict remaining weather parameters, such as linear regression, time series, local learners, artificial neural networks, among others. The weather parameters can be probabilistically modelled based on their probability distribution functions, or Monte Carlo simulations.

3.1 *Fourier Series Model*

Historical weather data, such as ambient temperature (T_a) and wind speed (W_s), are fitted into a Fourier series to predicted future measures as given by (5):

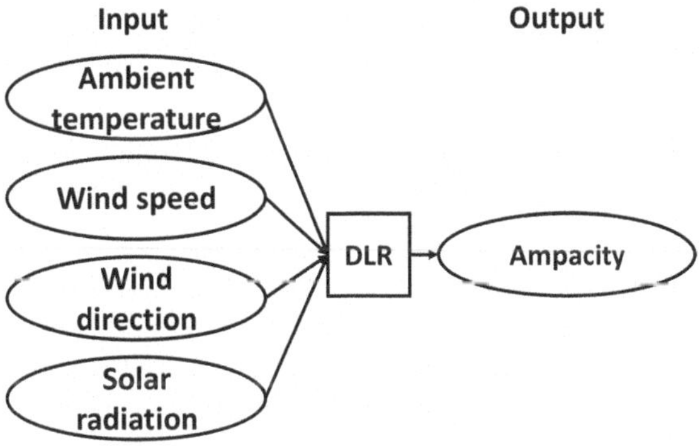

Fig. 2 NWP model parameters

$$Y_t = A_0 + k\sum_{i=1}^{n} C_i \mathrm{Sin}(\omega_i t) + \sum_{i=1}^{n} B_i \mathrm{Cos}(\omega_i t) \tag{5}$$

where: Y_t: is the predicted weather parameter at time t, A_0, C_i, B_i: are the model coefficients, ω: is the fundamental frequency equals to $2\pi/24h$, k: is an adjusting factor calculated based on the forecasts of the national weather providers as given in (6):

$$k = \frac{Y_f(t_{\max}) - Y_f(t_{\min})}{Y(t_{\max}) - Y(t_{\min})} \tag{6}$$

where: $Y_f(t_{\max})$: is the daily maximum value of the weather parameter by the national provider, $Y_f(t_{min})$: is the daily minimum value of the weather parameter by the national provider, $Y(t_{\max})$, and $Y(t_{\min})$ are estimated using Fourier series when $t = t_{\max}$, and $t = t_{\min}$ respectively.

The model parameters (A_0, C_i, B_i) are determined using the least square regression method [11].

3.2 Time Series Model

The auto-regressive (AR) process is a time series predictor that forecasts weather parameters from historic measures obtained by weather stations [12]. The AR model or order p forecasts and estimation of the weather parameter Y_t from a combination of p historic values Y_t; $i = 1, \ldots, p$. As given in (7):

$$Y_t = \beta_0 + \sum_{i=1}^{p} \beta_i Y_{t-i} + \varepsilon_t \tag{7}$$

where: β_0 is the regression constant, β_i are the regression coefficients, and ε_t is the Gaussian noise error. The β_i coefficients are usually estimated for time horizon $(t = 1, \ldots, n)$ using ordinary least square method as given in (8):

$$\begin{aligned} Y_n &= \beta_0 + \beta_1 Y_{n-1} + \beta_2 Y_{n-2} + \cdots + \beta_p Y_{n-p} + \varepsilon_n \\ Y_{n-1} &= \beta_0 + \beta_1 Y_{n-2} + \beta_2 Y_{n-3} + \cdots + \beta_p Y_{n-p-1} + \varepsilon_{n-1} \\ &\vdots \\ Y_1 &= \beta_0 + \beta_1 Y_0 + \varepsilon_{n-1} \end{aligned} \tag{8}$$

The system presented in (8) can be solved as a matrix system $(Y = XB + E)$ by minimizing the sum squares of the residuals (ε) using (9):

$$B = \left(X'X\right)^{-1}X'Y \tag{9}$$

where:

$$Y = \begin{bmatrix} Y_n \\ Y_{n-1} \\ \vdots \\ Y_1 \end{bmatrix},\ B = \begin{bmatrix} \beta_0 \\ \beta_1 \\ \vdots \\ \beta_p \end{bmatrix},\ X = \begin{bmatrix} 1 & Y_{n-1} & Y_{n-2} & \cdots & Y_{n-p} \\ 1 & Y_{n-2} & Y_{n-3} & \cdots & Y_{n-p-1} \\ 1 & \vdots & \vdots & \ddots & \vdots \\ 1 & Y_0 & Y_{-1} & \cdots & Y_{1-p} \end{bmatrix},\ E = \begin{bmatrix} \varepsilon_n \\ \varepsilon_{n-1} \\ \vdots \\ \varepsilon_1 \end{bmatrix}$$

As discussed earlier, the AR forecasting model relay on a historic time series of weather parameters at the same weather station or measuring location only. In order to consider historic time series at other surrounding locations, a vector auto-regressive (VAR) model can be used.

The VAR model provides a Spatio-temporal correlation between the target location and other locations. Given that weather variables are highly correlated between different locations, the use of a VAR model may have higher forecast accuracy over an AR model [13].

The victor autoregression model provides a weighted estimation of the historic time series in $(k-1)$ measuring locations around the sampling weather provider.

$$Y_t = A_0 + \sum_{i=1}^{p} A_i Y_{t-i} + \mathcal{E}_t \tag{10}$$

where: Y_t is a $(k \times 1)$ vector of weather parameters at a certain location K.

A_0: is a vector of constant coefficients,

A_i: is a $(k \times k)$ matrix of model coefficients at a certain horizon i,

$\mathcal{E}_t$: is the Gaussian error $(k \times 1)$ vector.

Similarly to the AR model, the VAR model given in (10) is solved by minimizing the sum square of residuals in the matrix system $Y = AX + \mathcal{E}$ where:

$$Y = \begin{bmatrix} Y_{1t} \\ Y_{2t} \\ \vdots \\ Y_{kt} \end{bmatrix},\ A = \begin{bmatrix} A_0 \\ A_1 \\ \vdots \\ A_k \end{bmatrix},\ X = \begin{bmatrix} A_{11}^i & A_{12}^i & \cdots & A_{1k}^i \\ A_{21}^i & A_{22}^i & \cdots & A_{2k}^i \\ \vdots & \vdots & \ddots & \vdots \\ A_{k1}^i & A_{k2}^i & \cdots & A_{kk}^i \end{bmatrix},\ \mathcal{E} = \begin{bmatrix} \varepsilon_{1t} \\ \varepsilon_{2t} \\ \vdots \\ \varepsilon_{kt} \end{bmatrix}$$

4 Modelling Dynamic Line Rating Using Artificial Intelligence

Artificial intelligence is the machine simulation of human intelligence and learning mechanisms. Artificial intelligence methods are widely used in industrial domains as they can represent huge datasets sufficiently and accurately as well. In this chapter, two classes of artificial intelligence algorithms are presented, which are machine learning local learners and Artificial neural networks as general learners.

4.1 Machine Learning Local Learners

Local learners are a class of machine learning that generate predictions based on the vicinity of the query point. It assumes that the neighbour points that are closer to the query point have the most information about it [3]. The vicinity of a certain query is usually determined by the points with the shortest distance to it [14]. There are two main functions used in detecting potential neighbors:

- Distance function: that determines the degree of vicinity between the query point and its neighbors based on a distance metric. The Euclidian distance is one of the simplest and most convenient functions used to measure the neighborhood of a certain query. A group of neighbors has similar characteristics based on a selected group of parameters. The Euclidian distance between two neighbor points (x^g, x^q) in a space of dimension i is estimated using (11):

$$d = \left\| x^g - x^q \right\| = \sqrt{\sum_{i=1}^{n} (x_i^g - x_i^q)^2} \tag{11}$$

- Weight function: that measures how much influence has a generic point on the query point. The weight of a certain neighbour point depends on its distance from the query point, as well as, on the radius of the neighbourhood under consideration. The simplest weight function is given in (12):

$$\omega_i = d_i \bigg/ \sum_{i=1}^{n} d_i \tag{12}$$

The weight function must gain its maximum value at zero distance from the interpolated point, and as the distance increases the function should decrease respectively [15]. These constraints are detailed as given in (13):

$$\begin{aligned} 0 < \omega\big(d\big(x^g,\ x^q\big),\ b\big) < 1 \\ \omega\big(d\big(x^q,\ x^q\big),\ b\big) = 1 \end{aligned} \tag{13}$$

Table 1 List of Kernel functions

Kernel	Parameters	References
$k_i = e^{(-d_i^2)}$	–	[17]
$k_i = \frac{1}{1+d_i^p}$	$p > 0$; is a power parameter chosen randomly, usually $p = 2$	[18]
$k_i = e^{-\lvert d_i \rvert}$	–	[18]
$k_i = \left(1 - d_i^2\right); \lvert d_i \rvert < 1$	–	[18]
$k_i = (1 - \lvert d_i \rvert^3)^3; \lvert d_i \rvert < 1$	–	[19]
$k_i = \frac{d_i - d_{\max}}{d_i - d_{\min} \times \frac{d_i + d_{\max}}{d_i + d_{\min}}}$	$\frac{d\max}{\min}$ s the distance between the interpolated point and furthest/nearest point in the set	[20]
$k_i = d_i^{-p}$	$p > 0$: is a power parameter chosen randomly	[20]
$k_i = \left[\frac{R - d_i}{R \times d_i}\right]^2$	R: is the distance from the interpolated point to the farthest point of the set of data	[21]

The simple weight function presented in (12) explains the influence of the neighbor points on the query point based only on the distance metrics within the vicinity. Therefore, complex weights can be assigned to the neighbor points using a Kernel function. The kernel function explains the relationship between the neighbor points as a complex form of the distance within the contiguity. Table 1 details some of the well-known Kernels used in local learning [16].

The prediction process of the local learning methods is categorized into three main behaviours:

- Defer: the data set is stored until predictions are required.
- Reply: data are trained using a local model to get an estimation that corresponds to the query point.
- Flush: once the prediction is obtained, the local model is discarded.

The local learner algorithms can be trained to correspond to the DLR estimation directly based on certain NWP inputs. The local learners can also be used to estimate the weather parameters individually and then the DLR will be estimated using heat balance equations. In this chapter, the direct prediction process is preferred. The implementation of the prediction step described in the 'Reply' mode is described in the following steps:

- The distance between the query section in the OHL and the sections with similar weather features is calculated using (14):

$$
\begin{aligned}
d_i &= \sqrt{D^2(T_a) + D^2(W_s) + D^2(W_d) + D^2(S_r)} \\
D(T_a) &= T_a(x)_i^g - T_a(x)_i^q \\
D(W_s) &= W_s(x)_i^g - W_s(x)_i^q \\
D(W_d) &= W_d(x)_i^g - W_d(x)_i^q \\
D(S_r) &= S_r(x)_i^g - S_r(x)_i^q
\end{aligned}
\tag{14}
$$

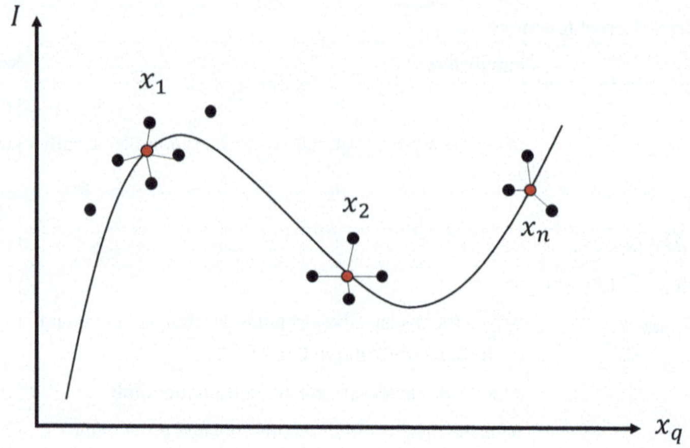

Fig. 3 Local learner prediction line

where: $D(T_a)$, $D(W_s)$, $D(W_d)$, $D(S_r)$ are the distance components for ambient temperature, wind speed, wind direction, and solar radiation, respectively.

- The weights of each generic point in the neighborhood are estimated using one of the kernel functions given in Table 1. The weight assigned to each kernel is calculated using (Eq. 15):

$$\omega_i = k_i / \sum_{i=1}^{n} k_i \tag{15}$$

- The DLR of the query section of the OHL is estimated by substituting all information form (Eq. 14) and (15) in

$$\hat{I}_i = \sum_{i=1}^{n} \omega_i I_i \tag{16}$$

where: $\hat{I}_i$ is the predicted ampacity of the query section. I_i is the ampacity of the neighbor points (Fig. 3).

4.2 *Artificial Neural Network*

Artificial neural network (ANN) is a global machine learning technique that is known also as the black box predictor. The neural network is an information managing tool that mimics the learning mechanism inside the human brain. The human brain manages the learning process using a network of connected nerve cells called neurons.

The neurons work as data processing units that transmit signals through the axon. The axon develops an association between its neuron and at least one other neuron in the network using several nods. Each neuron can receive signals (inputs), process them (capture), and send them to another neuron (output). This process forms the learning framework of the human brain [22].

Mathematically, the connection between different neurons is evaluated by a weight coefficient that reflects the degree of importance of each neuron. The neural network is able to find hidden information in the data, however, cannot extract them. The process of 'capturing' the unknown information is called 'learnin' or 'training' the neural network. In a mathematical formalism, to learn means to adjust the weight coefficients in such a way that some conditions are fulfilled [23]. The neural network consists of neurons that are ordered in several layers as depicted in Fig. 4. The input layer, the output layer, and the layers in between are called the hidden layers.

Each neuron in a certain hidden layer is connected to all neurons in the next layer. The connection between the i'th and j'th neuron is characterized by the weight coefficient ω_{ij}, and the bias of the i'th node ϑ_i. The output value (activity) of the i'th neuron x_i is determined using Eq. (17):

$$\begin{aligned} x_i &= f(\xi_i) \\ \xi_i &= \vartheta_i + \sum_j \omega_{ij}\, x_j \end{aligned} \quad (17)$$

where: where ξ_i is the potential of the i'th neuron, and $f(\xi_i)$ is the transfer function that holds:

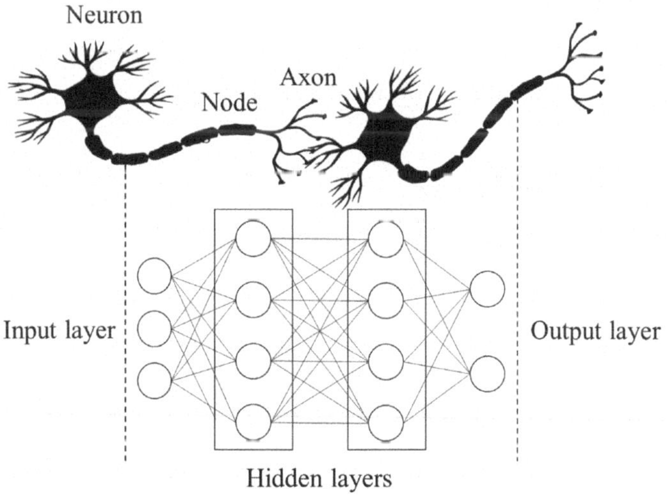

Fig. 4 Projection of human neuron on neural network

$$f(\xi) = (1 + \exp(-\xi))^{-1} \tag{18}$$

The weights and biases of the network are adjusted during the training process in such a way to minimize a cost function L such as the mean square error function presented in (19):

$$L = \frac{1}{n} \sum_{i=1}^{n} (\hat{y}_i - y_i)^2 \tag{19}$$

where: $\hat{y}_i$ is the estimated value of the output y_i.

The neural network presented in Fig. 4 is called Feed Forward Back Propagation (FFBP) net. The Feed Forward indicates that the nodes are connected in a straight direction and never form a cycle. Back propagation is the method of fine-tuning the weights of the nodes based on the error rate obtained in the previous iteration.

$$\omega_{ij}^{(k+1)} = \omega_{ij}^{(k)} + \lambda \left(\frac{\partial L}{\partial \omega_{ij}} \right)^{(k)}$$
$$\vartheta_i^{(k+1)} = \vartheta_i^{(k)} + \lambda \left(\frac{\partial L}{\partial \vartheta_i} \right)^{(k)} \tag{20}$$

where: λ is the positive rate of learning. The solution of (20) is obtained by finding the derivatives $\partial L/\partial \omega_{ij}$ and $\partial L/\partial \vartheta_i$ through the following steps:

- The error between the output vector $\hat{Y}$ and the target vector Y is calculated for the first output layer:

$$L = \frac{1}{n} \sum_{i=1}^{n} (\hat{x}_i^1 - x_i^1)^2 = \frac{1}{n} \sum_{i} e_i^2 \tag{21}$$

where: $e_i = 0$ for $i \notin$ output layer.

- Minimize the error in the weights of the first iteration:

$$\begin{aligned} \frac{\partial L}{\partial \omega_{ij}} &= \frac{\partial L}{\partial x_i} \cdot \frac{\partial x_i}{\partial \omega_{ij}} = \frac{\partial L}{\partial x_i} \cdot \frac{\partial f(\xi_i)}{\partial \omega_{ij}} = \frac{\partial L}{\partial x_i} \cdot \frac{\partial f(\xi_i)}{\partial \xi_i} \cdot \frac{\partial \xi_i}{\partial \omega_{ij}} \\ &= \frac{\partial L}{\partial x_i} \cdot f'(\xi_i) \cdot \frac{\partial \left(\vartheta_i + \sum_{j \in \Gamma_o} \omega_{ij} x_j \right)}{\partial \omega_{ij}} \\ &= \frac{\partial L}{\partial x_i} \cdot f'(\xi_i) x_j \end{aligned} \tag{22}$$

- Minimize the error in the bias of the first iteration:

$$\frac{\partial L}{\partial \vartheta_i} = \frac{\partial L}{\partial x_i} \cdot \frac{\partial x_i}{\partial \vartheta_{ij}} = \frac{\partial L}{\partial x_i} \cdot f'(\xi_i) \cdot l \tag{23}$$

- By solving (22) and (23) the following relation concludes:

$$\frac{\partial L}{\partial \omega_{ij}} = \frac{\partial L}{\partial \vartheta_i} \cdot x_j \tag{24}$$

- The final step is to find $\partial L / \partial \vartheta_i$ that fulfills (25) for $i \in$ output layer, and (26) for $i \in$ hidden layer considering (23):

$$\frac{\partial L}{\partial x_i} = e_i \tag{25}$$

$$\begin{aligned} \frac{\partial L}{\partial x_i} &= \sum_{l \in \Gamma_i} \frac{\partial L}{\partial x_l} \frac{\partial x_l}{\partial x_i} = \sum_{l \in \Gamma_i} \frac{\partial L}{\partial x_l} \frac{\partial f(\xi_l)}{\partial x_i} \\ &= \sum_{l \in \Gamma_i} \frac{\partial L}{\partial x_l} \frac{\partial f(\xi_l)}{\partial \xi_l} \frac{\partial \xi_l}{\partial x_i} = \sum_{l \in \Gamma_i} \frac{\partial L}{\partial \vartheta_l} \omega_{li} \end{aligned} \tag{26}$$

- The above-mentioned steps are repeated for all iterations until the sum square of errors is minimized.

Neural networks are labeled as hidden box algorithms because all the calculations are done in the hidden layers. A neural network with a sufficient number of hidden layers (neurons) is capable of predicting any set of continuous data. There is no simple rule to estimate the number of hidden layers required for a certain neural network. The number of hidden layers is usually proportional to the size of the dataset and the number of parameters included. A large subset of independent parameters requires a larger number of neurons. However, increasing the number of hidden layers may compromise the network's ability to be generalized, and may result in an overfitted estimation. Overfitting is the case when the neural network is unable to learn the general pattern and mimics the individual points (local minima) An overfitted prediction is caused when computational nodes calculate small observations leading to a decrease in the degree of freedom. Therefore, a neural network that has a generalized prediction performance requires a balanced number of weights and training subsets [24].

The advantages of ANN algorithms can be summarized in:

- Adaptability: ANNs can be used for a broad range of applications, due to their ability to capture subtle relationships even if they are difficult to describe.
- Generalization: they can handle imperfect or incomplete data without adding unacceptable errors in the predictions.
- Nonlinearity: they can map highly nonlinear relationships between inputs and outputs.

- Parallel computation: the elementary operations of an ANN can be performed independently and executed simultaneously.
- Robustness: ANN generates accurate predictions for complex systems with a large number of inputs and parameters [25].

The implementation of ANN in predicting the dynamic line ratings for an OHL can avoid the neglected factors of the physical models. The real-time thermal rating of the conductor can be estimated using neural networks in 2 steps. In the first step, a neural network is used to determine the real-time conductor temperature, based on four weather input parameters (ambient temperature T_a, solar radiation S_r, wind speed W_s, and wind direction W_d) and a load parameter (real-time current value I_r). Once the neural network is trained, the predicted conductor temperature is used in a thermal differential equation to estimate the transfer capacity of the OHL. Each conductor has a maximum temperature that cannot be exceeded due to material factors and critical sag. The thermal differential equation can estimate the surplus current that can be passed through the transmission lines to achieve this maximum temperature. The addition of the current and the surplus load represent the dynamic line rating.

Alternatively, the neural network can be trained to predict the dynamic line ratings directly by implementing the two steps at once. However, this requires the DLR measurements to be known for certain weather parameters in the training dataset. Thus, depending on the available inputs, the line ratings can be estimated using neural networks as illustrated in Fig. 5.

5 Discussion

Dynamic line ratings allow the transmission line to reach its full capacity by maximizing the allowable current passing through the conductor while maintaining the safety of the system. The estimation of DLR requires access to the real-time weather parameters in order to achieve accurate predictions. The DLR estimation for a certain OHL can be calculated using direct or indirect methods. The direct methods require the installation of measuring devices (sensors) on the conductor to provide real-time readings. The installed sensor provides live measures of the conductor temperature, sag, tension, and ground clearance. Although direct DLR methods provide accurate estimations, they are costly equipment and hard to install. They also require periodic maintenance and calibration along all sections of transmission line spans, which results in additional maintenance and installation costs. Besides, the direct DLR sensors are affected by the surrounding weather conditions that may affect the accuracy of the predictions. Large errors in dynamic rating are likely to result from lightly loaded lines when the line rating is calculated by tension and sag sensors. Furthermore, direct monitoring devices lack accuracy when the difference between the conductor temperature and ambient temperature is small. Similarly, in normal wind environments, the tension and sag sensors achieve poor measuring accuracy

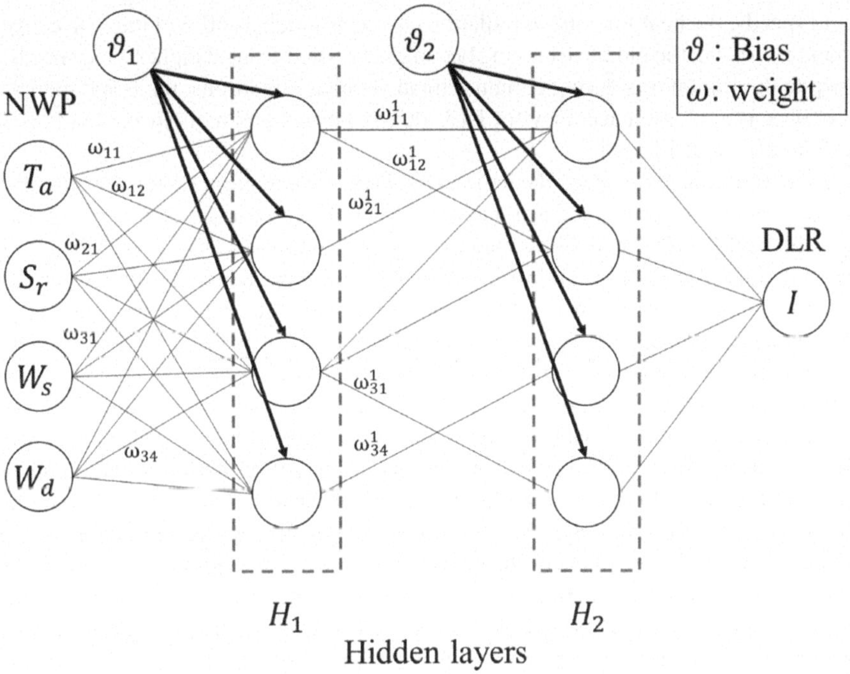

Fig. 5 Neural network for DLR forecasting

during low current load. On the contrary, the sensors have higher accuracy in the case of high current load during windy or low wind atmosphere [5].

Weather-based DLR modelling is the alternative to the direct DLR measuring methods. The numerical weather predictions are the indirect methods for estimating dynamic line ratings. Weather models are simplest to implement as weather monitoring devices are not required to be installed on the conductor itself. They are comparatively less expensive to direct sensors, highly reliable and do not require any special calibrations. The weather measurements can be obtained from local stations to be used for future predictions. The numeric weather predictions can be used for line rating estimation for a horizon of one day to one week ahead.

Although weather parameters are easy to obtain and costly effective, these measurements have large error margins. The weather measuring devices may not accurately represent the worst weather conditions along the transmission line. Thus, the line ampacity is highly affected by the uncertainty in NWP models as they relate the conductor temperature to the measured data.

For low-loaded lines, convective cooling of wind is the dominant factor to determine the line ampacity. For heavily loaded lines, the impact of ambient temperature and solar radiation is less significant and Joule heating from the line current is the principal contributor to conductor temperature. At elevated conductor temperatures, the temperature and line ampacity are not very sensitive to solar radiation. At low wind speeds, the heat loss due to radiation can be as much as 40% of the convective cooling term, but the radiative cooling becomes less significant at higher wind speeds. Despite the limitations during certain critical weather conditions, the NWP models have decent prediction accuracy for DLR during normal and high-speed wind rates with highly loaded lines [5].

In computational prospect, the indirect methods of estimating DLR have different performances depending on the simulated dataset. The physical models represented by IEEE and CIGRE models are simple to implement, have decent versatility, and they explain the physical relationships clearly. On the other hand, they ignore certain influencing factors and have difficulties in obtaining few parameters accurately. The statistical weather models overcome the problem of insufficient and inaccurate parameters in the physical models. However, they are only applicable to lines with similar characteristics, they are sensitive to abnormal data, and require a consistent amount of historical data for fitting. On the contrary, the artificial intelligence models can be generalized, they overcome the accuracy limitation in the parameters, and they can solve the problem of incomplete factors. However, they require a huge amount of training data, hence, a huge computational power. Local learners depend only on the neighbors of the query point, hence, they ignore the trend of the whole dataset in the estimation. Besides, they are highly affected by the extreme values, resulting in an overfitted model. On the contrary, artificial neural networks are general learner algorithms in which they use the whole dataset in the prediction. Also, they can optimize the proper weight for each point in the prediction curve.

6 Summary

This chapter presented the concepts of estimating the dynamic line ratings of an electric transmission line. The methods discussed in this chapter were divided into two categories, direct and indirect methods. The indirect methods have two main categories, physical and mathematical models. The physical models such as IEEE and CIGRE models estimate the dynamic line ratings based on the concept of heat balance. The mathematical models use weather parameters to reflect their effect on the conductor temperature. Mathematical models are classified as statistical weather models (Fourier Series time series and autoregression) and artificial intelligence algorithms (local learners and artificial neural networks). The mathematical models based on numeric weather prediction can estimate the DLR of a transmission line for a time horizon between one day to one week ahead.

References

1. Zainuddin, N.M., Rahman, M.A., Kadir, M.A., Ali, N.N., Ali, Z., Osman, M.: Review of thermal stress and condition monitoring technologies for overhead transmission lines: issues

and challenges. IEEE Access **8**, 120053–120081 (2020)
2. Aziz, M.H.A., Miswan, N.S., Kasran, F.A.M., Shokri, M.S.M., Noran, M.N., Abd Rahim, A.: Pilot project on dynamic line rating (DLR) system for optimal use of Tenaga Nasional Berhad (TNB) grid capacity. In: 2020 International Conference on Technology and Policy in Energy and Electric Power (ICT-PEP), pp. 77–81, 2020
3. Pepiciello, A., Coletta, G., Vaccaro, A., Villacci, D.: The role of learning techniques in synchrophasor-based dynamic thermal rating. Int. J. Electr. Power Energy Syst. **115**, 105435 (2020)
4. Hou, Y., Wang, W., Wei, Z., Deng, X., Ji, Q., Wang, T.: Research and application of dynamic line rating technology. Energy Rep. **6**, 716–730 (2020)
5. Karimi, S., Musilek, P., Knight, A.M.: Dynamic thermal rating of transmission lines: a review. Renew. Sustain. Energy Rev. **91**, 600–612 (2018)
6. Douglass, D., Reding, J.: IEEE Standard for Calculating the Current-Temperature of Bare Overhead Conductors, pp. 738–2006. IEEE Standard (2007)
7. Iglesias, J., Watt, G., Douglass, D., Morgan, V., Stephen, R., Bertinat, M.: Guide for Thermal Rating Calculations of Overhead Lines: Cigré (2014)
8. Simms, M., Meegahapola, L.: Comparative analysis of dynamic line rating models and feasibility to minimise energy losses in wind rich power networks. Energy Convers. Manage. **75**, 11–20 (2013)
9. Rácz, L., Gabor, G., Németh, B.: Different approaches of dynamic line rating calculations. In: 2019 7th International Youth Conference on Energy (IYCE). pp. 1–6, 2019
10. Alberdi, R., Albizu, I., Fernandez, E., Fernandez, R., Bedialauneta, M. T.: Overhead line ampacity forecasting with a focus on safety. IEEE Trans. Power Deliv. (2021)
11. Deb, A.K.: Powerline Ampacity System: Theory, Modeling and Applications. CRC Press (2017)
12. Fan, F., Bell, K., Infield, D.: Probabilistic weather forecasting for dynamic line rating studies. In: 2016 Power Systems Computation Conference (PSCC). pp. 1–7, 2016
13. Fan, F., Bell, K., Hill, D., Infield, D.: Wind forecasting using kriging and vector auto-regressive models for dynamic line rating studies. In 2015 IEEE Eindhoven PowerTech. pp. 1–6, 2015
14. Hamed, Y., Alzahrani, A.I., Mustaffa, Z., Ismail, M.C., Eng, K.K.: Two steps hybrid calibration algorithm of support vector regression and K-nearest neighbors. Alex. Eng. J. **59**, 1181–1190 (2020)
15. Hamed, Y., Shafie, A., Mustaffa, Z., Rusma, N.: Error-reduction approach for corrosion measurements of pipeline inline inspection tools. Measur. Control **52**, 28–36 (2019)
16. Hamed, Y., Shafie, A., Mustaffa, Z.B., Idris, N.R.B.: The effect of different weight functions on calibrating corrosion measurements using KNN technique. In: AIP Conference Proceedings. vol. 1787, p. 080027, 2016
17. Hinton, G.E., Roweis, S.T.: Stochastic neighbor embedding. Adv. Neural. Inf. Process. Syst. **15**, 833–840 (2002)
18. Fan, J., Hall, P.: On curve estimation by minimizing mean absolute deviation and its implications. Ann. Stat. **22**, 867–885 (1994)
19. Wang, Z., Isaksson, T., Kowalski, B.R.: New approach for distance measurement in locally weighted regression. Anal. Chem. **66**, 249–260 (1994)
20. Gou, J., Du, L., Zhang, Y., Xiong, T.: A new distance-weighted k-nearest neighbor classifier. Inf. Comput. Sci. **9**, 1429–1436 (2012)
21. Dumitru, P.D., Plopeanu, M., Badea, D.: Comparative study regarding the methods of interpolation. In: 1st European Conference of Geodesy & Geomatics Engineering GENG'13, Antalya, Turkey, pp. 45–52, 2013
22. Barham, A., Ismail, M.S., Hermana, M., Padmanabhan, E., Baashar, Y., Sabir, O.: Predicting the maturity and organic richness using artificial neural networks (ANNs): a case study of Montney Formation, NE British Columbia, Canada. Alexandria Eng. J. **60**, 3253–3264 (2021)
23. Svozil, D., Kvasnicka, V., Pospichal, J.: Introduction to multi-layer feed-forward neural networks. Chemom. Intell. Lab. Syst. **39**, 43–62 (1997)

24. Sabir, O., Ya, T.T.: Cell by cell artificial neural network model for predicting laminar, incompressible, viscous flow. ARPN J. Eng. Appl. Sci. **11**, 12084–12089 (2006)
25. Abiodun, O.I., Jantan, A., Omolara, A.E., Dada, K.V., Mohamed, N.A., Arshad, H.: State-of-the-art in artificial neural network applications: a survey. Heliyon **4**, e00938 (2018)

Estimating Marine CSEM Responses Using Gaussian Process Regression Based on Synthetic Models

Muhammad Naeim Mohd Aris, Hanita Daud, Khairul Arifin Mohd Noh, and Sarat Chandra Dass

Abstract Seabed logging (SBL) is an application of controlled-source electromagnetic (CSEM) waves to discover marine hydrocarbon-filled reservoirs based on the resistivity contrast of subsurface underneath the seabed. Current practice for processing marine CSEM responses utilizes meshes-based algorithms. The ad hoc algorithms require high computational time to solve the integrals and linear equations. Therefore, this synthetic-based study proposes Gaussian process regression (GPR) to estimate marine CSEM responses at various resistivities of target layer. Synthetic multifrequency SBL responses with target depth of 500 m from the seabed are modelled by finite element (FE) method using computer simulation technology (CST) software. As the prior information to the GPR, the target layer is parameterized with resistivity of 30–510 Ωm with an increment of 60 Ωm. By using MATLAB software, a two-dimensional (2D) GPR model is developed to estimate the marine CSEM responses at unobserved resistivities. For the validation, the mean absolute deviation (MAD), mean squared error (MSE) and root mean squared error (RMSE) between the 2D GP model and the CST outputs (i.e., true values) at the unobserved resistivities are calculated. The computational time for evaluating the marine CSEM using GPR and FE are computed and compared. The resulting error measurements and

M. N. Mohd Aris (✉) · H. Daud
Department of Fundamental and Applied Sciences, Universiti Teknologi PETRONAS, 32610 Seri Iskandar, Perak, Malaysia
e-mail: muhammad_16000991@utp.edu.my

H. Daud
e-mail: hanita_daud@utp.edu.my

K. A. Mohd Noh
Department of Geosciences, Universiti Teknologi PETRONAS, 32610 Seri Iskandar, Perak, Malaysia
e-mail: khairula.nmoh@utp.edu.my

S. C. Dass
School of Mathematical and Computer Sciences, Heriot-Watt University Malaysia, 62200 Putrajaya, Malaysia
e-mail: s.dass@hw.ac.uk

© Institute of Technology PETRONAS Sdn Bhd 2022

S. A. Abdul Karim and A. Shafie (eds.), *Towards Intelligent Systems Modeling and Simulation*, Studies in Systems, Decision and Control 383,
https://doi.org/10.1007/978-3-030-79606-8_17

the computational time revealed that GPR can estimate the marine CSEM responses efficiently and at par to the current methods.

Keywords Gaussian process regression · Controlled-source electromagnetic · Seabed logging · Simulation · Finite element

In this study, Gaussian process regression (GPR) acts as the alternative forward modelling technique to evaluate the marine controlled-source electromagnetic (CSEM) responses. The synthetic marine CSEM responses with target resistivity of 30–510 Ωm (with a gap of 120 Ωm each) and target depths of 400, 800 and 1200 m from the seafloor evaluated by finite element (FE) method using computer simulation technology (CST) software are treated as the GPR prior information. Then, a two-dimensional (2D) GPR model at the unobserved resistivities of target layer is developed for every target depth. The performance of the GPR models is quantified by calculating the mean absolute deviation (MAD), mean square error (MSE) and root mean squared error (RMSE) between the estimates and the computer simulation responses. The computational time taken by the GPR and the simulation software for evaluating the marine CSEM responses are compared as well. The resulting error measurements and the computational time revealed that GPR can estimate the marine CSEM responses efficiently and at par to the current methods.

1 Introduction

The application of controlled-source electromagnetic (CSEM) technique for hydrocarbon exploration in deep seawater environment is also known as seabed logging (SBL). SBL was introduced by Eidesmo et al. [1] in 2002 to remotely detect the offshore high-resistive bodies beneath the seafloor. This application is highly reliant on the resistivity (i.e., inversely proportional to conductivity) contrasts of subsurface. The target reservoirs (i.e., hydrocarbon) are known to have very high electrical resistivity (~30–500 Ωm) compared to its surroundings such as seawater (~0.5–2 Ωm) and sediment (~1–2 Ωm) [2]. The marine CSEM technique uses low-frequency electromagnetic (EM) signal (typically 0.1–10 Hz) emitted by a powerful EM transmitter to characterize the hydrocarbon-saturated reservoirs underneath the seafloor. The purpose of using low-frequency EM signal in the SBL is to obtain high wavelength and decrease the signal attenuation [3]. High-frequency EM signal can influence the attenuation rate and subsequently affect the distance of the signal propagation along the high-resistive bodies. The SBL has become the complementary technique to the conventional imaging survey (e.g., seismic technique) for the offshore hydrocarbon exploration.

In the EM problems, forward modelling of the EM responses is one of the most crucial tasks to reconstruct the resistivity distribution of the subsurface from the responses. The current practice for processing the marine CSEM responses utilizes mesh-based algorithm. Cai et al. [4] mentioned integral equation (IE), finite difference (FD) and finite element (FE) are the most famous computational methods used for the EM forward modelling. FE method is highly preferable compared to the other two numerical methods due to its potential of modelling the EM responses in an intricate geoelectrical structure [4]. It also supports the unstructured meshes in the modelling. However, using mesh-based algorithm methods can increase the computational efforts in terms of time and cost. Li and Key [5] and Bakr et al. [6] stated that the most time-consuming task in the EM forward modelling is solving the integrals and linear systems. Besides, the SBL also generates huge amount of data (i.e., EM responses), and processing those data has become a challenge to geophysicists [2]. Therefore, this study proposes Gaussian process regression (GPR) as the alternative forward modelling technique to effectively estimate the EM profiles with low computational efforts. The GPR models are developed based on very general input domains and computer simulation outputs, as is typical in the SBL application, without significantly increase the computational complexity. This study uses a software which was developed for the numerical simulation of EM fields, namely computer simulation technology (CST) software.

The main objectives of this study are:

(a) To develop seabed logging models and generate electromagnetic responses using computer simulation technology software
(b) To develop two-dimensional Gaussian process regression models based on the computer simulation outputs
(c) To determine the performance of Gaussian process regression in estimating the electromagnetic responses.

This chapter consists of six sections. Section 1 presents the introduction of the study as well as the related literatures. While Sect. 2 describes the statistical background of Gaussian process regression (GPR). Sections 3 and 4 explain the acquisition of the synthetic marine controlled-source electromagnetic (CSEM) data using computer simulation technology (CST) software and the estimation of the marine CSEM responses using GPR, respectively. Section 5 is dedicated for results and discussion, and lastly Sect. 6 is for the summary.

2 Statistical Background—Gaussian Process

Gaussian process regression (GPR) can be defined as an infinite collection of random variables where its finite subsets of the random variables are normally distributed [7, 8]. A GPR model can be specified by a mean function $m(x)$ and a covariance function $k(x, x')$, i.e., $f(x) \sim N[m(x),\ k(x, x')]$. GPR is a probabilistic and non-parametric method. It has convenient properties for numerous modelling tasks in statistics problems and machine learning [9]. Rasmussen and Nickisch [10] mentioned GPR can be utilized to express distributions over functions without having to carry out to a particular functional form. This probabilistic method places "a priori" on the function space and produces "a posteriori" over the function. Harari and Steinberg [11] stated GPR is commonly employed as a "substitute" model for relatively complicated mathematical model that needs high computational time. It also performs well in the non-linear modelling problems. GPR model provides predictive mean and variance. The mean represents the most likely outputs, and the predictive variance (i.e., predictive error-bar) is the confidence measure of the estimations.

In geophysics problems, Asante-Okyere et al. [12] employed GPR to evaluate the porosity and permeability of reservoir. Furthermore, Mohd Aris et al. [13] utilized the predictive variance of the GPR to detect the existence of thin hydrocarbon layer. Besides, GPR also was used by Li et al. [14] for seismic signal detection. In the SBL application, Mohd Aris et al. [15] and Daud et al. [16] used GPR to evaluate the electromagnetic (EM) profiles with different hydrocarbon depths. But, in all applications reported in the literatures, there has not been an attempt to use the GPR to estimate the EM profiles based on computer simulation outputs with various target resistivities for different target depths. The details of the methodology are described in the following sections.

3 Synthetic Marine CSEM Data Acquisition

In the context of this study, computer simulation technology (CST) software is used to replicate the seabed logging (SBL) models. The CST Studio Suite® utilizes the finite element (FE) method to solve the electromagnetic (EM) simulation in low-frequency applications [17]. All the SBL models consist of four different layers which are air, seawater, hydrocarbon (i.e., target) and sediment. Each layer is parameterized with different thicknesses such as 300 m for air, 1000 m for seawater, 200 m for hydrocarbon and 3500 m for total thickness of the sediment layer. Figure 1 shows the SBL models developed by using the CST software with three different target depths. The thickness of the overburden layer (i.e., target depth) is varied at depths of 400, 800 and 1200 m.

Based on Fig. 1, the hydrocarbon layer is positioned within the sediment layer. The total height, width and length of the SBL models are 5000 m, 20,000 m and 20,000 m, respectively. The EM source (i.e., transmitter) is located at the centre of the models and 30 m above the seafloor. For every depth, five different target resistivities are observed which are 30, 150, 270, 390 and 510 Ωm. Hence, five CST datasets are acquired for every target depth. The other input parameters used for the simulation of the SBL models are tabulated in Table 1.

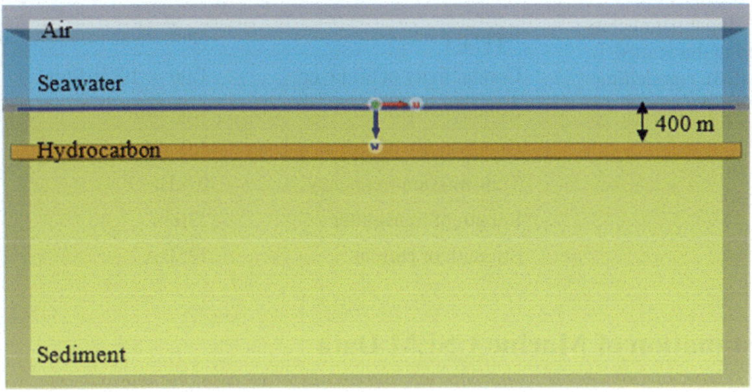

(a)

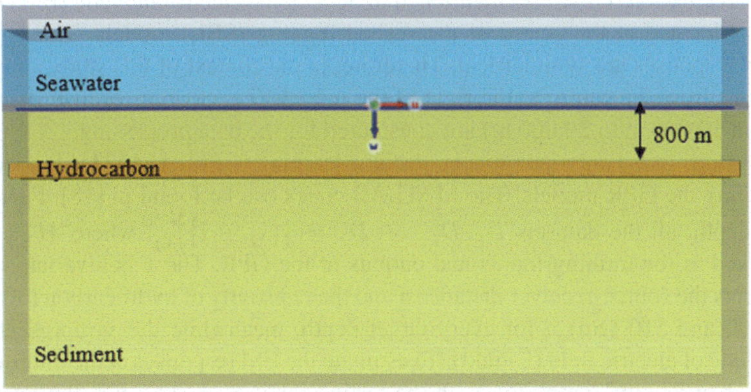

(b)

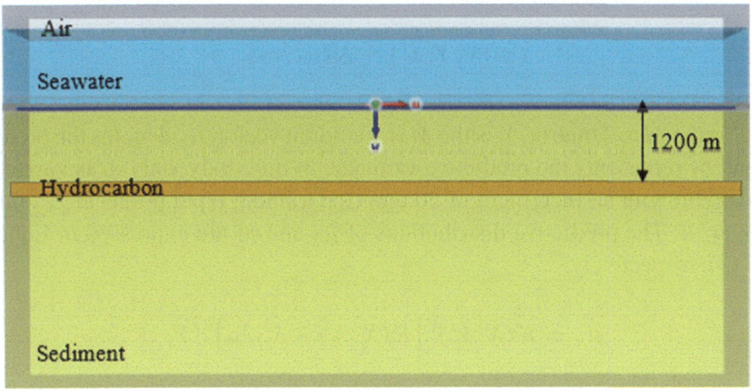

(c)

Fig. 1 Seabed logging (SBL) models replicated by the computer simulation technology (CST) software at three different target depths: **a** 400 m, **b** 800 m and **c** 1200 m

Table 1 Input parameters exercised in the seabed logging (SBL) modelling

Input parameters	Numerical values (in unit)
Conductivity of air layer	1.00×10^{-11} S/m
Conductivity of seawater layer	1.63 S/m
Conductivity of sediment layer	1.00 S/m
Transmission frequency	0.5 Hz
Length of transmitter	270 m
Strength of current	1250 A

4 Estimation of Marine CSEM Data

All the generated datasets are pre-processed before being utilized by the Gaussian process regression (GPR). As mentioned in Sect. 3, the electromagnetic (EM) transmitter is located at the centre of the seabed logging (SBL) models. This implies the EM responses are symmetrical. Therefore, in the context of this study, only EM responses from the centre to the right of the models (i.e., source-receiver distances are from 10,000 m to 20,000 m) are considered for the data processing.

This study uses Gaussian process machine learning (GPML) MATLAB codes for developing the GPR models. The MATLAB codes can be found in [18]. For every target depth, all the datasets $D_1, D_2, \ldots, D_5 = \{(x_j, y_j)\}_{j=1}^{M}$, where $M = 101$, are served as the training inputs and outputs of the GPR. The x is bivariate which represents the source-receiver distance u and the resistivity of hydrocarbon (30, 150, 270, 390 and 510 Ωm) v for every target depth, meanwhile the y represents the magnitude of electric field (E-field). To estimate the EM responses at the unobserved inputs of resistivity complying with the testing input $x_* = (u_*, v_*)$, the posterior distribution of the y_* is defined in Equation (1).

$$(y_* | X, Y, x_*) \sim N(\mu_*, \sigma_*), \tag{1}$$

where X is the $M \times 2$ matrix, Y is the $M \times 1$ column vector, μ_* denotes the predictive mean and σ_* represents the predictive variance. In this study, only v_* is varied from 30 to 510 Ωm with an increment of 20 Ωm (the training input points are excluded), while $u_* = u$. The predictive distributions of μ_* and σ_* are expressed in Equations (2) and (3), respectively.

$$\mu_* = K(X, x_*)^T \left[K(X, X) + \delta_M^2 I_M\right]^{-1} Y, \tag{2}$$

and

$$\sigma_* = k(x_*, x_*) - K(X, x_*)^T \left[K(X, X) + \delta_M^2 I_M\right]^{-1} K(X, x_*), \tag{3}$$

where $K(X, X)$ is the $M \times M$ covariance matrix of X, $K(X, x_*)$ is the $M \times 1$ covariance matrix of X and x_*, $k(x_*, x_*)$ is the covariance of x_* and $\delta_M^2 I_M$ is the M-dimensional covariance matrix of noise. Each element of the covariance matrices is formed of squared exponential (SE) covariance function. The SE covariance function is expressed in Equation (4).

$$k(x, x') = \gamma_f^2 \left[\exp\left(\frac{-|x - x'|^2}{2\tau^2} \right) \right], \tag{4}$$

where γ_f and τ are the hyper-parameters of the SE,$\theta = \{\gamma_f\ ,\ \tau\}$, which denote the signal variance and the length scale, respectively.

In the GPR, the θ is estimated by optimizing the negative-log marginal likelihood. The marginal likelihood is defined as follows,

$$-\log[p(Y|X, \theta)] = \frac{1}{2}\Big[M \log(2\pi) + \log\big|K(X, X) + \delta_M^2 I_M\big| + Y^T \big[K(X, X) + \delta_M^2 I_M\big]^{-1} Y\Big]. \tag{5}$$

The estimates of EM responses and its variances are then computed by using Equations (2) and (3) after the θ has successfully been optimized. The μ_* is the estimate of the magnitude of E-field at x_*, while the σ_* is the confidence measure of the estimate. In the context of this study, two standard deviations are chosen as the confidence interval. The general methodological flow of estimating the EM responses at the unobserved target resistivity using the GPR is depicted in Fig. 2.

5 Results and Discussion

All the seabed logging (SBL) models were developed using the computer simulation technology (CST) software. All the pre-processed datasets (five datasets per target depth) are plotted in Fig. 3.

In Fig. 3, the x-axis denotes the source-receiver distance and the y-axis is the log10 of magnitude of electric field (E-field). Based on the figure, the behaviour of the datasets with target resistivities of 150, 270, 390 and 510 Ωm is almost similar. Meanwhile the blue points (i.e., dataset with resistivity of 30 Ωm) deviate from those four datasets. This is because the high-resistive layers (e.g., 150, 270, 390 and 510 Ωm) guide the electromagnetic (EM) signal along the layers over longer distance with lower attenuation rate compared to the layer that has lower resistivity (e.g., 30 Ωm). The total time taken by the CST software to generate all the 15 datasets was approximately 45 min (~15 min for five datasets per depth). All datasets presented in Fig. 3 were fully utilized by the Gaussian process regression (GPR) for estimating the EM responses at the unobserved target resistivities. A two-dimensional (2D)

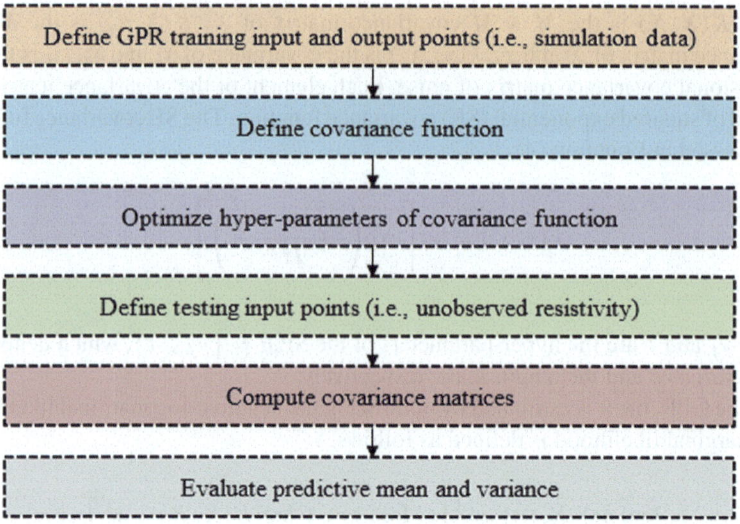

Fig. 2 Gaussian process regression (GPR) methodological flow to estimate the electromagnetic (EM) responses

GPR model for every target depth was developed based on the computer simulation outputs. The models are depicted in Fig. 4.

Based on Fig. 4, the x-axis is the source-receiver distance, the y-axis is the unobserved target resistivity from 30 to 510 Ωm with a gap of 20 Ωm each (the training input points were excluded) and the log10 of magnitudes of E-field are presented in colour codes. These 2D GPR models provide new information of the EM responses for 20 unobserved target resistivities for every target depth with low computational efforts. In average, only approximately 0.17 min (~10 s) was needed by the GPR to produce the 2D GPR model for every depth. In other words, approximately 45 min was needed for the GPR prior information acquisition using the CST software and approximately 0.5 min was needed for developing the 2D GPR models. Meanwhile, the CST software could consume approximately 225 min to generate all the datasets with information of 25 target resistivities (30–510 Ωm with a gap of 20 Ωm each) for every target depth (i.e., 75 datasets for three different target depths).

Besides comparing the time consumption, the performance of the GPR modelling in terms of accuracy also was determined. Three different error measurements which are mean absolute deviation (MAD), mean squared error (MSE) and root mean squared error (RMSE) were calculated using Equations (6), (7) and (8), respectively.

$$MAD = \frac{1}{M}\sum |y_{FE} - y_{GPR}|, \tag{6}$$

$$MSE = \frac{1}{M}\sum (y_{FE} - y_{GPR})^2, \tag{7}$$

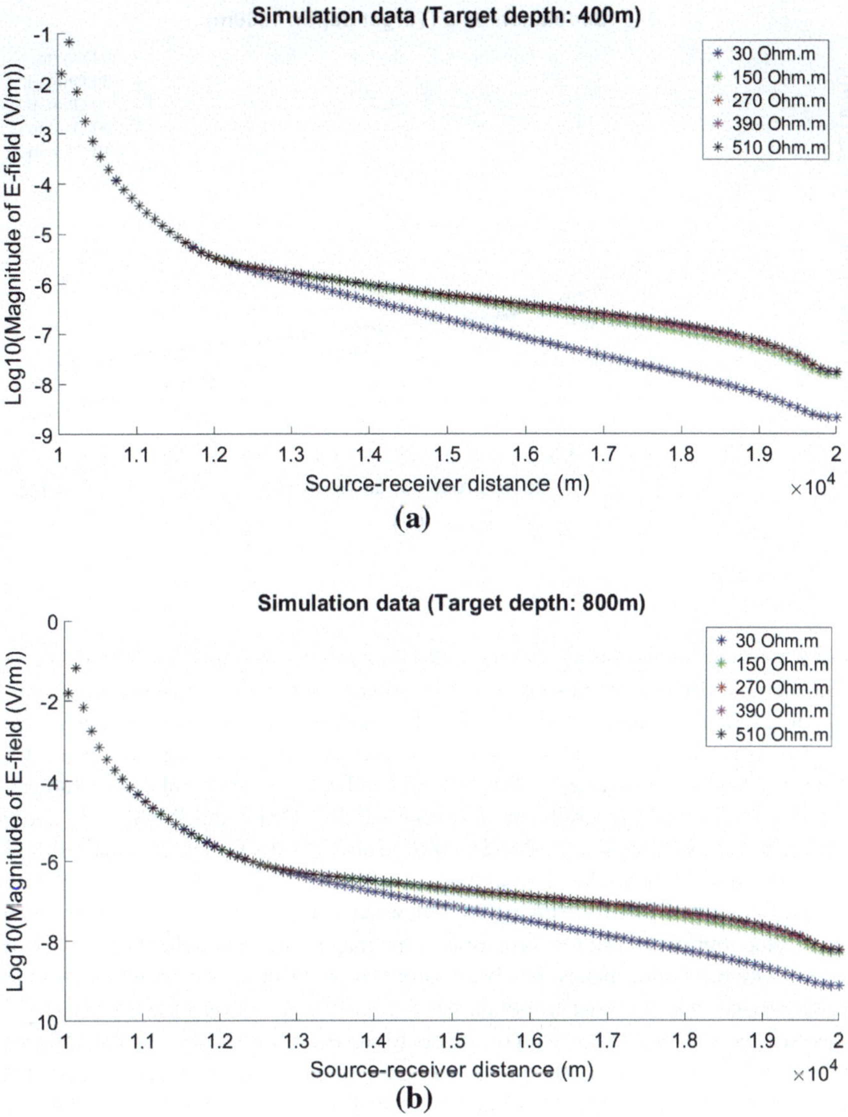

Fig. 3 Computer simulation outputs of five different target resistivities for three different target depths: **a** 400 m, **b** 800 m and **c** 1200 m

$$RMSE = \sqrt{\frac{1}{M}\sum(y_{FE} \quad y_{GPR})^2}, \tag{8}$$

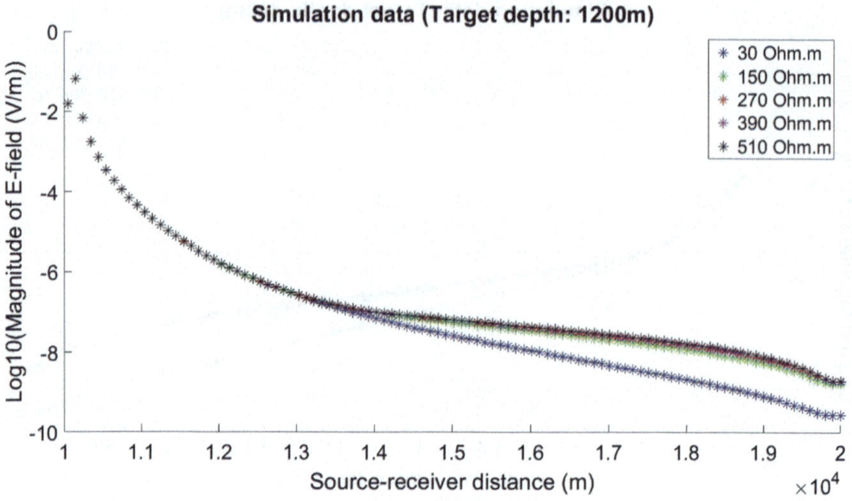

(c)

Fig. 3 (continued)

where y_{FE} represents the true value of the magnitude of E-field evaluated by the finite element (FE) method using the CST software and y_{GPR} represents the magnitude of E-field estimated by the GPR. For demonstration purposes, four unobserved target resistivities were randomly selected. The resulting error measurements for unobserved target resistivities of 90, 210, 330 and 450 Ωm are tabulated in Table 2.

Based on Table 2, generally the resulting MADs, MSEs and RMSEs for every target depth are very small. The MADs, MSEs, and RMSEs for target resistivities of 210, 330 and 450 Ωm are lesser than 0.05. From the table, the errors for dataset with target resistivity of 90 Ωm are bit higher compared to the others. This is highly related to the behaviour of the prior information of the magnitude of E-field at resistivity of 30 Ωm. The behaviour makes the GPR requires more prior information that close to the resistivity value in order to obtain better accuracy for lower target resistivities. However, the differences between the errors for resistivities of 210, 330 and 450 Ωm and errors for resistivity of 30 Ωm are not significant. This reveals that the 2D GPR models can significantly provide reliable information of EM responses at the untried input parameters with low errors.

6 Summary

Seabed logging (SBL) models with three different target depths were developed, and the electromagnetic (EM) responses were generated using the computer simulation technology (CST) software. Three different target depths were tested, and five target resistivities for every depth were observed. All the generated datasets were utilized

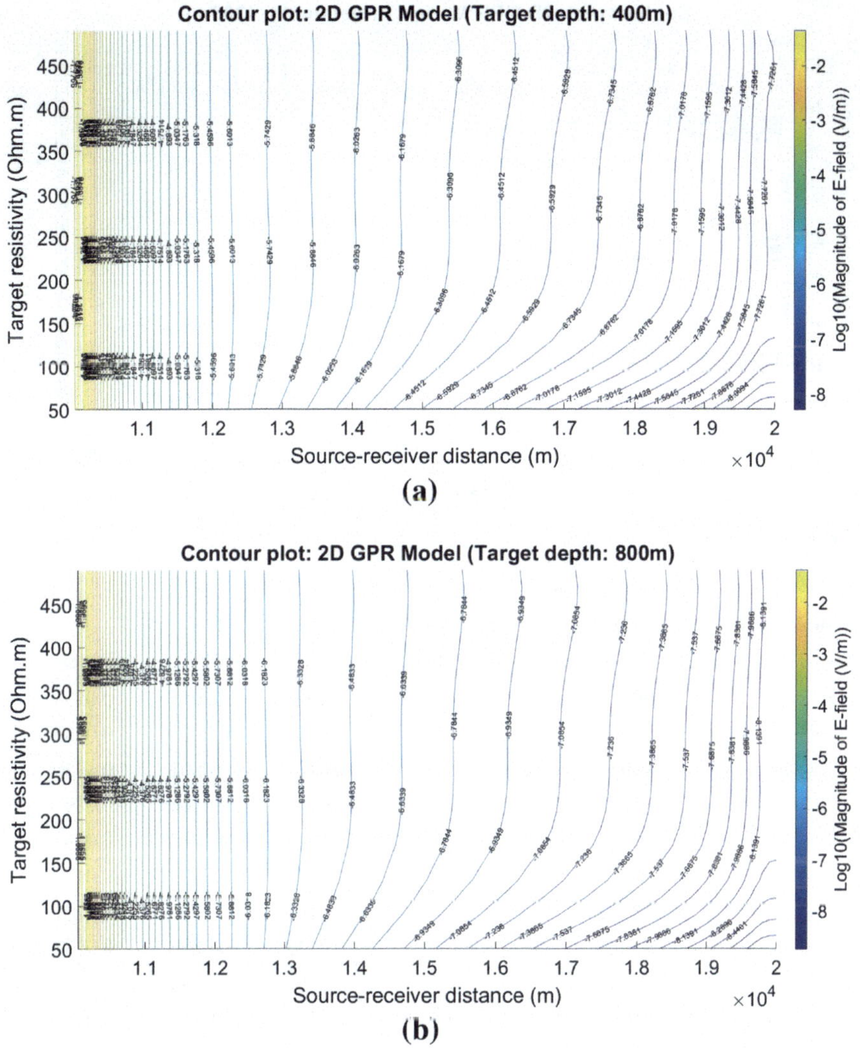

Fig. 4 Two-dimensional (2D) Gaussian process regression (GPR) model for three different target depths: **a** 400 m, **b** 800 m and **c** 1200 m

by the Gaussian process regression (GPR) as its prior information. For every target depth, a two-dimensional (2D) GPR model was successfully developed using the MATLAB software. EM profiles with 20 unobserved target resistivities (30–510 Ωm with an increment of 20 Ωm) were modelled by the GPR. Simulation of all 25 datasets (observed and unobserved) using the CST software could consume approximately 225 min (for three different target depths), while GPR took approximately 0.5 min to evaluate the EM responses for 20 unobserved target resistivities and needed 45 min for the prior data acquisition using the CST software. Besides, the mean absolute

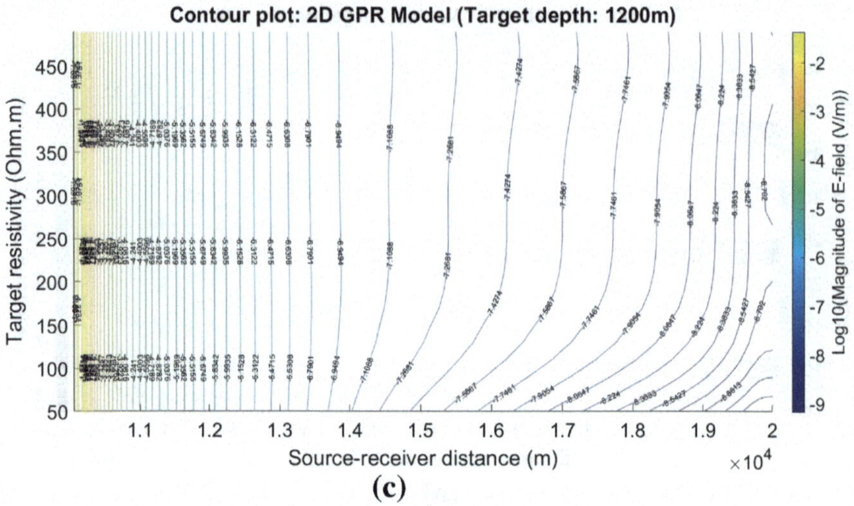

Fig. 4 (continued)

Table 2 Error measurements for four random unobserved target resistivities

Target depth (m)	Unobserved resistivity (Ωm)	MAD	MSE	RMSE
400	90	0.0618	0.0055	0.0738
	210	0.0154	0.0003	0.0185
	330	0.0111	0.0002	0.1450
	450	0.0159	0.0004	0.0192
800	90	0.0564	0.0048	0.0692
	210	0.0144	0.0003	0.0178
	330	0.0105	0.0002	0.0141
	450	0.0149	0.0003	0.0185
1200	90	0.0499	0.0040	0.0634
	210	0.0132	0.0003	0.0169
	330	0.0097	0.0002	0.0136
	450	0.0136	0.0003	0.0175

deviation (MAD), mean squared error (MSE) and root mean squared error (RMSE) between the true values of the magnitude of electric field (modelled by the finite element method) and the magnitudes estimated by the GPR also were computed, and all the computed deviations were very small. This proved that the performance of the GPR in evaluating the EM responses was at par to the current method but with significant improvement on the computational efforts.

Acknowledgements The authors would like to thank Universiti Teknologi PETRONAS (UTP) for the supports.

References

1. Eidesmo, T., Ellingsrud, S., MacGregor, L.M., Constable, S., Sinha, M.C., Johansen, S., Kong, F.N., Westerdahl, H.: Sea bed logging (SBL), a new method for remote and direct identification of hydrocarbon filled layers in deep water areas. First Break **20**, 144–152 (2002)
2. Ellingsrud, S., Eidesmo, T., Johansen, S.: Remote sensing of hydrocarbon layers by seabed logging: results from a cruise offshore Angola. Lead. Edge **21**, 972–982 (2002)
3. Daud, H., Yahya, N., Sagayan V., Talib, A.M.: Magnitude vs low frequency EM waves for sea bed logging applications. In Proceeding of 2010 IEEE Asia-Pacific Conference on Applied Electromagnetics (APACE), 9–11 Feb 2010, Port Dickson, Malaysia, pp. 1–6, 2010
4. Cai, H., Xiong, B., Han, M., Zhdanov, M.: 3D controlled-source electromagnetic modeling in anisotropic medium using edge-based finite element method. Comput. Geosci. **73**, 164–176 (2014)
5. Li, Y., Key, K.: 2D marine controlled-source electromagnetic modeling: part 1—an adaptive finite-element algorithm. Geophysics **75**, WA51–WA62 (2007)
6. Bakr, S.A., Pardo, D., Mannseth, T.: Domain decomposition Fourier finite element method for the simulation of 3D marine CSEM measurements. J. Comput. Phys. **255**, 456–470 (2013)
7. Fang, D., Zhang, X., Yu, Q., Jin, T.C., Tian, L.: A novel method for carbon dioxide emission forecasting based on improved Gaussian processes regression. J. Clean. Prod. **173**, 143–150 (2018)
8. Kong, D., Chen, Y., Li, N.: Gaussian process regression for tool wear prediction. Mech. Syst. Signal. Process. **104**, 556–574 (2018)
9. Rasmussen, C.E., Williams, C.K.I.: Gaussian Processes for Machine Learning. MIT Press, Cambridge, MA, USA (2006)
10. Rasmussen, C.E., Nickisch, H.: Gaussian processes for machine learning (GPML) toolbox. J. Mach. Learn. Res. **11**, 3011–3015 (2010)
11. Harari, O., Steinberg, D.M.: Optimal designs for Gaussian process models I via spectral decomposition. J. Stat. Plan. Inference **154**, 87–101 (2014)
12. Asante-Okyere, S., Shen, C., Ziggah, Y.Y., Rulegeya, M.M., Zhu, X.: Investigating the predictive performance of Gaussian process regression in evaluating reservoir porosity and permeability. Energies **11**, 3261 (2018)
13. Mohd Aris, M.N., Daud, H., Mohd Noh, K.A., Dass, S.C., Mukhtar, S.M.: Modelling 1-D synthetic seabed logging data for thin hydrocarbon detection: an application of Gaussian process. In: AIP Conference Proceedings 2266, 090002 (2020): https://doi.org/10.1063/5.0018105
14. Li, J., He, M., Cui, G., Wang, X., Wang, W., Wang, J.: A novel method of seismic signal detection using waveform features. Appl. Sci. **10**, 2919 (2020)
15. Mohd Aris, M.N., Daud, H., Dass, S.C., Mohd Noh, K.A.: Gaussian process for hydrocarbon depth estimation in forward modelling of seabed logging. J. Environ. Eng. Geophys. **24**, 399–408 (2019)
16. Daud, H., Mohd Aris, M.N., Mohd Noh, K.A., Dass, S.C.: A novel methodology for hydrocarbon depth prediction in seabed logging: Gaussian process-based inverse modelling of electromagnetic data. Appl. Sci. **11**, 1492 (2021)
17. Electromagnetic Simulation Solvers (CST Studio Suite): Available online: https://www.3ds.com/products-services/simulia/products/cst-studio-suite/solvers/. Accessed 02 Nov 2020
18. Documentation for GPML Matlab Code Version 4.2: Available online: http://www.gaussianprocess.org/gpml/code/matlab/doc/. Accessed 9 Sept 2020

Stochastic Process Algebra Model of Amyloidogenic Processing in Alzheimer's Disease

Angelyn Lao, Michelle Andrea Dela Cruz, Grazielle Joy Tenio, Polly Sy, and Eduardo Mendoza

Abstract Studies based on the amyloid hypothesis have shown that the proteolytic breakdown of amyloidogenic processing is the main pathology of Alzheimer's disease (AD). In this study, we focused on the effect of receptor SORLA, a 230 kDa type-1 transmembrane glycoprotein, after binding with the amyloid precursor protein (APP) in the Trans-Golgi Network (TGN). We built a Stochastic Process Algebra (SPA) model to capture receptor SORLA's behavior and understand its influence in the APP processing. Through the SPA modeling approach, the amyloidogenic processing is modeled as concurrent systems in continuous time Markov chains. We fitted the simulations of our model to the data published by Schmidt and colleagues in 2012. We built a model that initially considers the amyloidogenic processing to occur only in the endosomes. Then, we extended it so that the amyloidogenic processing also occurs in the TGN. In our study, we are able to validate the hypothesis proposed in Willnow and Spoelgen's study that there might be an indirect interaction between SORLA and—secretase. This indirect interaction takes place when APP and the receptor SORLA reverses, the APP that unbinds with SORLA may be cleaved by

A. Lao (✉) · M. A. Dela Cruz · G. J. Tenio · E. Mendoza
Mathematics and Statistics Department, De La Salle University, 2401 Taft Avenue, 0922 Manila, Philippines
e-mail: angelyn.lao@dlsu.edu.ph

M. A. Dela Cruz
e-mail: michelle_andrea_delacruz@dlsu.edu.ph

G. J. Tenio
e-mail: grazielle_tenio@dlsu.edu.ph

E. Mendoza
e-mail: eduardo.mendoza@dlsu.edu.ph

P. Sy
Institute of Mathematics, University of the Philippines Diliman, 1101 Diliman, Quezon City, Philippines

A. Lao · E. Mendoza
Systems and Computational Biology Research Unit, CENSER, De La Salle University, 2401 Taft Avenue, 0922 Manila, Philippines

© Institute of Technology PETRONAS Sdn Bhd 2022

S. A. Abdul Karim and A. Shafie (eds.), *Towards Intelligent Systems Modeling and Simulation*, Studies in Systems, Decision and Control 383,
https://doi.org/10.1007/978-3-030-79606-8_18

the—secretase in the TGN. Our SPA model brought new insights about SORLA's effect on the amyloidogenic processing, particularly to—secretase in the TGN.

Keywords Alzheimer's disease · APP processing · Amyloidogenic processing · SPiM · Stochastic process algebra · SORLA

Studies based on the amyloid hypothesis have shown that the proteolytic breakdown of amyloidogenic processing is the main pathology of Alzheimer's disease (AD). This study is focused on the effect of SORLA after binding with the amyloid precursor protein (APP). A Stochastic Process Algebra (SPA) model is established to capture SORLA's behavior and understand its influence in the APP processing. Through SPA modeling approach, the amyloidogenic processing is modeled as concurrent systems in continuous time Markov chains. The model is fitted to the experimental data. The simulation results validated the hypothesis regarding the indirect interaction between SORLA and β-secretase. This interaction takes place when APP and SORLA reverses, the APP that unbinds with SORLA may be cleaved by the β-secretase in the Trans Golgi Network (TGN). The SPA model brought new insights about SORLA's effect on the amyloidogenic processing, particularly to β-secretase in the TGN.

1 Introduction

Alzheimer's disease (AD), a neurodegenerative disorder that leads to cognitive function impairment [1]. Several studies have been made to understand this disease and stop it from progressing. However, new discoveries lead to more questions that required further research.

Amyloid Precursor Protein (APP), a type-1 transmembrane protein expressed in neurons, is closely linked to the pathogenesis of AD. In an APP processing (Fig. 1a), APP travels from the Trans Golgi Network (TGN) to the cell surface where most are cleaved by α-secretase producing soluble APP alpha (sAPP α) and a membrane-anchored fragment C83. APP that are not cleaved by the α-secretase continue to travel to the endosomes where APP is further cleaved by the β-secretase forming soluble APP beta (sAPP β) and a membrane-anchored fragment C99 [6, 11]. Both products of the two secretases undergo further cleavage by γ-secretase but, the cleavage of APP by the β-secretase is the only one that produces amyloid beta (A β) peptides which causes the formation of plaques and eventually results to neurodegeneration [3, 14].

Sorting protein-related receptor (SORLA), a 230 kDa type-1 transmembrane glycoprotein found in the brain, is a member of the VPS10-domain receptor family which adopts the structure of a large tunnel involved in peptide ligands binding. SORLA is a pro-receptor with 53 amino acid pro-peptide produced in the cell. Its

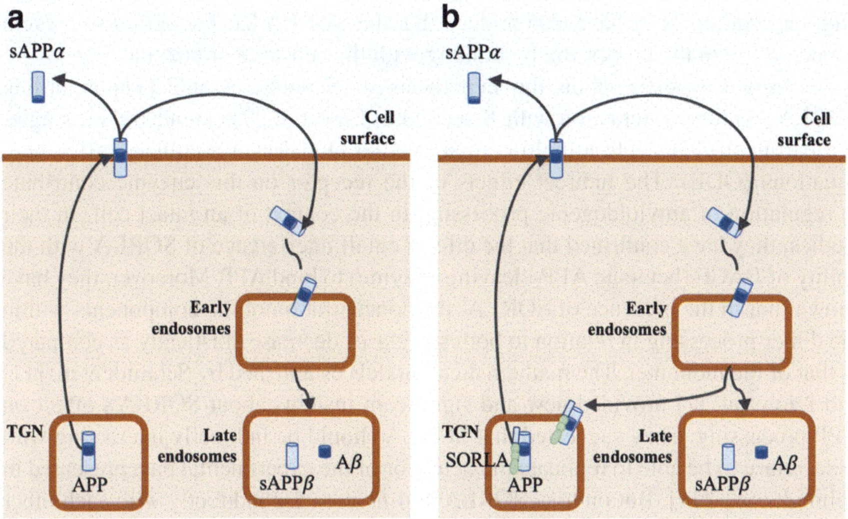

Fig. 1 A comprehensive diagram of APP processing in the absence (**a**) and presence (**b**) of SORLA. **a** From the TGN, APP travels to the cell surface (1) where it can be cleaved by α-secretase to produce sAPP α. The APP not cleaved by α-secretase would then continue to travel to the early endosome (2) where it can be cleaved by β-secretase and advance to the late endosomes. **b** From the TGN, APP binds with SORLA (1) only 10% or less are able to travel to the cell surface (2) where it can be cleaved by α-secretase to produce sAPP α. The APP not cleaved by α-secretase would then continue to travel to the early endosomes (3) where it can either be cleaved by β-secretase and advance to the late endosomes (Retromer) or revert to the TGN (4)

cleavage by the convertases found in the TGN activates the receptor that allows it to bind with APP [13].

Under the influence of SORLA (shown in Fig. 1b), APP is bound in the TGN. Only 10% or less of the free APP is transported to the cell surface where most are cleaved by the α-secretase. Henceforth, only a small amount of APP will reach the endosomes (early to late), which reduces the probability of forming the A β peptides [6, 7]. There is also the possibility that the small amount of APP in the early endosome returns to the TGN where the reverse of the interaction of APP and SORLA occurs. The APP that unbinds with SORLA (or free APP) may then proceed to amyloidogenic processing [13].

The first kinetic model for the amyloidogenic processing under the influence of SORLA was developed by Schmidt et al. [11]. In their study, they have shown that the secretases are allosteric enzymes that require cooperation from APP oligomers. Allosteric enzymes have the ability to respond differently in accordance to their environment, that is why each secretase yield different products. Cooperativity between APP and the secretases lead to prompt adaptive changes in enzyme activity making room for efficient processing. They have learned that in the presence of SORLA the secretases change from allosteric to non-allosteric enzymes, thus blocking APP oligomerization. They have also noticed that the SORLA has the ability to prevent cooperativity in the processing by binding with the substrate or enzyme.

To further investigate on the hypothesis of Schmidt et al.'s [11] regarding SORLA's indirect interaction with β-secretase, Lao et al. [7] extended their single-compartment model into a multi-compartmental model using ordinary differential equations (ODE). The indirect effects of the receptor on this enzyme contribute to regulation of amyloidogenic processing in the context of an intact cell. In their studies, they have confirmed that the effects entail interference of SORLA with the ability of BACE (beta-site APP-cleaving enzyme) to bind APP. Moreover, they have shown that in the presence of SORLA, the concentration of the components within the dimer processing in relation to both secretases decrease drastically as compared to that of the monomer. The mathematical models established by Schmidt et al. [11] and Lao et al. [7] provided new and significant insights about SORLA's effect on APP processing. They suggested that SORLA should be indirectly interacting with β-secretase to be able to replicate the behavior of the experimental data presented in Schmidt et al. [11]. But on how SORLA and β-secretase indirectly with each other interact needs further investigations.

Interesting hypothesis regarding the interactions between SORLA and β-secretase are proposed [12, 13]. Spoelgen et al. [12] suggested the possibility of SORLA interacting with β-secretase through coimmunoprecipitation (use to analyze protein–protein interaction using an antibody), predominantly in the TGN; thus the competition between APP and β-secretase for binding with SORLA in the TGN is probable. Similarly, Willnow and Andersen [13] also suggested that the amyloidogenic processing does not only occur in the endosomes but can also be generated in the TGN. The APP that was not cleaved in the early endosome may be transported to the late endosome or revert to the TGN; either way those APP may be cleaved by β-secretase which results to the increase of A β peptides (see Fig. 2).

With SORLA in the system, APP is prevented from travelling to different compartments and reduces the amyloidogenic processing in the TGN.

This work extended the studies of Schmidt et al. [11] and Lao et al. [7] by integrating the hypotheses proposed by Willnow and Andersen [13] and Spoelgen et al. [12]. Instead of using a deterministic ODE model, the relationship of SORLA and β-secretase is elucidated using a Stochastic Process Algebra (SPA) modeling approach.

In this study, SPA is chosen as the modeling approach because of its known strength to formally define compositionality, i.e. the ability to model a system as the interaction of its subsystem [2]. This feature allows a system to be modelled explicitly and systematically making it clear and easier to understand. It also allows the mathematical models to be rigorously derived [2, 4]. Besides compositionality, SPA presents features such as formality and abstraction. Formality provides a precise definition to all the terms in the language. Abstraction on the other hand allows the construction of complex models using detailed components excluding internal behavior when possible [2].

The scope of this study revolves around the first cleavage of the two secretases in APP processing. The main objective of the study is to build a SPA model that

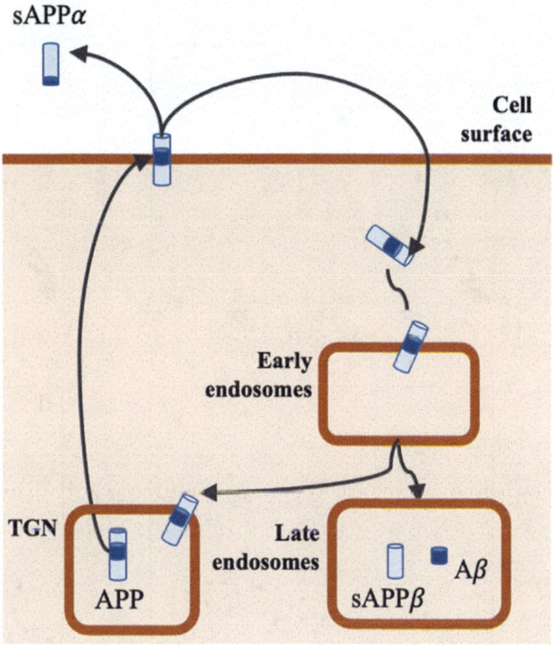

Fig. 2 APP Processing in the absence of SORLA with amyloidogenic processing in the TGN. From the TGN, APP travels to the cell surface where it can be cleaved by α-secretase to produce sAPP α. The APP not cleaved by α-secretase would then advance to the early endosome where it can be cleaved by β-secretase. Those that were not cleaved in the early endosome may continue to the late endosomes or revert to the TGN; either way both has the possibility to be cleaved by the by β-secretase

will capture the behavior of APP processing in the presence and absence of SORLA extending the studies of Schmidt et al. [11] and Lao et al. [7]. Modelling the APP processing under the influence of SORLA using SPA would bring forth, if not new, a better understanding of the amyloidogenic processing and its relationship with SORLA.

2 Experimental Data

The data used in this study is the dose response data published by Schmidt et al. [11]. They utilized the Tet-off mechanism, using enzyme-linked immunosorbent assay (ELISA), to elucidate APP and sAPP products in the presence and absence of SORLA. In the absence of SORLA, both sAPP α (Fig. 3a) and sAPP β (Fig. 3b) exhibits an increasing sigmoidal behavior. The maximum value of sAPP α and sAPP β is approximately 225 and 10 femtomole (10^{-15}), respectively. In the presence of SORLA, the production of sAPP α (Fig. 3c) reduces by 95%, while sAPP β (Fig. 3d) reduces only by 50%.

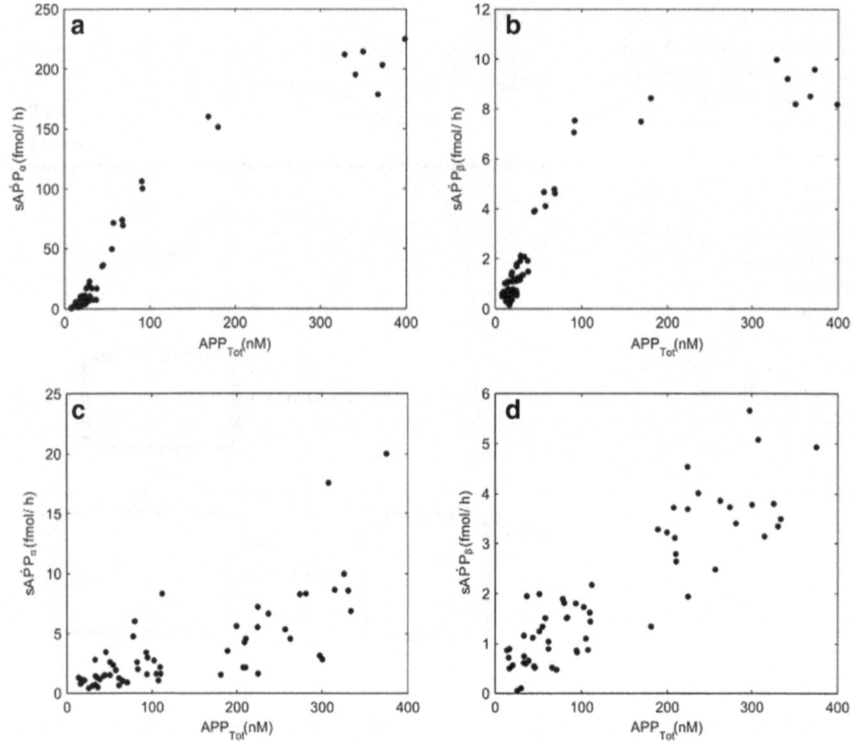

Fig. 3 The data generated from the dose–response experimental design. In the absence of SORLA (**A**, **B**), sAPP α and sAPP β exhibit an increasing sigmoidal curve wherein curves rise up to approximately 225 fmol/nM and 10 fmol/nM, respectively. In the presence of SORLA (**C**, **D**), sAPP α decreased by about 95.56% while sAPP β went down by 50%. Data are taken from [11]

3 Stochastic Process Algebra

We used Visual SPiM (Stochastic Pi Machine), an advanced programming language for designing and simulating computer models of biological systems as a collection of modular processes that communicates with each other through message-passing. It all started with pi-calculus, developed by Milner, Parrow and Walker [8], with the aim of modeling computer networks that can dynamically remodel themselves over time. Due to the complexity of the network, stochasticity was introduced [9, 10]. It has now evolved to what is known as SPA. A stochastic model possesses randomness which estimates the probability distributions and produces different outputs given the same set of parameter values and initial conditions by allowing random variation in one or more inputs over time.

SPA offers the concept of randomness and explicitly shows the behavior of each component in the system. One unique feature of SPA modeling is that it allows compositionality that offers several benefits for model construction. The systems can be modelled independently enabling direct composition of simple components that can be reconstructed to larger models. Moreover, it also enables mathematical analysis of systems which could possibly provide comprehension about the underlying properties of biological systems [8].

We ran 100 iterations to each model simulation, both in the presence and absence of SORLA, to see if it would fit the data generated by Schmidt et al. [11]. We considered it a good fit in the absence of SORLA, if the steady state values of sAPP α and sAPP β are within the range of 160–240 and 7–15, respectively. In the presence of SORLA, a fit is good if it is within the range of 5–12 for sAPP α and 3–6 for sAPP β. We utilized the z-test for one proportion at significance level of 5% to analyze if at least 60% of sAPP α and sAPP β simulations are of good fit, i.e. H_o: $\Pi \geq 0.6$ versus H_1: $\Pi < 0.6$.

4 SPA Model of Amyloidogenic Processing Influenced by SORLA

There are several studies that suggest possible interaction between SORLA and β-secretase [7, 11, 13]. In this study, investigation of the hypotheses proposed by Spoelgen et al. [12] and Willnow and Andersen [13] are conducted through SPA modeling. As mentioned earlier, this approach will allow us to easily reconstruct the amyloidogenic processing model under the influence of SORLA, by combining the components represented by the TGN, cell surface and endosomal compartments.

The SPA models of amyloidogenic processing influenced by SORLA are built under the SPiM environment. The initial SPA model, an amyloidogenic processing model (shown in Fig. 4), is built based on the model of Schmidt et al. [11].

To investigate on the proposed interaction of β-secretase and SORLA by Willnow and Andersen [13], we modified and extended the initial model (shown in Fig. 5) by including the possibility that some APP in the early endosome returns to the TGN where the reverse interaction of APP and SORLA occurs so that the free APP may proceed to amyloidogenic processing.

In our initial SPA model (Fig. 4), we first set the directive sample to 0.3 and specified the maximum number of plots to 3000 with minimum time interval of 0.0001 (encoded in Line (1)). The initial value of **APP**, **alpha**, **beta** and **SORLA** are set to 225, 180, 5 and 215, respectively (see Line (22)). Then, the components to be plotted at the end of each simulation are declared in Line (2).

In SPiM, delay actions and outputs of communicating actions occur after an exponentially distributed time with constant rate. Thus we can say that these actions are equivalent to the transitions in a continuous-time Markov chain (CTMC). It also computes for the probability of transitions of APP processing in the presence and absence of SORLA using conditional probability with a constant rate [5].

The four channels that represent the transition rate from one state to another state are as follows: **alkene_a**, **share_AppSorla**, e, and **alkene_b.** The transition

```
(1)   directive sample 0.3 3000
(2)   directive plot sAPPalpha(); sAPPbeta(); APP()
(3)
(4)   new alkene_a@0.10:chan
(5)   new share_AppSorla@10000000.0:chan(chan)
(6)   new e@0.1:chan()
(7)   new alkene_b@0.10:chan
(8)
(9)   let alpha() = ?alkene_a; alpha()
(10)  let beta() = ?alkene_b; beta()
(11)  let SORLA() = ?share_AppSorla(e);SORLA_bound(e)
(12)  and SORLA_bound(e:chan) = ?e; SORLA()
(13)
(14)  let APP() =
(15)  do !alkene_a; sAPPalpha()
(16)  or !alkene_b; sAPPbeta()
(17)  or !share_AppSorla(e);APP_Bound(e)
(18)  and APP_Bound(e:chan) = !e;APP()
(19)  and sAPPalpha() = ()
(20)  and sAPPbeta() = ()
(21)
(22)  run(225 of APP() | 180 of alpha() | 5 of beta() | 215 of SORLA())
```

Fig. 4 Codes of initial amyloidogenic processing model. Process **APP** may send on **share_AppSorla**, **alkene_a** or **alkene_b** that would execute the process **sAPPalpha, sAPPbeta** or **APP_Bound(e)**, respectively. When sAPPalpha or **sAPPbeta** is chosen, the process terminates. While when **APP_Bound(e)** is chosen, the process sends on channel e to perform process **APP**

rates for all channels are assumed to be equivalent to 0.10. Unlike in the absence of SORLA, the model in the presence of SORLA did not fit the data. Henceforth, we systematically incremented the transition rate of **share_AppSorla** until the model fitted the data.

After a series of systematic calibration, the transition rates of the four channels namely, **alkene_a**, **share_AppSorla**, **e**, and **alkene_b** are now set to 0.10, 10000000.0, 0.10 and 0.10, respectively (see Lines (4) to (7)). These channels receive from their respective processes to execute their designated process. For example, **alpha** will receive command signal from **alkene_a** channel and execute the **alpha** process (shown in Line (9)). Similar command applies to **beta** (shown in Line (10)) and **SORLA** (shown in Line (11)).

From Lines (14) to (20), **APP** undergoes one of the three different actions. APP could either be bound to SORLA yielding $C_{APPSORLA}$ (see Fig. 4 Line (17)), be cleaved by α-secretase forming sAPP α (see Fig. 4 Line (15)) or be cleaved by β-secretase creating sAPP β (see Fig. 4 Line (16)). APP and SORLA have a reversible interaction which means they could either go back to original state (see Fig. 4 Line (18)) or remain being bound with SORLA.

```
(1)   directive sample 0.05 5000
(2)   directive plot sAPPalpha(); sAPPbeta(); APP()
(3)
(4)   new alkene_a@0.10:chan
(5)   new share_AppSorla@10000000.0:chan(chan)
(6)   new e@0.1:chan()
(7)   new alkene_b@0.10:chan
(8)   new tgn_b@0.0003247:chan
(9)
(10)  let alpha() = ?alkene_a; alpha()
(11)  let beta() = ?alkene_b; beta()
(12)  let SORLA() = ?share_AppSorla(e);SORLA_bound(e)
(13)  and SORLA_bound(e:chan) = do ?e; SORLA()
(14)  or ?tgn_b; SORLA()
(15)
(16)  let APP() =
(17)  do !share_AppSorla(e);APP_Bound(e)
(18)  or !alkene_a; sAPPalpha()
(19)  or !alkene_b; sAPPbeta()
(20)  and APP_Bound(e:chan) =
(21)  do !e;APP()
(22)  or !tgn_b; sAPPbeta()
(23)  and sAPPalpha() = ()
(24)  and sAPPbeta() = ()
(25)
(26)  run(225 of APP() | 180 of alpha() | 6 of beta() | 215 of SORLA())
```

Fig. 5 Codes of modified amyloidogenic processing model. Process **APP** can send on: **share_AppSorla, alkene_a or alkene_b** to perform process **APP_Bound(e), sAPPalpha** or **sAPP-beta. When APP_Bound(e)** is executed, it may send on e to execute process **APP** or send on **tgn_b** to implement process **sAPPbeta.** When either **sAPPalpha or sAPPbeta** is chosen, the process will terminate

In relation to the objective of the study that is, to further investigate on the hypothesis proposed by Willnow and Andersen [13] and Lao et al. [7], we indicated the possibility that SORLA can indirectly interact with β-secretase in the TGN compartment (shown in Line (8) Fig. 5). Those APP that unbound with SORLA could either be bound again or be cleaved by β-secretase in the TGN compartment (encoded in Line (21)–(22)).

5 Results and Discussion

The simulation results of our initial SPA model in the absence of SORLA demonstrate that APP exponentially decreases from its initial value of 225 to 0 (Fig. 6a), while both sAPP α (Fig. 6b) and sAPP β (Fig. 6c) undergo sigmoidal increase to approximately 225 and 10, respectively. In the presence of SORLA (shown in Fig. 6d–f), as APP decreases (Fig. 6d), both sAPP α (Fig. 6e) and sAPP β (Fig. 6f) also decreases by

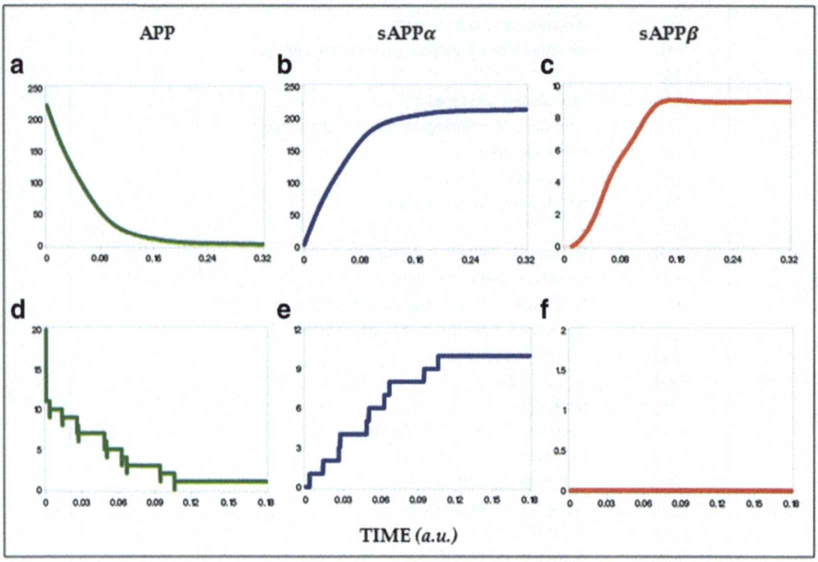

Fig. 6 Simulation results of amyloidogenic processing in the absence (**a–c**) and presence (**d–f**) of SORLA. In the absence of SORLA, **a** APP exponentially decreases from its initial value of 225 to 0; **b** sAPP α is an increasing sigmoidal curve from 0 to approximately 225; **c** sAPP β is also an increasing sigmoidal curve from 0 to roughly around 10. In the presence of SORLA, **d** APP exhibits a downward trend that eventually approaches 0; **e** sAPP α is decreased by 95.37%; **f** sAPP β is decreased by 100%

95.37% and 100%, correspondingly. Note that the amount of APP reaches zero but increases minimally over time (Fig. 6d).

After running 100 iterations of the SPA model (defined in Fig. 4) without SORLA, 100 and 60 of the simulation results fitted the sAPP α and sAPP β data, respectively. As for SPA model with SORLA, although all sAPP α simulations fitted the data but none for sAPP β simulations.

Due to the poor fit of sAPP β in the presence of SORLA, we calibrated the initial values of **APP, alpha** and **SORLA** corresponding to the changes done in **beta** (shown in Line (22) of Fig. 4), to improve the fit of the model to the data. By doing so, the closest we got is a good fit for APP, sAPP α and sAPP β in the presence of SORLA (see Fig. 7d–f); a good fit for APP and sAPP α in the absence of SORLA (see Fig. 7a, b); but a poor fit for sAPP β in the absence of SORLA (see Fig. 7c).

After modifying the initial values of APP, alpha, beta and SORLA (encoded in Line (22) Fig. 4), sAPP β in the absence of SORLA did not fit the data. In the absence of SORLA (Fig. 7a–c), APP exponentially decreases from 250 to 0, both sAPP α and sAPP β exhibits an increasing sigmoidal curve with steady state value at 200 and 50, respectively. While in the presence of SORLA (Fig. 7d–f), APP fluctuates decreasingly without reaching zero. As APP decreases, both sAPP α and sAPP β also decreases. sAPP α decreases by about 95% which is similar to the data of Schmidt

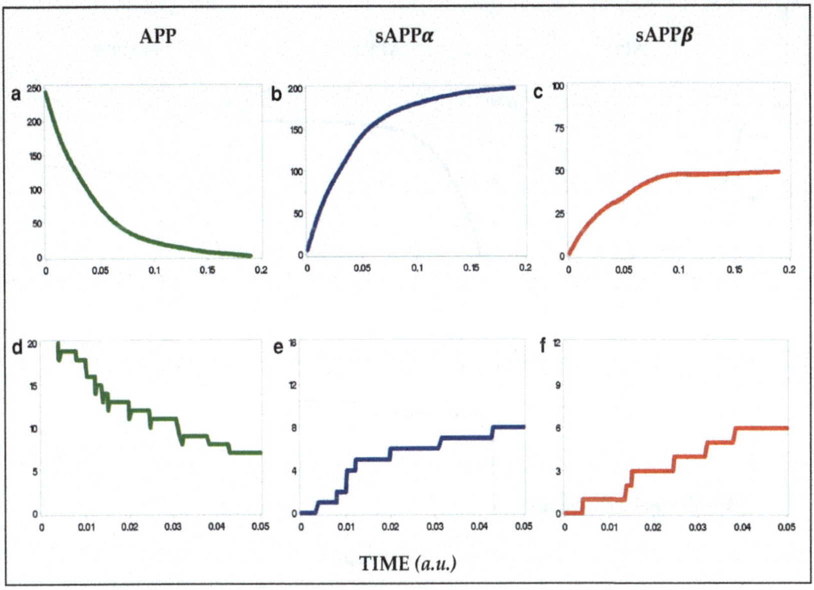

Fig. 7 Simulation results of changed of initial values in amyloidogenic processing in the absence (**a–c**) and presence (**d–f**) of SORLA. In the absence of SORLA, **a** APP exponentially decreases from 250 to 0; **b** sAPP α shows an increasing sigmoidal curve to approximately 200; **c** sAPP β has an increasing sigmoidal curve of approximately 50. In the presence of SORLA, **d** at time 0.05, APP does not reach 0 and fluctuates decreasingly; **e** sAPP α decreased by 95%; **f** sAPP β decreased by 86%

et al. [11]. However, sAPP β decreases significantly by 86% which is quite larger compared to the 50% decrease observed in the data of Schmidt et al. [11].

Using the initial model (Fig. 4), we were not able to get a good fit for sAPP β, as shown in Figs. 6 and 7. It either fitted sAPP β in the absence of SORLA and not in the presence of SORLA (Fig. 6) or vice versa (Fig. 7). Hence, in the next step, we try to improve the fit of our model to sAPP β data by calibrating the rates (Line (4)–(7) of Fig. 4) and initial value (Line (22) of Fig. 4) of the variables related to sAPP β.

After changing the rates and initial values of beta, the model simulations of sAPP β (without SORLA) still do not fit the data. We reduced the rates of **alkene_a** from 0.10 to 0.90 (Fig. 4 Line (4)) and **share_AppSorla** from 10000000.0 to 100000.0 (Fig. 4 Line (5)); increased the rate of **alkene_b** from 0.10 to 0.70 (Fig. 4 Line (7)) and remained the rate of **e** as is (Fig. 4 Line (6)). As for the initial value of beta, we increased it from 5 to 100 (Fig. 4 Line (22)).

Similar to Figs. 6 and 7, in the absence of SORLA, APP is exponentially decreasing over time from 225 to 0 (Fig. 8a); both sAPP α and sAPP β exhibits an increasing sigmoidal curve of approximately 200 (Fig. 8b) and 70 (Fig. 8c), respectively. While in the presence of SORLA, APP decreasingly fluctuates (Fig. 8d); sAPP α has decreased at about 93.79% (Fig. 8e); and sAPP β has decreased significantly by 92.19% (Fig. 8f).

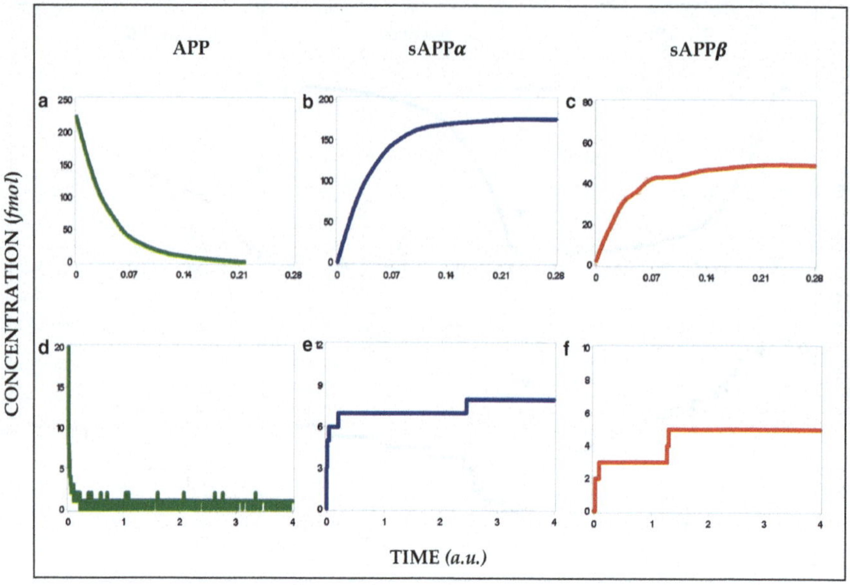

Fig. 8 Simulation results of changed rates and initial value of beta in amyloidogenic processing in the absence (**a–c**) and presence (**d–f**) of SORLA. In the absence of SORLA, **a** APP is decreasing exponentially from 225 to 0; **b** sAPP α has an increasing sigmoidal curve of approximately 200; **c** sAPP β has an increasing sigmoidal curve of approximately 70. In the presence of SORLA, **d** APP is decreasing and fluctuates; **e** sAPP α decreased by 93.79%; **f** sAPP β decreased by 92.19%

Even after several calibrations of the variables and rates, the model simulations of sAPP β do not have a good fit to the data. It either has a good fit in the absence of SORLA or in the presence of SORLA, but not both at the same time. It can be hypothesized that there should be some other interaction(s) among APP, SORLA and β-secretase that is present in the system, and is responsible for the dynamical behavior of sAPP β observed in the data.

The result is consistent with what Willnow and colleagues [7, 11–13] showed in their study that "*there must be an indirect interaction between SORLA and the β-secretase*". With this, the initial model (shown in Fig. 4) can be improved and modified by integrating the proposed interaction of SORLA and β-secretase at the TGN. A new channel is added. It serves as the β-secretase in the TGN so that when the reverse interaction of APP and SORLA occurs, the free APP may be cleaved by it (added lines include Lines (8), (14) & (22) of Fig. 5). Applying z-test to our modified SPA model (Fig. 5) confirms that at least 60% of sAPP α and sAPP β simulations in the presence and absence of SORLA are of good fit to Schmidt et al. [11] data (see Table 1).

In the absence of SORLA, APP exponentially decreases from its initial value of 225 to 0 (Fig. 9a); while both sAPP α (Fig. 9b) and sAPP β (Fig. 9c) has sigmoidal increase with steady state value at 225 and 10, respectively. In the presence of SORLA, APP decreases faster and more drastically (Fig. 9d) as compared to that

Table 1 Summary of the z test for one proportion

	Observed proportion of good fit	Z statistics
Without SORLA		
sAPP α	1	6.17
sAPP β	0.60	0.00
With SORLA		
sAPP α	1	8.16
sAPP β	1	8.16

In modified amyloidogenic processing model. At $\alpha = 0.05$, we fail to reject that at least 60% of sAPP α and sAPP β simulations arc of good fit

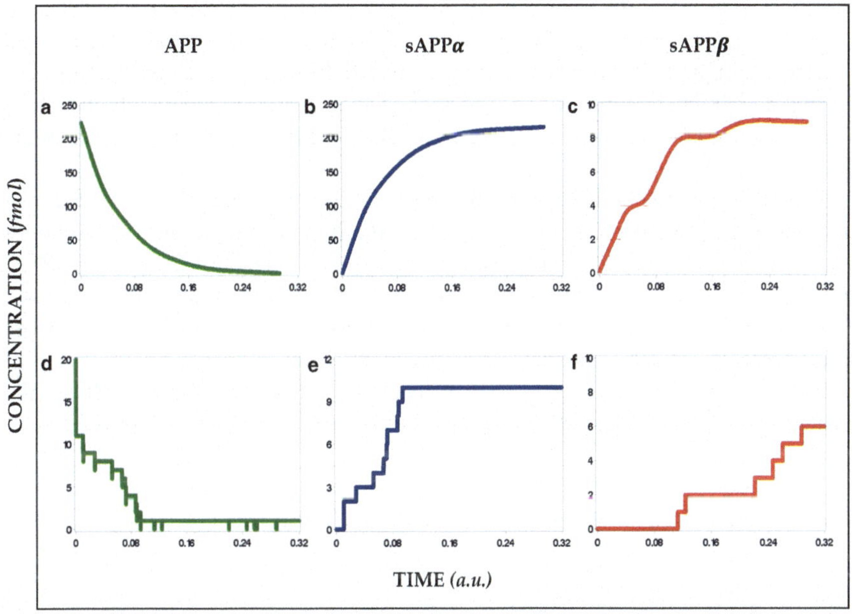

Fig. 9 Simulation results of modified amyloidogenic processing in the absence (**a**–**c**) and presence (**d**–**f**) of SORLA. In the absence of SORLA, **a** APP is exponentially decreasing from 225 to 0; **b** sAPP α undergoes a sigmoidal increase of approximately 225; **c** sAPP β undergoes a sigmoidal increase of approximately 10. In the presence of SORLA, **d** APP decreasingly fluctuates and at some point interval would reach 0; **e** sAPP α decreases by 95%; **f** sAPP β decreases by 50%

in the absence of SORLA. It decreases continuously and fluctuates minimally over time (Fig. 9d). As APP decreases, both sAPP α and sAPP β also decrease by about 95% (Fig. 9e) and 50% (Fig. 9f), correspondingly. The decrease of sAPPα and sAPP β that we have observed in the presence of SORLA (Fig. 9) is consistent to what was observed in Schmidt et al. [11] data (shown in Fig. 4).

6 Summary and Future Works

Several studies have been made in trying to understand the pathology of AD through a series of experiments, or mathematical and computational modeling [7, 11–13]. In this study, an SPA based modelling approach is used to capture the behavior suggested by the data, which are gathered from the study of Schmidt et al. [11]. Using an SPA modeling approach, it becomes possible to integrate the cleavage of APP by β-secretase at the TGN where SORLA is located. Interestingly, the simulations of this new model show good fit to the experimental data, both in the absence and presence of SORLA.

When SORLA and β-secretase do not interact at the TGN, none of the generated simulations fitted the data (shown in Fig. 4). Either sAPP β in the presence of SORLA will fit or that in the absence of SORLA, but not both at the same time. Hence, it can be the case that there should exist some kind of interaction between SORLA and β-secretase in order to improve the fitting. This led the direction of this work to what Spoelgen et al. [12], Willnow and Andersen [13] discussed in their study that there might be an indirect interaction between SORLA and β-secretase that occurs in the TGN.

The SPA model is built based on the works of Spoelgen et al. [12] and Willnow and Andersen [13]. It did not only validated the hypothesis presented in Schmidt et al. [11] and Lao et al. [7] that there is an indirect interaction between SORLA and β-secretase; but also gave a better understanding of the proteolytic breakdown of the amyloidogenic processing. Particularly, it is once again showed that SORLA and β-secretase possibly interacts indirectly at the TGN.

For future studies, we suggest the inclusion of the γ-secretase to replicate the formation of A β peptides the leading cause of AD to fully understand the effectiveness of SORLA in the proteolytic breakdown of amyloidogenic processing. Moreover, we recommend future researchers to include compartmental details in the SPA model where at each compartment, the amount of APP is accounted for.

References

1. Ali, G.-C., et al.: World Alzheimer Report 2015 The Global Impact of Dementia An Analysis of Prevalence, Incidence, Cost and Trends. Alzheimer's Disease International, London (2015)
2. Clark, A., et al.: Stochastic process algebras. In: Proceedings of the 7th International Conference on Formal Methods for Performance Evaluation, pp. 132–179. Springer, Berlin, Heidelberg (2007)
3. De Strooper, B., Annaert, W.: Proteolytic processing and cell biological functions of the amyloid precursor protein. J. Cell Sci. **113**(Pt 11), 1857–1870 (2000)
4. Gilmore, S., et al.: Stochastic process algebras: from individuals to populations. Comput. J. **55**(7), 866–881 (2011)
5. Goldstein, L.: A stochastic π-calculus model of MHC class I antigen (2005)
6. Lao, A.: A Systems Biology Approach to Understand the Influence of SORLA on Amyloidogenic Processing in Alzheimer's Disease. Rostock University (2013)

7. Lao, A., et al.: Multi-compartmental modeling of SORLA's influence on amyloidogenic processing in Alzheimer's disease. BMC Syst. Biol. **6**, 74 (2012). https://doi.org/10.1186/1752-0509-6-74
8. Phillips, A.: A visual process calculus for biology. Microsoft Res. (2009)
9. Priami, C.: Stochastic π-calculus. Comput. J. **38**(7), 578–589 (1995). https://doi.org/10.1093/comjnl/38.7.578
10. Regev, A., et al.: Representation and simulation of biochemical processes using the pi-calculus process algebra. In: Pacific Symposium on Biocomputing, pp. 459–470 (2001)
11. Schmidt, V., et al.: Quantitative modelling of amyloidogenic processing and its influence by SORLA in Alzheimer's disease. ResearchGate **31**(1), 187–200 (2012). https://doi.org/10.1038/emboj.2011.352
12. Spoelgen, R., et al.: Interaction of the cytosolic domains of sorLA/LR11 with the amyloid precursor protein (APP) and β-Secretase β-Site APP-cleaving enzyme. J. Neurosci. **26**(2), 418–428 (2006). https://doi.org/10.1523/JNEUROSCI.3882-05.2006
13. Willnow, T.E., Andersen, O.M.: Sorting receptor SORLA—a trafficking path to avoid Alzheimer disease. J Cell Sci. **126**(13), 2751–2760 (2013). https://doi.org/10.1242/jcs.125393
14. Zhang, Y., et al.: APP processing in Alzheimer's disease. Mol. Brain. **4**, 3 (2011). https://doi.org/10.1186/1756-6606-4-3

Soliton, Rogue Wave and Double Layer in an Unmagnetized Collisionless Plasma

Md. Golam Hafez and Samsul Ariffin Abdul Karim

Abstract This chapter deals with the ion-acoustic (IA) solitons, rogue waves (RWs), and double layers (DLs) in multi-component plasmas consisting of positively charged ion fluid and (α, q)-distributed electrons as well as positrons. The nonlinear evolution equations, namely, Korteweg–de Vries equation (KdVE), modified KdV equation (mKdVE), nonlinear Schrödinger equation (NLSE) and standard Gardner equation (SGE) along with their solutions are derived using the reductive perturbation method. It is examined that KdV, and SG solitons, RWs and DLs are supported both of positive and negative electrostatic potential structures, but mKdV supports only negative electrostatic potential structure by depending on the related plasma parameters. The outcomes presented may be very helpful in understanding the behavior of nonlinear coherent structures in many astrophysical and space environments and the future experimental studies in laboratory.

Keywords Nonlinear evolution equations · Wave phenomena · Multi-component plasma

1 Introduction

Numerous theoretical and experimental studies [1–15] confirmed that the linear properties of several kinds of low-frequency electrostatic and electromagnetic modes in the **multi-component plasmas** are now well understood. It has been reported that the linear theories are only supported for small amplitude via diverse evolution. How-

Md. G. Hafez (✉)
Department of Mathematics, Chittagong University of Engineering and Technology, Chattogram 4349, Bangladesh
e-mail: hafez@cuet.ac.bd

S. A. Abdul Karim
Fundamental and Applied Sciences Department, Centre for Systems Engineering (CSE), Institute of Autonomous System, Universiti Teknologi PETRONAS, Bandar Seri Iskandar, 32610 Seri Iskandar, Perak DR, Malaysia
e-mail: samsul_ariffin@utp.edu.my

© Institute of Technology PETRONAS Sdn Bhd 2022
S. A. Abdul Karim and A. Shafie (eds.), *Towards Intelligent Systems Modeling and Simulation*, Studies in Systems, Decision and Control 383,
https://doi.org/10.1007/978-3-030-79606-8_19

ever, the evolution processes indicate that unstable mode can saturate and obtain large amplitudes. For this case, **nonlinearity** cannot be ignored. Such situation contribute to the formation of waves by balancing with dispersive and or dissipative media and leads to various types of cohesive structure, i.e. solitons, shocks, double layer, vortices etc. in many **space and astrophysical environments** [16–25]. But, it is very difficult to examine the basic features of plasmas without **tedious mathematical procedures**. One can study the basic characteristics, (e.g. amplitude and width of electrostatic potential, polarity, wave structures, rogue waves (RWs), double layer (DL), etc.) by the **nonlinear evolution equations (NLEEs)** and their stationary solutions. However, a huge amount of authors have just solved the NLEEs with nonlinear and dispersive/dissipative coefficients either unity or any arbitrary numbers by implementing various solution methods [26–40]. Whereas the nonlinear and dispersive/dissipative coefficients of the NLEEs are strongly dependent on the physical parameters, e.g. density and temperature ratio's, nonthermality parameter, nonextensvity parameter, etc. [41–60]. So, it is therefore important to required a proper guideline about NLEEs and their solutions for not only plasma physics and applied mathematics communities but also NLEEs solver.

This chapter aims to explore how to examine the features of **ion acoustic waves (IAWs)** in the plasmas by deriving NLEEs from **the theoretical model equations**. To do so, a plasma system is proposed by composing of cold ion fluid and (α, q)-distributed electrons and positions. It is noted that the (α, q)-distribution is very useful for describing not only nonthermal and nonextensive but also nonthermal lighter particles with the presence of nonextensivity. Several types of NLEEs are formulated from the model equation via **the reductive perturbation technique (RPT)**. The effect of some of the plasma parameters on the nonlinear wave structures are investigated with physical interpretation by deriving **Korteweg–de Vries equation (KdVE)**, **modified KdV equation (mKdVE)**, **nonlinear Schrödinger equation (NLSE)** and **standard Gardner equation (SGE)**.

2 Model Equations

A one-dimensional unmagnetized collisionless multi-component plasma is proposed by the mixture of inertial positively charged ions fluid and inertialess (α, q) distributed electrons as well as positrons along with the neutrality condition $N_{i0} + N_{p0} = N_{e0}$. One can easily define respectively the normalized densities for nonthermal nonextensive electrons and positrons (see details in [61]) as

$$\begin{aligned} N_e &= N_{e0}\,[1 + (q-1)\psi]^{\frac{(q+1)}{2(q-1)}} \times \left[1 + k_1\psi + k_2\psi^2\right], \\ N_p &= N_{p0}\,[1 - (q-1)\sigma\psi]^{\frac{(q+1)}{2(q-1)}} \times \left[1 - k_1\sigma\psi + k_2\,(\sigma\psi)^2\right], \end{aligned} \tag{1}$$

where $k_1 = -16q\alpha/\left(3 - 14q + 15q^2 + 12\alpha\right)$ and $k_2 = -B_1\,(2q-1)$. To study the basic features of IAWs by deriving NLEEs, one can formulate the normalized model equations by taking the mass and momentum conservation low for ions fluid in the following forms:

$$\frac{\partial N_i}{\partial T} + \frac{\partial (N_i U_i)}{\partial X} = 0, \tag{2}$$

$$\frac{\partial U_i}{\partial T} + U_i \frac{\partial U_i}{\partial X} = -\frac{\partial \psi}{\partial X}, \tag{3}$$

With the help of Eq.(1) and neutrality condition, the plasma system is closed by the following:

$$\frac{\partial^2 \psi}{\partial X^2} = 1 + c_1 \psi + c_2 \psi^2 + c_3 \psi^3 + ... - N_l = -\rho, \tag{4}$$

where,

$$c_1 = \frac{1}{1-p}\left(k_1 + \frac{q+1}{2}\right)(1 + p\sigma),$$

$$c_2 = \frac{1}{1-p}\left[k_2 + \frac{q+1}{2}k_1 + \frac{(q+1)(3-q)}{8}\right]\left(1 - p\sigma^2\right),$$

$$c_3 = \frac{1}{1-p}\left[\frac{q+1}{2}k_2 + \frac{(q+1)(3-q)}{8}k_1 + \frac{(q+1)(3-q)(5-3q)}{48}\right]\left(1 + p\sigma^3\right).$$

It is noted that the meaning of the physical variables along with their scaling and parameters are included in the Table 1.

3 Formation of KdVE with Solution

To derive KdVE form Eqs.(2)–(4), one considers the new starching coordinates as

$$x = \varepsilon^{1/2}(X - V_p T), t = \varepsilon^{3/2} T \tag{5}$$

and expansions for physical quantities as

$$\begin{bmatrix} N_i \\ U_i \\ \psi \\ \rho \end{bmatrix} = \begin{bmatrix} 1 \\ 0 \\ 0 \\ 0 \end{bmatrix} + \varepsilon \begin{bmatrix} N_i^{(1)} \\ U_i^{(1)} \\ \psi^{(1)} \\ \rho^{(1)} \end{bmatrix} + \varepsilon^2 \begin{bmatrix} N_i^{(2)} \\ U_i^{(2)} \\ \psi^{(2)} \\ \rho^{(2)} \end{bmatrix} + \varepsilon^3 \begin{bmatrix} N_i^{(3)} \\ U_i^{(3)} \\ \psi^{(3)} \\ \rho^{(3)} \end{bmatrix} + ... \tag{6}$$

based on RPT(see detail in [43]). If one applies Eqs.(5) and (6) into Eqs. (2)–(4), then Eqs. (2)–(4) can easily transform by involving of ascending order of ε, that is $O(\varepsilon)$. From the $O(\varepsilon^{3/2})$ equations, one obtains

Table 1 List of variables, parameters and their abbreviations

Physical variables/ parameters	Abbreviations
N_{i0}	Unperturbed ion density
N_{e0}	Unperturbed electron density
N_{p0}	Unperturbed positron density
N_i	Ion density normalized by N_{e0}
U_i	Ion fluid velocity normalized by $\sqrt{k_B T_e/m_i}$
ψ	Electrostatic potential normalized by $\psi \to (k_B T_e/e)\psi$
$T_e(T_p)$	Electron (positron) temperature
m_i, e, k_B	Ion mass, electric charge, Boltzmann constant
α	Population of non-thermal electrons and positrons
$q = 1, \alpha = 0$	Isothermal electrons and positrons
$q = 1, \alpha \neq 0$	Nonthermal electrons and positrons
$0 < q < 1, \alpha = 0$	Superthermal electrons and positrons
$q > 1, \alpha = 0$	Subthermal electrons and positrons
X	Space coordinate normalized by Debye length $\sqrt{k_B T_e/4\pi N_{i0} e^2}$
T	Time coordinate normalized by plasma frequency $\sqrt{4\pi N_{e0} e^2/m_i}$
$p = N_{p0}/N_{e0}$	Density ratio
$\sigma = T_e/T_p$	Temperature ratio

$$N_i^{(1)} = \frac{1}{V_p^2}\psi^{(1)},\ U_i^{(1)} = \frac{1}{V_p}\psi^{(1)},\ V_p = \sqrt{\frac{1}{c_1}}. \tag{7}$$

For $O(\varepsilon^{5/2})$, one can determine the following:

$$\frac{\partial N_i^{(1)}}{\partial t} - V_p\frac{\partial N_i^{(2)}}{\partial x} + \frac{\partial U_i^{(2)}}{\partial x} + \frac{\partial (N_i^{(1)}U_i^{(1)})}{\partial x} = 0, \tag{8}$$

$$\frac{\partial U_i^{(1)}}{\partial t} - V_p\frac{\partial U_i^{(2)}}{\partial x} + U_i^{(1)}\frac{\partial U_i^{(1)}}{\partial x} = -\frac{\partial \psi^{(2)}}{\partial x}, \tag{9}$$

$$\frac{\partial^2 \psi^{(1)}}{\partial x^2} = c_1\psi^{(1)} + c_2\left[\psi^{(1)}\right]^2 - N_i^{(2)}, \tag{10}$$

Eliminating $N_i^{(2)}$ and $U_i^{(2)}$ from Eqs. (9)-(10), the following KdVE is obtained:

$$\frac{\partial \psi^{(1)}}{\partial t} + N_{cf}\psi^{(1)}\frac{\partial \psi^{(1)}}{\partial x} + D_{cf}\frac{\partial^3 \psi^{(1)}}{\partial x^3} = 0, \tag{11}$$

where

$$N_{cf} = \frac{V_p}{2}\left[\frac{3}{V_p^2} - \frac{2c_2}{c_1}\right], D_{cf} = \frac{V_p^3}{2}. \tag{12}$$

To study the basic features of the consider plasma, one can use the following well established and very useful stationary solution of Eq. (11):

$$\psi^{(1)} = \psi_{ap} \sec h^2\left(\frac{x - V_{rf}t}{\psi_w}\right), \tag{13}$$

where $\psi_{ap} = (3V_{rf}/N_{cf})$, and $\psi_w = \sqrt{(4D_{cf}/V_{rf})}$ are respectably the amplitude and width of soliton and V_{rf} is the constant velocity of the considered moving reference frame. Equation (13) clearly indicates that the ideal plasma is only supported compressive (rarefactive) soliton for the case of $(3/V_p^2) > (2c_2/c_1)$ $((3/V_p^2) < (2c_2/c_1))$. Besides, the considered theory is not valid for $N_{cf} \to 0$ which indicates the amplitudes approach infinity. For instance, the following critical value (q_c) for q leads to $\psi_{ap} \to \infty$ with the presence of nonextensive distributed electrons and positrons, that is, $\alpha = 0$.

$$q = q_c \to \frac{3p(\sigma^2 + 2\sigma + 1)}{p\sigma^2(1 - 4p) + p(1 - 6\sigma) - 4}. \tag{14}$$

Equation (14) clearly indicates the higher order correction of RPT is still required to describe the polarity around q_c.

4 Formation of mKdVE with Solution

To study soliton formation around q_c, one can derive mKdVE by taking higher order correction into account. To do so, the stretched coordinates are defined as

$$x = \varepsilon(X - V_pT), t = \varepsilon^3 T. \tag{15}$$

Using Eq. (15) and the expansion for perturbed quantities as in Eqs. (5) into Eqs. (2)–(4), the similar expressions as in Eq. (7) are obtained by taking $O(\varepsilon^2)$ equations. For $O(\varepsilon^3)$ yields

$$-V_p\frac{\partial N_i^{(2)}}{\partial x} + \frac{\partial U_i^{(2)}}{\partial x} + \frac{\partial (N_i^{(1)}U_i^{(1)})}{\partial x} = 0, \tag{16}$$

$$-V_p\frac{\partial U_i^{(2)}}{\partial x} + U_i^{(1)}\frac{\partial U_i^{(1)}}{\partial x} = -\frac{\partial \psi^{(2)}}{\partial x}, \tag{17}$$

$$0 = c_1\psi^{(2)} + c_2\left[\psi^{(1)}\right]^2 - N_i^{(2)}, \tag{18}$$

Solving $O(\varepsilon^3)$ equations, one obtains

$$N_i^{(2)} = \frac{1}{V_p^2}\left[\psi^{(2)} + \frac{3}{2V_p^2}\left\{\psi^{(1)}\right\}^2\right], U_i^{(2)} = \frac{1}{V_p}\left[\psi^{(2)} + \frac{1}{2V_p^2}\left\{\psi^{(1)}\right\}^2\right]$$
$$-\frac{1}{2}N_{cf}\left\{\psi^{(1)}\right\}^2 = 0 = \rho^{(2)}. \tag{19}$$

For $O(\varepsilon^4)$ gives

$$\frac{\partial N_i^{(1)}}{\partial t} - V_p\frac{\partial N_i^{(3)}}{\partial x} + \frac{\partial U_i^{(3)}}{\partial x} + \frac{\partial}{\partial x}\left(N_i^{(1)}U_i^{(2)} + N_i^{(2)}U_i^{(1)}\right) = 0, \tag{20}$$

$$\frac{\partial U_i^{(1)}}{\partial t} - V_p\frac{\partial U_i^{(3)}}{\partial x} + \frac{\partial}{\partial x}\left(U_i^{(1)}U_i^{(2)}\right) = -\frac{\partial \psi^{(3)}}{\partial x}, \tag{21}$$

$$\frac{\partial^2\psi^{(1)}}{\partial x^2} = c_1\psi^{(3)} + 2c_2\left[\psi^{(1)}\psi^{(2)}\right] + c_3\left[\psi^{(1)}\right]^3 - N_i^{(3)}, \tag{22}$$

Simplifying Eqs.(20)–(21) by considering Eqs.(7) and (19), the following mKdVE is obtained:

$$\frac{\partial\psi^{(1)}}{\partial t} + D_{cf}N'_{cf}\left\{\psi^{(1)}\right\}^2\frac{\partial\psi^{(1)}}{\partial x} + D_{cf}\frac{\partial^3\psi^{(1)}}{\partial x^3} = 0, \tag{23}$$

where

$$N'_{cf} = \left[\frac{15}{2V_p^6} - \frac{3c_3}{c_1V_p^2}\right]. \tag{24}$$

To study the basic features of the consider plasma around the critical value, one can use the following well established and useful stationary solution of Eq.(23):

$$\psi^{(1)} = \psi'_{ap}\sec h\left(\frac{x - V_{rf}t}{\psi'_w}\right), \tag{25}$$

where $\psi'_{ap} = \sqrt{\left(6V_{rf}/D_{cf}N'_{cf}\right)}$, and $\psi'_w = \psi'_{ap}\sqrt{\left(N'_{cf}/6\right)}$ are respectably the amplitude and width of soliton.

5 Formation of NLSE with Solution

It is well confirmed that the envelope soliton is theoretically studied by deriving NLSE from mKdVE [62–64] for the determination of critical value with some plasma parameter, which is also observed experimentally in plasma laboratory [65]. In addition, the unstable soliton, so called the RW is examined in the plasma system by determining the critical value based on the certain parameters. For investigating the behavior of unstable soliton, one employs the following transformation in Eq. (23):

$$\psi^{(1)}(x,t) = \sum_{m=1}^{\infty} \varepsilon^m \sum_{l=-m}^{m} \left[\Psi_l^m(x,t) \exp\{il(kx - \omega t)\}\right], \chi = \varepsilon(x - V_g t), \tau = \varepsilon^2 t. \tag{26}$$

For $m = 1$ and $l = 1$(first harmonic), one can easily formulate the dispersion relation for electrostatic wave as $\omega = -D_{cf}k$ (ω (k) is the frequency (wave number)). For $m = 2$ and $l = 1$ gives the group velocity as $V_g = -3D_{cf}k^2$. Finally, the following NLSE is derived by taking $m = 3$ and $l = 1$ with the comparability condition:

$$\frac{\partial \Psi}{\partial \tau} + \frac{1}{2} P_{cf} \frac{\partial^2 \Psi}{\partial \chi^2} + Q_{cf} \Psi |\Psi|^2 = 0, \tag{27}$$

where $P_{cf} = -6D_{cf}k$ and $Q_{cf} = -D_{cf}N_{cf}^{'}k$. One can study the stability or instability of pulse envelope for external perturbations by depending on $P_{cf}/Q_{cf} = 6/N_{cf}^{'}$. It is clearly indicated that the stable (unstable) plane waves are produced based on $N_{cf}^{'} < 0\,(N_{cf}^{'} > 0)$. The following well established rational function solution can be used to demonstrate the shape of RWs within the modulational unstable region:

$$\Psi(\chi, \tau) = \sqrt{\frac{P_{cf}}{Q_{cf}}} \left[\frac{4(1 + 2iP_{cf}\tau)}{1 + 4P_{cf}^2\tau^2 + 4\chi^2} - 1 \right] \exp(iP_{cf}\tau). \tag{28}$$

6 Formation of SGE with Soliton and DL Solution

It is noted that Eq. (23) has a soliton solution around q_c, but not DLs solution. To study both of soliton and DLs around $q_c (N_{cf} = 0)$, one can derive next higher order NLEE, so called SGE. The details derivation of this equation and its solution are given below.

For q around its q_c, $N_{cf} = N_{cf}^0$ is defined as

$$N_{cf}^0 = h \left(\frac{\partial N_{cf}}{\partial q} \right)_{q=q_c} |q - q_c| = hg\varepsilon, \tag{29}$$

where $|q - q_c|$ is equivalent to ε (because $|q - q_c|$ is small quantity), $h = 1(-1)$ for $q > q_c(q < q_c)$ and $g = \left(\frac{\partial N_{cf}}{\partial q}\right)_{q=q_c}$ is constant and strongly dependent on the related parameters. As a result, one can define $\rho^{(2)} \equiv -\varepsilon^3 \frac{1}{2} hg\psi^2$ and Eq. (22) yields

$$\frac{\partial^2 \psi^{(1)}}{\partial x^2} + \frac{1}{2} hg\psi^2 = c_1 \psi^{(3)} + 2c_2 \left[\psi^{(1)} \psi^{(2)}\right] + c_3 \left[\psi^{(1)}\right]^3 - N_i^{(3)}. \tag{30}$$

Now, simplifying Eqs. (16), (17) and (30), one obtains the following SG equation:

$$\frac{\partial \psi^{(1)}}{\partial t} + hgD_{cf}\psi^{(1)} \frac{\partial \psi^{(1)}}{\partial x} + D_{cf} N_{cf}^{'} \left\{\psi^{(1)}\right\}^2 \frac{\partial \psi^{(1)}}{\partial x} + D_{cf} \frac{\partial^3 \psi^{(1)}}{\partial x^3} = 0. \tag{31}$$

Equation(31) is also known as the mixed mKdVE. It can easily converted to KdVE by ignoring $\left\{\psi^{(1)}\right\}^2$ and insetting $hgD_{cf} = N_{cf}$. The nonlinear term of KdVE vanishes at $q \equiv q_c$ which indicate this equation is unable to describe basic feature of wave structures around at $q \equiv q_c$. But Eq. (31) is supported such phenomena for $q \equiv q_c$ which is the main advantage of this equation. It can also be used to describe not only soliton but also DL structures formation by determining the useful solution of Eq. (31). One can use the well established soliton and DL solution of this equation to study physical issues of the considered plasma. The soliton solution of Eq. (31) is obtained as

$$\psi^{(1)} = \left\{\frac{1}{\psi_{ap1}} - \left(\frac{1}{\psi_{ap2}} - \frac{1}{\psi_{ap1}}\right) \cos h^2 \left(\frac{x - V_{rf} t}{\psi_{gw}}\right)\right\}^{-1}, \tag{32}$$

where $\psi_{ap1,2} = G_0\left[1 \mp \sqrt{1 + V_{rf}/G_1}\right]$, $V_{rf} = (hgD_{cf}/3)\psi_{ap1,2} + (N_{cf}^{'} D_{cf}/6)$ $\psi_{ap1,2}$, $G_0 = -(hg/N_{cf}^{'})$, $G_1 = h^2 g/6N_{cf}^{'})$, and $\psi_{gw} = 2/\sqrt{-N_{cf}^{'} \psi_{ap1} \psi_{ap2}/6}$. Besides, the DL solution of Eq. (31) is obtained as

$$\psi^{(1)} = \frac{\psi_{ga}}{2} \left\{1 + \tan h \left(\frac{x - V_{rf} t}{\psi_{gw}}\right)\right\}, \tag{33}$$

where $\psi_{ga} = 6V_{rf}/hD_{cf}$ ($\psi_{gw} = 2/\psi_{ga}\sqrt{-N_{cf}^{'}/6}$) is the hight (thickness) of DL and $V_{rf} = -h^2 gD_{cf}/6N_{cf}^{'}$.

7 Results and Discussion

Based on the obtained NLEEs and their solutions, the effect of some plasma parameters on nonlinear propagation of IAWs for the considered plasmas are displayed graphically with the physical description. In the present investigation, the ranges of

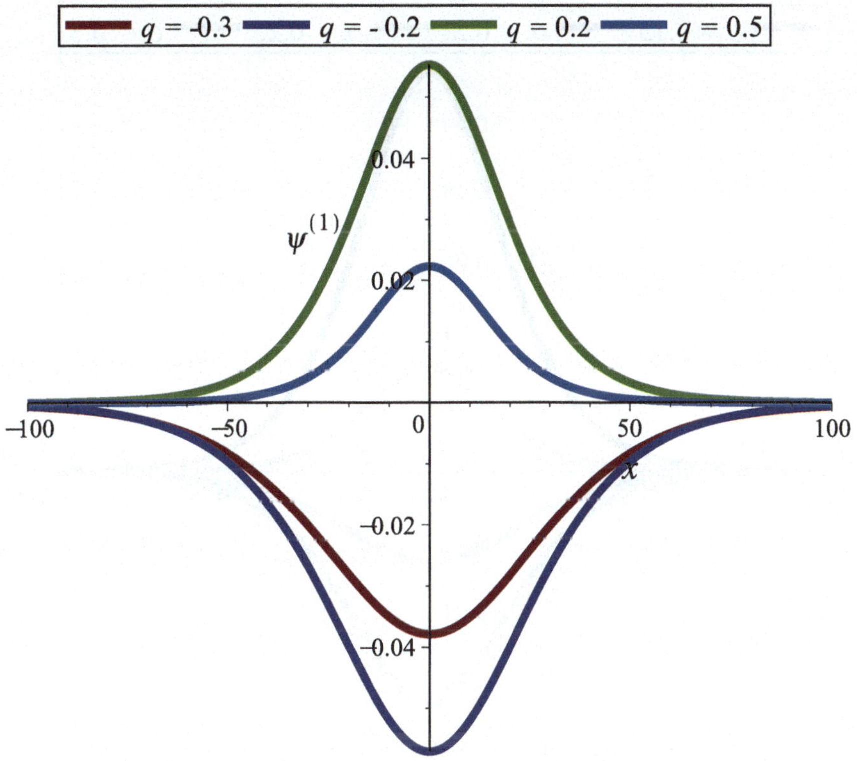

Fig. 1 KdV soliton for different values of q with $p = 0.001$, $\sigma = 1$, $V_{rf} = 0.0075$ and $\alpha = 0$

relevant parameter are assumed based on the [41–48, 59, 61] which are obviously related to astrophysical and space plasma environments.

Figures 1 and 2 displays the KdV soliton for different values of q and α, respectively by taking the remaining parameter constants. It is observed from Figs. 1 and 2 that the amplitude and width of electrostatic potential gains (losses) energies with increased α (q). In addition, Figs. 1 and 2 are indicated both positive and negative electrostatic potential are produced in the considered plasmas with the presence of strength of nonextensivity as well as number of nonthermal lighter particles population. The mathematical expression for the amplitude of KdV soliton is also clearly indicated that there exits no soliton at the critical value as well as around the critical value by depending on the plasma parameters. As a result, one can study the behavior of electrostatic potential around the critical value dy only deriving mKdVE. Figures 3 and 4 demonstrates the mKdV soliton structures around the critical value for q and α, respectively by choosing the remaining parameter constant. Consequently, one can study the unstable soliton by deriving NLS equation from mKdV equation. Because

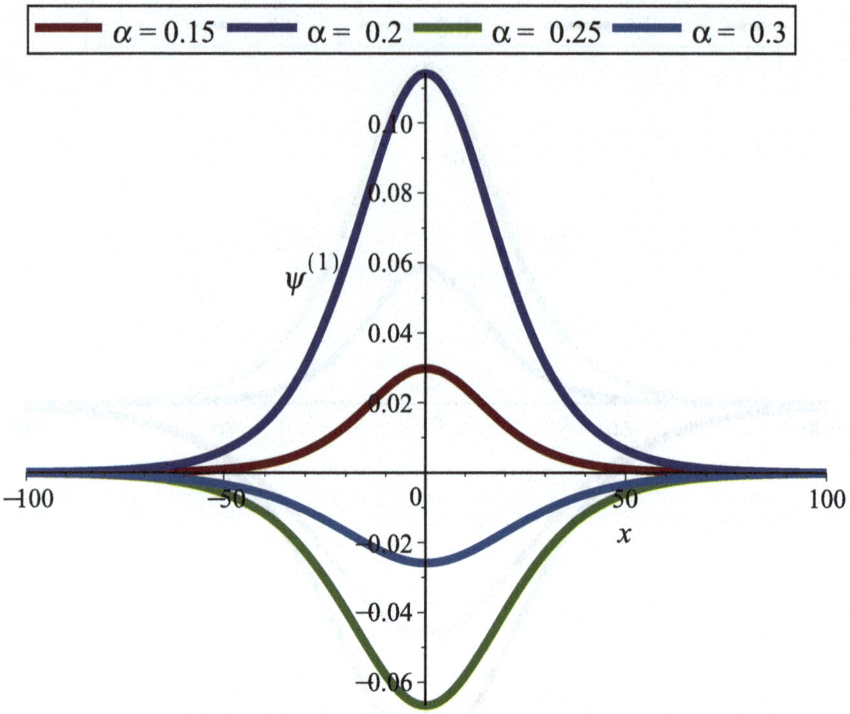

Fig. 2 KdV soliton for different values of α with $p = 0.1$, $\sigma = 1$, $V_{rf} = 0.0075$ and $q = 1$

the weakly nonlinear theory are only supported modulationally stable soliton. This means that the NLS equation that obtained from the KdV equation is not supported RW solution. Figures 5 and 6 show the shape of RW structure and its contour plot for better vitalization of space and time variation with the constant values of the other parameters, respectively. However, mKdVE is unable to examine the basic features of nonlinear propagation of ion acoustic not only soliton but also DLs at the critical values. In such situations, SGE is only an arena to over such difficulty. Because, this equation is supported soliton at the critical value and DL structures. Figures 7 and 8 show both of positive and negative soliton at the critical value and DLs by considering constant values of the parameters. The effect of plasma parameters based on various types of plasma assumption on the nonlinear propagation of several kind of acoustic waves are still not clear and a considerable progress in this direction can be achieved by undertaking a systematic study.

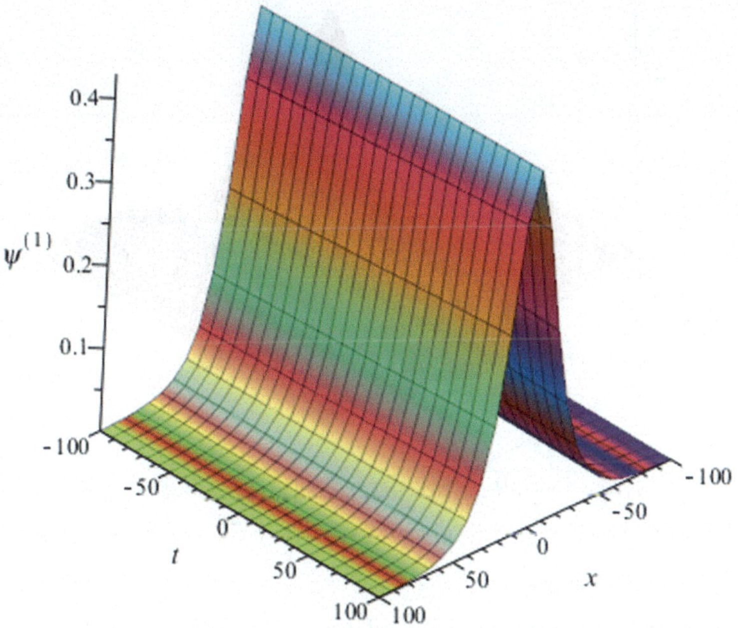

Fig. 3 mKdV soliton around the critical value for $q = -0.4$ with $p = 0.25$, $\sigma = 1$, $V_{rf} = 0.0075$ and $\alpha = 0$

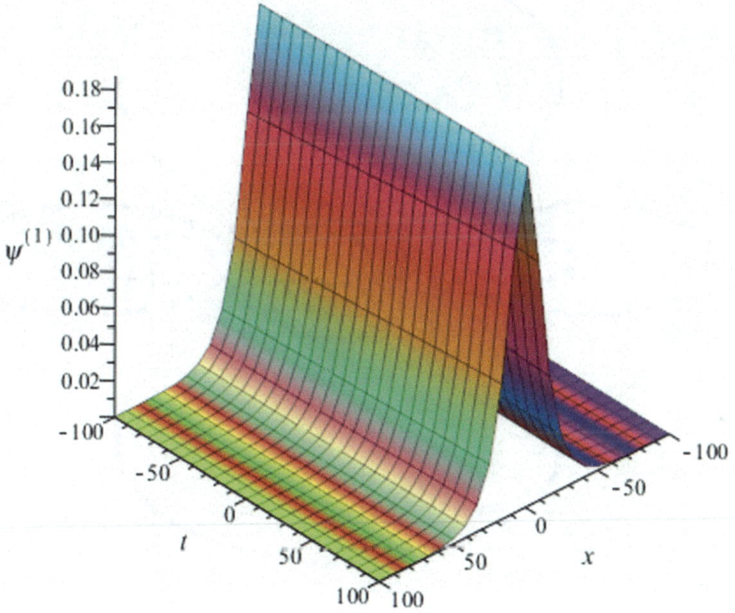

Fig. 4 KdV soliton for different values of α with $p = 0.1$, $\sigma = 1$, $V_{rf} = 0.0075$ and $q = 1$

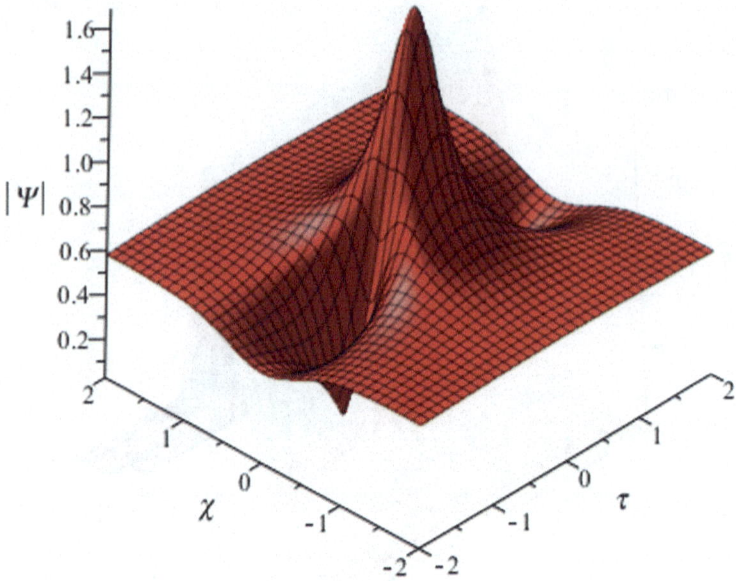

Fig. 5 RW structure for $q = 0.5$, $p = 0.3$, $\sigma = 1$, $k = 1$, $V_{rf} = 0.075$ and $\alpha = 0$

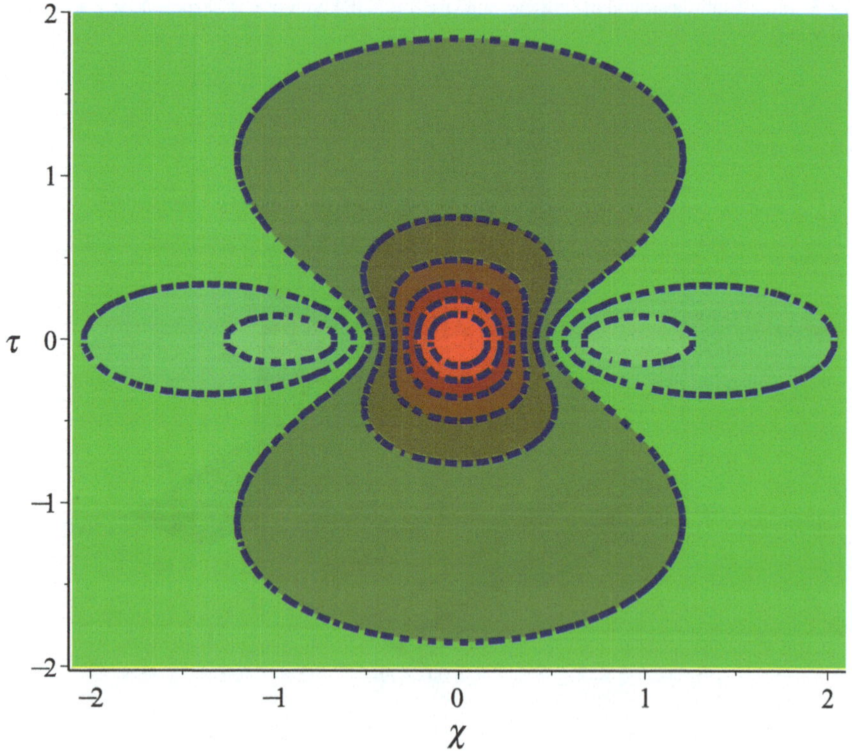

Fig. 6 Counter plot of RW for $q = 0.5$, $p = 0.3$, $\sigma = 1$, $k = 1$, $V_{rf} = 0.075$ and $\alpha = 0$

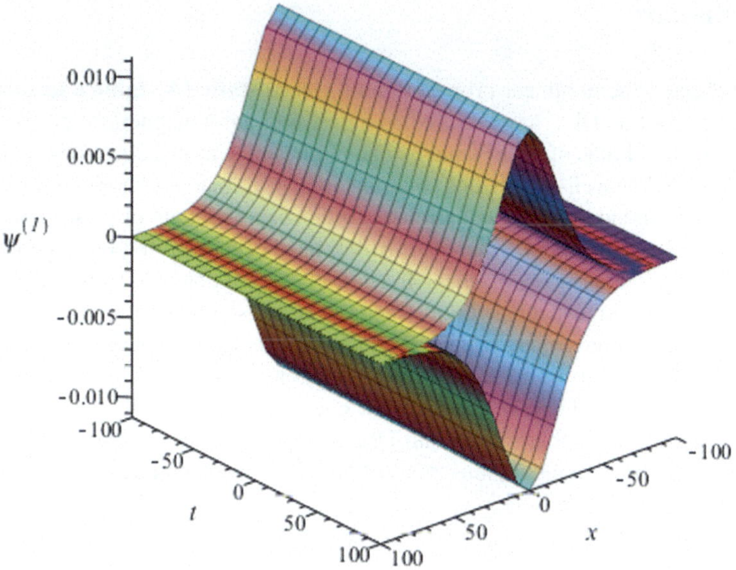

Fig. 7 Positive (upper surface) and negative (upper surface) Gardner soliton at $q_c = -0.5714285715$ for $h = 1$ and $h = -1$, respectively with $p = 0.25, \sigma = 1, V_{rf} = 0.075$ and $\alpha = 0$

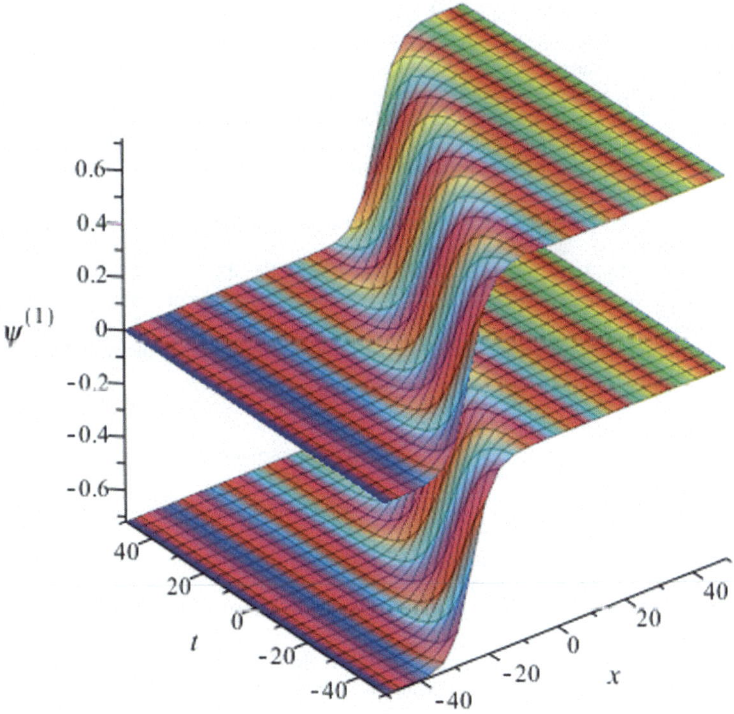

Fig. 8 Positive (upper surface) and negative (lower surface) potential DL structures for $h = 1$ and $h = -1$, respectively with $p = 0.1, \sigma = 1, \alpha = 0.5$ and $q = 1$

8 Summary

In this chapter, the nonlinear propagation of **electrostatic IA stable and unstable soliton**, as well as DLs, have studied by proposing an unmagnetized collisionless plasma. Several kinds of NLEEs have obtained from the model equations by implementing RPT. The well-established solutions for these equations have used to archive this goal. It is found that KdV equation has been supported both of positive and negative electrostatic potential, whereas mKdVE has been supported only the positive electrostatic potential in the plasmas. In addition, NLSE obtained from mKdVE has only been produced unstable soliton structures. It is also remarkable that KdVE and mKdVE are not applicable to study the basic features of soliton at the critical values. But, SGE is not only useful for describing soliton formation at the critical values but also DLs. It would be concluded that one can systematically study the basic features of stable/unstable soliton and DLs formation in various astrophysical and space plasmas by deriving various types of NLEEs. The outcomes will be useful for future laboratory verification.

References

1. Mendis, D.A.: The role of field emission in the electrostatic disruption of cosmic dust. Astrophys. Space Sci. **176**, 163 (1991)
2. Mendis, D.A.: Advances in Dusty Plasmas. In: Shukla, P.K., Mendis, D.A., Desai, T. (eds.) World Scientific, Singapore (1997)
3. Shukla, P.K., Mamun, A.A.: Introduction to Dusty Plasma Physics. Institute of Physics, Bristol (2002)
4. Mendis, D.A., Horanyi, M.: Cometary Plasma Processes. In: Johnston, A.J. (ed.) Geophys, Washington, DC (1991)
5. Rao, N.N., Shukla, P.K., Yu, M.Y.: Dust-acoustic waves in dusty plasmas. Planet. Space Sci. **38**, 543 (1990)
6. Shukla, P.K., Silin, V.P.: Dust ion-acoustic wave. Phys. Scr. **45**, 508 (1992)
7. Verheest, F.: Nonlinear dust-acoustic waves in multispecies dusty plasmas. Planet. Space Sci. **40**, 1 (1992)
8. Kaw, P.K., Sen, A.: Low frequency modes in strongly coupled dusty plasmas. Phys. Plasmas **5**, 3552 (1998)
9. Thompson, C., Barkan, A., Merlino, R.L., D'Angelo, N.: Video imaging of dust acoustic waves. IEEE Trans. Plasma Sci. **27**, 146 (1999)
10. Popal, S.I., Vladimirov, S.V., Shukla, P.K.: Ion-acoustic solitons in electron–positron–ion plasmas. Phys. Plasmas **2**, 716 (1995)
11. Shen, M.B., Meyer-ter-Vehn, J.: Pair and γ-photon production from a thin foil confined by two laser pulses. Phys. Rev. E **65**, 016405 (2001)
12. Barkan, A., Merlino, R.L., D'Angelo, N.: Experiments on ion-acoustic waves in dusty plasmas. Planet. Space Sci. **44**, 239 (1996)
13. Shukla, P.K., Marklund, M.: Dust acoustic wave in a strongly magnetized pair-dust plasma. Phys. Scr. **113**, 36 (2004)
14. Shukla, P.K., Rosenberg, M.: Boundary effects on dust-ion-acoustic and dust-acoustic waves in collisional dusty plasmas. Phys. Plasmas **6**, 1038 (1999)
15. Shukla, P.K., Varma, R.K.: Convective cells in nonuniform dusty plasmas. Phys. Fluids B **5**, 236 (1993)

16. Michel, F.C.: Theory of pulsar magnetospheres. Rev. Mod. Phys. **54**, 1 (1982)
17. Miller, H.R., Witta, P.J.: Active Galactic Nuclei, p. 202. Springer, Berlin (1987)
18. Tandberg-Hansen, E., Emshie, A.G.: The Physics of Solar Flares, p. 124. Cambridge University Press, Cambridge (1988)
19. Michel, F.C.: Theory of Neutron Star Magnetosphere. Chicago University Press, Chicago (1991)
20. Burns, M.L.: Positron-Electron Pairs in Astrophysics. American Institute of Physics, Melville NY (1983)
21. Rees, M.J.: In the Very Early Universe. In: Gibbons, G.W., Hawking, S.W., Siklas, S. (eds.) Cambridge University Press, Cambridge (1983)
22. Chen, F.F.: Introduction to Plasma Physics and Controlled Fusion, 2nd edn. Plenum Press (1984)
23. Gurnett, D.A., et al.: Radio emission from the heliopause triggered by an interplanetary shock. Science **262**, 199 (1993)
24. Kivelson, M.G., Russell, C.T.: Introduction to Space Physics. Cambridge University Press (1995)
25. Krall, N.A., Trivelpiece, A.W.: Principles of Plasma Physics. San Fransisco Press (1986)
26. Uddin, M.F., et al.: Traveling wave with beta derivative spatial temporal evolution for describing the nonlinear directional couplers with metamaterials via two distinct methods. Alexandria Eng. J. **60**, 1055 (2021)
27. Uddin, M.F., Hafez, M.G.: Interaction of complex short wave envelope and real long wave described by the coupled Schrödinger–Boussinesq equation with variable coefficients and beta space fractional evolution. Results Phys. **19**, 103268 (2020)
28. Iqbal, S.A., Hafez, M.G., Karim, S.A.A.: Bifurcation analysis with chaotic motion of oblique plane wave for describing a discrete nonlinear electrical transmission line with conformable derivative. Results Phys. **18**, 103309 (2020)
29. Hafez, M.G., Iqbal, S.A., Akther, S., Uddin, M.F.: Oblique plane waves with bifurcation behaviours and chaotic motion for resonant nonlinear Schrodinger equations having fractional temporal evolution. Results Phys. **15**, 102778 (2019)
30. Akther, S., Hafez, M.G., Ferdous, F.: Oblique resonance wave phenomena for nonlinear coupled evolution equations with fractional temporal evolution. Eur. Phys. J. Plus **134**, 473 (2019)
31. Akther, S., Hafez, M.G., Rezazadeh, H.: Resonance nonlinear wave phenomena with obliqueness and fractional time evolution via the novel auxiliary ordinary differential equation method. SN Appl. Sci. **1**, 567 (2019)
32. Ferdous, F., et al.: Oblique resonant optical solitons with Kerr and parabolic law nonlinearities and fractional temporal evolution by generalized exp $(-\Phi(\xi))$-expansion. Optik **178**, 439–449 (2019)
33. Ferdous, F., Hafez, M.G.: Nonlinear time fractional Korteweg-de Vries equations for interaction of wave phenomena in fluid-filled elastic tubes. Eur. Phys. J. Plus **133**, 384 (2018)
34. Ferdous, F., Hafez, M.G.: Oblique closed form solutions of some important fractional evolution equations via the modified Kudryashov method arising in physical problems. J. Ocean Eng. Sci. **3**, 244 (2019)
35. Hafez, M.G.: Exact solutions to the (3+1)-dimensional coupled Klein-Gordon-Zakharov equation using Exp-expansion method. Alexandria Eng. J. **55**, 1635 (2016)
36. Hafez, M.G., Talukder, M.R., Ali, M.H.: New analytical solutions for propagation of small but finite amplitude ion-acoustic waves in a dense plasma. Waves Random Complex Media **26**, 68 (2016)
37. Hafez, M.G., Sakthivel, R., Talukder, M.R.: Some new electrostatic potential functions used to analyze the ion-acoustic waves in a Thomas fermi plasma with degenerate electrons. Chin. J. Phys. **53** (2015)
38. Hafez, M.G.: New travelling wave solutions of the (1+1)-dimensional cubic nonlinear Schrodinger equation using novel (G'/G)-expansion method. Beni-Suef Uni. J. Basic. Appl. Sci. **5**, 109 (2016)

39. Akter, S., et al.: Analytic wave solutions of beta space fractional Burgers equation to study the interactions of multi-shocks in thin viscoelastic tube filled. Alexandria Eng. J. **60**, 877 (2021)
40. Uddin, M.F., Hafez, M.G., Hammouch, Z., Baleanu, D.: Periodic and rogue waves for Heisenberg models of ferromagnetic spin chains with fractional beta derivative evolution and obliqueness. Waves Random Complex Media **279** (2020). https://doi.org/10.1080/17455030.2020.1722331
41. Hafez, M.G.: Nonlinear ion acoustic solitary waves with dynamical behaviours in the relativistic plasmas. Astrophys. Space Sci. **365**, 78 (2020)
42. Hafez, M.G.: Nonlinear Schamel Korteweg-de Vries-Burgers equation to report ion acoustic waves in the relativistic plasmas. IEEE Trans. Plasma Sci. **47**, 5314 (2019)
43. Hafez, M.G., et al.: Dust ion acoustic multi-shock wave excitations in the weakly relativistic plasmas with nonthermal nonextensive electrons and positrons. AIP Adv. **10**, 065234 (2020)
44. Hafez, M.G., et al.: Face to face collisions of ion acoustic multi-solitons and phase shifts in dense plasma. Braz. J. Phys. **49**, 221 (2019)
45. Hafez, M.G., Talukder, M.R., Sakthivel, R.: Ion acoustic solitary waves in plasmas with nonextensive distributed electrons, positrons and relativistic thermal ions. Indian J. Phys. **90**, 603 (2016)
46. Hafez, M.G., Talukder, M.R., Ali, M.H.: Effects of trapped electrons on the oblique propagation of ion acoustic solitary waves in electron-positron-ion plasmas. Phys. Plasmas **23**, 082904 (2016)
47. Hafez, M.G., Roy, N.C., Talukder, M.R., Ali, M.H.: Oblique propagation of ion acoustic shock waves in weakly and highly relativistic plasmas with nonthermal electrons and positrons. Astrophys. Space Sci. **361**, 312 (2016)
48. Hafez, M.G., Roy, N.C., Talukder, M.R., Ali, M.H.: Ion acoustic shock and periodic waves through Burgers equation in weakly and highly relativistic plasmas with nonextensivity. Plasma Sci. Technol. **19**, 015002 (2017)
49. Alam, M.S., Hafez, M.G., Talukder, M.R., Ali, M.H.: Head-on collision of ion acoustic shock waves in electron-positron-ion nonextensive plasmas for weakly and highly relativistic regimes. Phys. Plasmas **25**, 072904 (2018)
50. Hafez, M.G., Talukder, M.R., Ali, M.H.: Comment on "Electrostatic compressive and rarefactive shocks and solitons in relativistic plasmas occurring in Polar Regions of pulsar". Astrophys. Space Sci. **361**, 154 (2016)
51. Hafez, M.G., Talukder, M.R., Ali, M.H.: Nonlinear propagation of ion-acoustic waves through the burgers equation in weakly relativistic plasmas. Plasma Phys. Rep. **43**, 499 (2017)
52. Alam, M.S., Hafez, M.G., Talukder, M.R., Ali, M.H.: Head-on collision of ion acoustic solitary waves in electron-positron-ion nonthermal plasmas for weakly and highly relativistic regimes. Phys. Plasmas **24**, 072901 (2017)
53. Alam, M.S., Hafez, M.G., Talukder, M.R., Ali, M.H.: Head-on collision between positron acoustic waves in homogeneous and inhomogeneous plasmas. Astrophys. Space Sci **363**, 102 (2018)
54. Alam, M.S., Hafez, M.G., Talukder, M.R., Ali, M.H.: Effects of two-Temperature ions on head-on collision and phase shifts of dust acoustic single-and multi-solitons in dusty plasma. Phys. Plasmas **24**, 103705 (2017)
55. Alam, M.S., Hafez, M.G., Talukder, M.R., Ali, M.H.: Interactions of ion acoustic multi-soliton and rogue wave with Bohm quantum potential in degenerate plasma. Chin. Phys. B **26**, 095203 095203 (2017)
56. Hafez, M.G., Talukder, M.R., Ali, M.H.: Two-dimensional nonlinear propagation of ion acoustic waves through KPB and KP equations in weakly relativistic plasmas. Adv. Math. Phys. **2016**, 12 (2016). (Article ID 9352148)
57. Hafez, M.G., Roy, N.C., Talukder, M.R., Ali, M.H.: Ion acoustic shock and solitary waves in highly relativistic plasmas with nonextensive electrons and positrons. Phys. Plasmas **23**, 012902 (2016)
58. Hafez, M.G., Talukder, M.R., Ali, M.H.: Nonlinear propagation of weakly relativistic ion-acoustic waves in electron-positron-ion plasma. Pramana J. Phys. **87**, 70 (2016)

59. Hafez, M.G., Talukder, M.R.: Ion acoustic solitary waves in plasmas with nonextensive electrons, Boltzmann positrons and relativistic thermal ions. Astrophys. Space Sci. **359**, 27 (2015)
60. Hafez, M.G., Akter, P., Karim, S.A.A.: Overtaking collisions of ion acoustic N-shocks in a collisionless plasma with pair-ion and (α, q) distribution function for electrons. Appl. Sci. **10**, 6115 (2020)
61. Tribeche, M., Amour, R., Shukla, P.K.: Ion acoustic solitary waves in a plasma with nonthermal electrons featuring Tsallis distribution. Phys. Rev. E **85**, 037401 (2012)
62. El-Tantawy, S.A., Wazwaz, A.M., Schlickeiser, R.: Solitons collision and freak waves in a plasma with Cairns-Tsallis particle distributions. Plasma Phys. Control. Fusion **57**, 125012 (2015)
63. El-Tantawy, S.A., El-Awady, E.I., Tribeche, M.: On the rogue waves propagation in non-Maxwellian complex space plasmas. Phys. Plasmas **22**, 113705 (2015)
64. El-Tantawy, S.A.: Rogue waves in electronegative space plasmas: The link between the family of the KdV equations and the nonlinear Schrödinger equation. Astrophys. Space Sci. **361**, 164 (2016)
65. Bailung, H., Nakamura, Y.: Observation of modulational instability in a multi-component plasma with negative ions. J. Plasma Phys. **50**, 231 (1993)

Sparse Closed-Loop Surface Related Multiple Elimination

Abdul Halim Abdul Latiff and Hazwan Hashim

Abstract A new approach for multiple removal were developed using the closed-loop surface related multiple elimination (CL-SRME) method. Instead of directly updating the estimated primaries, the impulse response and the source wavelet are updated together during inversion as both are used to defines the primaries estimate. Through imposing the sparsity constraint toward the update of the primary impulse, the spiky primary impulse can be obtained and reduced the time for its convergence. Through example from both synthetic and field data of Malay Basin, it shows that this new approach works on both shallow and complex geological area. One advantage of using the new approach is it were able to recover the primaries continuity which are often missing due to primaries-multiples overlap. The faster rate of convergence also proves that this new approach could be an alternative to existing multiple elimination process such as SRME, EPSI or CL-SRME.

Keywords Multiples elimination · SRME · Seismic processing

1 Introduction

Multiples in its simplest term can be described as a repetition of a seismic events. As seismic reflections are used to transfer the information of the subsurface, the repetition from the same reflection may cloud the primary reflections and cause further confusion during processing and imaging. In seismic exploration, the signal from the primary reflection events is used to convey the information of the subsurface. Primary reflection or usually known as primaries is a seismic wave that have been propagated through the earth subsurface with a single bounce (reflected once) without

A. H. Abdul Latiff (✉) · H. Hashim
Centre for Subsurface Imaging, Universiti Teknologi PETRONAS, Bandar Seri Iskandar, 32610 Seri Iskandar, Perak Darul Ridzuan, Malaysia
e-mail: abdulhalim.alatiff@utp.edu.my

H. Hashim
e-mail: hazwan.hashim@utp.edu.my

© Institute of Technology PETRONAS Sdn Bhd 2022
S. A. Abdul Karim and A. Shafie (eds.), *Towards Intelligent Systems Modeling and Simulation*, Studies in Systems, Decision and Control 383,
https://doi.org/10.1007/978-3-030-79606-8_20

bouncing against the surface. This seismic wave travels in downwards and upwards direction only once.

In contrast to the primaries, multiples are the seismic events that have been reflected more than once both upward and downward before it reaches the receiver. Multiples have a lower amplitude and usually had a longer travel times than its corresponding primaries. The amplitude of the multiples depends on the product of the reflection coefficients for each reflection, therefore only strong impedance contrasts will generate multiples that can recognized as an event [1].

The strength of the multiple reflections depends on two factors, the strength of the reflectors its originated from and the lateral continuity of that reflector [2]. Due to this factor, multiples are easily recognized at the strong and coherent reflection interfaces such as the layer between the water bodies, salt structure, carbonate platforms and basalt layers. Since conventional seismic reflection process were based solely on the primary signal to describe the subsurface information, the other unwanted signals must be removed beforehand including the multiple reflections.

1.1 Categories of Multiples

The interface at which the multiples had its downward bounces can be used to determine whether it was a surface-related multiple or internal multiple. Surface-related multiples is a multiple reflection that had at least one downward bounce against the water surface. If the seismic reflection had at least one downward bounce at the water-bottom and below, but does not reflected at the water surface, this multiple reflection can be considered as internal multiple. Figure 1 below shows the difference ray path took by the surface-related multiples and internal multiples when it travels in the subsurface.

The types of multiples can be further categorized based on their travel ray path. Figure 2 depicts different type of multiples according to their ray path [3, 4]. It is important to know the type of multiples as some multiple removal techniques are only valid for certain type of multiples. The order of the multiples described the number of downward bounces that the multiples have before it reaches the receiver.

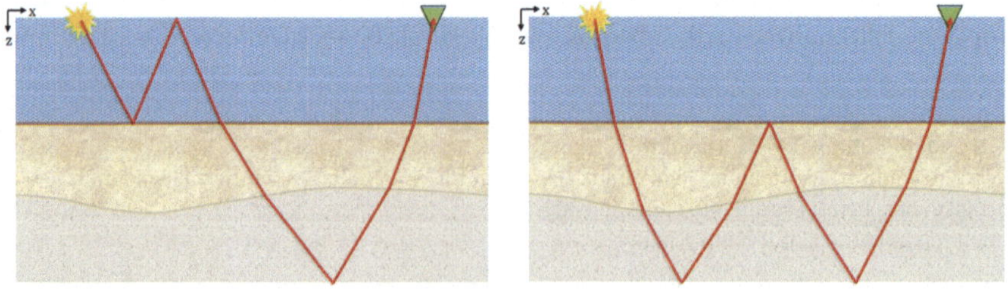

Fig. 1 Raypath for surface-related multiple (**a**) and internal multiple (**b**)

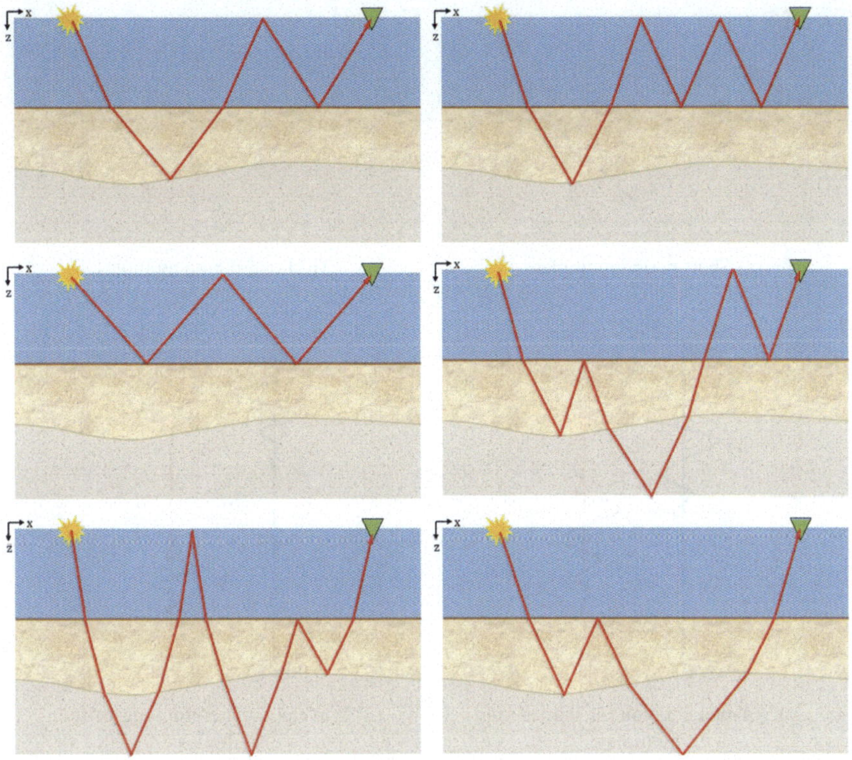

Fig. 2 Classification of multiples according to their ray paths; **a** first-order surface multiples, **b** second-order surface multiples, **c** water-bottom multiples, **d** water-layer reverberations, **e** other surface-related multiples and **f** internal multiples

For example, first-order multiples had only one downward bounce (Fig. 2a), second-order multiples had two downward bounce (Fig. 2b), third-order multiples had three downward bounces and so on. If the multiples only propagate within the water layer without travelling below it, as in Fig. 2c, these multiples are known as the water-bottom multiples. In Fig. 2d, e, some multiples can contain both surface and internal multiples. If this multiple had an upward and downward bounces within the water layer, it was considered as water-layer reverberation (Fig. 2d). Whereas if the multiples do not bounce within the water-layer, they are considered as other surface-related multiples (Fig. 2e).

Besides the interfaces and ray paths, the delay time between the primaries and the multiples are used to determine the type of multiples. Short-period multiples are the multiples where its delay time is in the order of the primary wavelet duration. These types of multiples are difficult to recognised as separate events in the seismic data and it is usually generated in a fine layering medium. As for the long-period multiples, these types of multiples had a delay time which are longer than the primary wavelet

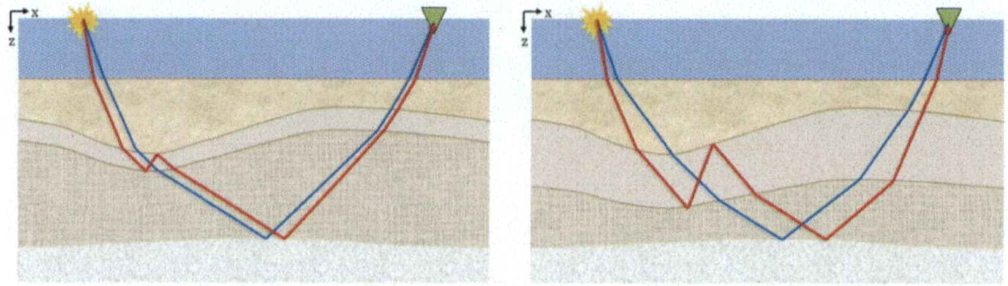

(a) Short-period multiples (b) Long-period multiples

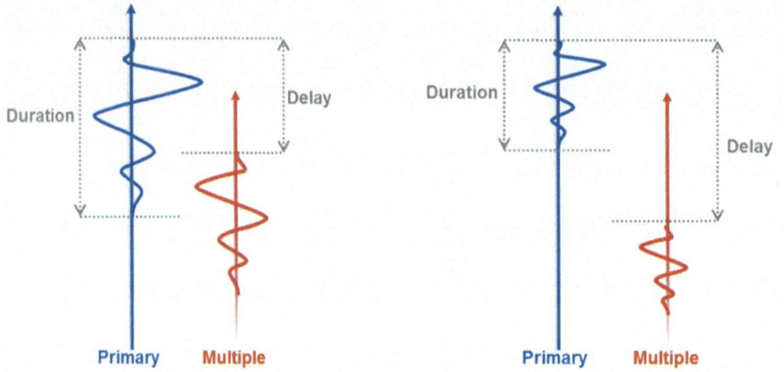

(c) Wavelet duration for short-period multiples (d) Wavelet duration for long-period multiples

Fig. 3 Raypaths for short and long period multiples

duration and it can be recognised as separate events in the seismic data. Figures 3 shows the ray paths of these two types of multiples and its wavelet duration.

1.2 Characteristic of Multiples

Since multiples are basically a repetition of the primary events, it is important to know how to identify them in the seismic data. According to [5], here are some of the common characteristics of multiples found in seismic data (Fig. 4). One of the recognizable features of multiples in the seismic data is its periodicity. Since multiples are periodic in nature, it can be observed in the seismic data as a repeating event in a regular interval with its amplitude increasing or decreasing regularly.

Every time multiples bounce against a reflector, it will carry a structural imprint of the reflectors. If this reflector contained a dipping structure, the dip of this reflectors will be added to the dip of the primary path which comes from its first downward bounce. In a case where the primaries had already dipping, the multiples that comes

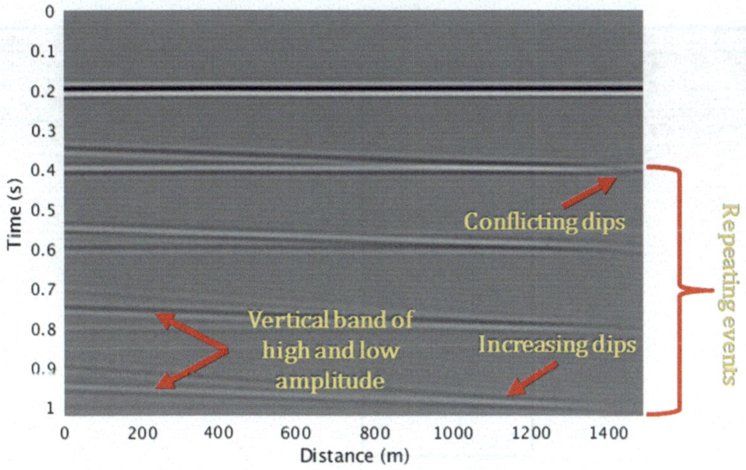

Fig. 4 Some characteristic of multiples commonly found in seismic data

from this primary will had more increasing dips. Whereas for a case where the multiples overlap with a deeper primary, it can be recognized if it's contained a conflicting dip.

If the reflectors where the multiples originated contains a small structure variation such as anticline or syncline, it will cause a geometric effect (focusing/defocusing) to the reflection events in terms of travel times and amplitudes. This effect will be further amplified with the increase order of the multiples. Due to the repeated focusing and defocusing effects, it will create a laterally incoherence higher-order multiples [6]. A higher-order multiple can also increase the amplitude variations of the multiple generating reflectors. As the multiples bounces along several reflector as it propagates through the subsurface, it creates a vertical band of high and low energy along the seismic section. This high and low amplitude variations depend on the strength of the reflectors where the bounce occurs.

1.3 Effect of Multiple Reflection in Seismic Data

One of the main challenges in processing the seismic data from Malay Basin is the presence of multiples in the dataset [7–8]. The presence of multiples in seismic dataset reduced the quality of the seismic data, therefore making its unreliable for accurate interpretation. Since multiples replicate the same primary's structure, it could be easily mistaken as a primary in the seismic section. Figure 5 shows an example of multiples found in Malay Basin. Noticed that the multiple reflectors shown by the black arrows could be easily mistaken as a primaries structure.

Beside reducing the reliability of seismic interpretation, it could also affect the quality of the AVO inversion. Multiples are considered as coherent noise which

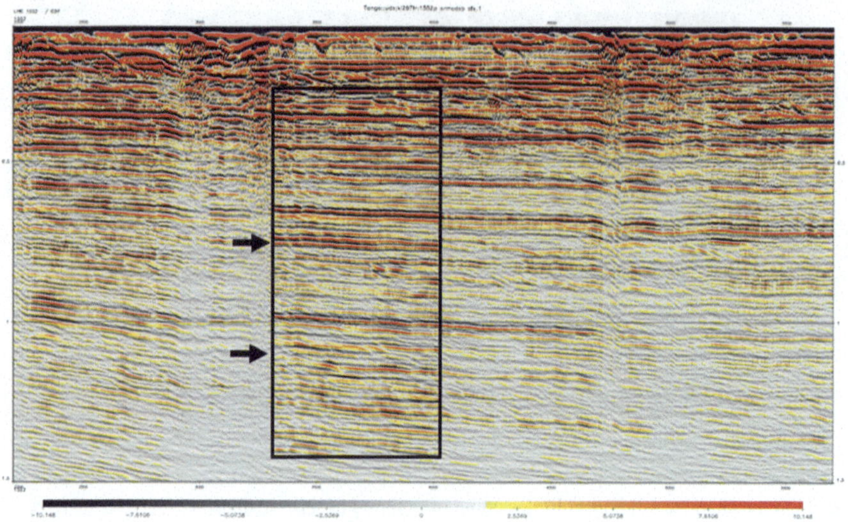

Fig. 5 Examples of multiples from Malay Basin [9]. The black box shows the seismic data before any multiple's reflection were removed and the two black arrows indicate the multiples events in the seismic data

could reduce the quality of the inversion, therefore creating an unreliable quantitative interpretation [10–11].

2 WRW Model

Using the two-dimensional version of the Rayleigh II integral, the propagation operator can be rewritten as a function W of the lateral coordinates as:

$$P(x_A, z_A, \omega) = \int_x P(x, z_S, \omega) W(x_A, x, z_A, z_S, \omega) dx \quad (1)$$

or:

$$P(x_A, z_A, \omega) = P(x, z_S, \omega) * W(x_A, x, z_A, z_S, \omega) \quad (2)$$

Acording to Berkhout [12], by rewriting the equation this way, the forward extrapolation in the frequency-space domain can be formulated in terms of convolution. The operator W depends on the output points or the points source. For example, given a point source at depth $z = z_0$, then the response from the depth level z_m will be given by the $W(x_A, x, z_A, z_s, \omega)$.

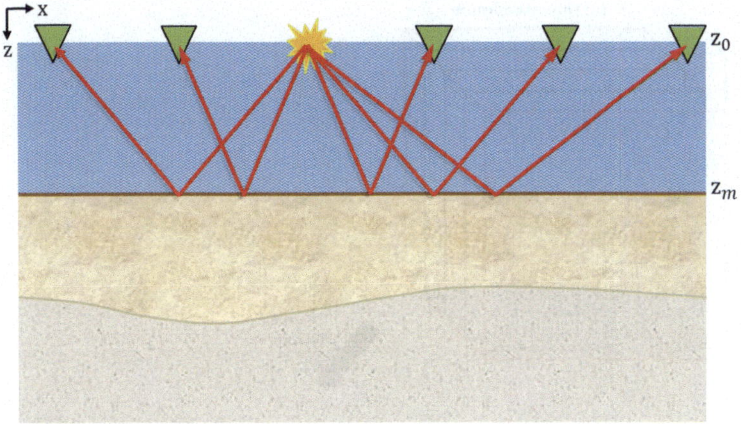

Fig. 6 Examples of seismic acquisition in fixed-spread configuration

2.1 Matrix Representative of Seismic Data

Consider a seismic dataset cotaining N shot records in a fixed-spread configuration with N detectors as shown in Fig. 6. The source location of this seismic dataset are the same as the detector location and this dataset are called $p(x_r, x_s, t)$ with x_r denotes the different receiver-stations, x_s is the different source-stations and t stands for the time for which the response are related. The recorded seismic were later placed in a matrix as shown in Fig. 7a, where each column represents one recorded trace for a specific shot receiver gather.

Using temporal Fourier transform, this data matrix is transformed from the time–space domain to the frequency-space domain, where each sample of this matrix cube (shown in Fig. 7b) is now a complex number containing the amplitude and phase information for one frequency and trace. The dataset $p(x_r, x_s, t)$ now becomes $P(x_r, x_s, \omega)$, with ω represents the frequency component. By taking the horizontal slice of these data cube as shown by the green line in Fig. 7c, it allowed us to calculate for each frequency component individually which are called as the data matrix, **P**. Each column and row of this matrix represents a monochromatic shot and receiver gather respectively. The matrix multiplication which is done in this WRW model was equivalents to the spatial integrals representing the convolution, which are discussed earlier.

2.2 Detail Notation and Conversion

Below are the detail-hiding notation and convention used by Berkhout [12] to represents the wavefields:

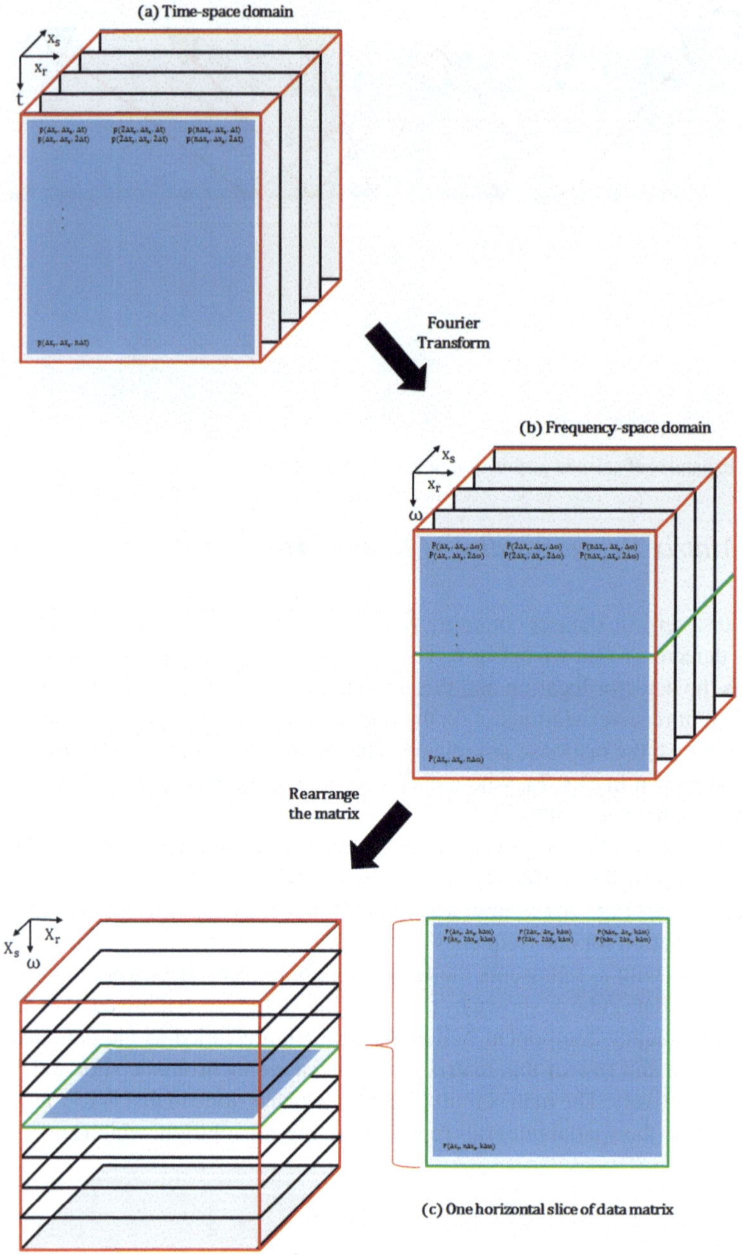

Fig. 7 The matrix representation of seismic dataset in the **a** time–space domain and **b** frequency-space domain. Δx_r, Δx_s, Δt and $\Delta\omega$ represents the sample interval of the receiver, source, time and frequency component. The horizontal slice (green line) of the data matrix **c** represents the kth frequency component in the frequency space domain

- **P**: Data matrix which represents the seismic measurement for one frequency component. Each column of this matrix represents the shot gather while each row represents the receiver gather.
- **S**: Source matrix contained the source wavefield which are generated by the sources in the subsurface. This wavefield are represented by the dipole sources which includes the source-ghost.
- **W**: This propagation operator based on the Green's function (as mentioned in Eqs. 1 and 2) describes the wavefield propagation from z_0 depth level to the z_m depth level.
- **D**: Describe the detector matrix which converts the upgoing wavefield $\mathbf{P}^-$ into the measured wavefield **P**. This matrix also contains the receiver-ghost.
- **R**: This reflectivity matrix can convert the up-going wavefield at the surface into a down-going wavefield. In marine case, at $z = z_0$ it can be approximated $\mathbf{R}^{\cap}(z_0) = -\mathbf{I}$.
- **X**: This matrix represents the impulse response generated by the dipole sources. It contained the information of the downward propagation into the subsurface and upward propagation to the receivers after reflections.
- $\mathbf{P}_0$,$\mathbf{X}_0$: The subscript "0" indicate there is no surface-related multiples presents in the matrix. This matrix contained all the primary reflection and all the internal multiples generated in the subsurface.
- W^+, W^-: The superscripts "+" and "−" indicate the downward and upward wavefield propagation.
- z_0, z_m: Represents the depth level at which the data is measured. $\mathbf{P}(z_m)$ indicate that the data are obtained at depth level z_m, $\mathbf{P}(z_0)$ indicate the data are obtained at the surface and $\mathbf{P}(z_m, z_n)$ indicate the wavefields originated from the source at depth level z_m and measured at depth level z_n.
- $\mathbf{P}_i$: The subscripts "i" denotes the number of iterations for that given matrix and its property are the results after ith iterations.
- $\widehat{\mathbf{P}}$: If the matrix contained the "^" symbol, it means that the property of that matrix is an estimation.
- $\Delta\mathbf{P}$: The "Δ" symbol indicate that this matrix contained only the primary reflection.

2.3 *Formulation of WRW Model*

To describe the construction of the wavefield using the WRW-model, the data-matrix must be properly rearranged as shown in Fig. 7c where the columns and rows of the data matrix **P** denotes the shot and receiver gathers respectively. For a single shot gather, the measured primary wavefield are given as:

$$\Delta\mathbf{P}_s(z_0, z_m) = \mathbf{D}(z_0)\Delta\mathbf{P}_s^-(z_0, z_0) \tag{3}$$

where $\Delta\mathbf{P}_s^-(z_0, z_0)$ denotes the reflected wavefield, and it was convolved with the receiver-properties, $\mathbf{D}(z_0)$ as it was measured by the receiver at the surface. This reflected wavefields can be further elaborated as:

$$\Delta\mathbf{P}_s^-(z_0, z_0) = \sum_{m=1}^{\infty} \mathbf{W}(z_0, z_m)\mathbf{R}(z_m, z_m)\mathbf{S}_s^+(z_m, z_0) \quad (4)$$

where the source wavefield, $\mathbf{S}_s^+(z_m, z_0)$ is scaled by the reflectivity operator, $\mathbf{R}(z_m, z_m)$ and forward propagated to the surface by the propagation operator, $\mathbf{W}(z_0, z_m)$ as a reflected wavefield. The source wavefield are generated by the source located at the surface (z_0) and it can be further expanded as:

$$\mathbf{S}_s^+(z_m, z_0) = \mathbf{W}(z_m, z_0)\mathbf{S}_s(z_0) \quad (5)$$

where $\mathbf{W}(z_m, z_0)$ is the downgoing propagation operator which brings the source-signal into the subsurface depth-level (z_m) as the source wavefields. Following Eqs. 3–5, the measured primary wavefields can be written as:

$$\Delta\mathbf{P}_s(z_0, z_m) = \mathbf{D}(z_0)\Delta\mathbf{X}(z_0, z_0)\mathbf{S}_s(z_0) \quad (6)$$

which are frequently referred as the 'WRW-model' [12]. The transfer matrix, $\Delta\mathbf{X}(z_0, z_0)$ represents the Earth's impulse responses for primary reflection in the subsurface:

$$\Delta\mathbf{X}(z_0, z_0) = \sum_{m=1}^{\infty} \mathbf{W}(z_0, z_m)\mathbf{R}(z_m, z_m)\mathbf{W}(z_m, z_0) \quad (7)$$

2.4 Extension for Surface Related Multiple

To include the surface-related multiple inside the WRW-model, the model was extended through adding a feedback path at the surface. The Eq. 6 now becomes:

$$\Delta\mathbf{P}_s(z_0, z_m) = \mathbf{D}(z_0)\Delta\mathbf{X}(z_0, z_0)\left[\mathbf{R}^\wedge(z_0, z_0)\mathbf{P}_s^-(z_0, z_0) + \mathbf{S}_s(z_0)\right] \quad (8)$$

In this equation, the boundary operator, $\mathbf{R}^\wedge(z_0, z_0)$ transforms the reflected wavefield, $\mathbf{P}_s^-(z_0, z_0)$ into the down going wavefield and the transfer matrix, $\Delta\mathbf{X}$ transforms the down going wavefield into the reflected wavefield again. Figure 8 shows the workflows and the visualization of the WRW-model and the feedback-model in terms of ray path. Rewriting Eq. 8 in terms of primaries and multiples, the following equation were obtained:

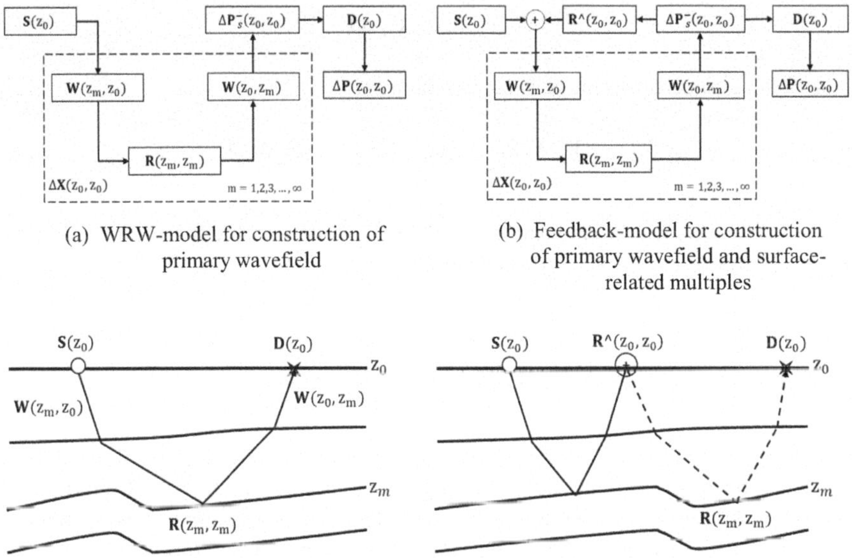

(a) WRW-model for construction of primary wavefield

(b) Feedback-model for construction of primary wavefield and surface-related multiples

(c) Visualization of WRW-model in terms of raypath.

(d) Visualization of Feedback-model in terms of raypath.

Fig. 8 WRW and feedback model for primary wavefield construction and surface-related multiples [13]

$$\mathbf{P} = \mathbf{D}\mathbf{X}_0\mathrm{S}^+ + \mathbf{D}\mathbf{X}_0\mathbf{R}^\cap\mathbf{P}^- \tag{9}$$

and after applying receiver deghosting, Eq. 9 becomes:

$$\mathbf{P}^- = \mathbf{X}_0\mathrm{S}^+ + \mathbf{X}_0\mathbf{R}^\cap\mathbf{P}^- \tag{10}$$

In practical term, due to the material discontinuity at the surface, the primary wavefield will be reflected into the subsurface for an additional round-trip, where the surface itself becomes the secondary sources for this secondary wavefield. This secondary wavefield will then be reflected again at the free surface for another extra round-trip and this process will be repeated again until the multiples lose its energy. Therefore, the wavefield measured at the surface can be described as an infinite sum of multiple contributions [14]:

$$\mathbf{P}^- = \mathbf{X}_0\mathrm{S}^+ + (\mathbf{X}_0\mathbf{R}^\cap)\mathbf{X}_0\mathrm{S}^+ + (\mathbf{X}_0\mathbf{R}^\cap)^2\mathbf{X}_0\mathrm{S}^+ + (\mathbf{X}_0\mathbf{R}^\cap)^3\mathbf{X}_0\mathrm{S}^+ + \cdots \tag{11}$$

with $\mathbf{X}_0\mathrm{S}^+$ represent the primaries, $(\mathbf{X}_0\mathbf{R}^\cap)\mathbf{X}_0\mathrm{S}^+$ represent the first-order multiples, $(\mathbf{X}_0\mathbf{R}^\cap)^2\mathbf{X}_0\mathrm{S}^+$ represent the second-order multiples and so on. By defining the surface operator, $\mathbf{A}$ as $\mathbf{A} = \mathbf{S}^{-1}\mathbf{R}^\cap$ and $\mathbf{P}_0^- = \mathbf{X}_0\mathrm{S}^+$ Eq. 10 can be written as:

$$\mathbf{P}^- = \mathbf{P}_0^- + \mathbf{P}_0^- \mathbf{A}\mathbf{P}^- \tag{12}$$

with the geometric series of Eq. 11 written as:

$$\mathbf{P}^- = \mathbf{P}_0^- - (\mathbf{P}_0^-\mathbf{A})\mathbf{P}_0^- + (\mathbf{P}_0^-\mathbf{A})^2\mathbf{P}_0^- - (\mathbf{P}_0^-\mathbf{A})^3\mathbf{P}_0^- + \ldots \tag{13}$$

where $\mathbf{P}_0^-$ now represent the primaries, $(\mathbf{P}_0^-\mathbf{A})\mathbf{P}_0^-$ represent the first-order multiple, $(\mathbf{P}_0^-\mathbf{A})^2\mathbf{P}_0^-$ represent the second-order multiple and so on.

3 Multiple Elimination Theoretical Review

To removes the multiple reflections, several different methods had been developed such as using move-out discrimination method (F-K and Radon Transforms), filter method (predictive deconvolution) and wavefield extrapolation method. The focused of this thesis is in using the data-driven demultiple methods based on the Feedback-model introduced by Berkhout [12]. Below are some of the data-driven demultiple method that were previously created based on the Feedback model.

From here onwards, it is assumed that the data had been applied a receiver deghosting beforehand. Therefore the superscript "–" and "+" will be dropped from the matrix notation.

4 Closed-Loop SRME: CL-SRME Theory

Following the CL-SRME algorithm [15] and the detail hiding notation [13], the measured wavefield at the surface, $\mathbf{P}$ can be described by the primaries, $\mathbf{P}_0$ and the multiples, $\mathbf{P}_0 A\mathbf{P}$ as a product of up-going wavefield, $\mathbf{P}_0$ and down-going wavefield, $(\mathbf{I} + A\mathbf{P})$:

$$\mathbf{P} = \mathbf{P}_0(\mathbf{I} + A\mathbf{P}) \tag{14}$$

According to Berkhout [15], the total wavefields can also be described using the primary impulse response, $\mathbf{X}_0$, source wavelet, $\mathbf{S}$ and surface reflectivity, $\mathbf{R}^\cap$:

$$\mathbf{P} = \mathbf{X}_0\mathbf{S} + \mathbf{X}_0\mathbf{R}^\cap\mathbf{P} \tag{15}$$

This formulation can also be used to described the measured wavefields in terms of up-going and down-going wavefields where Eq. 15 now becomes:

$$\mathbf{P} = \mathbf{X}_0\left(\mathbf{S} + \mathbf{R}^\cap\mathbf{P}\right) \tag{16}$$

Instead of describing the measured wavefields in terms of primaries and multiples, this formulation described the wavefields in terms of its direction of propagation, where **P** represent the up-going part, and $(\mathbf{S} + \mathbf{R}^{\cap}\mathbf{P})$ represent the down-going part, with the primary impulse, $\mathbf{X}_0$ are used to connect these two parts. Following Eq. 15, the primaries can be described as matrix multiplication results between the primaries impulse, $\mathbf{X}_0$ and the source wavelet, **S**, where $\mathbf{P}_0 = \mathbf{X}_0\mathbf{S}$. Replacing this in Eq. 14, the measured wavefields now becomes:

$$\mathbf{P} = \mathbf{X}_0\mathbf{S}(\mathbf{I} + A\mathbf{P}) \tag{17}$$

This equation can also be described in terms of primaries and multiples wavefield as:

$$\mathbf{P} = \mathbf{X}_0\mathbf{S} + (\mathbf{X}_0\mathbf{S})A\mathbf{P} \tag{18}$$

The main goal of this inversion process is to find an objective function, J that will be minimized as the correct value for the unknowns are found, which accurately describe the input dataset. The objective function for this method were based on Eq. 17, and it was written as:

$$J = \sum_{\omega} \left\| \mathbf{P} - \widehat{\mathbf{X}}_0\widehat{\mathbf{S}}\left(\mathbf{I} + \widehat{A}\mathbf{P}\right)^2 \right\| \tag{19}$$

The variables $\widehat{\mathbf{X}}_0$ and $\widehat{\mathbf{S}}$ represent the real value estimate of $\mathbf{X}_0$ and **S**. Following Eq. 19, if this value accurately described the measured wavefields, **P**, the objective function will goes to zero. The solution for this objective function must be obtained over all the frequencies ω. Similar to the CL-SRME approach, the initial value for A that were used in the first iteration were $\widehat{A} = A_{SRME}$. This value of $\widehat{A}$ were obtained by calculating A as a result of matching filter between the wavefield **P** and **PP**. The reason for this initial value of $\widehat{A}$ is to ensure that a correlation type of update was obtained for $\mathbf{P}_0$ during its first iteration [14].

5 Application of Sparsity Constraint

In order to reach the minimum in the objective function, a condition or constraint must be applied to the inverse problem. Through employing the sparsity constraint in the regularization of the inverse problem, it reduces the amount of possible solution it needed to reach its convergence. Sparse signal is a high dimensional signal which contained a small number of non- zero elements [16]. The sparsity of a signal can be defined by its number of a non-zero elements. Primary impulse is an example of a sparse signal, where it can be defined as a signal with a series of spikes in the time domain, where each spikes represents the reflection events in the subsurface [1]. The sparsity constraint can be imposed using the $\boldsymbol{L}_0$ norm or $\boldsymbol{L}_1$ norm depending of the type of signal that were recovered. Through incorporating an extra constraint into the algorithm, the new objective function are now written as:

$$J = \sum_{\omega} J_{\omega}^{(LS)} + \sum_{t} J_{t}^{(reg)} = \sum_{\omega} \left\| \mathbf{P} - \widehat{\mathbf{X}}_0 \widehat{\mathbf{S}} \left(\mathrm{I} + \widehat{A}\mathbf{P} \right) \right\|^2 + \lambda \sum_{t} \| \Delta\mathbf{x}_0 \| s \tag{20}$$

where J is the new objective function, $J_{\omega}^{(LS)}$ is the original objective function and $J_{t}^{(reg)}$ is the objective function with the extra constraint.

The goal of this inversion is to minimize the objective function shown by the above equation. In this equation, $\Delta\mathbf{x}_0$ is the update of the primary impulse response in the time domain and $\boldsymbol{\lambda}$ is the regularization constant that were defined by the user. The norm.s represents the sparsity-norm that were applied to the update of the primary impulse, $\Delta\mathbf{x}_0$ in the time domain. Since the regularization norm were applied in the time domain as shown in Eq. 23, sparsity norm were also executed in that domain. Following the new objective function, the update direction for the impulse response, $\Delta\mathbf{x}_0$ were given as:

$$\Delta\mathbf{X}_0 = \Delta\mathbf{X}_0^{(LS)} + \Delta\mathbf{X}_0^{(reg)} \tag{21}$$

with:

$$\Delta\mathbf{X}_0^{(LS)} = \left[\mathbf{P} - \widehat{\mathbf{P}}_0 \left(\mathbf{I} + \widehat{A}\mathbf{P} \right) \right] \left(\mathbf{I} + \widehat{A}\mathbf{P} \right)^H \widehat{\mathbf{S}}^H \tag{22}$$

and:

$$\Delta\mathbf{x}_0^{(reg)} = \lambda \| \Delta\mathbf{x}_0 \| s \tag{23}$$

where $\Delta\mathbf{X}_0$ is the total primary impulse update, $\Delta\mathbf{X}_0^{(LS)}$ and $\Delta\mathbf{X}_0^{(reg)}$ is the portion of the $\Delta\mathbf{X}_0$ which are related to the least-square part and regularized part of J. Note that $\Delta\mathbf{X}_0^{(reg)}$ is the Fourier transformed of the $\Delta\mathbf{x}_0^{(reg)}$. The symbol λ represent the overall scaling factor which handle the strength of the sparsity regularization.

The residual $\mathbf{V}$ are calculated using the steepest descent method through the following equations:

$$\mathbf{V} = \left[\mathbf{P} - \widehat{\mathbf{P}}_0 \left(\mathbf{I} + \widehat{A}\mathbf{P} \right) \right] \left(\mathbf{I} + \widehat{A}\mathbf{P} \right)^H \tag{24}$$

where $\left(\mathbf{I} + \widehat{A}\mathbf{P} \right)^H$ represents the complex adjoint of $\left(\mathbf{I} + \widehat{A}\mathbf{P} \right)$. The term $\left[\mathbf{P} - \widehat{\mathbf{P}}_0 \left(\mathbf{I} + \widehat{A}\mathbf{P} \right) \right]$ can be considered as the remaining residual obtained from the results of the previous iteration. Through the matrix multiplication between

$\left[\mathbf{P}-\widehat{\mathbf{P}}_0\left(\mathbf{I}+\widehat{A}\mathbf{P}\right)\right]$ and $\left(\mathbf{I}+\widehat{A}\mathbf{P}\right)^H$, it allows the information from the multiple to be contribute to the primaries estimation. The initial values for the primaries estimate $\widehat{\mathbf{P}}_0$ were set to be zero in the first iteration.

The update of the primary impulse $\Delta\mathbf{X}_0$ will then be calculated from the residual as:

$$\Delta\mathbf{X}_0 = (\mathbf{V})\mathbf{S}^H \tag{25}$$

where $\mathbf{S}^H$ represents the complex adjoint of the source wavelet that were applied to the residual, $\mathbf{V}$ in order to obtained the update direction $\Delta\mathbf{X}_0$ of the primary impulse. In the first iteration, the Gaussian wavelet are used as the initial source wavelet. This source wavelet will then be replaced in later iteration with the new estimated source wavelet.

Since it was assumed that he primary impulse was sparse in the time domain, the sparsity-promoting regularization norm.s were applied to the update of the primary impulse $\mathbf{X}_0$ in the time domain. This extra constraint will have allowed us to obtain a sparse representation of the primary impulse while removing the aliasing noise during the update of the primary impulse. Using the recursion formula, the updates $\Delta\mathbf{X}_0$ are used to obtain the new estimate of $\mathbf{X}_0$ in every iteration i as:

$$\widehat{\mathbf{X}}_0^{(i+1)} = \widehat{\mathbf{X}}_0^{(i)} + \alpha\Delta\mathbf{X}_0^{(i)} \tag{26}$$

The scaling factor α are obtained through line search method such that the value of the objective function $J\left(\widehat{\mathbf{X}}_0^{(i+1)}\right) = J\left(\widehat{\mathbf{X}}_0^{(i)} + \alpha\mathbf{X}_0^{(i)}\right)$ will be minimized after every iteration.

Following Eq. 24 the update of $\widehat{\mathbf{S}}$ can be obtained through the matrix multiplication between the residual $\mathbf{V}$ with the complex adjoint of primary impulse update $\Delta\mathbf{X}_0$:

$$\Delta\mathbf{S} = (\mathbf{V})\mathbf{X}_0^H \tag{27}$$

This method is assumed that the source array used in sparse CL- SRME are originated from one source, and a single source wavelet are estimated for all shots. This updated source wavelets are then brought in to the time domain to ensure that the length of the wavelet are limited. To reduce the objective function value in the following iteration, a scale factor is then applied to the update of the source wavelet in the frequency domain. The primary estimate $\widehat{\mathbf{P}}_0$ can then be obtained directly through the matrix multiplication between the estimated primary impulse, $\widehat{\mathbf{X}}_0$ with the estimated source wavelet, $\widehat{\mathbf{S}}$ in the frequency domain.

$$\widehat{\mathbf{P}}_0 = \sum_{\omega}\widehat{\mathbf{X}}_0\widehat{\mathbf{S}} \tag{28}$$

These two parameters are estimated in an alternate manner, where the primary impulse are estimated first, followed by the update of the source wavelet in every iterations as both of them are strongly dependent on each other. As for the surface-operator, A, the update for this parameter are strongly dependent on the primaries estimate, $\widehat{\mathbf{P}}_0$. This parameter can be estimated as a matched filter obtained through least-square matching with the estimated primaries in the time domain. The estimation of $\widehat{A}$ were done similar to the method introduced by Lopez [15], where it was proposed that the estimation are done in two stages. For early iteration, the surface operator were estimate using the minimum energy constraint [17]:

$$\widehat{A} = A(\omega) \text{ s.t } \mathbf{P} - \widehat{\mathbf{P}}_0 \widehat{A} \mathbf{P}^2 \rightarrow \min \quad (29)$$

Through obtaining the surface-operator using the minimum energy criterion, the trivial solution of $\hat{A} = 0$ and $\mathbf{P}_0 = \mathbf{P}$ are avoided. In later iteration, this minimum energy constraint are dropped and replaced by estimating the surface-operator through minimizing the residual:

$$\hat{A} = A(\omega) \text{ s.t } \mathbf{P} - \hat{\mathbf{P}}_0 - \hat{\mathbf{P}}_0 \hat{A} \mathbf{P}^2 \rightarrow \min \quad (30)$$

This way a more accurate estimate of $\widehat{A}$ can be obtained and the residual will be resolved much better compared to the used of minimum energy criterion.

6 Synthetic Data Examples: Malay Basin Model

This model was created using the finite-difference algorithm based on the velocity model shown in Fig. 9. The structure of this model was based on the structures of Malay Basin particularly at the northern area of the basin. The geological structure of this model is slightly dipping with sudden change to horizontally flat at 500 m with a small anticline at the bottom. To obtain a fixed-spread data, the source and the receiver were placed at the same location. This model contained 150 shots and 150 traces with a 10 m spacing between each interval. Each traces contained a 256 time samples with a 4 ms sample interval. The maximum frequency used during the forward modelling is set to be around 70 Hz.

6.1 Shot Gather Display

The shot gather for the input dataset and the results of various multiple attenuation process were displayed in Figs. 10, 11, 12 and 13. In the shot gather for the input data in Fig. 10a, some multiples interference can be observed particularly at 0.3 and 0.5 s (marked by the blue arrow) where the primaries continuity appears to be missing.

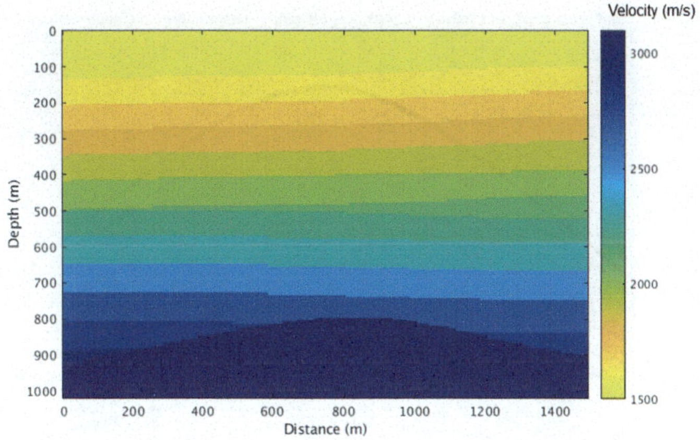

Fig. 9 Velocity model for Malay Basin model

Furthermore, there were also some presences of multiple reflections at 0.7 s and at the bottom of the shot gather at 0.95 s. From Figs. 10b and 11b, it appears that both SRME and CL-SRME were unable to properly recover the missing primaries reflector after the multiples were removed. The predicted multiples that were generated by these two methods (Figs. 12b and 13b) also shows that they were unable to properly construct the overlapping multiples at 0.5 s (marked by the blue arrow). Sparse CL-SRME (Fig. 11a, b) were able to properly recover the missing primaries reflector at 0.3 s (shown by the green arrow). The predicted multiples generated by these two methods also appears to be able to construct the overlapping multiples shows as shown in Fig. 13a, c.

6.2 *Zero Offset Gather Display*

Figures 14, 15, 16 and 17 shows the zero-offset gather for the input data and the results of various multiple attenuation process. In Fig. 14, it can be observed that there were several multiple interference in the input data. Marked by the blue arrows, the multiples reflections interfere with the primary reflector resulting in reducing the primaries continuity at the 0.3 and 0.5 s. The multiples can also be observed at 0.55 and 0.95 s.

Due to the strong multiple interference at 0.3 s, the SRME were unable to recover the missing primaries continuity as shown by the blue arrow in Fig. 15a. Due to its inability to properly recovered the primaries reflectors, this missing primaries reflectors were replicated by the subsequent order of multiples which appear in the SRME predicted multiples (blue arrow at 0.5 s) in Fig. 15b. Therefore, the corresponding

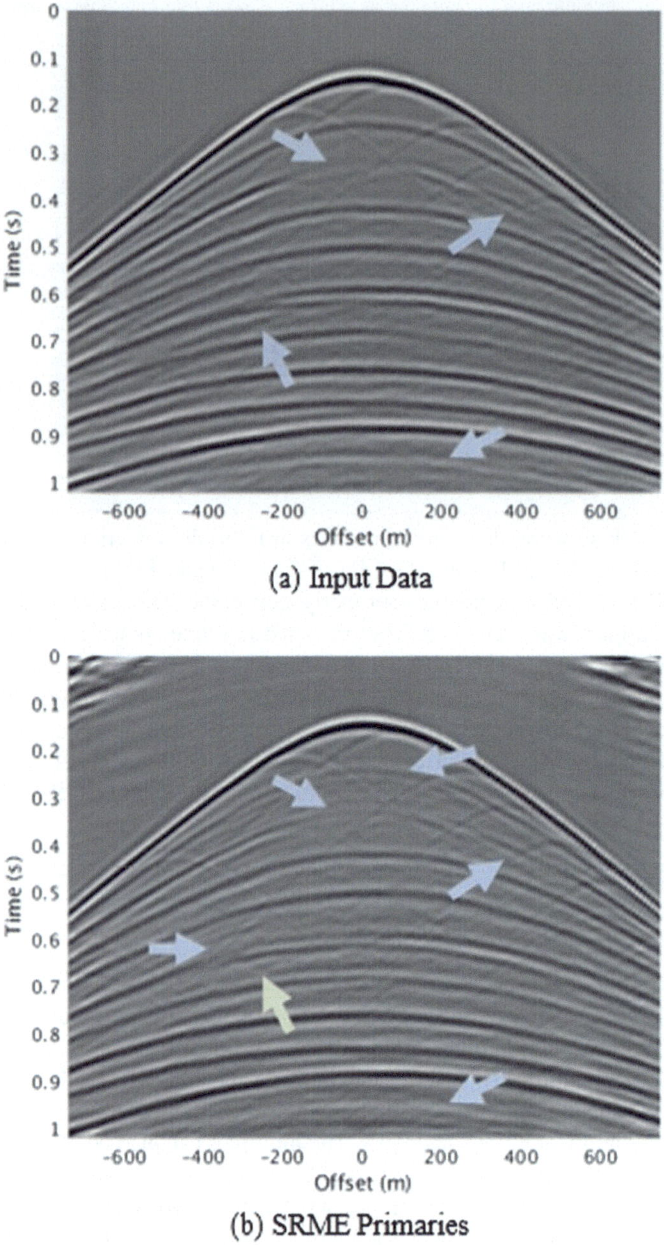

Fig. 10 Primaries shot gather of Malay Basin model: **a** input data and **b** SRME primaries

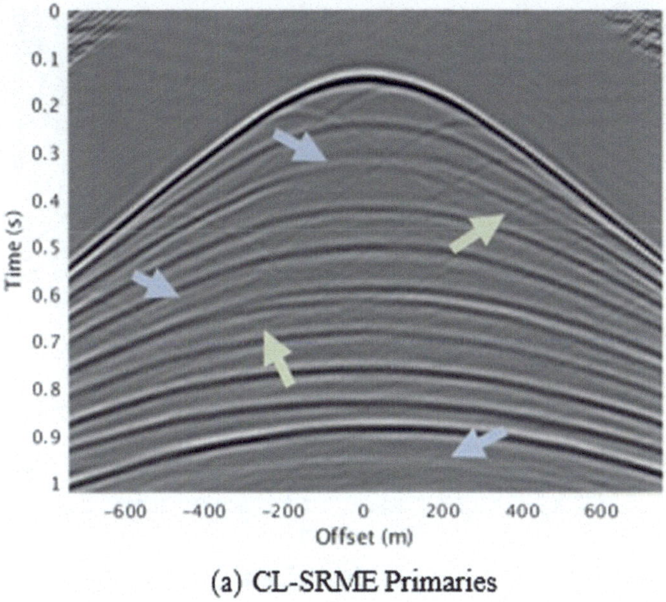

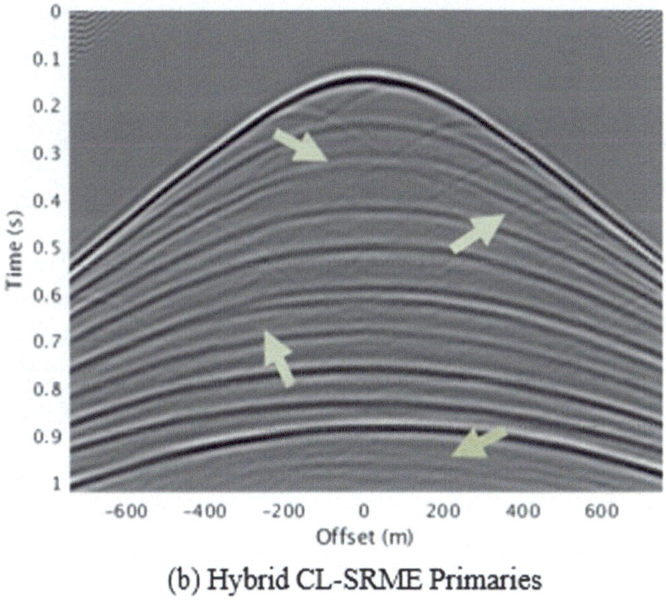

Fig. 11 Primaries shot gather of Malay Basin model: **a** CL-SRME primaries and **b** sparse CL-SRME primaries

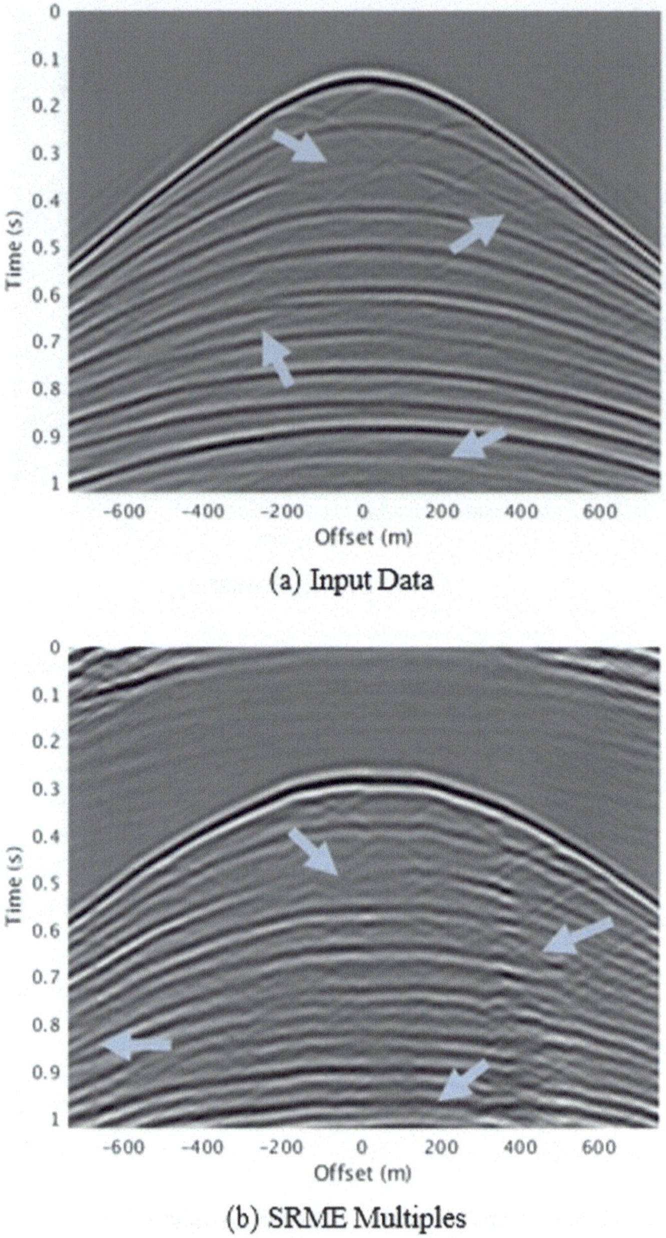

Fig. 12 Multiples shot gather of Malay Basin model: **a** input data and **b** SRME multiples

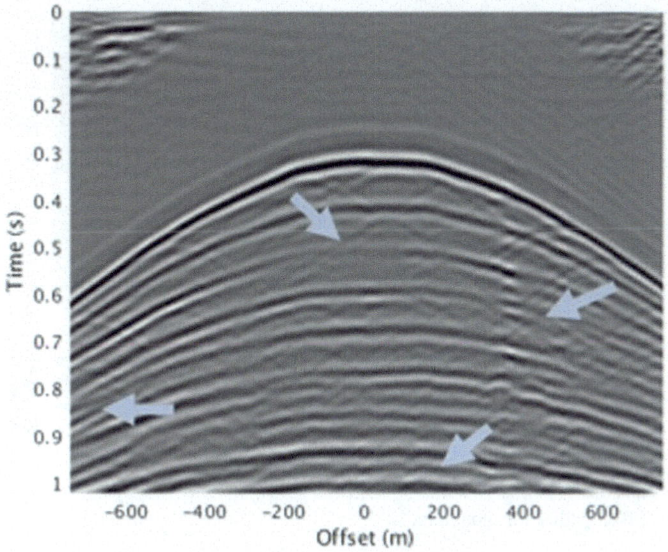

(a) CL-SRME Multiples

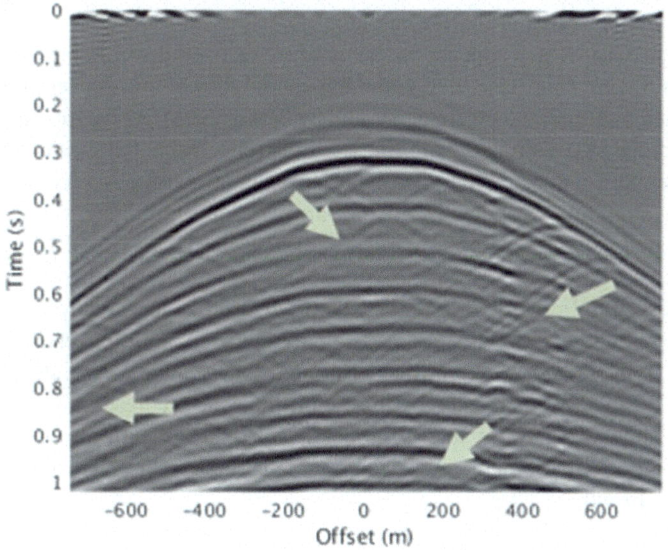

(b) Hybrid CL-SRME Multiples

Fig. 13 Multiples shot gather of Malay Basin model: **a** CL-SRME multiples and **b** sparse CL-SRME multiples

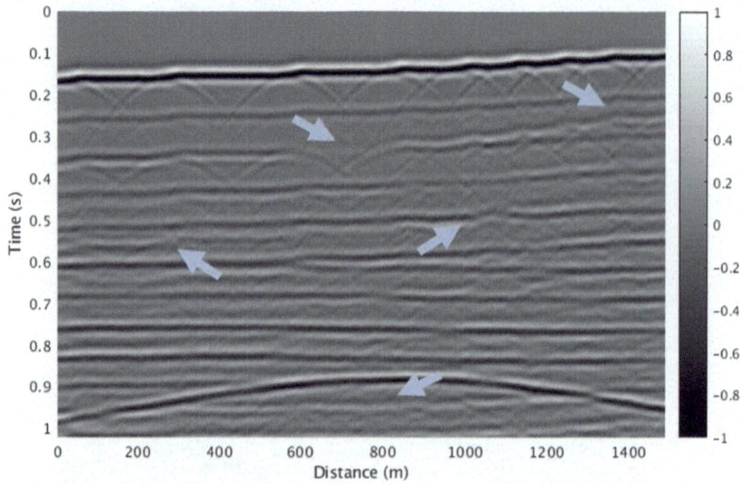

Fig. 14 Zero-offset gather for input data

multiple model created by the SRME were already innacurate to begin with. Furthermore, it appears that there were also some multiple residuals can be observed at 0.55 and 0.95 s in the SRME primaries output.

Although CL-SRME appears to be able to recover the missing overlapping primaries reflector (Fig. 15a), due to the low amplitudes of the recovered primaries, the predicted multiples which are generated by the CL-SRME (Fig. 16b) appear to be unable to properly construct the missing multiple reflectors at 0.5 s (marked with the blue arrow). It also appears there was some ringing artefacts observed at the top of the predicted multiples as noted by the blue arrows. Some multiple residuals can also be observed at 0.3 and 0.25 s as (marked by the blue arrow in CL-SRME primaries output).

As observed in Fig. 17a, the sparse CL-SRME were able to recover the missing primary continuity at 0.3 s. It were also able to reduce the multiple energy at 0.95 s (noted by the green arrow). However, as noted by the blue arrow at 0.3 s, there appears to be an artefacts created by this algorithm. The predicted multiples section (Fig. 17b) show that it were able to properly construct the multiples model with a good continuity. Although, there were some small artefacts observed at 0.75 s (noted by the blue arrow).

6.3 Auto-Correlation Section

Figures 18, 19 and 20 shows the auto-correlation section of the input data and the primaries output for all the multiple attenuation process. Based on the auto-correlation section of the input data shown in Fig. 18, there were some multiples

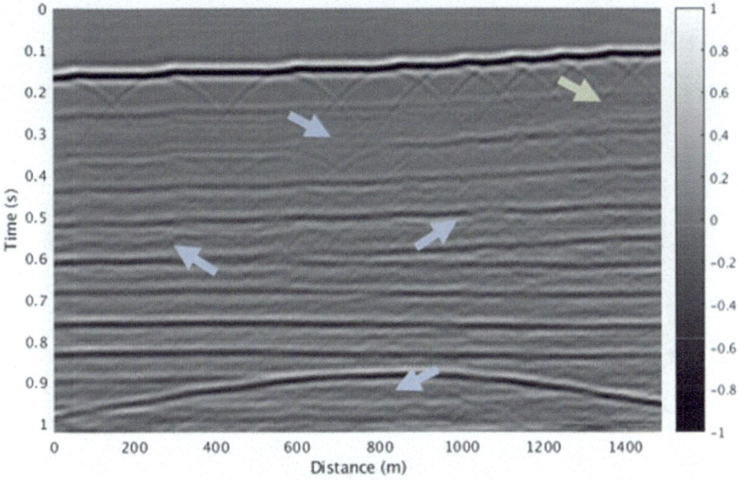

(a) SRME Primaries

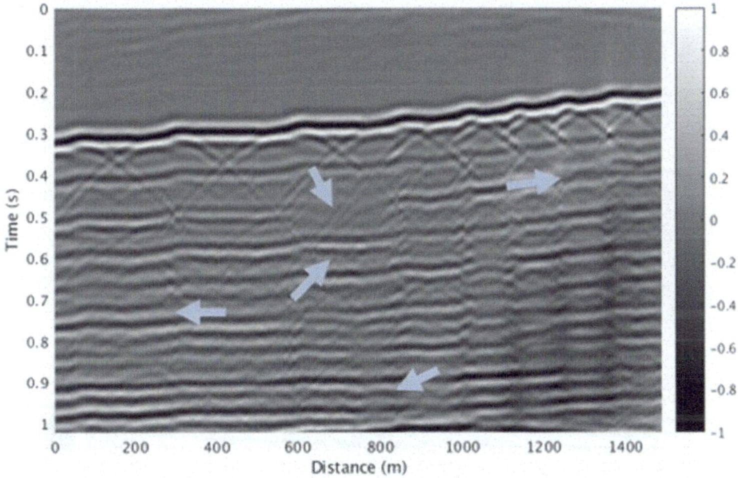

(b) SRME Multiples

Fig. 15 Zero-offset gathers for: **a** SRME primaries and **b** SRME multiples

events can be observed in the auto-correlation section as marked by the blue arrow. The strongest events can be noted at 0.2 s which clearly shows a line of multiple reflectors. After the application of SRME (Fig. 19) and CL-SRME (Fig. 20a), it can observe that these two methods were unable to properly removes this multiple reflector, as there were some presence of multiples residuals left (marked by the blue arrow at 0.2 s) in the auto-correlation section of this two primaries output. Through applying the sparse CL-SRME (Fig. 20b) algorithm, the line of multiples was now

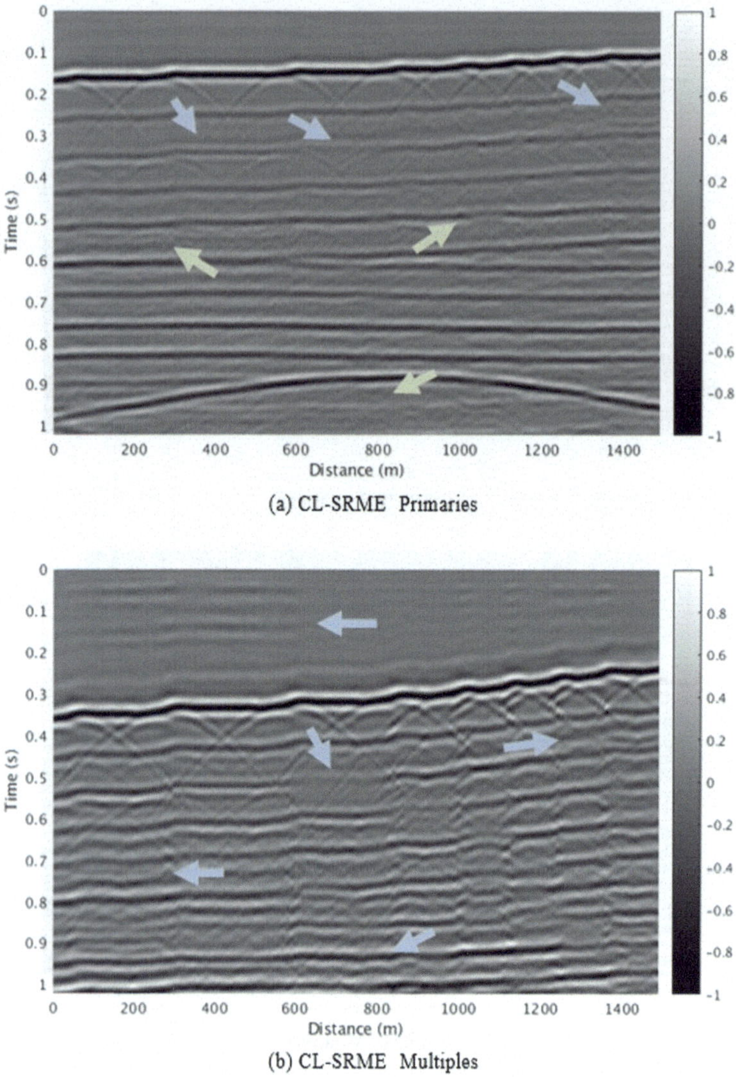

Fig. 16 Zero-offset gathers for: **a** CL-SRME primaries and **b** CL-SRME multiples

removed, however it can be also noted that there were still some small residuals left in the primaries output as noted by the blue arrows.

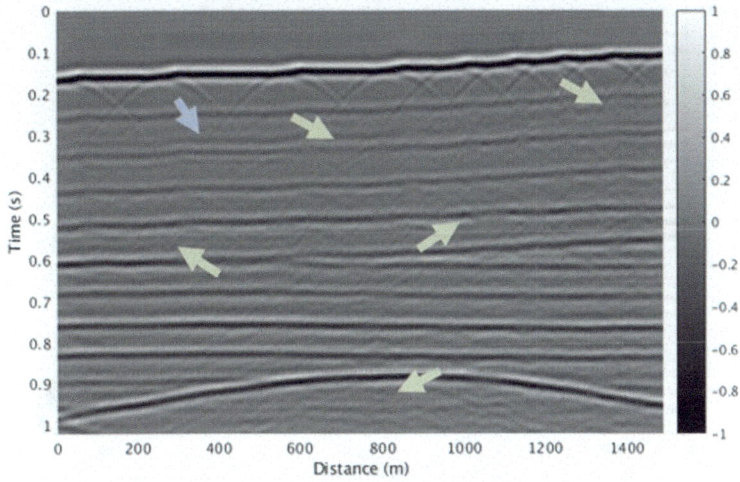

(a) Hybrid CL-SRME Primaries

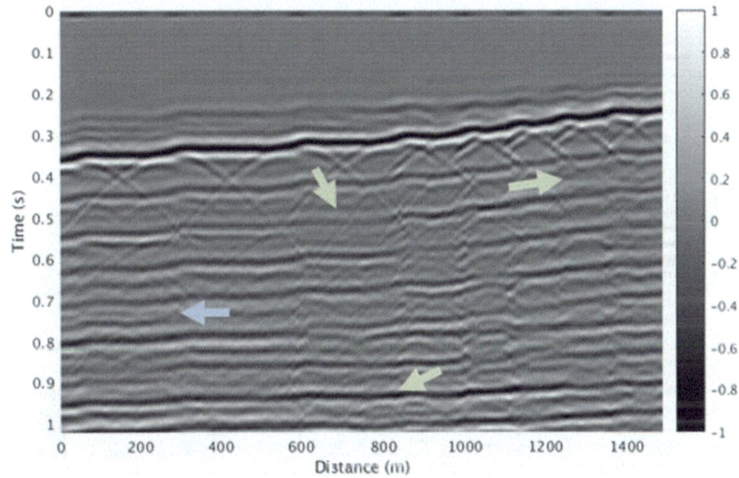

(b) Hybrid CL-SRME Multiples

Fig. 17 Zero-offset gathers for: **a** sparse CL-SRME primaries and **b** sparse CL-SRME multiples

6.4 *Spectral Analysis*

The amplitude spectrum for the input data and primaries output of various multiple attenuation methods are shown in Figs. 21, 22 and 23. When compared with the input data, the amplitude spectrum generated from the SRME primaries (Fig. 21b) shows a significant increase in the spectral power at the low and high frequency range. Similar to the case of the Complex Salt model, it was observed that the CL-SRME

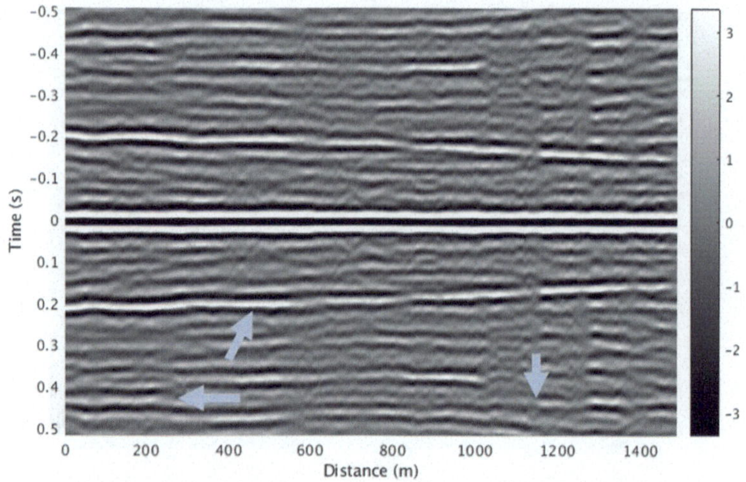

Fig. 18 Autocorrelation of input data

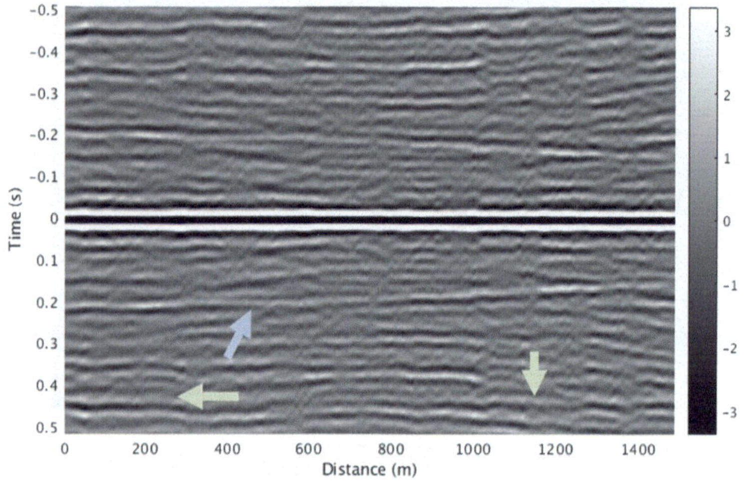

Fig. 19 Auto-correlation section for SRME primaries

(Fig. 22b) and sparse CL-SRME (Fig. 23a) show a good match of the spectral power in the middle frequency range between 5 and 60 Hz, and there is a sudden reduction of spectral power noticed from 60 Hz and above. Moreover, the amplitude spectrum for the sparse CL-SRME also shows a sudden increase in the amplitude spectrum at the low frequency section as noted in Fig. 23a. As for overall amplitude spectrum (Fig. 23b), it shows that the CL-SRME method gave the best amplitude spectrum compared to the other methods, as the amplitude spectrum it produces are closer to the input data.

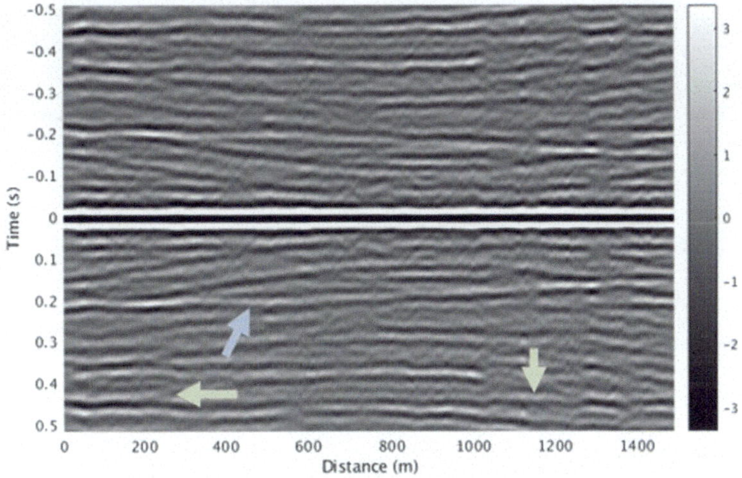

(a) Auto-correlation of CL-SRME Primaries

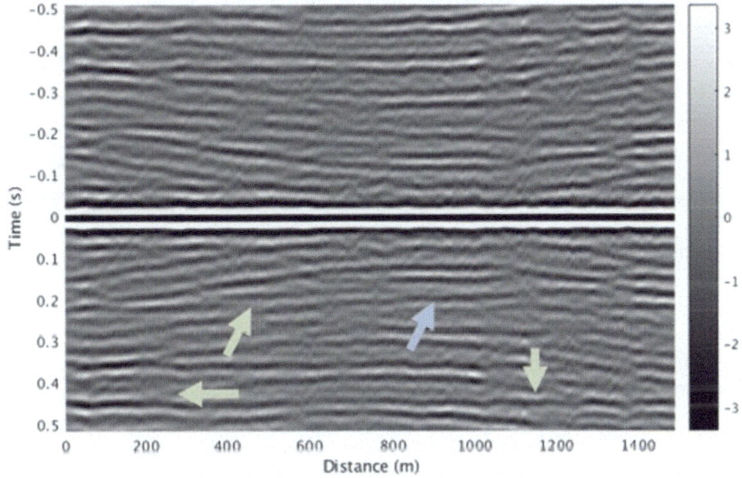

(b) Auto-correlation of Hybrid CL-SRME Primaries

Fig. 20 Auto-correlation section for: **a** CL-SRME primaries and **b** sparse CL-SRME primaries

7 Field Data Examples: Offshore Malay Basin

This field data were acquired from offshore Malay Basin, at the northern region area (Fig. 24). This area had a shallow water depth which makes it prone to short-period multiples. The structure of this field data is almost flat with a slight dipping but it contained many pockets of high amplitude anomalies which might indicate a shallow gas. According to Ghazali [18], gas cloud was one of the prominent multiple

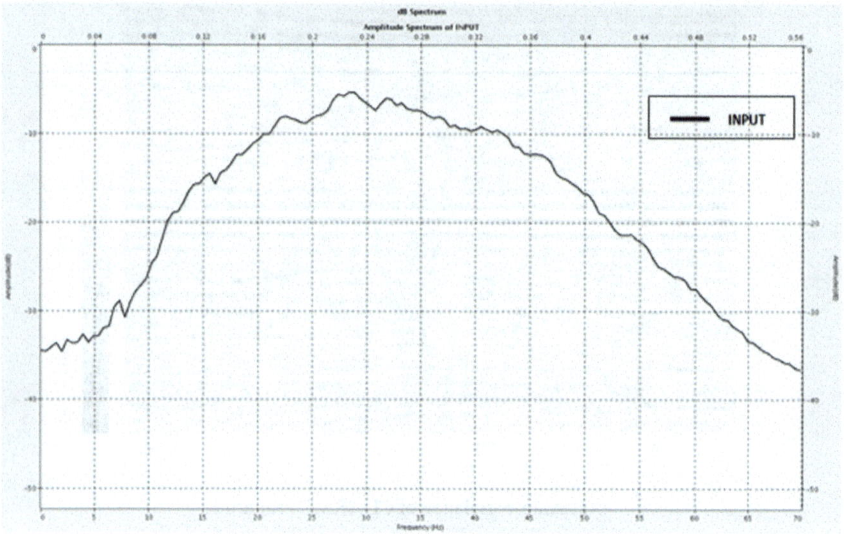

(a) dB Spectrum of Input Data

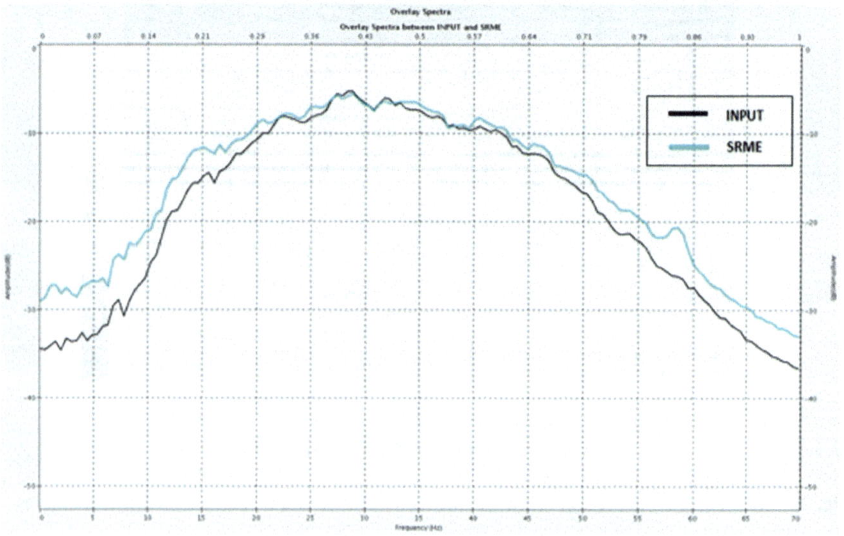

(b) Spectrum Overlay between Input Data and SRME

Fig. 21 Amplitude spectrum for: **a** input data and **b** SRME primaries

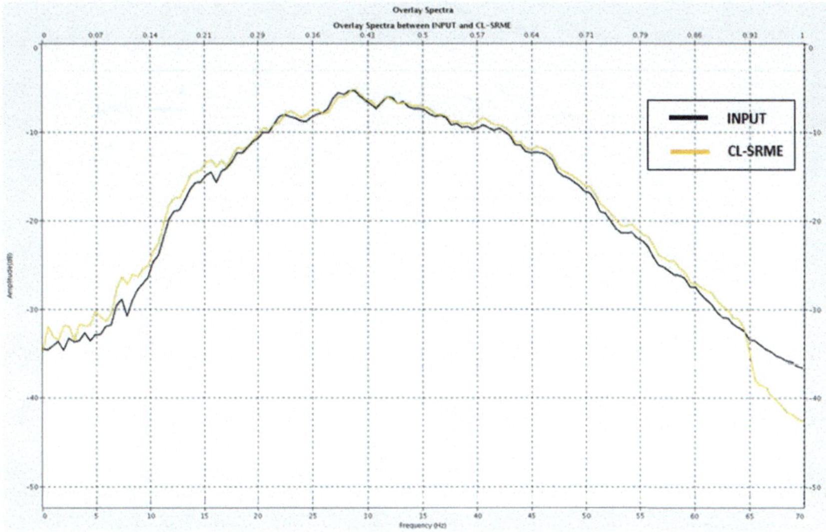

Fig. 22 Amplitude spectrum for CL-SRME primaries

generators in Malay Basin. This field data was previously used by Arshad et.al [19] for velocity estimation using the Joint Migration Inversion.

Before the multiple attenuation were apply, this data had been subjected to several conditioning to removes any unwanted signal such as the swell noise and the direct wave. This data was resampled from 37.5 to 12.5 m so that it will be suitable for the multiple attenuation process. To recover the missing near-offset gap, the spatial trace interpolation was applied. As the data was acquired as a moving-spread, the reciprocity was applied to turn it into a fixed-spread dataset of 150 shots with 150 traces. After successful implementation of the multiple attenuation process, the data had been applied with the Auto-Gain Control (AGC) to increase its amplitude for better observation (Fig. 25).

7.1 Shot Gather Display

Figures 26, 27, 28 and 29 shows the shot gather for the input data, primaries and multiples generated by the various multiple attenuation technique. As shown by Fig. 26a, the input data contained several multiple reflectors which are noted by the blue arrows. This multiple reflector results in the missing primary continuity, unwanted artefacts and also a change in polarity. The results of SRME primaries (Fig. 26b) shows that it was unable to attenuate the multiples properly. As noted by the blue arrows in at 0.4 s, SRME were unable to properly construct the overlapping primaries. There were also some multiples leakage observed at 0.8 and 0.9 s.

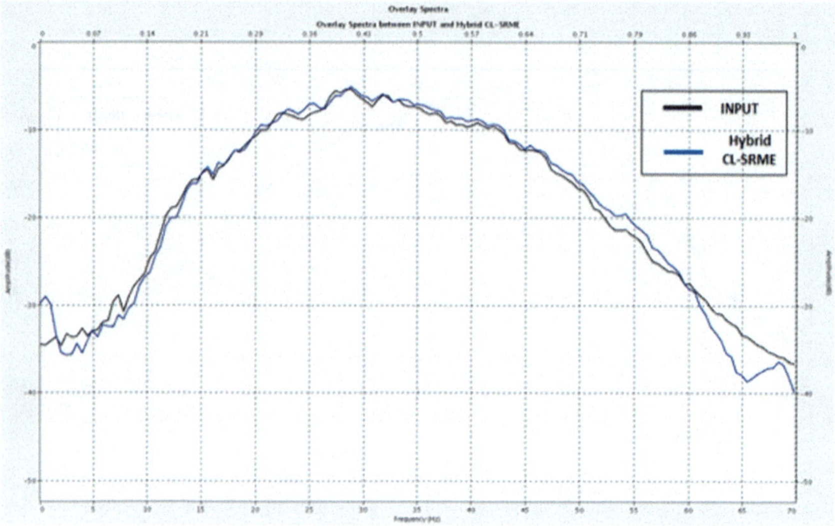

(a) Spectrum Overlay between Input Data and Hybrid CL-SRME

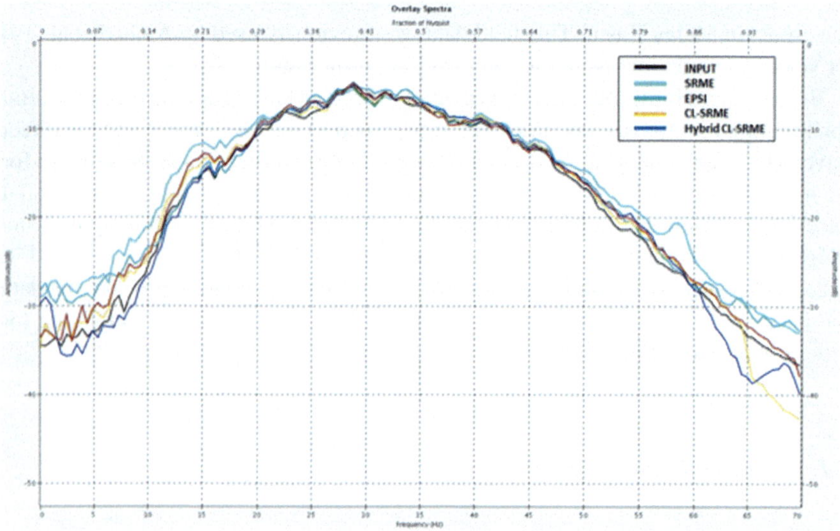

(b) Spectrum Overlay for all

Fig. 23 Amplitude spectrum for: **a** sparse CL-SRME primaries and **b** all methods

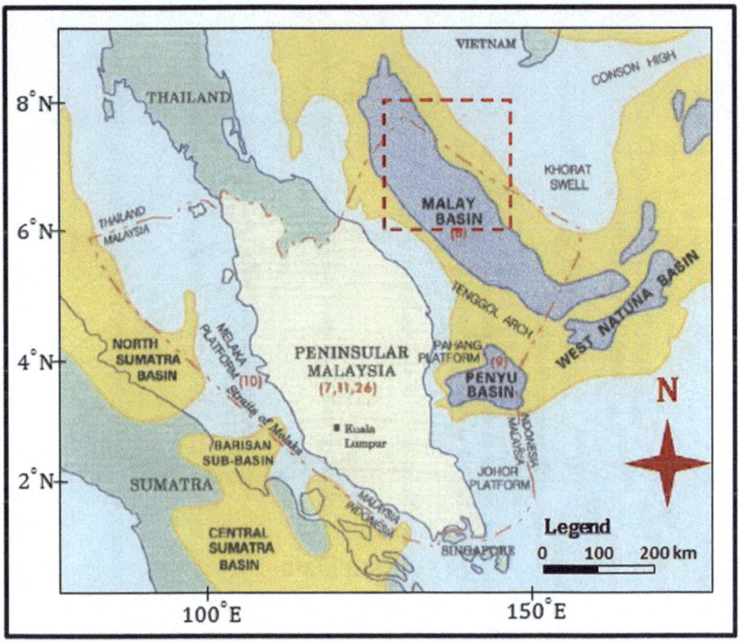

Fig. 24 The location for the Malay Basin. The red box indicates the location for the field data, offshore Malay Basin. Taken and modified from [20]

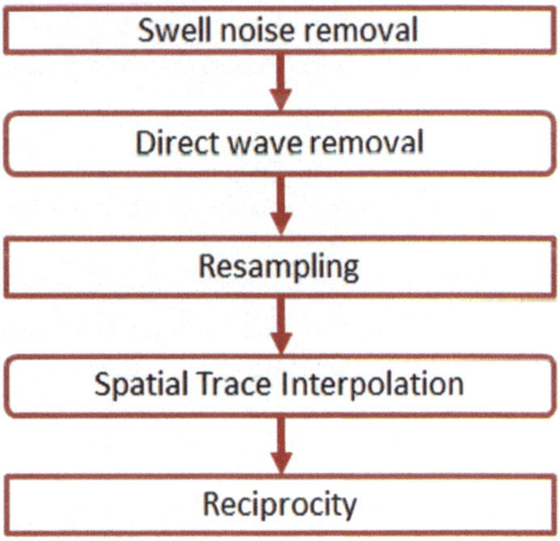

Fig. 25 Pre-conditioning workflow for input dataset

The predicted multiples created through also shows that it were unable to properly construct the multiple reflections at 0.9 s.

Both CL-SRME and sparse CL-SRME appears to be able to recovered the missing overlapping primaries as shown by the green arrow at Fig. 27a, b. However, it also appears that both of these methods were unable to properly removed the multiples

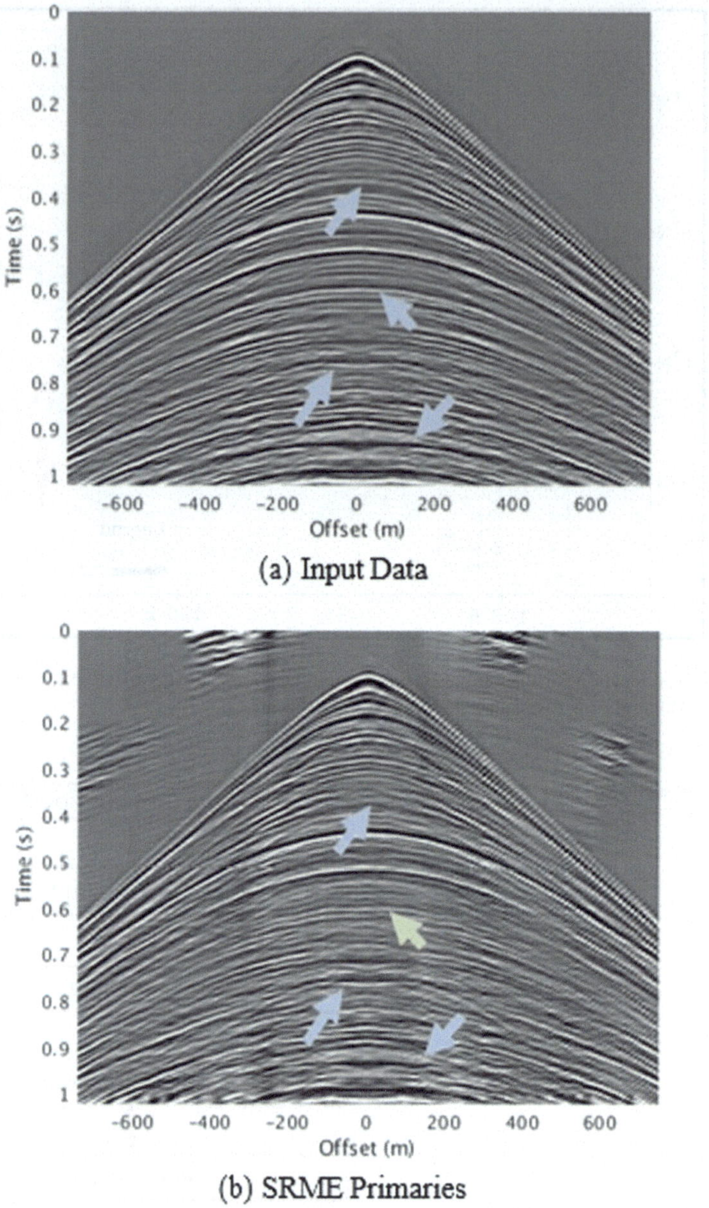

Fig. 26 Primaries shot gather of offshore Malay Basin: **a** input data and **b** SRME primaries

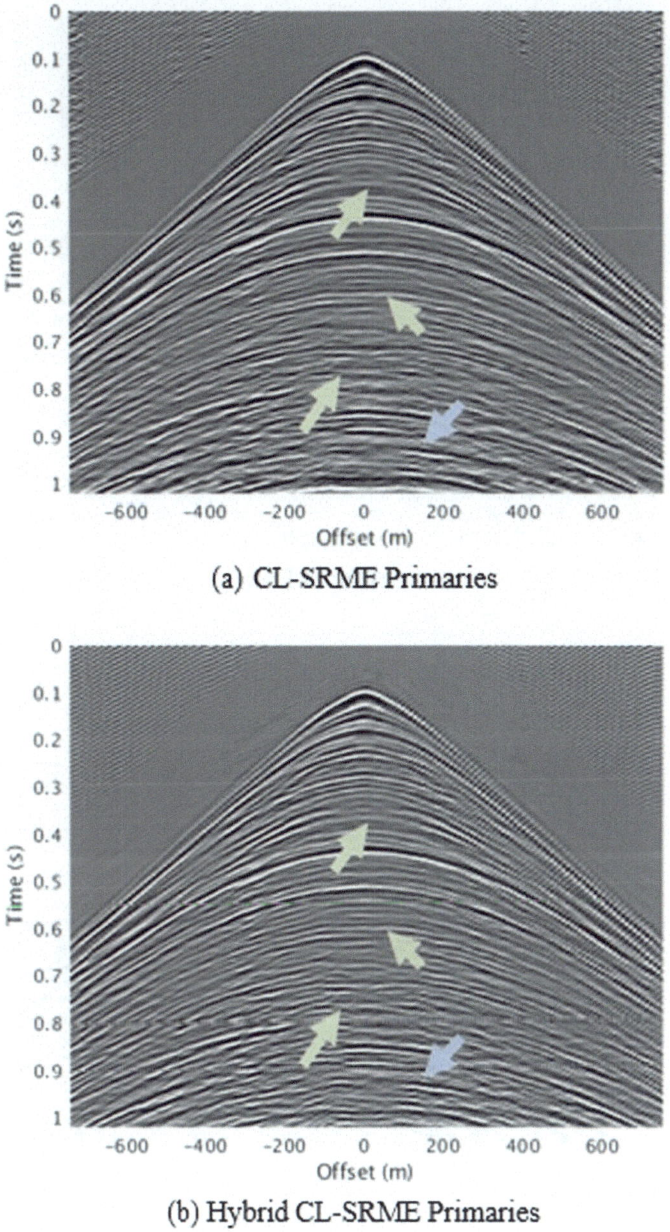

Fig. 27 Primaries shot gather of offshore Malay Basin: **a** CL-SRME primaries and **b** sparse CL-SRME primaries

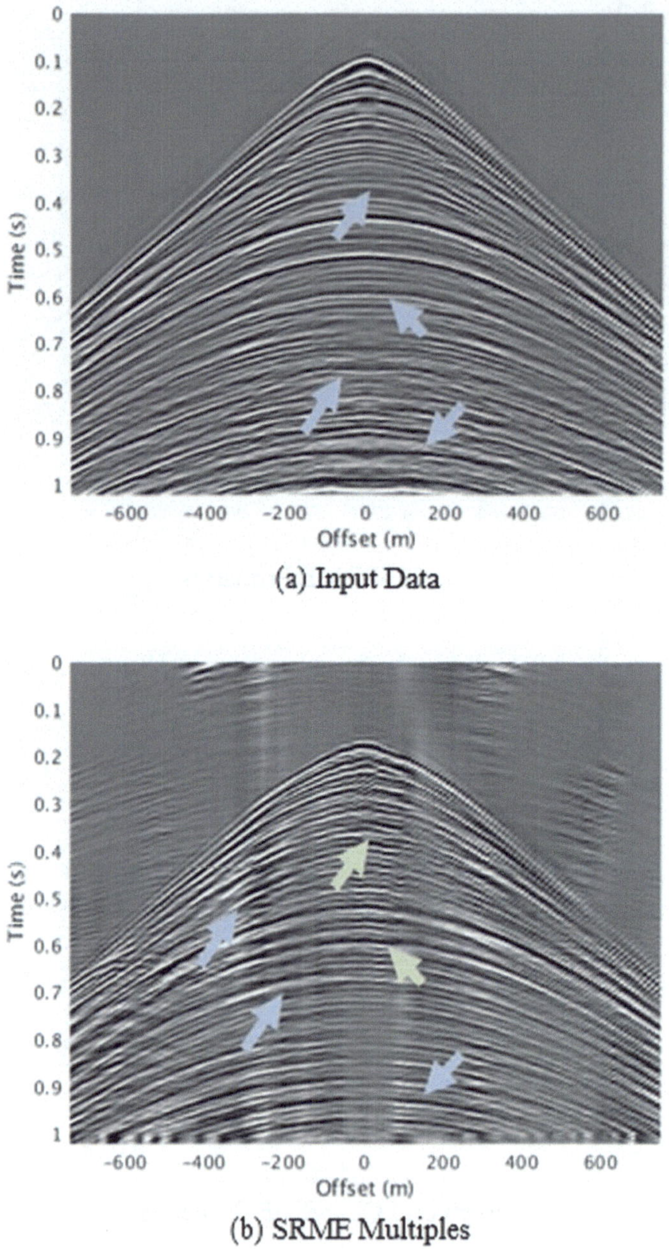

Fig. 28 Multiples shot gather of offshore Malay Basin: **a** input data and **b** SRME multiples

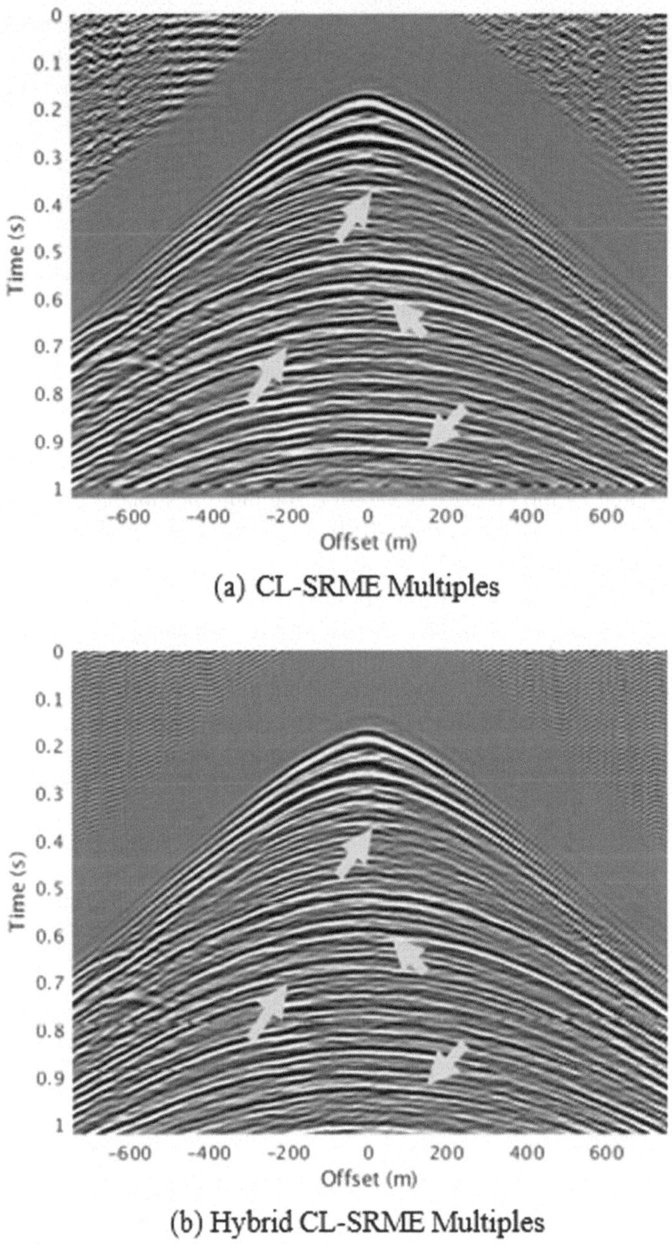

Fig. 29 Multiples shot gather of offshore Malay Basin: **a** CL-SRME multiples and **b** sparse CL-SRME multiples

reflector shown by the blue arrow at 0.9 s. The predicted multiples generated by these two methods (Fig. 29a, b) appears to be properly constructed even for the case of overlapping primaries and multiples.

7.2 Zero-Offset Gather Display

Figures 30, 31, 32, 33 shows the zero-offset gathers for the input data, primaries and multiples obtained from various multiple attenuation methods. From Fig. 30, some prominent multiples reflector can be observed in the input data as marked by the blue arrows. These multiples most probably generated by the water-bottom area and through the gas cloud seepage at 0.45 s.

After the application of SRME (Fig. 31a), it managed to remove most of the multiples as marked by the green arrow. However, it appears there are some multiples residual as noted by the blue arrow at 0.9 s. The multiples model generated by the SRME in Fig. 31b also shows that SRME were unable to properly construct the multiples reflector which are noted by the blue arrows.

CL-SRME (Fig. 32a) shows a good primaries estimation, except at the area mark by the blue arrow at 0.75 and 0.9 s. The predicted multiples generated by the CL-SRME shows a good construction of the prominent multiples. However, through observation at the shallow area of the predicted multiples (Fig. 32b), it can be observed that there was a presence of ringing artefacts. Compared to the other methods, the result produce by sparse CL-SRME (Fig. 33a) shows the most promising results in the zero-offset gather. It manages to obtained a good primaries estimate especially

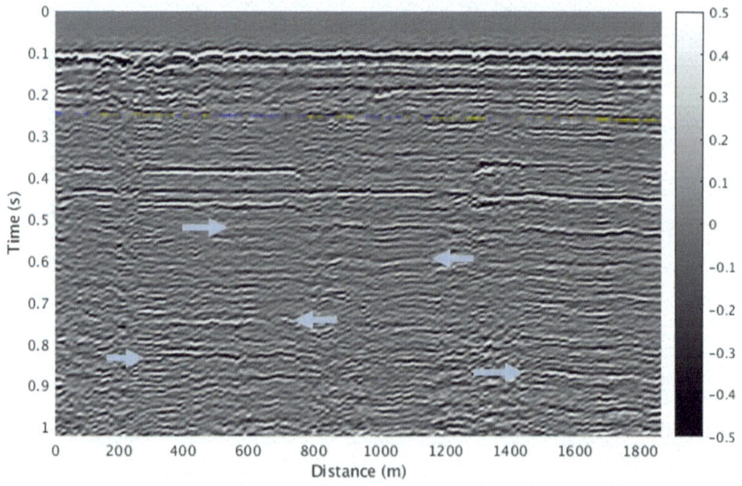

Fig. 30 Zero-offset gather for input data

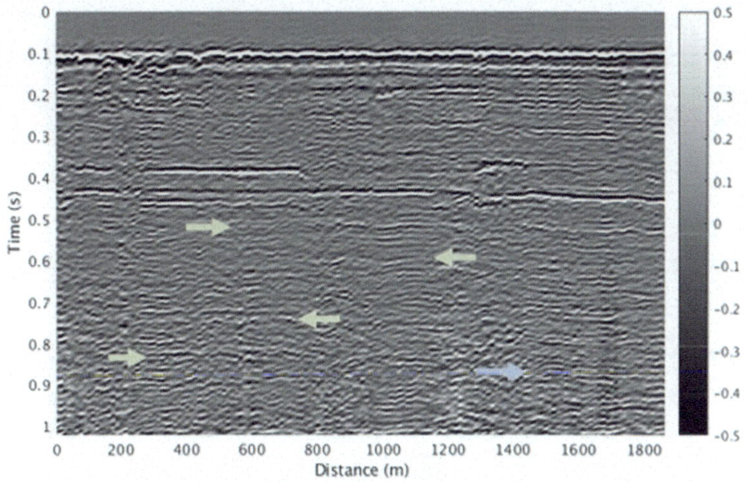

(a) SRME Primaries

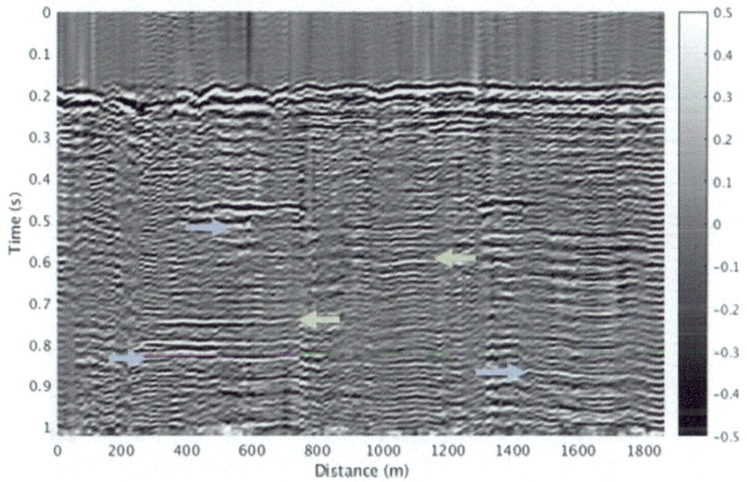

(b) SRME Multiples

Fig. 31 Zero-offset gathers for: **a** SRME primaries and **b** SRME multiples

at the area dominated by the gas clouds multiples. It also shows a good predicted multiples (Fig. 33b) without any ringing artefacts as shown by the CL-SRME.

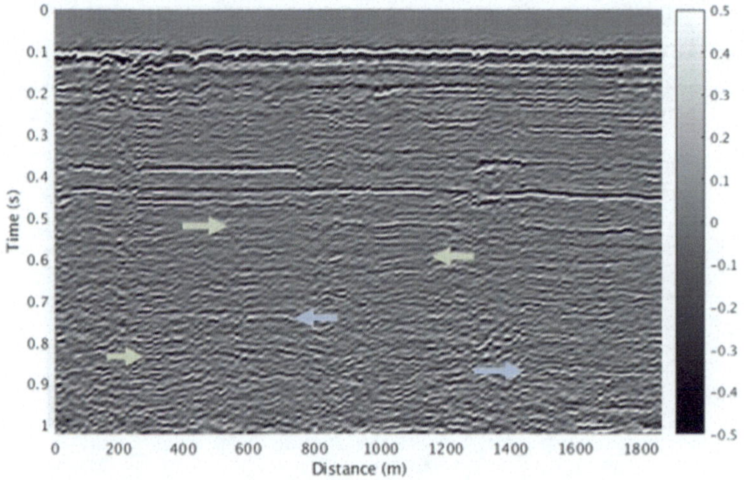

(a) CL-SRME Primaries

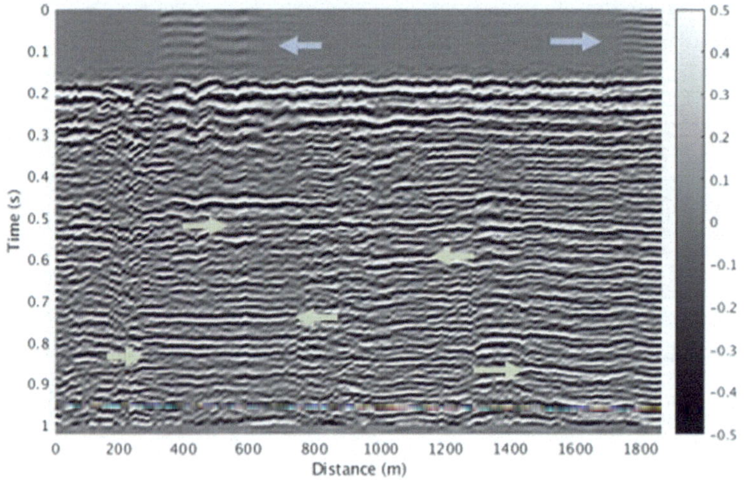

(b) CL-SRME Multiples

Fig. 32 Zero-offset gathers for: **a** CL-SRME primaries and **d** CL-SRME multiples

7.3 Auto-Correlation Section

Figures 34, 35 and 36 shows the auto-correlation section of the input data and the primaries output for all the multiple attenuation process. From the input data shown in Fig. 34, the multiples energy presence in the input data can be observed as noted by the blue arrows at 0.4, 0.35 and 0.45 s. After the application of SRME (Fig. 35), a reduction of multiples energy can be observed in the primaries output. However, it

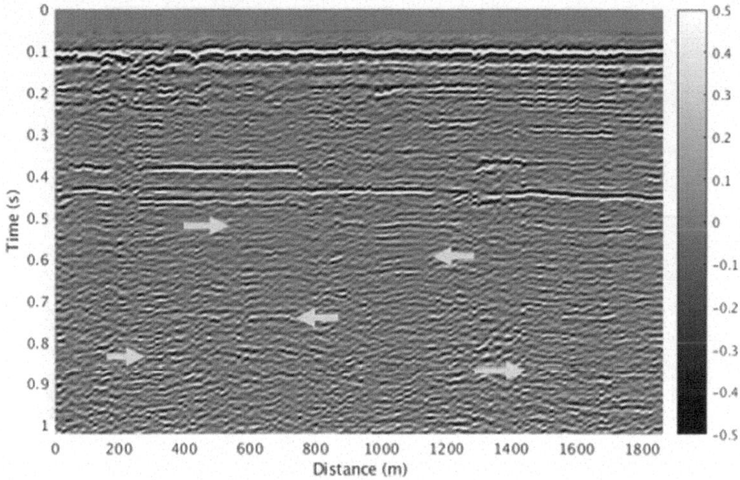

(a) Hybrid CL-SRME Primaries

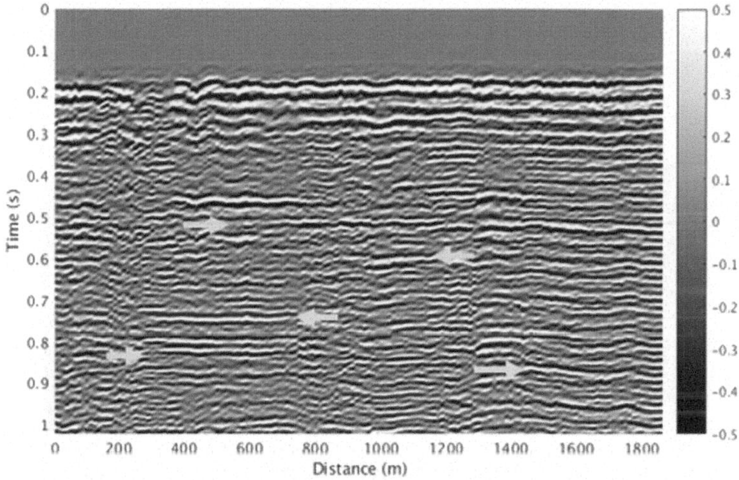

(b) Hybrid CL-SRME Multiples

Fig. 33 Zero-offset gathers for: **a** sparse CL-SRME primaries and **b** sparse CL-SRME multiples

can also observed there was some residuals of the multiples energy left in the SRME primaries which are noted by the blue arrow at 0.45 s. Some reduction of primaries and multiples energy can also be observed in the lower section of the primaries output between 0 and 0.2 s.

Both auto-correlation section for CL-SRME (Fig. 36a) and sparse CL-SRME (Fig. 36b) shows a good reduction in multiple energy in the primaries output as

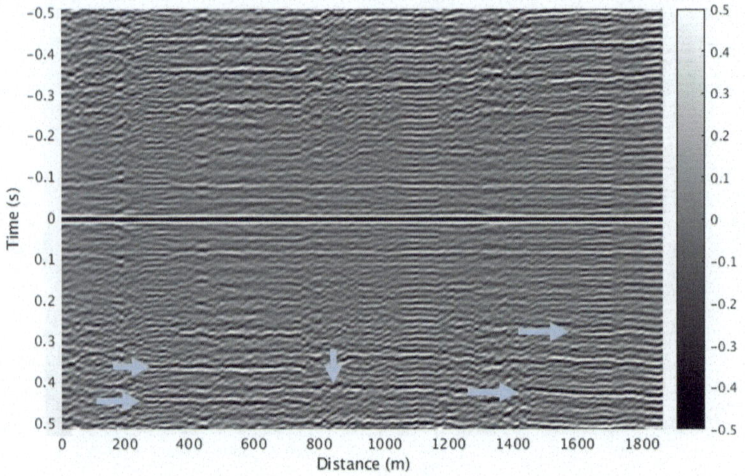

Fig. 34 Auto-correlation of input data

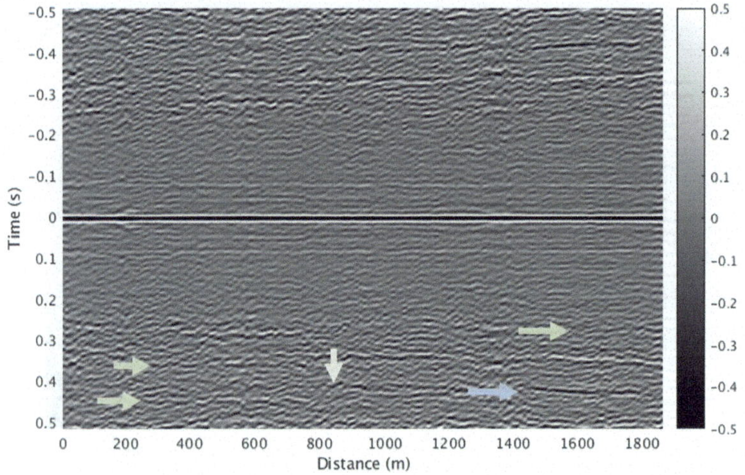

Fig. 35 Auto-correlation section for SRME primaries

mark by the green arrow. There were also no significant artefacts shown by this two algorithm.

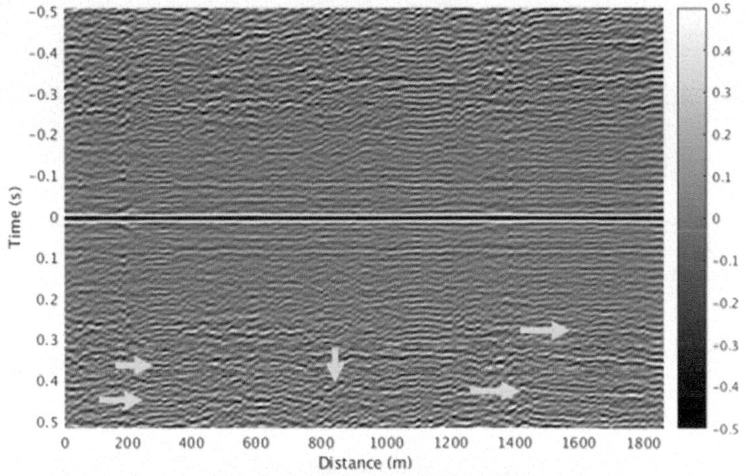

(a) Auto-correlation of CL-SRME Primaries

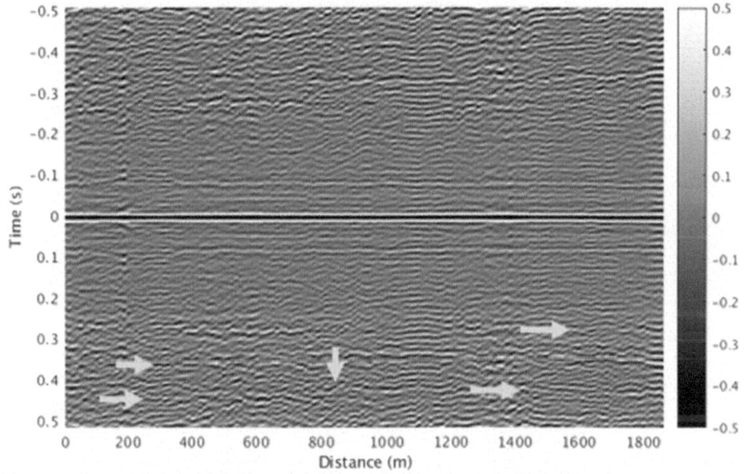

(b) Auto-correlation of Hybrid CL-SRME Primaries

Fig. 36 Auto-correlation section for: **a** CL-SRME primaries and **b** sparse CL-SRME primaries

7.4 Spectral Analysis

The amplitude spectrum for the input data and the primaries output of various multiple attenuation methods were shown in Figs. 37 and 38. Based on Fig. 37b, it appears there was an increase in the spectral power for the SRME primaries in the low (0–10 Hz) and high frequency (90–125 Hz) range. CL-SRME produced an amplitude spectrum which are consistent with the input data, although it appears there was a significant reduction of spectral power at 90 Hz and above. This reduction of spectral

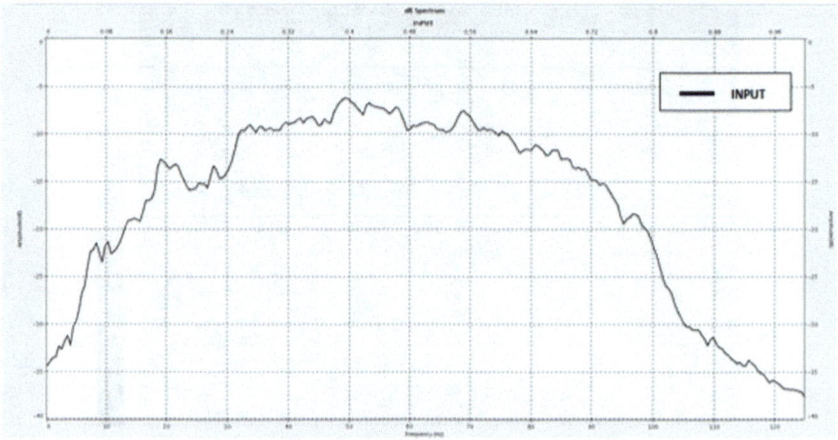

(a) dB Spectrum of Input Data

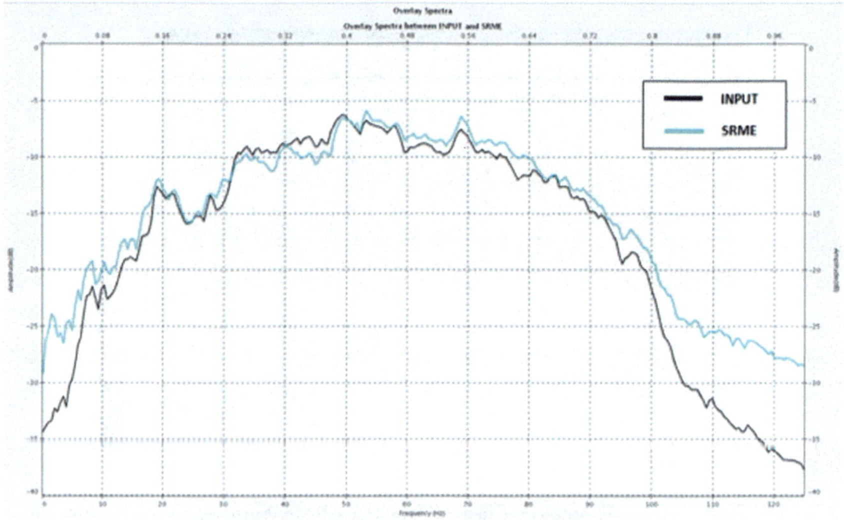

(b) Spectrum Overlay between Input Data and SRME

Fig. 37 Amplitude spectrum for: **a** input data and **b** SRME primaries

power were also observed in the sparse CL-SRME (Fig. 38a), although there were also a significant reduction of spectral power from 0 to 40 Hz. The reason for the drop in frequency at the low frequency section of sparse CL-SRME might due to the fact that sparse CL-SRME uses constant source wavelet throughout the inversion process. This results in the loss of low frequency events as the source wavelet may vary with time due to the propagation effects. Based on the overall spectrum in Fig. 38b, the amplitude spectrum generated by EPSI shows the most consistent frequency

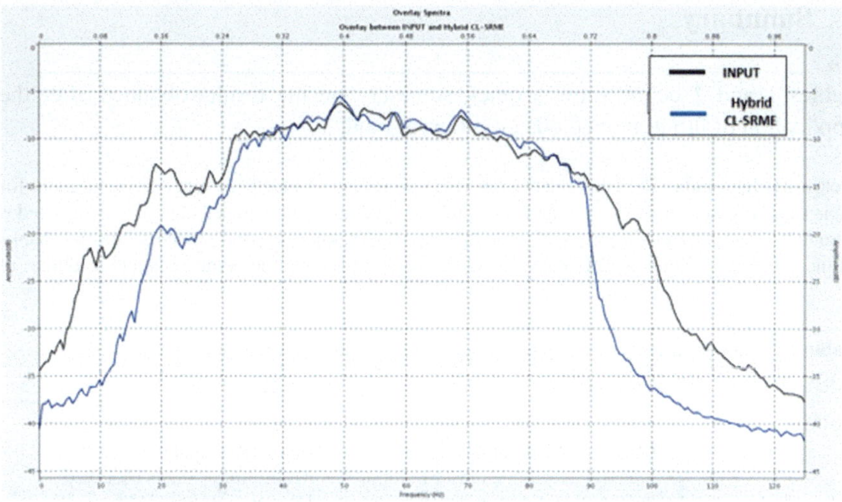

(a) Spectrum Overlay between Input Data and Hybrid CL-SRME

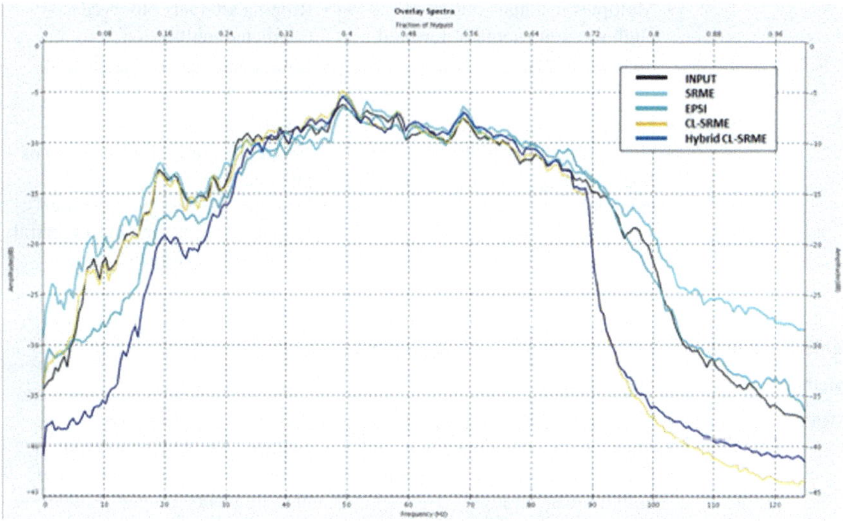

(b) Spectrum Overlay for all

Fig. 38 Amplitude spectrum for: **a** sparse CL-SRME primaries and **b** all methods

energy distribution when compared with the input data, although there was a sudden reduction of energy at the low frequency range.

8 Summary

Tables 1 and 2 below show a quick summary on the results obtained after the application of different methods on different models.

Acknowledgements The authors are grateful to all members of the Geoscience Department for their support and as well as the members of the above consultancy model's team, which includes engineers, associate professors, lecturers, laboratory technologists, and students. Special thanks to the Universiti Teknologi PETRONAS, Malaysia for providing the permission and resources to

Table 1 Summary and comparison between different methods and model

Method	Two-layer dipping model	Complex salt model
SRME	• Missing primaries not recovered • Multiples residuals observed • Spectral amplitude closer to input data with less notches	• Poor primary continuity • Some artefacts observed • Multiples residuals observed • Spectral amplitude close to input data
CL-SRME	• Missing primaries recovered • Multiples residuals observed • High spectral energy at low and high frequency	• Good primary continuity • Ringing artefacts observed (predicted multiples) • Low spectral energy at the high frequency range
Sparse CL-SRME	• Missing primaries recovered • No significant multiples residual • Spectral analysis closer to input data with less notches	• Good primary continuity • No significant multiples residuals and artefacts • Low spectral energy at the high frequency range, and high spectral energy at low frequency range

Table 2 Summary and comparison between different methods and models

Method	Malay Basin model	Offshore Malay Basin
SRME	• Missing primaries not recovered • Multiples residuals observed • High spectral energy at low and high frequency range	• Poor primary continuity • Multiples residuals observed • High spectral energy at low and high frequency range
CL-SRME	• Missing primaries recovered • Multiples residuals observed • Ringing artefacts observed (predicted multiples) • Low spectral energy at high frequency range	• Good primary continuity • Ringing artefacts observed (predicted multiples) • Low spectral energy at high frequency range
Sparse CL-SRME	• Missing primaries recovered • Some artefacts observed • Low spectral energy at the high frequency range, and high spectral energy at low frequency range	• Good primary continuity • No significant artefacts and multiples residuals • Low spectral energy at low and high frequency range

carry out the laboratory demonstration. Without the kind cooperation this study would have been impossible.

References

1. Telford, W.M., Geldart, L.P., Sheriff, R.E.: Applied Geophysics. Cambridge University Press, Cambridge (1990)
2. Zhou, H.W.: Practical Seismic Data Analysis, vol. 1. Cambridge University Press (2014)
3. Dragoset, W.H., Jericevic, Z.: Some remarks on surface multiple attenuation **63**(2), 772–789 (1998)
4. Verschuur, D.J., Berkhout, A.J., Wapenaar, C.: Adaptive surface related multiple elimination. Geophysics **57**(9), 1166–1177 (1992)
5. Verschuur, D.J.: Estimation of primaries by sparse inversion including the ghost. In: SEG Technical Program Expanded Abstracts 2013, pp. 4094–4100. Society of Exploration Geophysicists (2013)
6. Riley, D.C., Claerbout, J.F.: 2-D multiple reflections. Geophysics **41**(4), 592–620 (1976)
7. Reilly, J.M., Pitcher, D., Ghosh, D.: SEG applied research workshop: geophysical challenges in southeast Asia exploration. Lead. Edge **27**(10), 1282–1299 (2008)
8. Phan, Y.P., Wang, X.: Re-acquiring or re-processing seismic data? A successful re-processing case study offshore Malaysia. In SEG Technical Program Expanded Abstracts 2014, pp. 4325–4329. Society of Exploration Geophysicists, Aug 2014
9. Hydal, S., Alai, R., Soh, N.N.H.C., Tang, W.H., Rahman, N.N., F. Khan, A., Ramlan, M.D.M., Seymour, N., Ngo, L.S.: High-resolution Earth model building and imaging strategies in a pre-carbonate play. In: SEG Technical Program Expanded Abstracts 2015, pp. 5217–5221. Society of Exploration Geophysicists (2015)
10. Ramlan, M.D.M., Seymour, N., Ngo, L.S.: High-resolution Earth model building and imaging strategies in a pre-carbonate play. In: SEG Technical Program Expanded Abstracts 2015, pp. 5217–5221. Society of Exploration Geophysicists (2015)
11. Cambois, G.: Can P—wave AVO be quantitative? Lead. Edge **19**(11), 1246–1251 (2000)
12. Kumar. D.: Multiples or residual moveout: the presence of which noise type in seismic is worse for wide angle AVO inversion. In: 7th International Conference & Exposition on Petroleum Geophysics, Hyderabad, 2008
13. Berkhout, J.: Seismic Migration Imaging of Acoustic Energy by Wave Field Extrapolation, A: Theoretical Aspects. Developments in Solid Earth Geophysics, 2nd edn. Elsevier (1984)
14. Iverson, A.: The impact of interbed multiples on the inversion and interpretation of pre-stack data. CSEG Rec. **39**(1), 36–41 (2014)
15. Lopez, G.A.: Closed-loop surface-related multiple estimation. Ph.D. thesis, Delft University of Technology (2016)
16. Berkhout, A.J., Verschuur, D.J.: Removal of internal multiples with the common-focus-point (CFP) approach: part 1 explanation of the theory. Geophysics **70**(3), V45–V60 (2005)
17. Lopez, G.A., Verschuur, D.J.: Closed-loop SRME with data-reconstruction. In: 77th Conference and Exhibition, (June):Tu N114 03, 2015
18. De Almeida, L.A.: Seismic data interpolation using sparsity constrained inversion. Master's thesis, Colorado School of Mines (2017)
19. Long, A.S., van Borselen, R., Fountain, L.: Surface related multiple elimination—applications to an offshore Australia data set. ASEG Ext. Abstr. **2001**(1):1 (2001)
20. Ghazali, A.R., Verschuur, D.J., Gisolf, A.: Seismic imaging through gas clouds: a data-driven imaging strategy. In: SEG Technical Program Expanded Abstracts 2008, pp. 2302–2306. Society of Exploration Geophysicists, 2008

21. Alai, R., Rahman, N.N.A., Abd Mutalib, M.A.A., Abu Bakar, H.H., Beg, M.A., Phan Y.P., Wang X.: Re-acquiring or re-processing seismic data? A successful re-processing case study offshore Malaysia. In SEG Technical Program Expanded Abstracts 2014, pp 4325–4329. Society of Exploration Geophysicists (2014)
22. Hashim, H., Abdul Latiff, A.H.: Elimination of surface-related multiple in offshore Malay Basin—a case study. In: ICIPEG 2016, Aug 2016

Hybrid Illumination Analysis for Complex Overburden Geological Structure

Abdul Halim Abdul Latiff

Abstract The seismic data quality can be improved by evaluating two aspects: seismic data acquisition and seismic imaging. In recent years, a number of innovative acquisition layouts and methods have been introduced that produced better subsurface images compare to previous dataset. However, the seismic technology in the industry has been largely on an ad-hoc basis without any qualitative analysis of the subsurface geology and complexity and its influence on data quality. In this chapter, the poor images beneath shallow gas cloud were analysed by using two types of illumination methods, namely wave propagation and ray interpolation techniques. Meanwhile the data enhancement methodology was conducted through optimization of receiver positions on surface level. This will address the issue of subsurface illumination below complex overburden and shallow anomaly that cause wave field distortion. The outcome from this work aims at developing the new technique of getting a good and reliable seismic data underneath the near surface anomaly by integrating the existing illumination techniques and optimizing the source and receiver positions. The proposed integrated method of illumination analysis, leads to a new attribute known as Illumination Factor (IF), was formulated based on information from two forward modelling methods; (i) Amplitude distribution in spatial domain and (ii) Ray hit count at the target subsurface. Evaluation of the proposed Illumination Factor will set a basis for improving seismic acquisition design, by enhancing seismic data through better illumination analysis while providing an insight for accurate reservoir characterization.

Keywords Seismic illumination · Complex overburden · Illumination factor

The poor images beneath shallow gas cloud were analysed by using two types of illumination methods, namely wave propagation and ray interpolation techniques. The data enhancement methodology was conducted through optimization of receiver

A. H. Abdul Latiff (✉)
Centre for Subsurface Imaging, Universiti Teknologi PETRONAS, 32610, Bandar Seri Iskandar, Perak Darul Ridzuan, Malaysia
e-mail: abdulhalim.alatiff@utp.edu.my

© Institute of Technology PETRONAS Sdn Bhd 2022
S. A. Abdul Karim and A. Shafie (eds.), *Towards Intelligent Systems Modeling and Simulation*, Studies in Systems, Decision and Control 383,
https://doi.org/10.1007/978-3-030-79606-8_21

positions on surface level. This chapter will address the issue of subsurface illumination below complex overburden and shallow anomaly that cause wave field distortion. This chapter aims at developing the new technique of getting a good and reliable seismic data underneath the near surface anomaly by integrating the existing illumination techniques and optimizing the source and receiver positions. This was conducted by developing a new technique for illumination analysis, by taking into account both wave extrapolation and ray tracing interpolation methods for better subsurface guidance. The integrated method of illumination analysis known as Illumination Factor (IF), was formulated based on information from two forward modelling methods; (i) Amplitude distribution in spatial domain (focal beam propagation), and (ii) Ray hit count at the target subsurface (ray tracing interpolation). Evaluation of the proposed Illumination Factor will set a basis for improving seismic acquisition design, by enhancing seismic data through better illumination analysis while providing an insight for accurate reservoir characterization.

1 Introduction

For the last few years, several breakthroughs have been achieved in developing seismic imaging algorithm to solve both travel-time error and absorption effect, such as amplitude and bandwidth expansion in Q-migration technique [1] and Q-Reverse Time Migration (Q-RTM) technique [2]. Advancement in migration algorithm by incorporating the wave and ray theories such as Reverse Time Migration [3], Gaussian Beam Migration [4] and Wave Equation Migration [5] were also carried out and implemented with a certain degree of success. On the other hand, seismic data quality through innovative acquisition methods like wide-azimuth acquisition [6], Ocean Bottom Cable (OBC) on the seabed [7], coil acquisition with circular shooting [8] and blended acquisition with dispersed source arrays [9] are also being extensively tested in acquisition industry. Implementation the mentioned methods have been successful in producing better seismic image data, particularly in the area of complex overburden and shallow anomaly.

In general, the solutions discussed above follow the forward path of seismic value chain [10], where a cross-disciplinary connections were established between seismic acquisition, structural imaging and reservoir characterization (Fig. 1). Prior to a re-acquisition process, the potential hydrocarbon reservoirs need verification through a forward modelling process. However, the current processing and imaging procedures require extensive computational power, longer processing period and huge data space, before the subsurface data were evaluated. In addition, the acquired seismic data requires a tool to explain the shadow zones which detected after all processing stage was completed. With these constraints in mind, various illumination techniques based on subsurface velocity information were developed to enable faster evaluations of an acquisition design. These illumination analysis methods put a strong emphasis in forward modelling technique using either wave extrapolation or ray interpolations methods.

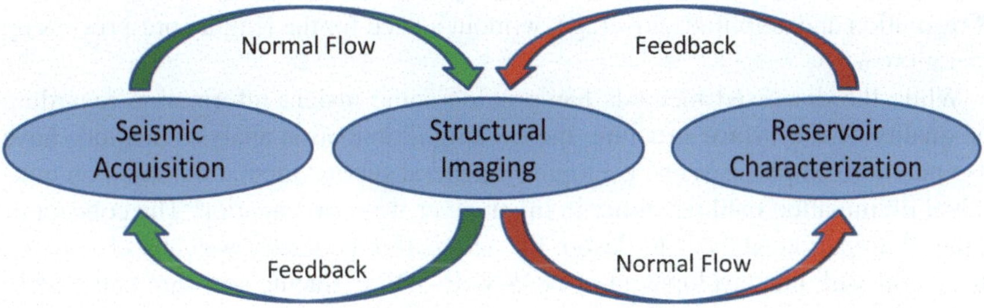

Fig. 1 The seismic value chain indicated the cross-disciplinary between different seismic section, from seismic acquisition to structural imaging and reservoir characterization. The cycle involves in this research is within the green arrow loop

All existing illumination analysis methods were reviewed [11] and they categorized into two major categories: (i) Global method, i.e. information distribution over a whole reflector horizon base on hit count, amplitude or offset attribute; and (ii) Local method, i.e. locating information at a specific subsurface area or point. Global illumination technique is commonly used in exploration industry due to its ability to have larger target horizon as well as faster implementation of ray tracing technique, thus enables subsurface illumination evaluation carried out instantly during acquisition stage. Contradictory to the global solution, local methods such as focal beam method [12–15] (Fig. 2) have the ability to estimate the amplitude distribution at a particular subsurface point by incorporate wave field propagation through inhomogeneous layers, although high performance computing system sometimes is needed. One major advantage of conducting illumination analysis in local method is that the information on seismic data quality can be obtained and analysed in term

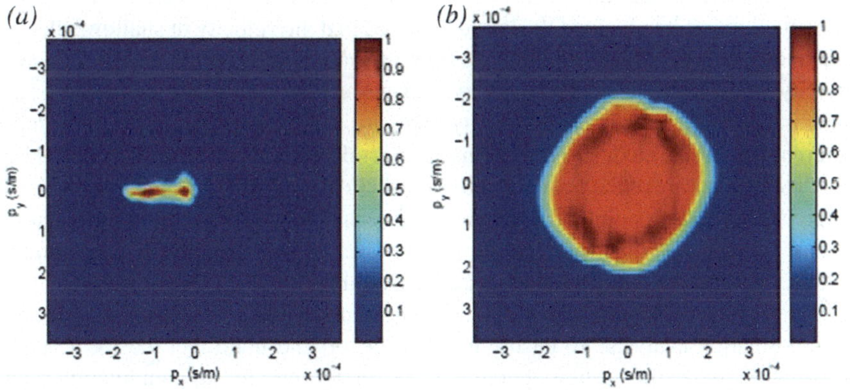

Fig. 2 The focal beam method application illustrated above indicated the various illumination data quality from different acquisition design. The illumination data obtained for the *Ziggy*[1] velocity model for **a** conventional parallel geometry **b** full 3D geometry [15]

of resolution and amplitude coverage, without a need for the lengthy pre-processing operation.

While the discussed methods has provide some insight information regarding the quality of subsurface structure, the existing illumination analysis methods have also put more emphasises on developing a general survey orientation based on individual illumination method, either in ray trace or wave propagation. The concept of using illumination analysis to design the acquisition geometry was already proven successful with the development of NORSAR-3D ray tracing software that widely used in the industry to conduct the acquisition feasibility study before a real acquisition took place. The software developed by NORSAR AS is based on the wave front construction method [16] which allows estimation the wave travel time and amplitude information. Meanwhile, most of other illumination analyses and survey design methods are solely base either on ray trace interpolation or wave field propagation theory. As a result, the outcomes from their acquisition configurations as well as their illumination analysis might be inconsistent due to different wave characteristic.

By taking into account of both wave and ray concepts, it is very crucial to have an integrated method of illumination analysis and exerts a greater accuracy in optimizing a seismic survey configuration, in term of its quality, cost and time constraints of seismic data acquisition. In the meantime, solution to the source and receiver positioning requires an optimization technique, with the aid of existing subsurface information, to fully optimize the source and receiver position.

2 Focal Beam Theoretical Review

In forward modelling process of focal beam method, several model parameters were interchangeable during the simulation, thus illumination quality of different subsurface situation can be obtained. It should be noted that the velocity of shallow subsurface anomaly is not definitive, for example in shallow gas cloud region, the wave velocity through the gas zone is related to its gas concentration, where a denser gas content will yield slower wave velocity compare to thin gas layer. Therefore, four different gas cloud models were evaluated with different gas cloud velocity while other parameters are remain constant. Beside shallow gas velocity value, the effect of different target depth point, different nominal seismic frequency as well as different surface geometry on the image quality will be investigated. In addition to gas cloud, a salt anomaly model was also developed and analyzed.

Choosing the suitable wave operator is important in producing the correct output for illumination analysis. To ensure external factors did not affect the focal beam propagation, the extrapolation operators were set constant for all computation with length of each operator was set at 19 points, its maximum propagation angle at 50° with tolerance of 1×10^{-6} and weighting factor of 1×10^{-5}. It should be noted that the operator coefficients chosen are the most suitable for non-uniform wave propagation, and it was based on several test runs conducted earlier.

2.1 Shallow Gas Cloud Velocity

The main objective of this work is to illustrate the difference in illumination strength for different gas cloud velocity profile. Depending on the concentration of gas accumulation, velocity of shallow gas cloud is differed from each field and basin. Thus, four velocity models were created, shown in Fig. 3 with each model contains different value of low velocity zone, represented by ellipsoid shape structure in the area between 800 and 1200 m deep, width of 200 m in both x and y direction and centered at coordinate [1000, 1000, 1000]. The total dimension for each model is approximately 2 km length, 2 km width and 4 km deep with thickness for each velocity layer at 15 m. Each layer possesses constant vertical velocity gradient of 0.56 m^2s^{-1} from water bottom (water bottom has fixed velocity at 1500 ms^{-1}) up till maximum velocity of 4000 ms^{-1}. The velocity chosen for shallow gas cloud area ranging from 1700, 1400, 1100 to 800 ms^{-1}, with maximum deviation for shallow gas cloud velocity compare to area outside the gas zone with constant velocity gradient trend (velocity position at coordinate [1201, 1000, 1000]) is 1.4% in 1700 ms^{-1} model, 18.8% in 1400 ms^{-1} model, 36.2% in 1100 ms^{-1} model and 53.6% in 800 ms^{-1} model. During different shallow gas cloud velocity analysis, nominal frequency of 35 Hz, fix reflector depth location at 4 km and full 3D source and receiver configuration parameters were used.

The illumination outputs at the reflector depth of 4 km after focal beam process for each velocity model are shown in Fig. 4. If the gas cloud velocity close to velocity outside the gas cloud, such as for velocity model of 1700 ms^{-1} (Fig. 4a) where maximum velocity deviation stands at 1.4%, a convolution between focal source beam and focal receiver beam yield a good resolution image, by producing a sharp and focus image at the centre as well as excellent AVP imprint attributed to uniform amplitude coverage obtained. The AVP imprint and resolution function for slightly slower gas cloud velocity of 1400 ms^{-1} (Fig. 4b), which contains moderate velocity deviation also yield good illumination coverage with high resolution image is achievable. For 1100 ms^{-1} (Fig. 4c) and 800 ms^{-1} (Fig. 4d) models, a sharp contrast surrounding the gas zone structure caused the wave field scattering and amplitude loss during upward and downward wave propagation. As a result, AVP imprint at the target point within these two models produce poor illumination coverage. However, the resolution function obtained remains identical as no aliasing effect occurred due to good sampling requirement for source and receiver configuration.

One potential solution in overcome poor AVP coverage in high velocity contrast is to smooth the velocity model particularly at the edge of gas cloud zone, preventing the gas zone from start and end abruptly. The gradual change of velocity gives better subsurface illumination quality, particularly in the area underneath the shallow anomaly. The application of dense sampling grid on the surface acquisition had prevented aliasing from occurred during wave field propagation. Another viable option for improving the resolution and illumination coverage that is yet to be explored is by optimizing operator function. In earlier test runs conducted on focal beam operator prior to this analysis show that by shortening the operator length, better resolution can be achieved, probably due to more wave field energy able to

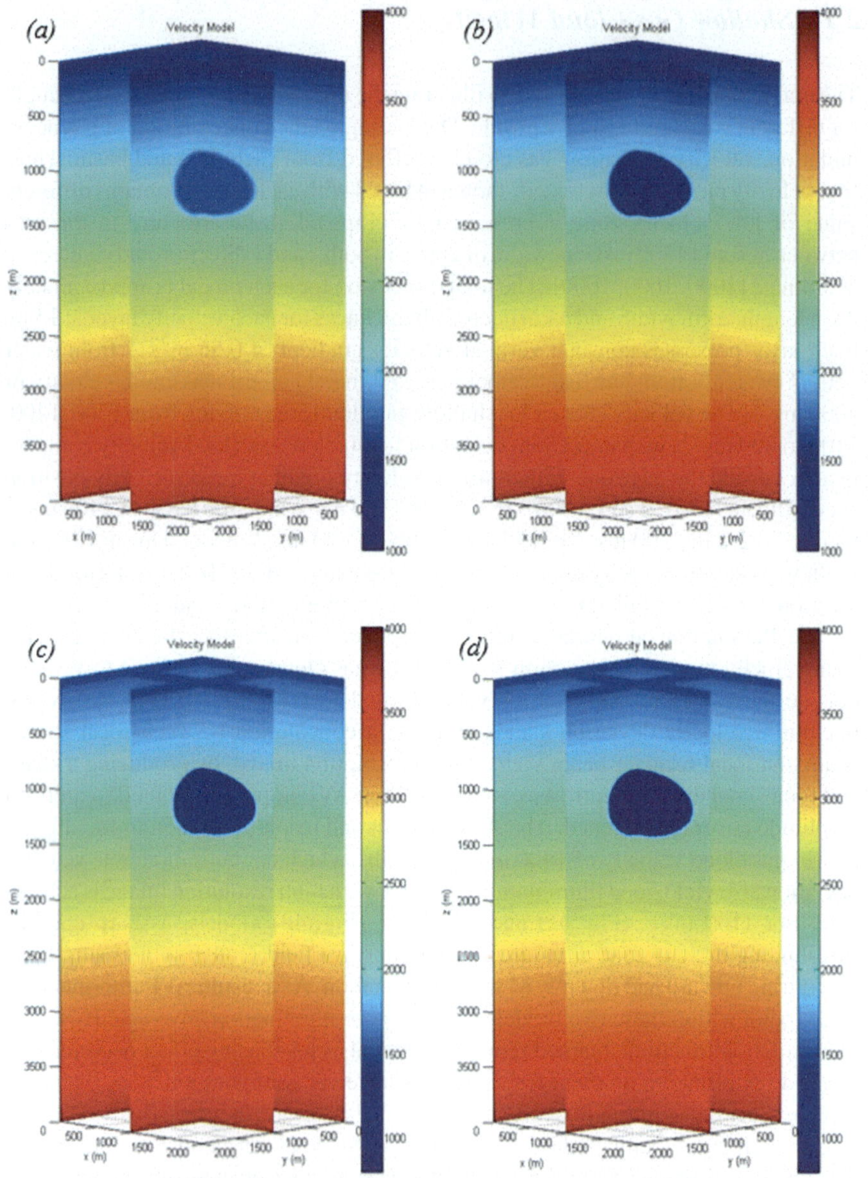

Fig. 3 Simple velocity model of 2 km × 2 km × 4 km dimension representing the shallow gas cloud model, where the gas zone marked by ellipsoid shape structure from 800 to 1200 m depth. The wave propagation velocity through shallow gas cloud was set as follows: **a** 1700 ms^{-1}, **b** 1400 ms^{-1}, **c** 1100 ms^{-1} and **d** 800 ms^{-1}. The color bar at the side of each model indicates the velocity scale

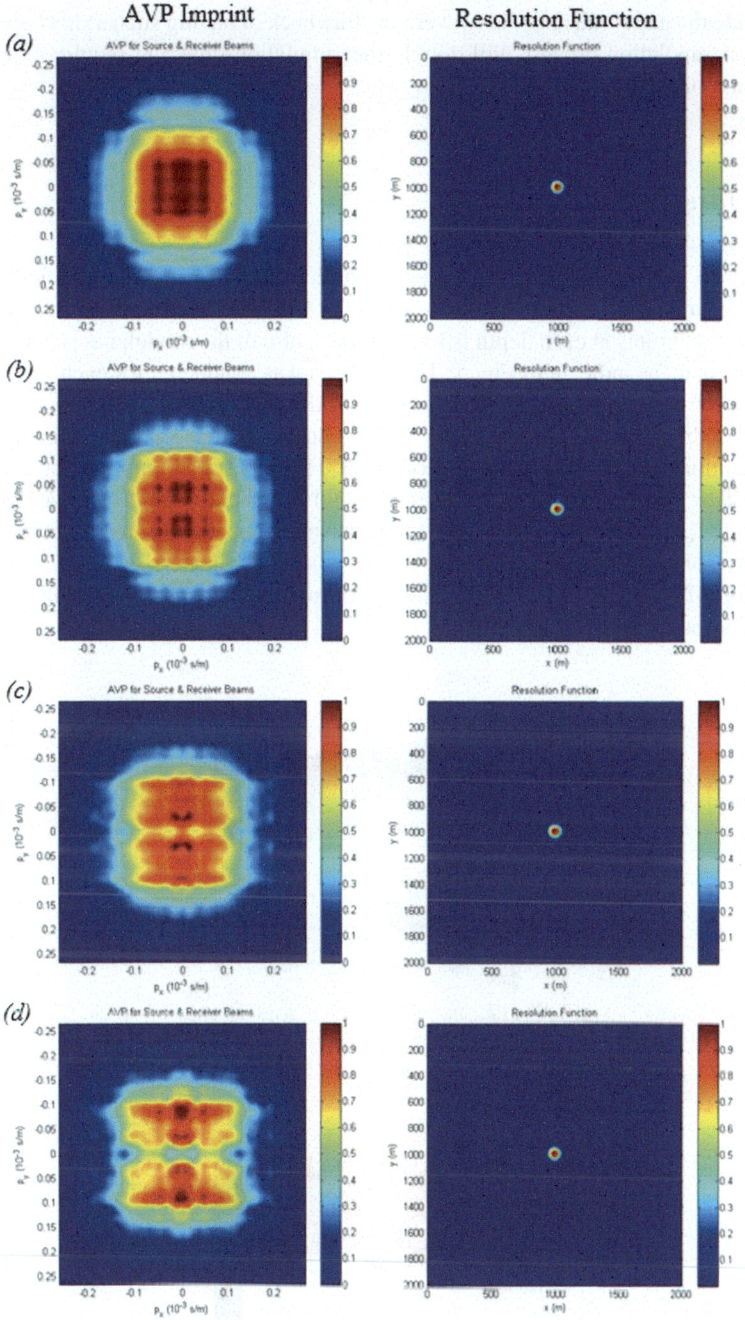

Fig. 4 The AVP imprint and resolution results from different shallow gas cloud velocity models: **a** 1700 ms^{-1}, **b** 1400 ms^{-1}, **c** 1100 ms^{-1} and **d** 800 ms^{-1}. Other parameters were remained constant: full 3D, 4 km depth and 35 Hz frequency

penetrate the gas cloud zone. However, the drawback of having such a short operator is the extrapolation process will double the simulation time and require extensive computational power.

2.2 *Target Depth Level*

In order to understand the illumination strength of individual target position, three different target depths, z, were chosen: 1500, 2000 and 4000 m, while lateral location of the target points at each depth is fixed at the centre of the model, i.e. [1000, 1000, z] coordinate, as indicate in Fig. 5. The analysis was conducted for each individual location on target reflector as energy level at different depth level is independent of each other, thus there will be no wave field interference between focal beam computations of 1500 m point reflector to 2000 and 4000 m point reflector and vice versa. The wave field was propagated through the shallow gas cloud velocity of 1400 ms^{-1}, which had a moderate effect on wave field quality due to presence of low velocity zone, as proven in previous section. Other parameters were left constant; with full 3D acquisition geometry and 35 Hz nominal frequency were used in all three computations (Fig. 5).

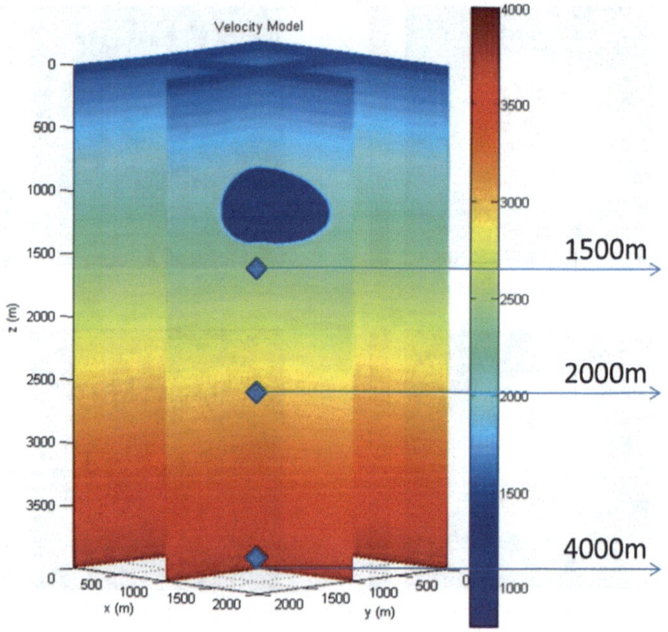

Fig. 5 A shallow gas cloud velocity model of 1400 ms^{-1} gas velocity with three different target depths for analysis. Full 3D acquisition geometry along with 35 Hz nominal frequency was used for all three different depth Positions

Varying the target depth producing expected results as target reflector closer to the surface has greater amplitude distribution and signal strength, whereas the response for target reflector at deeper depth is weaker with less energy detected, as demonstrated in Fig. 6. However, since attenuation factor, Q was not incorporated in the focal beam computation and analysis, the resolution strength, which based on the signal quality detected at a particular point, is identical for all three depth positions. Even though resolution outcomes in shallow gas cloud velocity model of 1400 ms^{-1} yield no significant difference between the lateral resolution for 1500, 2000 and 4000 m depth, by having closer depth position to the anomaly, larger part of wave field will be influenced by low velocity zone, as the propagation angle getting wider which in contrast, deeper target points will allow smaller range of propagation angles. This explains why AVP imprints at 4000 m depth are smaller than 2000 m depth. Throughout the varying depth analysis, emphasises was given on the observed amplitude coverage compare to resolution function.

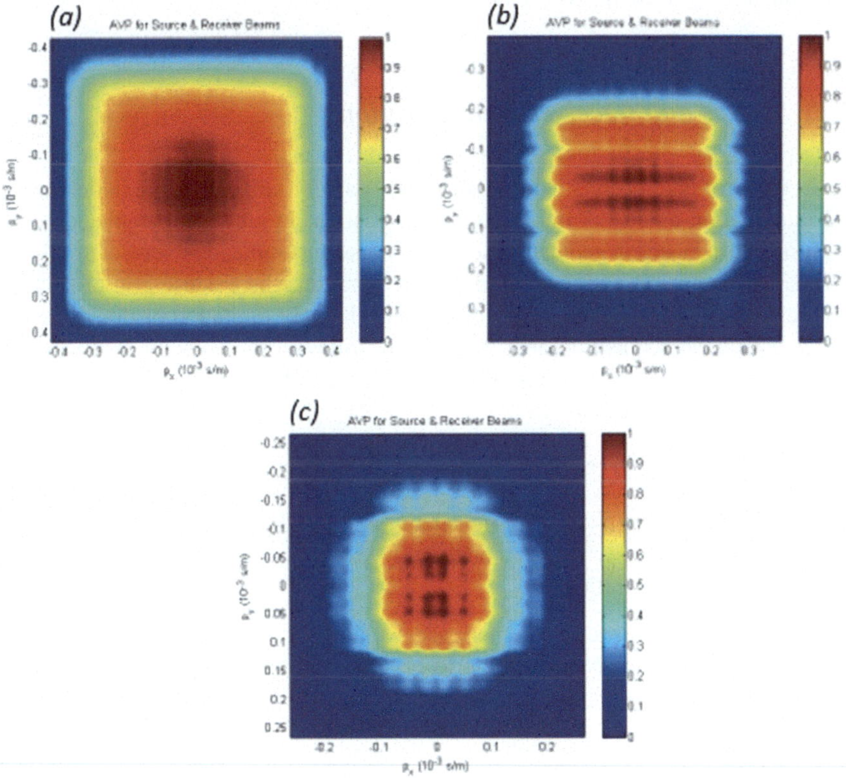

Fig. 6 The AVP's focal beam analysis for different target depth: **a** 1500 m **b** 2000 m **c** 4000 m. Other parameters were remained constant: shallow gas cloud velocity of 1400 ms^{-1}, full 3D acquisition and 35 Hz frequency

Additional simulations were carried out by varying the shallow gas velocity along with the depth of target reflector. In case of slower gas cloud velocity (as for 1100 and 800 ms^{-1} models), the presences of a strong velocity contrast yields poor resolution and amplitude coverage at all three depth points mainly due to limitation of extrapolation operators as the wave field propagation was badly affected by extreme velocity variation. In higher gas cloud velocity, it was observed that AVP imprints are less smooth at deeper point, contrary to spatial domain where no distinctive effect can be seen to resolution function at different depth level. These variations are due to the imperfect extrapolation operators which are not set to optimum. In addition, operator used in each focal beam computation applied to every grid point and depth step, means that for 2000 m depth location the same operator is applied 268 times (as 2000 m divided by 15 m for each depth step equal to 134 steps required for forward extrapolation while another 134 steps required for inverse extrapolation) for both focal detector beam and focal source beam. This imperfection is amplified by multiplication of the identical focal source beam and focal detector beam, which consequently squaring the artefacts.

2.3 *Source and Receiver Configurations*

Surface source and receiver configuration plays a very influential role in determine the illumination quality of the target depth location. In previous parameterization analysis, the acquisition geometry used in focal beam computation was set to full 3D acquisition geometry (Fig. 7a), an ideal configuration where dense grid of sources and detectors fully cover the acquisition area. The only limitation of this design is the extent of the acquisition area that strictly followed the velocity model build-up area. The following features describe the full 3D survey design used in the parameterization's analysis.

(i) Source location is same as receiver location.
(ii) Source (and receiver) lines have length of 2 km (x-direction) and 2 km (y-direction), same as velocity model area.
(iii) Minimum distance between consecutive sources (and receivers) are 10 m in both x and y direction.

Prior to choosing this set-up, numerical calculations have been carried out to determine the minimum geometry spacing to prevent any data aliasing. The minimum source and receiver spacing allowable were calculated by using the following formula:

$$\Delta x_s = \Delta x_r = \frac{V_{\min}}{2 f_{\max}} \tag{1}$$

The nominal frequency used of 35 Hz is the maximum frequency in the data, while minimum velocity in the model was taken from shallow gas cloud velocity value of

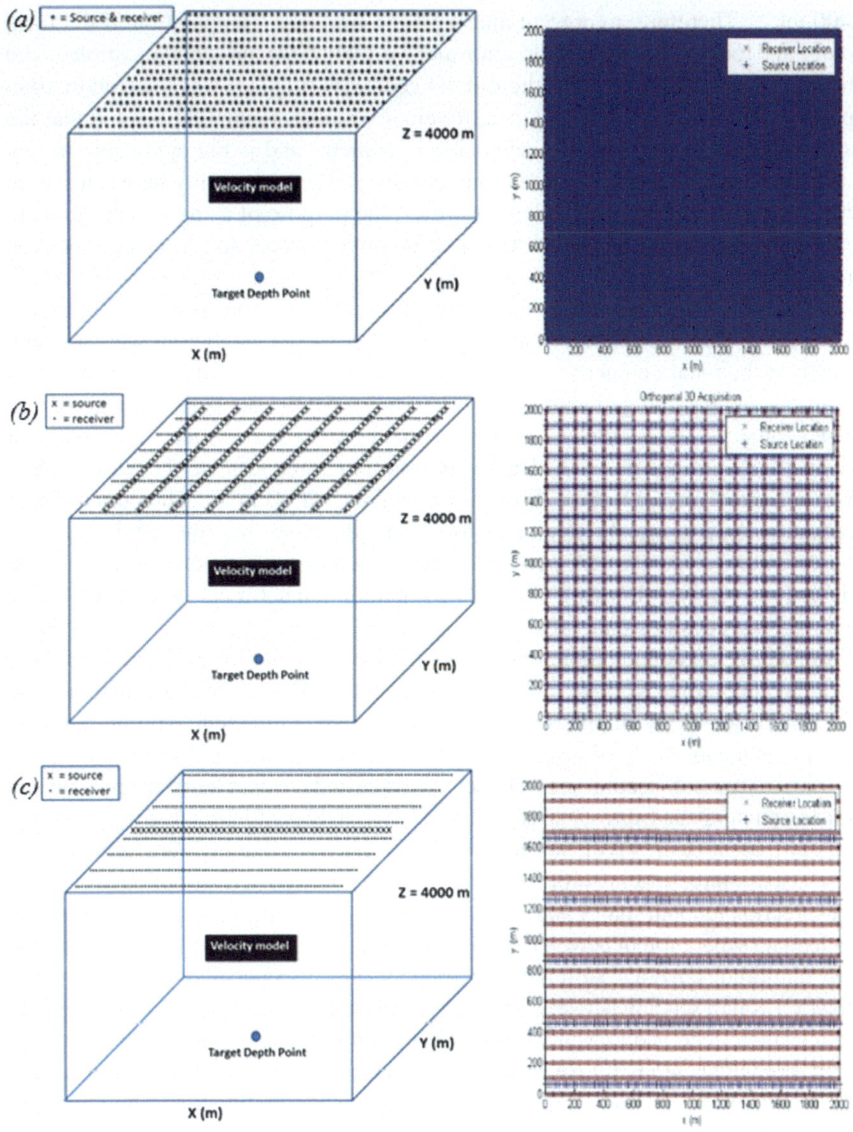

Fig. 7 Three types of 3D surface acquisition geometry used for surface geometry focal beam analysis; **a** Full 3D geometry where the source and receiver location is identical, **b** Orthogonal 3D geometry commonly used in land acquisition and **c** Parallel 3D geometry that commonly used in marine acquisition. Other parameters were remained constant: 1400 ms^{-1} gas cloud velocity, 4 km target depth and 35 Hz Frequency

1400 ms^{-1}. Therefore, to prevent aliasing from occurred, computation of the focal source and receiver beams in *x*-direction and *y*-direction for surface acquisition model should not exceed 20 m. Thus, the full 3D geometry (Fig. 7a) was lowered to 10 m spacing which equivalent to the grid of velocity model. It should be noted that the full 3D acquisition is the ideal acquisition geometry and is not applicable in real acquisition world even in land environment due to close proximity between source and receiver location in the design. However, the purpose of using this design is to understand illumination strength when all possible sources and receiver's location are available.

Meanwhile, for better understanding regarding illumination coverage for the area affected by shallow subsurface anomaly, two other survey geometries were applied, namely orthogonal geometry (Fig. 7b), to represent the standard land environment geometry and parallel geometry (Fig. 7c), to represent the standard marine environment geometry. The parallel acquisition geometry involves one source line and eight receiver lines arrange in swath. The swath was configured to cover the whole survey area with receiver spacing is 20 m in *x*-direction and 50 m in *y*-direction, source separation is 20 m in *x*-direction and 400 m in *y*-direction. In orthogonal geometry, source and receiver lines are perpendicular to each other, with spacing in between source (and receiver) lines are 50 m while separation in between consecutive source (and receiver) positions are 10 m.

In focal beam computation, the target illumination point for analysis in all three different geometries is chosen at 4 km depth located at the centre of acquisition area. The full 3D geometry amplitude distribution (Fig. 8a) is sufficient to produce a good seismic image as plenty of signals were detected on the surface. As shown here and in previous two sub-sections as well, dense source and receiver locations prevent any aliasing phenomena from occurred in full 3D design. In parallel acquisition geometry (Fig. 8b), focal beam's AVP suffered from very poor aliasing in *y*-direction due to a very sparse source spacing (100 m spacing) while pretty good distribution achievable in *x*-direction for both source and receiver focal beams. In orthogonal geometry, due to inadequate separation in between the adjacent source (and receiver) lines (50 m), the AVP amplitude coverage suffered from mild aliasing effect in all direction (Fig. 8c). However, small spacing between adjacent source (and receiver) location compensate the aliasing effect, and prevents from poorer amplitude coverage.

In acquisition geometry parameterization analysis, it was observed that in spatial domain, the better beam will dominate the final resolution function. The limitation in source sampling as in the case of parallel geometry will produce side lobes to source focal function, but the effect was offset by dense receiver positions, thus the resultant resolution function will not introduce any side lobes and aliasing effect. In radon domain, based on the parallel geometry, focal source beam will produce poor imprint while focal receiver beam produce acceptable imprint. Since the final AVP imprint will be dominated by poorer beams of two, the result obtained normally resembles the poorer image of the two, in this case is focal source beam.

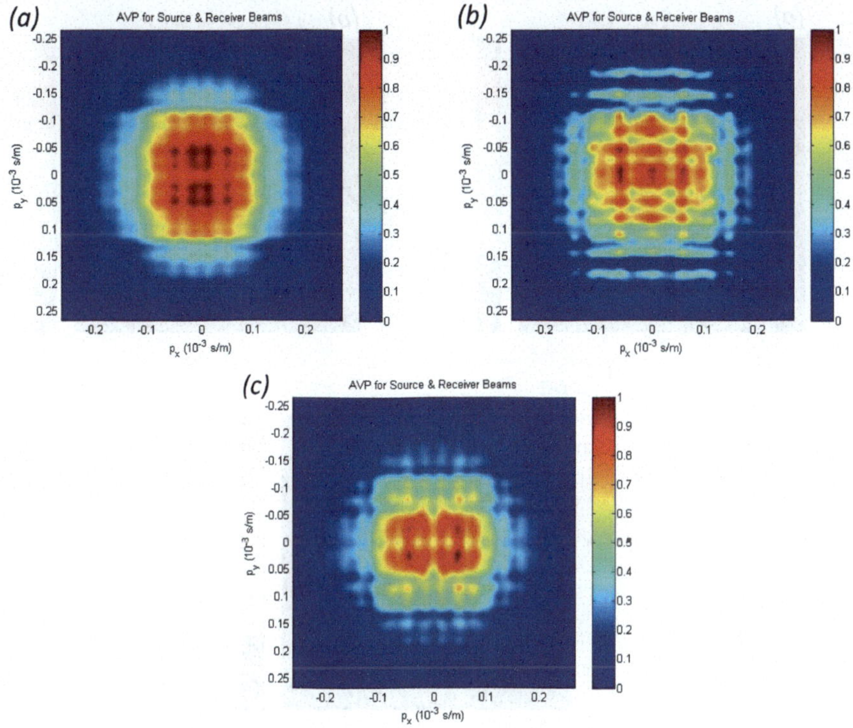

Fig. 8 The AVP results from different geometry configurations. The AVP results obtained after incorporating **a** full 3D geometry, **b** parallel geometry and **c** orthogonal geometry. Other parameters were remained constant: shallow gas cloud velocity of 1400 ms^{-1}, 4 km target depth and 35 Hz nominal frequency

2.4 *Nominal Seismic Frequency*

Nominal frequency used and detected in seismic exploration is important to determine the quality of subsurface image. In shallow gas accumulation environment, the frequency of the seismic signal was affected due to dispersion phenomena during seismic wave propagation. Generally, the useful seismic signal contains a broad range of frequency spectrum from 10 to 80 Hz. To understand the illuminations aspect of seismic data with respect to frequency, the seismic nominal frequency in focal beam parameterization analysis were changed as: 10 Hz for low frequency signal, 35 Hz for peak frequency signal and 60 Hz for high frequency signal. The AVP results as shown in Fig. 9 were obtained after applying full 3D geometry of 1400 ms^{-1} shallow gas cloud velocity model with fix illumination point at 4 km.

Initial assessments of the results indicate that all three nominal frequencies yield considerably good amplitude distribution in AVP with uniform coverage and no aliasing recorded which due to full 3D surface acquisition used in the analysis. However, in 10 Hz result, the image obtained is less focus compare to 35 and 60 Hz

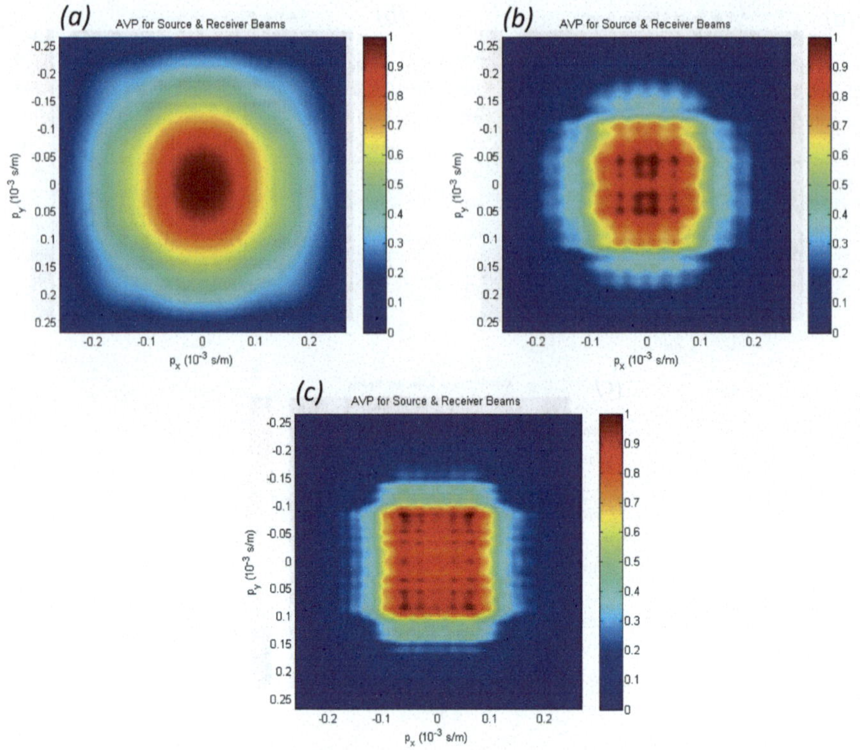

Fig. 9 Three different nominal frequencies used in focal beam parameterization analysis. The frequencies tested were **a** 10 Hz, **b** 35 Hz, **c** 60 Hz and each frequency yield the AVP result as shown. Other parameters were remained constant: 1400 ms^{-1} shallow gas cloud velocity, full 3D acquisition and 4 km target depth

frequency images, which have sharper images. Another difference noticed from the results is that lower frequency exhibited wider amplitude coverage compare to higher frequency, which indicates the lower frequency has better ability to illuminate deeper target. If the target is shallow, seismic data will be using higher frequency signal which undoubtedly will yield better resolution image. From the result of the analysis, it shows that as the frequency increase, the amplitude coverage decreases with better image resolution.

3 Ray Tracing Interpolations

Due to structural and seismic complexity of the earth subsurface, the forward modelling approaches to the wave equation using direct numerical methods such as finite difference and finite element sometimes are not desirable due to complexity and time consuming factors in solving the full wave equation. Alternatively, solving wave equation by using an approximation method in term of ray tracing interpolation [10] is considerably faster and simpler compare to full waveform solution. The basic of ray theory is based on wave equation solution using Eikonal and transport equations which will compute the travel time and amplitude of the wave respectively [18]. Based on 1D wave equation:

$$\frac{d^2 p}{d^2 z} = \frac{1 d^2 p}{c^2 d^2 t} \tag{2}$$

The approximation of Eq. 2 to high frequency leads to Eikonal equation:

$$\left|\nabla T^2\right| = \frac{1}{c^2} \tag{3}$$

where T is the wave travel time information. Since the frequency information has been approximated, the corresponding solution is non-dispersive and only contains phase velocity information. Propagating ray information over the velocity model instead of full waveform has a clear advantage in term of speed, cost and simplification of the forward process, though the resultant forward modelling are not as accurate as wave propagation and will not exhibit other wave phenomena such as interference and diffraction.

These are several approaches for ray interpolations in seismology, among them are bending method, pseudo-bending method, wave front construction and Euler-based method using numerical solution. Ray tracing methodology in this thesis will incorporate Euler-based approach by interpolating rays throughout the layers which will be determined by Runge-Kutta 4th order (RK4) method. The ray interpolation method works by specifying the source and receiver configuration on surface during the forward modelling process, then a group of rays were propagated through the velocity model and reflected at a target reflector.

The examples of ray tracing propagation through a shallow anomaly media is as illustrated in Fig. 10. Based on the propagation analysis and the number of detected rays on surface, it clearly demonstrated that rays will be easily reached the target subsurface, provided the rays were shot from the middle of the shallow gas zone (Fig. 10c). This is because the ray propagation obeys Snell's Law, whereas the slower media causes the ray bending inwards and towards a focus point. However, if the source position was located at the edge of the slower anomaly zone (Fig. 10b), less rays will be propagate towards the reflector as higher velocity media near the anomaly flank push the rays propagation away from the central zone (Fig. 11).

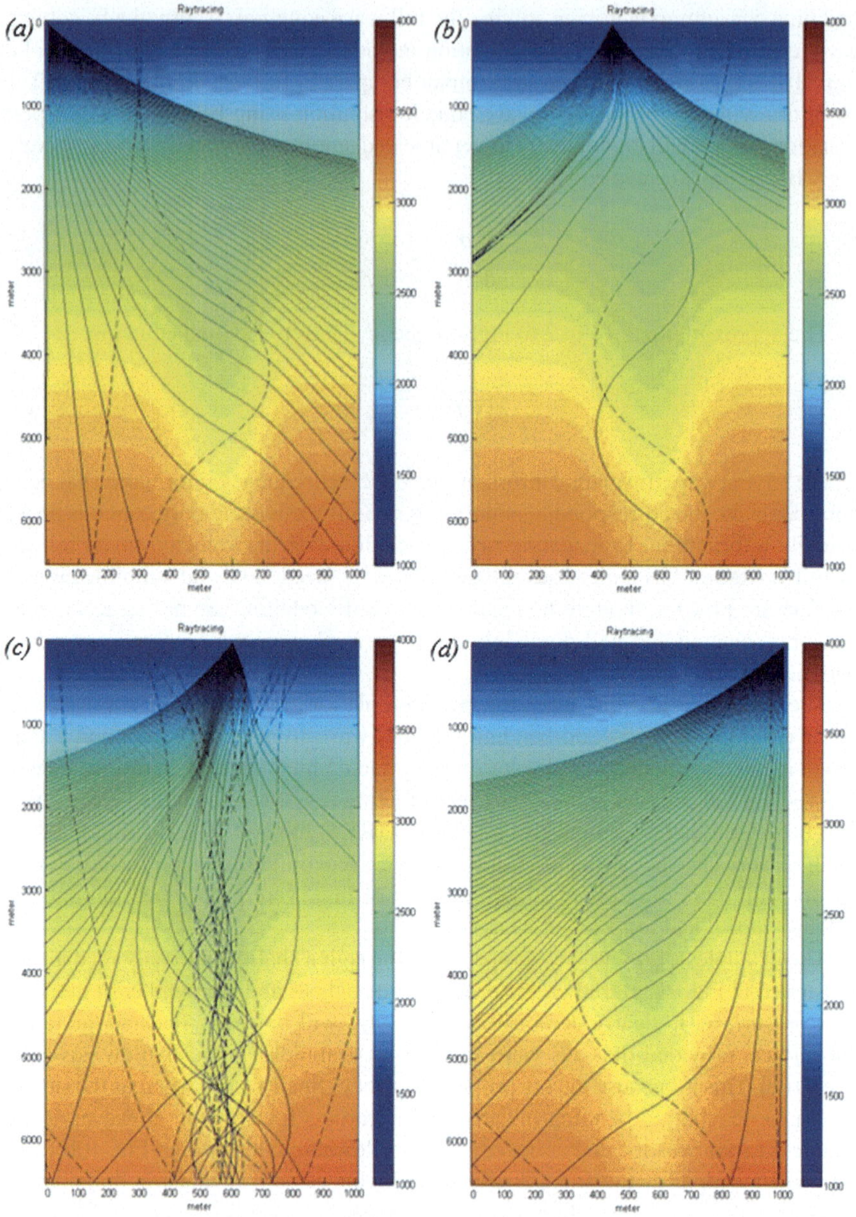

Fig. 10 The ray traces propagation through a shallow gas cloud media in four different surface positions within a 2D line. Downward rays (solid line) reflected at target subsurface produced reflected rays (dotted line)

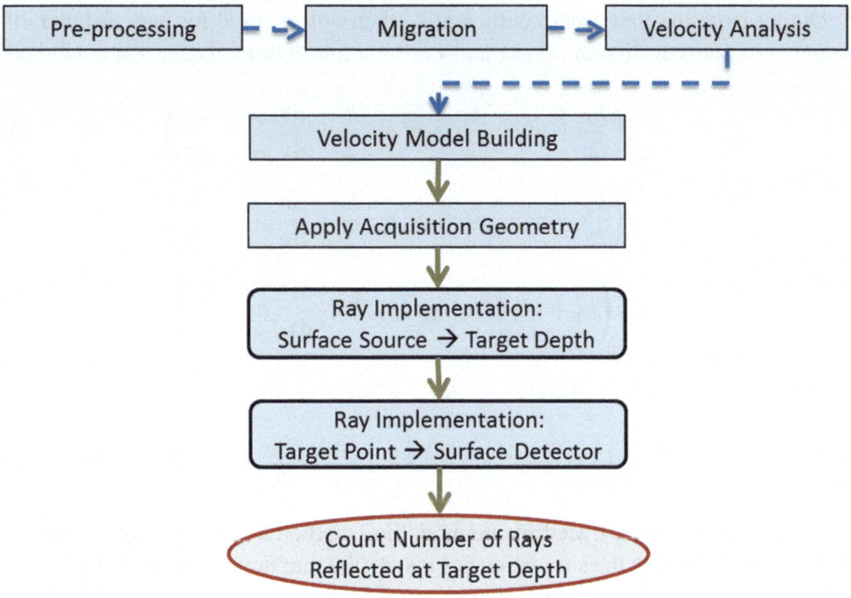

Fig. 11 Workflow for ray tracing interpolations through shallow gas cloud. Different model parameters related to ray trace interpolation were tested by follow the procedure describes in the workflow

3.1 Rung-Kutta 4th Order Methods

Eikonal equation in Eq. 3 can be represented in term of ordinary differential equation (ODE) and to solve the equation numerically, a solution to ODE's boundary problems is required. By formulate ray tracing based on shooting method, the kinematic ray equation can be solved as an initial value problem rather than using two points boundary value problems which is more complicated. The basic concept of shooting ray method lies in two critical parameters, namely the initial position of x, y and z coordinates, and its trajectory angle from the initial position. Based on these parameters, the consequent ray positions can be obtained from its initial position using Runge-Kutta 4th order (RK4) technique. To determine the next interpolation position, ray propagation by RK4 technique will go through four sets of positional evaluations where the derivative of the position was calculated four times for higher accuracy.

Using known information of original ray position with its trajectory angle, the first order RK4 coefficient, k_1 was obtained by multiplying the time step, h with velocity information, v and horizontal and vertical slowness, p_x and p_z, as well as with lateral and vertical velocity gradient, $\frac{dv}{dx}$ and $\frac{dv}{dz}$ respectively:

$$k_1 = h\left(v^2 p_x, v^2 p_z, \frac{dv}{dx}, \frac{dv}{dz}\right) \tag{4}$$

Output from the first order coefficient k_1 then will be used for 2nd, 3rd and 4th order coefficients evaluation, k_2, k_3 and k_4 of the fourth order Runge-Kutta method.

$$k_2 = h\left(\frac{k_1}{2} + v^2 p_x, \frac{k_1}{2} + v^2 p_z, \frac{h}{2} + \frac{dv}{dx}, \frac{h}{2} + \frac{dv}{dz}\right) \quad (5)$$

$$k_3 = h\left(\frac{k_2}{2} + v^2 p_x, \frac{k_2}{2} + v^2 p_z, \frac{h}{2} + \frac{dv}{dx}, \frac{h}{2} + \frac{dv}{dz}\right) \quad (6)$$

$$k_4 = h\left(k_3 + v^2 p_x, k_3 + v^2 p_z, h + \frac{dv}{dx}, h + \frac{dv}{dx}\right) \quad (7)$$

The next ray position can be obtained by combining Eqs. 4–7:

$$(x, y, z) = (x, y, z) + \frac{k_1}{6} + \frac{k_2}{3} + \frac{k_3}{3} + \frac{k_4}{6} \quad (8)$$

In order to get the ray trace forward modelling method as accurate as possible, the time step used through the simulation is 4 ms. As the maximum velocity in the model is 3750 ms^{-1}, the most suitable depth increment for this velocity model is every 15 m. The ray trace propagated from a source on surface level, with the trajectory defined from −45° to 45° in the z-direction. The rays will be reflected back once it hits the target reflector, located at specified depth level, and the number of rays detected at receivers will be counted. There are two conditions where ray trace will be discarded from the system:

(i) Any ray travels beyond the model's boundary.
(ii) Recorded rays without first reflected at the target reflector.

In illumination analysis based on ray tracing method, number of ray hits the target reflector will determine the subsurface quality, as information obtained on how much rays can propagate through the subsurface anomaly and recorded back at reflector. However, note that any information related to amplitude is not available in ray tracing method.

4 Parameterization Analysis

From the focal beam results, it clearly shown that the subsurface illumination analysis is highly dependent on the velocity model parameterizations, as a consequence of distorted wave propagation occurred and wave amplitude attenuation throughout the extrapolation process. However in ray tracing interpolations method, the subsurface illumination may differ from focal beam method due to contrasting characteristic between wave and ray theories. Therefore, before the integration of focal beam and ray tracing methods took off, ray tracing parameterization analyses were conducted

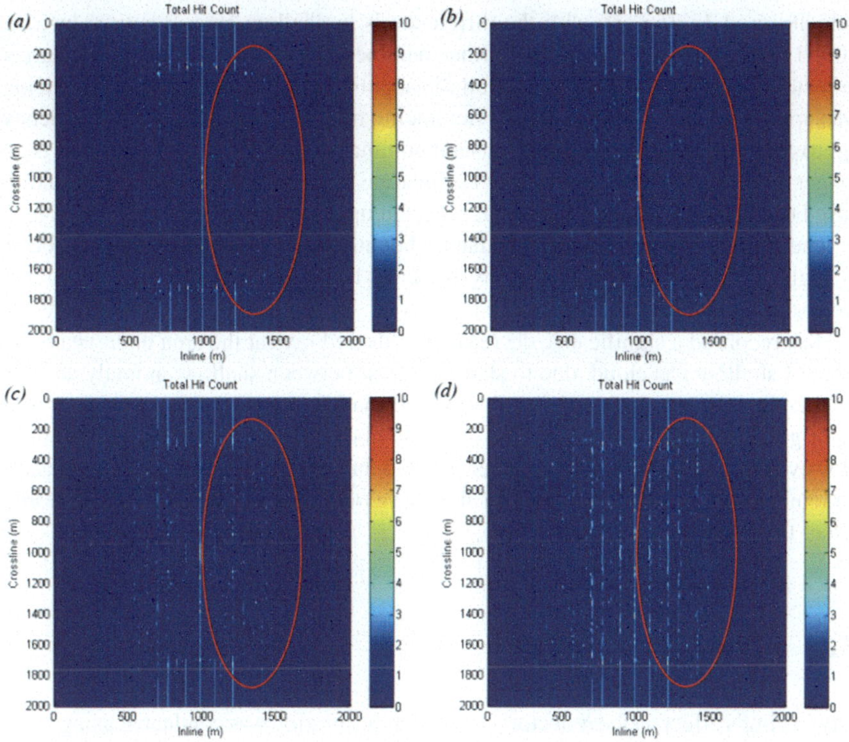

Fig. 12 Ray hit counts map for four different shallow gas cloud velocity; **a** 800 ms^{-1}, **b** 1100 ms^{-1}, **c** 1400 ms^{-1}, and **d** 1700 ms^{-1}. Target reflector was fixed at 4 km with full 3D geometry applied on the surface level. For better viewing, the area circled in red indicates the significance difference in number of ray hit count on the particular target reflector

to understand the influence of ray propagation and its illumination analysis in the shallow gas cloud area.

Unlike the focal beam method, conducting illumination analysis in ray tracing technique does not involve any amplitude strength and energy distribution, which appear constant throughout the interpolation process. Instead, in the ray tracing analysis, the hit count method will be used which is the number of rays hit the target horizon before reflected back to surface level. The reflected rays have to be detected by receivers at surface level as to be considered in illumination distribution. The workflow of the ray tracing work is described in Fig. 12.

4.1 Shallow Gas Cloud Velocity

Based on the similar shallow gas cloud velocity model developed for focal beam parameterization analysis in Sect. 2.1, rays were propagated through four different velocity models (Fig. 3), with the difference is in shallow gas velocity value, i.e. 800, 1100, 1400 and 1700 ms^{-1}. Other parameters were set constant with target reflector chosen at 4 km depth and full 3D acquisition geometry with 40 m spacing was applied on the surface level. As ray tracing method is already a high frequency approximation of wave equation, constant nominal frequency parameterization is no longer applicable. The results in Fig. 12 illustrate the subsurface illumination quality based on the number of ray hit count, reflected upon hit the target depth position.

The ray hit count results indicate that higher number of rays hit the target reflector in higher shallow gas cloud velocity models (1400 and 1700 ms^{-1}) compare to slower velocity models (800 and 1100 ms^{-1}). In particular, the slower velocity models recorded a significance decrease of reflected rays at the area underneath the edge of shallow gas cloud, due to sharp contrast between shallow anomaly and the surrounding. A drastic change in velocity information caused the ray to deviate from its original trajectory thus leading to lower number of rays detected at the receivers. However, a large number of ray hit recorded at the centre of the survey area (exactly underneath the centre of gas cloud) in all four velocity models illustrated the focusing effect from shallow gas cloud.

4.2 Target Depth Level

Ray tracing's effect on the reflector's target depth variation was conducted using identical velocity model with shallow gas velocity value of 1400 ms^{-1}. Three different reflector levels of 1500, 2000 and 4000 m were chosen, as describe in Fig. 5, while a full 3D geometry survey design with series of source and receivers of 40 m separations in between consecutive source and receivers were used in the simulation. The resultant ray hit count is illustrated in Fig. 13, where obvious reduction in number of rays recorded at the deeper target reflector. In general, the number of ray hit count obtained indicates shallower targets will have better illumination due to higher number of ray hit count, compared to deeper targets especially for the area underneath shallow gas cloud. In the shallower target, centre of the target horizon has the highest number of reflected rays because the shallow gas cloud perform as a lens, to focus all energy towards the centre of area underneath gas cloud.

5 Hybrid Illumination Method Development

The focal beam and ray tracing illumination methods discussed in Sects. 2 and 3 suffered from a few setbacks. For instance, the hit count's ray tracing technique will not took into account the Fresnel zone information, due to unknown amplitude coverage of the target area. The nature of ray characteristic also means that some of the wave phenomena were ignored. On the other hand, local approach in focal beam method give only limited illumination information for a particular target point. In

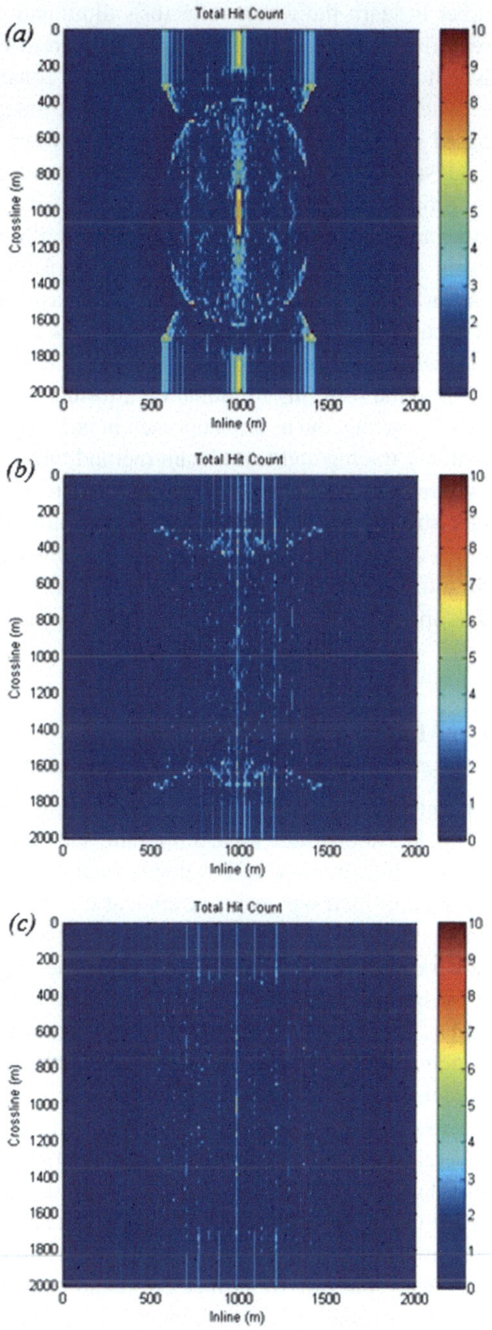

Fig. 13 Ray parameterization analysis for different reflector depth level; **a** 1500 m depth **b** 2000 m depth and **c** 4000 m depth. Other parameters were remained constant: 1400 ms^{-1} shallow gas cloud velocity and full 3D acquisition

order to learn the whole reflector's illumination coverage, the focal beam method requires multiple extrapolations from different target points within the target horizon, which is considerably expensive and time consuming.

The development of hybrid method to integrate both illumination methods, is based on work done by Bernitsas [19], where he patented a hybrid illumination analysis method of both wave equation and ray tracing techniques. His method works by propagating ray tracing through a velocity medium before measure the Fresnel zone around the target reflector. Then, the illumination information was calculated using wave equation for all frequencies involved, thus assess the quality of a particular subsurface by taking into account both wave propagation and ray tracing methods. It should be noted that the ray tracing work developed in Bernitsas [19] did not take into account the reflected ray from target subsurface to the receiver which eventually mislead the resulting illumination information.

Meanwhile, the hybrid approach in our work involved analysing and propagating both ray tracing and focal beam method through the velocity media independently, by introducing different numerical illumination information based on wave amplitude strength and number of ray hits at a particular location. Each of the illumination method had been analyses and discussed independently in Sects. 2 and 3. The workflow for hybrid illumination analysis was summarized in Fig. 14 and this method was applied to a field data from Malay Basin.

6 Malay Basin Case Study

The integration of hybrid illumination technique was applied on a field data to identify the field's subsurface wave amplitude and rays count. Based on previous migration velocity information, a 3D velocity model was built as shown in Fig. 15a. From the model and its respective time slice at 6 s (Fig. 15b), a shallow gas zone was easily spotted around inline 500 and crossline 1000, which indicated by a velocity pull-down effect. To validate the velocity model quality, the model was overlaid on top of seismic section (Fig. 16) which resulted in good match between the seismic data and its velocity information. Two 2D horizons were chosen for illumination analysis, which are the stretches along crossline 1000 with depth of 3 and 6 s of the velocity model.

Instead of localised target point, the revised focal beam method extrapolates several independent wave fields from multiple target points located within the specified horizons for focal analysis in spatial domain. The extrapolation points begin at inline 0 and end at inline 1000 with target point spacing of 20 m. Constant parameterization was used throughout the focal beam process, with 35 Hz nominal seismic frequency and full 3D surface acquisition Geometry.

The distributions of wave amplitude in spatial domain from two target reflectors were shown in Fig. 18. Both of the 3 and 6 s amplitude distribution along chosen 2D horizon indicates that adequate seismic energy fall on the reflector and good seismic image can be obtained, whilst shallower target appears to have sharper image

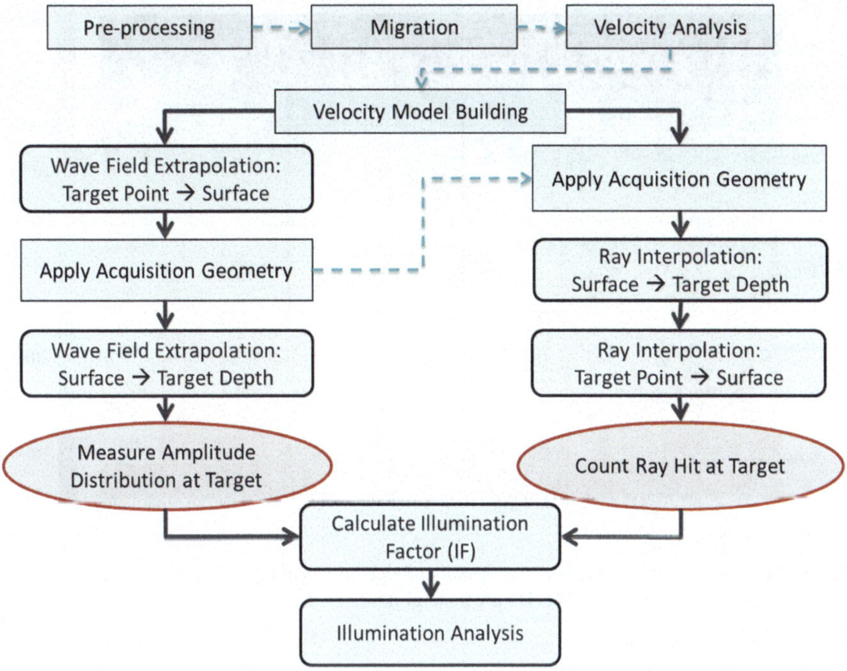

Fig. 14 Hybrid illumination analysis for shallow subsurface anomaly region. The method contains two separate analysis, namely ray tracing and focal beam techniques, and both methods integrate towards the end of the methodology

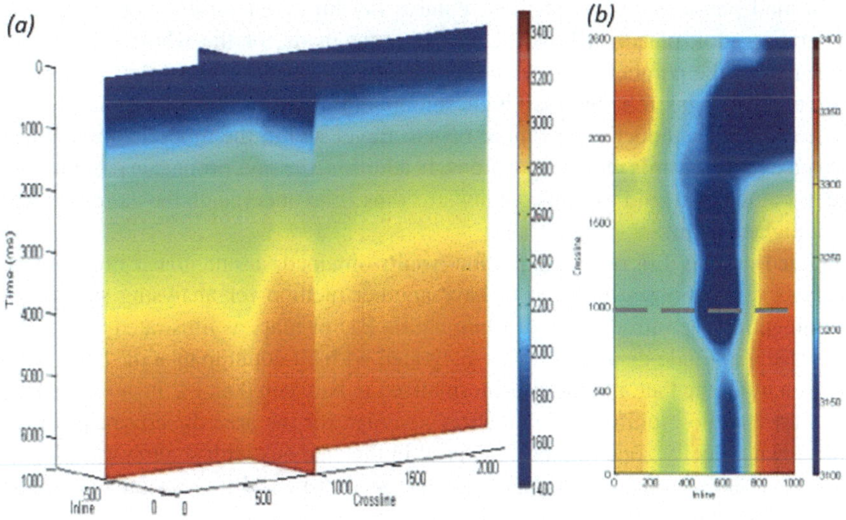

Fig. 15 The velocity model used for integrated illumination analysis. **a** Inline and crossline slices of the model with a velocity pull-down effect spotted at inline 500 and crossline 1000. **b** The time slice of the velocity model at 6 s data, with dash line indicates the 2D horizon to be used for analysis

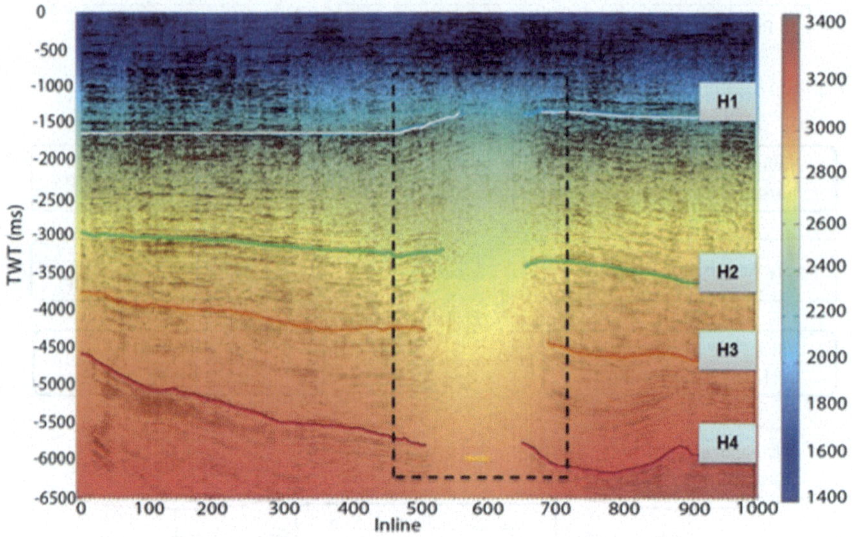

Fig. 16 Seismic section overlay with the velocity model built as in Fig. 15. The horizons continuity (area indicated in black box) for H1–H4 are general poor

resolution. However, it should be noted that the final seismic stack section (Fig. 17) at the chosen target horizon indicate otherwise, as poor image recorded with uncertainty subsurface layers, particularly at shallower target that located closer to shallow gas cloud.

Second phase of hybrid method requires ray hit count information before the integration took place. In Fig. 19, the hit count maps for the whole survey area indicate that the number of ray hit counts in 6 s reflector is higher than 3 s reflector, with middle of gas zone received highest number of ray hit. This is particularly true since the deeper model allows more rays to travel within the model boundary and reflected back towards receiver positions. In addition, the rays propagate past through the gas cloud also tend to converge inwards thus zone underneath the gas cloud will have higher ray hit.

Based on two different illumination results obtained, an integrated illumination analysis was proposed in order to have another method for analysing subsurface illumination. The introduction of Illumination Factor map, as illustrated in Fig. 20, the wave amplitude distribution was integrated with hit count map at the chosen 2D horizon, with the factor values was normalized in between 0 and 1. It shows that 6 s reflector has more uniform illumination coverage for the area directly beneath the pull-down velocity zone (2D horizon used in the analysis with shallow gas cloud). However, the 3 s reflector illumination coverage is wider as more points within the same reflector were illuminated compared to deeper target reflector.

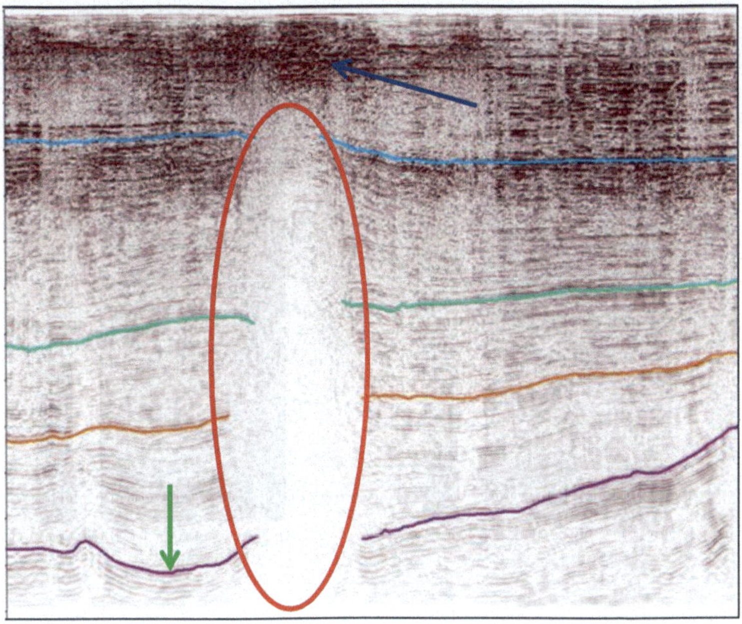

Fig. 17 Spatial domain focal beams illumination for a 2D reflector at **a** 3 s depth **b** 6 s depth. Several points on the target reflector were used to obtain these beams; begin from inline 0 until inline 1000 with increment of 20 m. The color bars show the normalize amplitude of the beams

7 Summary

The concept of wave extrapolation using focal beam method and ray tracing interpolations using Runge-Kutta 4th Order (RK4) method as explained in the beginning of this chapter were originally subject to individual illumination analysis. Table 1 summarizes its consequences, demonstrating the effectiveness of focal beam and ray tracing methods over various model parameters. However, it should be noted that focal beam method only allows a single point analysis even though the wave extrapolation process was carried out in 3D. In the ray tracing approaches, rays interpolation were in 2D, albeit its resulted in ray hit count at multiple points on the same reflector horizon level.

Therefore, a hybrid combination of both wave propagation and ray tracing was introduced and demonstrated to give more subsurface information than individual illumination analysis, yield a new illumination attribute called Illumination Factor, giving a greater level of confidence analysis. While these analyses did not directly enhance the subsurface image, a valuable insight and information on the acquisition geometry, velocity model parameter, wave and ray propagation phenomena and potential migration algorithm had been obtained.

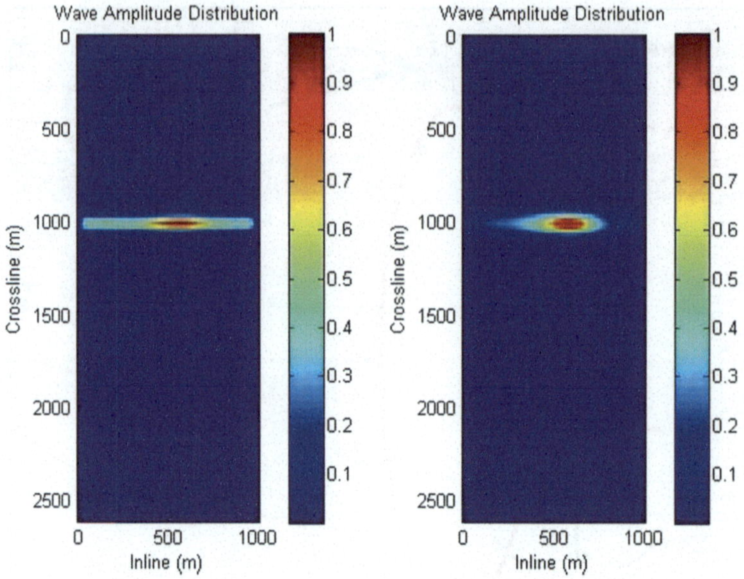

Fig. 18 Spatial domain focal beams illumination for a 2D reflector at **a** 3 s depth **b** 6 s depth. Several points on the target reflector were used to obtain these beams; begin from inline 0 until inline 1000 with increment of 20 m. The color bars show the normalize amplitude of the beams

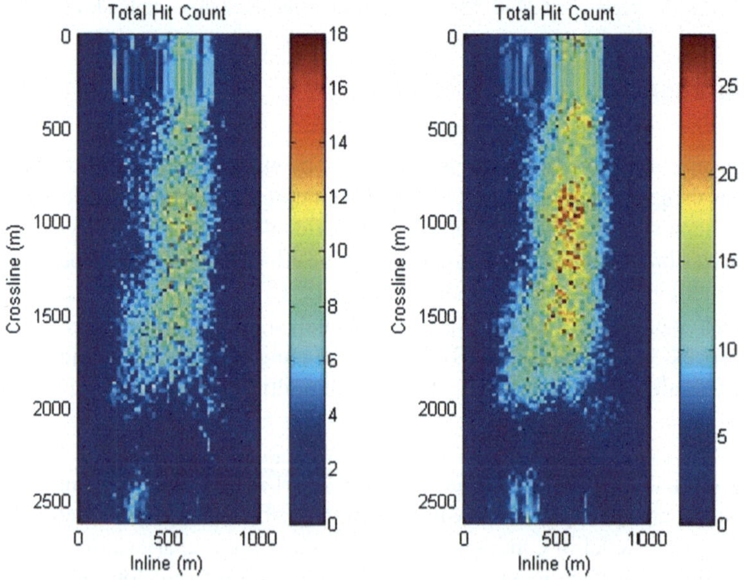

Fig. 19 Map of rays hit count for **a** 3 s horizon and **b** 6 s horizon. The rays were propagated from all source position on surface level according to full 3D acquisition geometry. The number of rays recorded back by receiver is indicates by color bar on the right

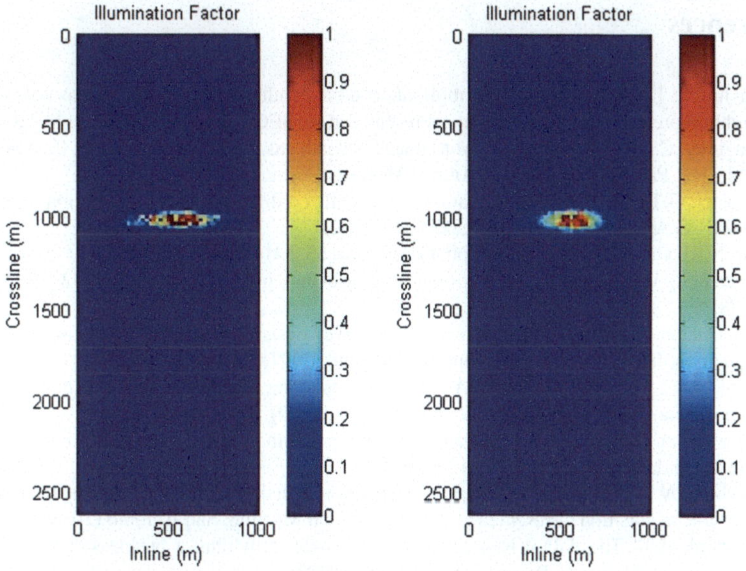

Fig. 20 Illumination Factor maps for [500, 1000] target point produce after integrating between spatial domain focal beam and rays hit count for **a** 3 s reflector and **b** 6 s reflector. The scales were normalized to give better illumination indicator

Table 1 Summary of parameterization effects on focal beam AVP analysis and number of ray trace hit count

Parameterization	Focal beam AVP analysis	Ray trace 'hit count'
Shallow gas cloud velocity	Higher velocity, wider coverage	Higher velocity, greater hit count
Depth of target point	Deeper target, smaller coverage	Deeper target, less hit count
Survey design	Full 3D—good coverage Parallel and Orthogonal—High coverage with aliasing problem	Not applicable
Nominal seismic frequency	Low frequency, wider coverage High frequency, better focusing	Not applicable

Acknowledgements The authors are grateful to all members of the Geoscience Department for their support and as well as the members of the above consultancy model's team, which includes engineers, associate professors, lecturers, laboratory technologists, and students. Special thanks to the Universiti Teknologi PETRONAS, Malaysia for providing the permission and resources to carry out the laboratory demonstration. Without the kind cooperation this study would have been impossible.

References

1. Traynin, P., Liu, J., Reilly, J.M.: Amplitude and bandwidth recovery beneath gas zones using Kirchhoff prestack depth Q-migration. Presented at the SEG Annual Meeting, Las Vegas (2008)
2. Yongping, C., Yan, H.: Use Q-RTM to image beneath gas hydrates in Alaminos Canyon, Gulf of Mexico. Presented at the SEG Annual Meeting, Denver (2010)
3. Schuster, G.T.: Reverse-time migration = Generalized diffraction stack migration. Presented at the SEG Annual Meeting, Salt Lake City (2002)
4. Hill, N.R.: Gaussian beam migration. Geophysics **55**, 1416–1428 (1990)
5. Gazdag, J.: Wave equation migration with phase-shift method. Geophysics **43**, 1342–1351 (1978)
6. Moldoveanu, N., Egan, M.: Some aspects of survey design for wide-azimuth towed-streamer acquisition. In SEG Annual Meeting, San Antonio (2007)
7. Granli, J.R., Arnsten, B., Sollid, A., Hilde, E.: Imaging through gas-filled sediments using marine shear-wave data. Geophysics **64**, 668–677 (1999)
8. Moldoveanu, N., Kapoor, J., Egan, M.: Full azimuth imaging using circular geometry acquisition. Lead. Edge **27**, 908–913 (2008)
9. Berkhout, A.J., Blacquière, G.: Blended acquisition with dispersed source arrays, the next step in seismic acquisition? Presented at the SEG Annual Meeting, San Antonio (2011)
10. Berkhout, A.J.: The data-driven seismic value chain, providing a business context for the velocity issue. Geophys. Prospect. **52**, 481–487 (2004)
11. Laurain, R., Gelius, L.J., Vinje, V., Lecomte, I.: A review of 3D illumination studies. J. Seism. Explor. **13**, 17–37 (2004)
12. Berkhout, A.J., Ongkiehong, L., Volker, A.W.F., Blacquiere, G.: Comprehensive assessment of seismic acquisition geometries by focal beams Part I: theoretical considerations. Geophysics **66**, 911–917 (2001)
13. Volker, A.W.F., Blacquiere, G., Berkhout, A.J., Ongkiehong, L.: Comprehensive assessment of seismic acquisition geometries by focal beams Part II: practical aspects and examples. Geophysics **66**, 918–931 (2001)
14. Volker, A.W.F.: Assessment of 3D seismic acquisition geometries by focalbeam analysis. Ph.D., Applied Physics, Delft University of Technology (2002)
15. Van Veldhuizen, E.J.: Integrated approach to 3D seismic acquisition geometry analysis. Delft University of Technology, PhD (2006)
16. Vinje, V., Iversen, E., Gjoystdal, H.: Traveltime and amplitude estimation using wavefront construction. Geophysics **58**, 1157–1166 (1993)
17. Gjøystdal, H., Iversen, E., Lecomte, I., Kaschwich, T., Drottning, Å., Mispel, J.: Improved applicability of ray tracing in seismic acquisition, imaging and interpretation. Geophysics **72**, 261–271 (2007)
18. Cerveny, V., Molotrov, I.A., Psencik, I.: Ray method in seismology. Univerzita Karlova, Praha (1977)
19. Bernitsas, N.: Subsurface illumination, a hybrid wave equation-ray tracing method. United States Patent (2004)

Sol-gel Synthesis of Silicadioxide/Organopolymer Nanocomposite for Potential Applications in EOR Nanotechnology

Hassan Soleimani, Mohammed Falalu Hamza, Zulkifli Merican Aljunid Merican, and Karl Dunbar Stephen

Abstract The application of enhanced oil recovery (EOR) nanotechnology has opened an important field of fundamental science and engineering studies offering a wide range of research interests. Owing to the physicochemical properties of some nanoparticles (NPs), several mechanisms have been employed to functionalize the surface of nanoparticles to improve the interface between organic and inorganic molecules. In this study, a β-1,3-glucan schizophyllan organopolymer (SPG-OH) was grafted onto hydroxyl activated silica NPs (SiO_2–OH NPs) via a sol-gel thermo-condensation polymerization process to further improve the surface properties of the SiO_2 NPs. The structure of the nanocomposite material (SPG-g-SiO_2 NPs) was characterized using different spectroscopic experiments such as FTIR, FESEM, XRD and TGA. From the spectroscopic experiments, the grafting bonding between the SPG-OH and the SiO_2–OH NPs was established in the FTIR spectra of the SPG-g-SiO_2 NPs due to the disappearance of the characteristic absorption wavelength corresponding to −OH at 3303–3500 cm^{-1} leading to strong chemical interactions between the SPG and SiO_2 NPs. Although, there was noticeable slight modification of the size and environment of the SPG-g-SiO_2 NPs composite due to surface activity in the FESEM imaging, however, a large proportion of these grafted particles were within 20–30 nm indicating minimum particle aggregations. The XRD experiments further confirmed that no surface morphology change from amorphous to crystalline was observed due to the appearance of diffraction shoulder at 20 2θ in the diffractogram. Similarly, the grafted material had a 53% residual mass after subjected to

H. Soleimani (✉) · Z. M. A. Merican
Fundamental and Applied Sciences Department, Universiti Teknologi PETRONAS, 32610 Seri Iskandar, Bandar Seri Iskandar, Perak Darul Ridzuan, Malaysia
e-mail: hassan.soleimani@utp.edu.my

Z. M. A. Merican
e-mail: zulkifli.aljunid@utp.edu.my

M. F. Hamza
Department of Pure & Industrial Chemistry, Bayero University Kano, Kano 3011, Nigeria

K. D. Stephen
Institute of Petroleum Engineering, Heriot-Watt University, Edinburgh E14 4AS, UK

© Institute of Technology PETRONAS Sdn Bhd 2022
S. A. Abdul Karim and A. Shafie (eds.), *Towards Intelligent Systems Modeling and Simulation*, Studies in Systems, Decision and Control 383,
https://doi.org/10.1007/978-3-030-79606-8_22

800 °C decomposition temperature, indicating thermal stability. Thus, this study suggests that a composite of the SPG-g-SiO_2 NPs could potentially be utilized in the EOR nanotechnology, such as in smart nanofluid injection or combined with surfactant in foam flooding for enhancing oil recovery.

Keywords EOR · Nanotechnology

1 Introduction

The application of nanotechnology has opened an important field of fundamental science and engineering studies offering a wide range of research interests in the areas of energy and fuel, biomedical, agriculture, and computing technology [1, 2]. To date, in the field of energy research, nano materials have been interestingly studied owing to their physical, chemical and mechanical properties. Several NPs have been used alone or mixed with surfactants to increase surface activity to mobilize the residual oil. In foam injection technology, the NPs are mainly combined with surfactant to generate stable foam to increase oil recovery [3]. However, due to the nature of the NPs surface properties, propagation in porous media, instability, surfactant/NPs compatibility are some of the major challenges faced during enhanced oil recovery (EOR) nanotechnology, leading to economic losses [4]. Consequently, surface modification offers a better alternative. Several literature studies reported various surface functionalized nanomaterials that involves the use of coating agent like inorganic polymers, silane coupling agents, etc. [5]. In the study of Xu et al. [6], the nano-SiO_2 was coated with two types coupling agents (KH550 and KH570), and the KH550 surface modified material was reported to be better than the KH570. To support this finding, Wang et al. [7] studied the conductivity effect of the composite solid polymer electrolytes containing KH550 functionalized nano-SiO_2, the authors discovered that due to the presence of functionalized SiO_2, the ionic conductivity increased by two order magnitude. Similarly, Park et al. [8] modified the surface of a bare SiO_2 NPs with hydrophilic polymers; poly(oxyethylene methacrylate-POEM) and poly(styrene sulfonic acid-PSSA) to improve the dispersion properties of the NPs. The results showed that the dispersion properties were improved significantly by the grafted materials. However, due to high sensitivity to salinity and temperature demonstrated by common inorganic polymers [9], there has been a growing research interest geared towards water soluble organic polymers such as Xhanthan gum and Schizophyllan (SPG). These organic polymers have been experimentally studied to show tolerance to many hindering factors in the field of energy and other technological studies [10].

In this study, the SPG polymer was used to modify the surface of activated nano-SiO_2 for application in EOR nanotechnology. The choice of this specific polymer in this study was based on the fact that the SPG being produced by fungus Schizophyllum commune displays characteristic helical water soluble complexes which could be an interesting functional material [11]. A comprehensive review on its structure, properties, bioactivities and recent developments reported by Zhang et al. [12] has given the authors of this paper a better understanding of its surface functionality that could offer a chemical benefit to participate in bonding with nanomaterials. A support of this claim was the study carried out by Abdel-Mohsen et al. [13] which developed a composite of SPG/Silver NPs for cytotoxicity activities, affirming the possibility to generate nanohybrid containing the SPG materials. To date, and as per as the objective of this research is concerned, limited work has been done on surface functionalization of nano-SiO_2 in the field of EOR nanotechnology, particularly using organic polymer. Although, for decades the SPG polymer field flooding EOR trial was conducted in Northern Germany oilfield [10] and the recent laboratory sensitivity to brine and EOR studies was investigated by Gao et al. [9]. However, in other studies such as those reported by Haraguchi et al. [11] it focused on the immobilization of polythiophene chirality induced by SPG using tetraethoxysilane as catalyst which did not gear towards the EOR. Consequently, in EOR foam technology [14], this research serves as a preliminary investigation to develop a grafted nanomaterial prior to its application to stabilize EOR foam. Nevertheless, the findings may open doors to other applications in the EOR technology including other material chemistry and industrial applications.

2 Materials and Methods

The amorphous mesoporous silica nanoparticles was purchased from US Research Nanomaterials, Inc. USA. The SiO_2 NPs have a diameter between 20 and 30 nm, surface area of 180–600 m^2/g, bulk and true density of < 0.10 and 2.4 g/cm^3, respectively. The Schizophyllan biopolymer was purchased from InvivoGen Limited, USA.

2.1 One Step Activation of Silonol Groups (−OH)

Activation of silanol group (−OH) of SiO_2 NPs was achieved via a one step treatment with 10 % HCl (Fig. 22.1). This treatment was done in order to activate the -OH on the surface of the SiO_2 NPs [8, 15]. Exactly 5 g of the SiO_2 NPs was transferred slowly into a beaker containing 50 mL of HCl with gentle stirring at 30 °C for 1 h. After the activation, the mixture was then treated with water and air dried under vacuum and analysed with Fourier transform infrared spectroscopy (FTIR).

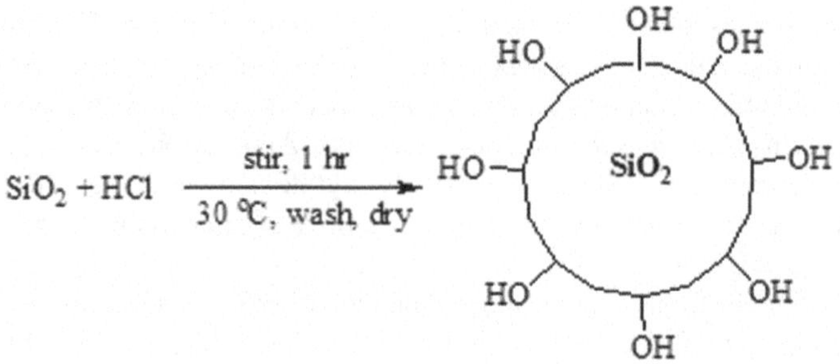

Fig. 22.1 Activation of silonol group (−OH)

Fig. 22.2 Chemical structure of SPG polymer

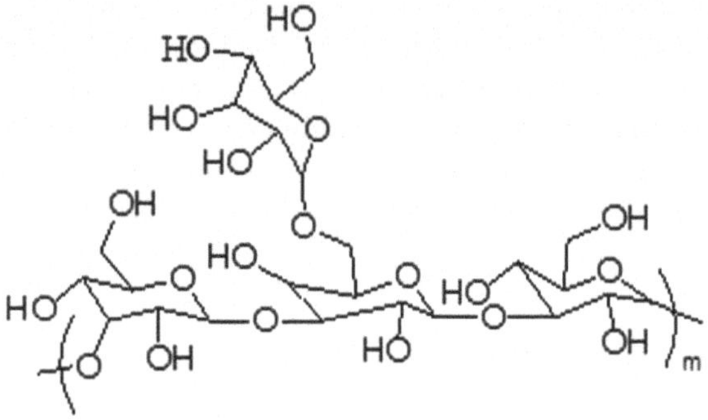

2.2 Sol-gel Synthesis of SPG-g-NPs Nanocomposite

The SPG-OH (0.1%), Fig. 22.2, was first treated with 10 mL of dimehylsulphoxide (DMSO) followed by 25 mL of distilled water and heated at 40 °C for an hour to ensure of a complete dissolution of the SPG polymer until the solution became homogenized. Thereafter, a predetermined mass of the activated SiO_2–OH NPs, was transferred gradually into the reaction beaker with continuous stirring and little amount of ammonium per-sulfate was added as a radical initiator. Thereafter, the reaction temperature was increased to 120 °C and maintained under stirring for 48 h (Fig. 22.3). After the reaction was completed, the mixture was centrifuged at 10,000 rpm for 15 min to obtain a sol-gel of SPG-g-NPs nanocomposite. The sol-gel SPG-g-NPs nanocomposite was dried in an oven at 80 °C for 24 h, ultrasonicated and characterized using spectroscopic instruments.

2.3 Spectroscopic Characterization

Fourier transform infrared spectroscopy FTIR (PerkinElmer Spectrum ISO AKX1300573 Nexus FTIR) was used to study the activation as well as surface

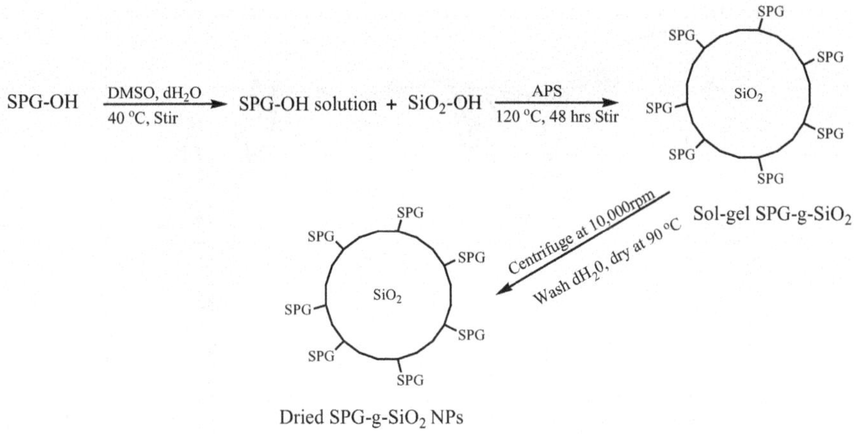

Fig. 22.3 Sol-gel synthesis reaction routes

chemical grafting between the polymer and nanoparticles by comparing their spectral absorptions before and after grafting. The transmittance was measured as a function of wave number between 600 and 4000 cm^{-1}.

The X-Ray diffraction (XRD) analysis was performed using an XPert Pro diffractometer used Cu Kα radiation at 40 kV and 40 mA. All scans were classically accomplished over a 2θ range from 10° to 99° at a 0.02/s speed, with an opening slit, an anti-scatter slit, and a receiving slit of 2 mm, 6 mm, and 0.2 mm, correspondingly.

Field Emission Scanning Electron Micrograph (FESEM) instrument (Carl Zeiss AG SUPRA 55VP) was used to study the nanocomposite particle morphology through which particles size and distribution was examined. The sample was coated with gold to minimize charges due to the SiO_2 to give proper viewing at different magnifications from 100–100,000X.

Thermogravimetric analysis (TGA) instrument (PerkinElmer Pyris 1) equipped with aluminum pans was used and measurements were carried out under air atmosphere and a controlled temperature program that automatically maintains constant temperature during thermal decompositions. About 8–10 mg sample mass was heated and the weight loss measurement was carried out between 25 and 800 °C at a heating rate of 10 °C min^{-1}.

3 Result and Discussion

3.1 *Characterization*

Figure 22.4 presents the FTIR spectra of the pure SiO_2 NPs, SPG, activated SiO_2–OH NPs and grafted nanocomposite SPG-g-SiO_2 NPs materials phased in one spectrum

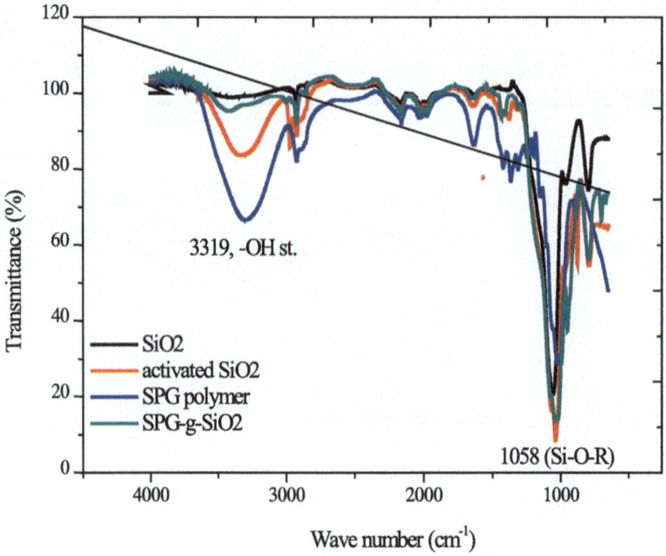

Fig. 22.4 The FTIR spectra of the activated SiO_2–OH NPs and SPG-g-SiO_2 NPs

for easy comparison. It can be clearly observed that the main absorption peaks at 3319 cm^{-1} that differentiates the spectra between the pure SiO_2 NPs and activated SiO_2–OH NPs corresponds to the absorption characteristic band of –OH groups, which typically indicated that the sylonation of the pure SiO_2 NPs with the silonol groups (−OH) was successful. This activation was carried out mainly to change the hydrophilic nature of the pure SiO_2 into hydrophobic for easy chemical interaction. Similarly, the high intensity and broad peaks around 3315 cm^{-1} in the SPG spectra matched with the numerous −OH groups embedded in the SPG structure. However, after the sol-gel process, all the broad peaks resonating at 3303–3500 cm^{-1} in the pure SPG and activated SiO_2–OH NPs spectra was noticeably reduced suggesting the formation of the SPG-g-SiO_2 NPs nanocomposite. This reduction of peak was attributed to the covalent bond between the −OH groups of the reactant species via a condensation reaction [16] leading to a strong chemical interaction between the SPG and SiO_2 NPs. Other characteristic peaks as evidence of Si-O-R in the composite material was the strong intensity at 1058 cm^{-1}. The assignments of these characteristic peaks conformed with the studies reported in [13].

The XRD analysis shown in Fig. 22.5 revealed the characteristic of amorphous peak diffraction shoulder appearing at 20 2θ in the diffractogram of both the SiO_2 NPs and SPG-g-SiO_2 NPs nanocomposite materials. This confirms the evidence that the morphological property of the SiO_2 NPs was not deformed to due the absence of crystalline peaks after surface grafting with the SPG biopolymer. Several studies have revealed that pure SiO_2 NPs and SPG exhibit the characteristic amorphous peaks at 20 2θ [13], thus, evidently supported this finding. Similarly, Park et al. [8] reported that the grafting of polymer did not alter the microstructure of nano-SiO_2.

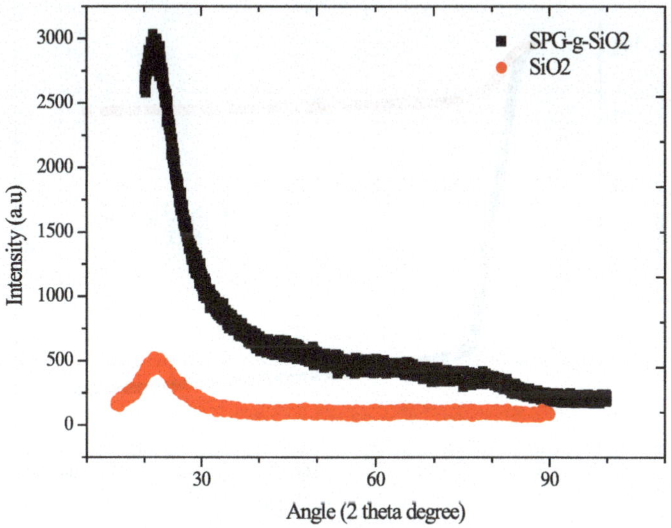

Fig. 22.5 The XRD spectra of the activated SiO_2–OH NPs and SPG-g-SiO_2 NPs

Figure 22.6 shows the image of the activated SiO_2 NPs and SPG-g-SiO_2 NPs. Typically, the SiO_2 NPs were within the average particle size of 20–30 nm, however, after the polymer grafting leading to the formation of the SPG-g-SiO_2 NPs composite, a slight modification of size, shape and environment of SPG-g-SiO_2 NPs composite had occurred indicating surface activity due to the presence of biopolymer. However, a large proportion of these grafted particles were within 20–30 nm. This size modification was also observed in the study of Numata et al. [17] and SPG was reported to suppresses particle aggregation [11] which may support the finding of this study.

From the thermogram of activated SiO_2 NPs and SPG-g-SiO_2 NPs in Fig. 22.7, it was observed that all the materials lost small masses during first decomposition occurring between 100 and 110 °C, this happened due to evaporation of the absorbed

Fig. 22.6 The FESEM image of the activated SiO_2–OH NPs and SPG-g-SiO_2 NPs

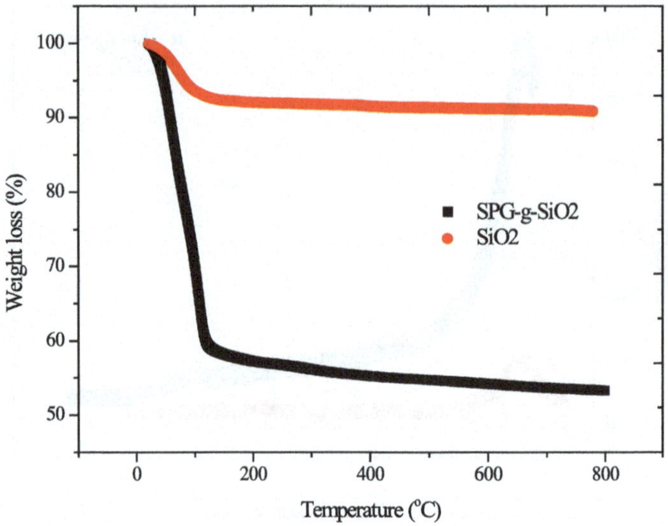

Fig. 22.7 The TGA thermogram of the activated SiO_2–OH NPs and SPG-g-SiO_2 NPs

water molecule in the form of moisture. However, the second decomposition after 110 °C suggested the cleavage of –OH bonds in the activated SiO_2 NPs and breakage of side chain monomer in the SPG-g-SiO_2 NPs, respectively. Subsequently, no further significant decomposition of materials was noticed after subjected to thermal decomposition temperature up to 800 °C. At the end of the decomposition process, more than 50 % of residual mass was obtained from the grafted material indicating thermal stability.

4 Summary

The synthesis of composite nanomaterial of the SPG-g-SiO_2 NPs was successful via a sol-gel synthesis process. Several spectroscopic experiments such as FTIR, FESEM, XRD, TGA have established the chemical bonding, particle size, morphology and thermal stability of the hybrid material, respectively, which could be of potential applications in the field of the EOR nanotechnology and other industrial applications.

Acknowledgements The authors of this paper have appreciated the support of the project (Grant: PRSB 513CB010) from the PETRONAS Research Sdn, Bhd, and Universiti Teknologi PETRONAS, Malaysia. Also, the Institute of Hydrocarbon Recovery, UTP, technical staff, especially, Mr. Iswadi Radzali (Technician, Coreflood system), and Mr Muhammad Annas Yahya (Technician, Spectroscopic instruments) were highly appreciated for their enormous supports in this project.

References

1. Soleimani, H., Latiff, A., Rasyada, N., Yahya, N., Sabet, M., Khodapanah, L., Guan, B.H.: Synthesis and characterization of Yttrium Iron Garnet (YIG) nanoparticles activated by electromagnetic wave in enhanced oil recovery. J. Nano Res. **38** (2016)
2. Hamza, M.F., Sinnathambi, C.M., Merican, Z.M.A., Soleimani, H., Karl, S.D.: An overview of the present stability and performance of EOR-Foam. Sains Malays. **46**(9), 1641–1650 (2017)
3. Hamza, M.F., Sinnathambi, C.M., Merican, Z.M.A.: Recent advancement of hybrid materials used in chemical enhanced oil recovery (CEOR): a review. IOP Conf. Ser. Mat. Sci. Eng. **206**(1), 012007 (2017)
4. ShamsiJazeyi, H., Miller, C.A., Wong, M.S., Tour, J.M., Verduzco, R.: Polymer-coated nanoparticles for enhanced oil recovery. J. Appl. Polym. Sci. **131**(15) (2014)
5. Protsak, I., Pakhlov, E., Tertykh, V., Le, Z.C., Dong, W.: A new route for preparation of hydrophobic silica nanoparticles using a mixture of poly (dimethylsiloxane) and diethyl carbonate. Polymers **10**(2), 116 (2018)
6. Xu, M.H., Cao, Y.Y., Gao, S.G.: Surface modification of nano-silica with silane coupling agent. Key Eng. Mater. Trans. Tech. Publ. (2015)
7. Wang, X.L., Mei, A., Li, M., Lin, Y., Nan, C.W.: Effect of silane-functionalized mesoporous silica SBA-15 on performance of PEO-based composite polymer electrolytes. Solid State Ionics **177**(15–16), 1287–1291 (2006)
8. Park, J.T., Seo, J.A., Ahn, S.H., Kim, J.H., Kang, S.W.: Surface modification of silica nanoparticles with hydrophilic polymers. J. Ind. Eng. Chem. **16**(4), 517–522 (2010)
9. Gao, C.: Application of a novel biopolymer to enhance oil recovery. J. Petrol. Explor. Prod. Technol. **6**(4), 749–753 (2016)
10. Leonhardt, B., Ernst, B., Reimann, S., Steigerwald, A., Lehr, F.: Field testing the polysaccharide schizophyllan: results of the first year. In: SPE Improved Oil Recovery Symposium. Society of Petroleum Engineers (2014)
11. Haraguchi, S., Numata, M., Kaneko, K., Shinkai, S.: Immobilization of polythiophene chirality induced by a helix-forming β-1, 3-glucan polysaccharide (schizophyllan) through sol-gel reaction. Bull. Chem. Soc. Jpn. **81**(8), 1002–1006 (2008)
12. Zhang, Y., Kong, H., Fang, Y., Nishinari, K., Phillips, G.O.: Schizophyllan: a review on its structure, properties, bioactivities and recent developments. Bioact. Carbohydr. Dietary Fibre **1**(1), 53–71 (2013)
13. Abdel-Mohsen, A.M., Abdel-Rahman, R.M., Fouda, M.M., Vojtova, L., Uhrova, L., Hassan, A.F., Al-Deyab, S.S., El-Shamy, I.E., Jancar, J.: Preparation, characterization and cytotoxicity of schizophyllan/silver nanoparticle composite. Carbohyd. Polym. **102**, 238–245 (2014)
14. Hamza, M.F., Sinnathambi, C.M., Merican, Z.M.A., Soleimani, H., Stephen, K.D.: Effect of SiO_2 on the foamability, thermal stability and interfacial tension of a novel nano-fluid hybrid surfactant. Int. J. Adv. Appl. Sci. **5**(1), 113–122 (2018)
15. Budiman, H.H., Sri, H., Setiawan, A.: Preparation of silica modified with 2-mercaptoimidazole and its Sorption. Properties of Chromium (III). J. Chem. **6**(1), 141–150 (2009)
16. Isoda, T., Maeda, R.: Development of an interaction assay between single-stranded nucleic acids trapped with silica particles and fluorescent compounds. J. Funct. Biomater. **3**(3), 601–614 (2012)
17. Numata, M., Li, C., Bae, A.H., Kaneko, K., Sakurai, K., Shinkai, S.: β-1, 3-Glucan polysaccharide can act as a one-dimensional host to create novel silica nanofiber structures. Chem. Commun. (37), 4655–4657 (2005)

Empirical and Neural Network Modelling of Oil Recovery in Nano Assisted Enhanced Oil Recovery

Afza Shafie, Sayed Ameenuddin Irfan, and Norsyahirah Aripin

Abstract In this chapter, an empirical model was developed to predict the oil recovery in enhanced oil recovery (EOR) application, based on rock permeability, rock wettability, particle size, and injection rate of nanofluids. The developed model efficiency is compared with several neural network models that are developed for the same purpose. A multi-layer feed forward artificial neural network (ANN) model trained with an error back-propagation algorithm was employed for developing a predictive model. The model has considered input variables of particle size, rock permeability, rock wettability, nanofluid injection rate, and temperature, while percentage of oil recovery is the output. The scaled conjugate gradient (SCG) optimization algorithm was used to train the ANN model. An ANN with six hidden neurons was highly accurate in predicting the oil recovery. The developed ANN model accuracy was determined using statistical measures such as R- square and Mean-square error (MSE). The values obtained for the developed model are 1 and 0.0009 for training, 1 and 0.0009 for validation and, 1 and 0.0005 for testing data sets, respectively. The comparison between the developed neural network model and the polynomial fitting method concluded that the ANN is better in terms of accuracy for predicting oil recovery in EOR applications.

Keywords Neural network · Oil recovery model · Scaled conjugate gradient and enhanced oil recovery

---

A. Shafie (✉)
Fundamental and Applied Sciences Department, Universiti Teknologi Petronas, 32610 Seri Iskandar, Perak, Malaysia
e-mail: afza@utp.edu.my

S. A. Irfan
Shale Gas Research Group, Universiti Teknologi Petronas, 32610 Seri Iskandar, Perak, Malaysia
e-mail: sayed.irfan@utp.edu.my

N. Aripin
Faculty of Science, Universiti Teknologi Malaysia, Johor Bahru 81310 Johor, Malaysia

© Institute of Technology PETRONAS Sdn Bhd 2022

S. A. Abdul Karim and A. Shafie (eds.), *Towards Intelligent Systems Modeling and Simulation*, Studies in Systems, Decision and Control 383,
https://doi.org/10.1007/978-3-030-79606-8_23

1 Introduction

Oil and gas remain as one of the critical aspects of energy in the world. Hence, it is very crucial to improve oil recovery by adopting the novel enhanced oil recovery methods. With the development in energy interest, and spending issue tracking down new oil wells, consideration has gone to Enhanced Oil Recovery (EOR) procedures because of their lower venture cost and higher benefit openings, Esfandari Bayat et al. [1]. Since EOR techniques have been seen to have a good advantage, improvement in the application of nanotechnology has provided new directions in the EOR process. This is due to their capability to help the fluid mobility by changing the wettability and interfacial tension of the reservoir along with cost-effective and environmentally friendly manner as well. Nanoparticles are significant materials since their properties allows for specialized applications because of their enormous surface area as highlighted in Irfan et al. [2, 3] and Yeken et al. [4].

According to Hendraningrat et al. [6], artificial neural Network (ANN) is a nonlinear measurable investigation method that is particularly reasonable for recreation of frameworks which are difficult to be depicted by a physical and substance models. The essential handling component of ANN is a neuron. It gets contributions from different sources, consolidates them, performs commonly a nonlinear procedure on the outcome, and afterward yields as end-product. An ANN gains from models and perceives designs in a progression of information and yield information with no earlier suppositions about their tendency and interrelations [4]. ANN takes out the restrictions of the traditional methodologies by extracting the ideal data based on the input information. Applying ANN to a framework needs adequate information and yield information rather than a numerical condition as mentioned in Khajeh et al. [5] and Yeken et al. [6]. ANN is a good alternative to conventional empirical models that are based on polynomial and linear regression. ANN consists of input, output and hidden layer and learns general rules based on calculation upon numerical data.

In this work, a multi-layer perceptron (MLP) based feed-forward ANN, is proposed for oil recovery. The aim is to investigate the effects of various boundaries, such as nanoparticle size, rock permeability, rock wettability, nanofluid infusion rate, and temperature on percentage of oil recovery. Section 2 looks at the methodology for data collection and extraction followed by the section on results and discussion. The conclusion will be provided at the end of the chapter.

2 Methodology

The research methodology used in this study is summarized in Fig. 1 and the architecture of the ANN model was presented in Fig. 2.

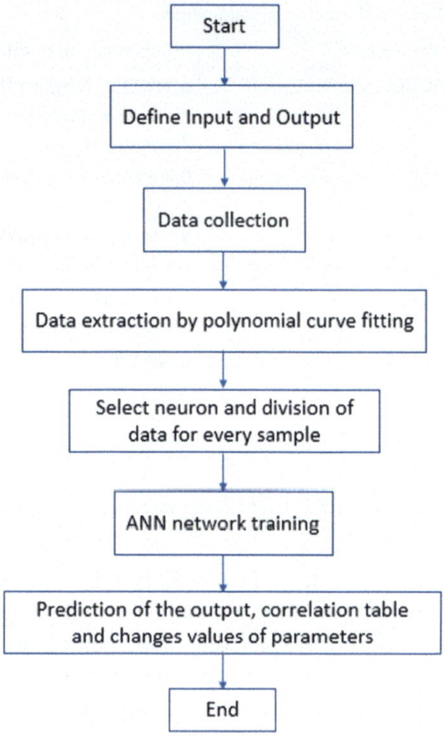

Fig. 1 Methodology of research

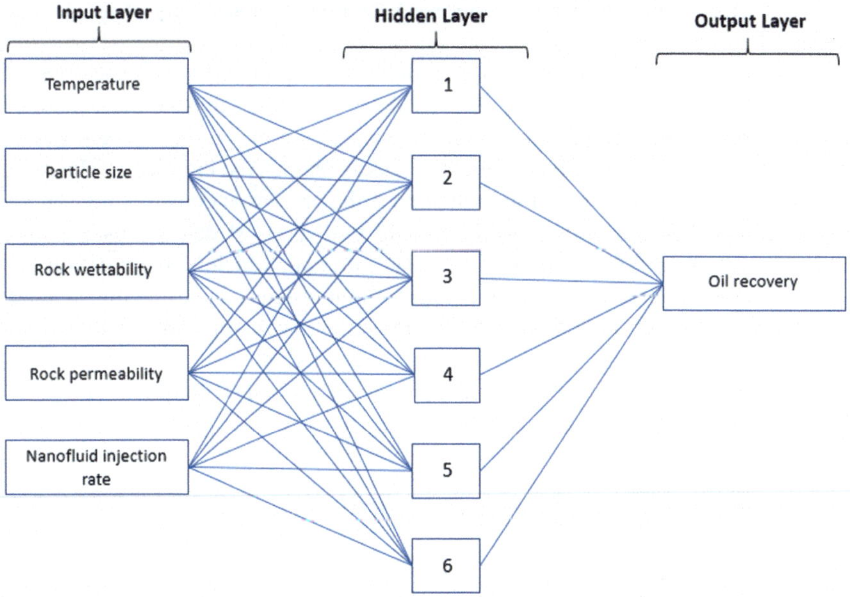

Fig. 2 Proposed ANN architecture

Table 1 Equation of parameters

Parameters	Polynomial curve fitting equation
Wettability	$\mathbf{0.37652}X_1 - \mathbf{0.0012763}X_2 + \mathbf{0.0061806}X_3 + \mathbf{0.48706}$ X_1 = *Nanofluid injection rate*, X_2 = *Permeability*, X_3= *Temperature*
Nanofluid injection rate	$\mathbf{0.000065769}X_1 + \mathbf{0.061856}X_2 - \mathbf{0.0016071}X_3 + \mathbf{0.25377}$ X_1 = *Permeability*, X_2 = *Wettability*, X_3= *Temperature*
Permeability	$\mathbf{8.27478}X_1 - \mathbf{26.3807}X_2 + \mathbf{1.70027}X_3 + \mathbf{55.1804}$ X_1 = *Nanofluid injection rate*, X_2 = *Wettability*, X_3 = *Temperature*
Temperature	$-\mathbf{11.9989X_1} + \mathbf{0.100896X_2} + \mathbf{7.58069X_3} + \mathbf{19.0234}$ X_1 = *Nanofluid injection rate*, X_2 = *Permeability*, X_3 = *Wettability*
Oil recovery	$\mathbf{0.0015781X_1} - \mathbf{4.4387X_2} + \mathbf{0.062598X_3} - \mathbf{6.1663X_4} + \mathbf{6.6265}$ X_1 = *Permeability*, X_2 =*Wettability*, X_3 = *Temperature*, X_4 = *Nanofluid injection rate*

2.1 Data Collection and Extraction

The exploratory data used in this work were arbitrarily divided into three sets consisting of 143, 31, and 31 points, which were utilized as training, validation, and testing respectively. The training data was utilized to calculate the network parameters while the validation data was used to determine robustness of the network parameters. To avoid over fitting of the model, the testing stage was utilized to control the error, which is the point at which it expanded, the training stops. Testing information collected were utilized to determine the predictive capacity of the created model [1].

The dataset for the study is extracted from literature using a digitizer software. 31 datasets were extracted from this method. With the 31 datasets that has been obtained from the extraction software, a polynomial curve fitting method have been used to increase the number of datasets, so that the results on the ANN model will be more accurate [6]. Polynomial curve fitting method is the process of constructing a curve, or mathematical function, that has the best fit to a series of data points, possibly subject to constraints. Curve fitting is divided into two, which is interpolation and extrapolation. The equation for the input and output is determine using this method. The equations are for wettability, nanofluid injection rate, permeability, temperature and oil recovery. The obtained equations are given in Table 1.

A total of 205 data sets are obtained from this extraction technique.

2.2 Scaled Conjugate Gradient Algorithm

In this work, a multi-layer perceptron (MLP) based feed-forward ANN, which uses back-propagation learning algorithm, was carried out to investigate the effects of particle size, rock permeability, rock wettability, nanofluid injection rate and temperature on the enhanced oil recovery. The inputs for the network are particle size, rock permeability, rock wettability, nanofluid injection rate and temperature and the output is the oil recovery. The ANN model is represented in Fig. 2. All the simulations are done using MATLAB.

Scaled conjugate gradient (SCG) is one the types of learning algorithm which can be used for the training of a network. The scaled conjugate gradient methods represent an important innovation for solving large-scale unconstrained optimization problems, min $f(x)$ where f: $R^n \rightarrow R$ is continuously differentiable and its gradient is available. These methods generate a sequence x_k of approximations to the minimum x^* of f, in which

$$x_{k+1} = x_k + \alpha_k d_k$$

$$d_{k+1} = -g_{k+1} + \beta_k d_k$$

where $g_k = \nabla f(x_k)$, α_k is selected to minimize $f(x)$ along $d_k \cdot \beta_k$ is a scalar parameter. An initial point x_0 and $d_0 = -g_0$ is selected to start the iterative process. When the function f is quadratic and α_k is selected to minimize $f(x)$ along the direction d_k, then all choices of β_k are equivalent, but for general nonlinear functions different choices of β_k give algorithms with very different convergence performances [7, 8] provides the details of the SCG algorithm.

3 Results and Discussion

A correlation investigation shown in Table 2 was done to decide the viability of a variable utilizing the recommended ANN model in this work.

The results in Table 2 showed that the correlation between oil recovery and temperature is 0.4262 and correlation between oil recovery and permeability is 0.479. The

Table 2 Correlation of parameters and output

	Temperature	Permeability	Nanofluid injection rate	Size	Wettability	Oil recovery
Temperature	1	0.9753	0.7391	0.9849	0.7397	0.4262
Permeability	0.9753	1	0.7537	0.9864	0.6713	0.479
Nanofluid injection rate	0.7391	0.7537	1	0.7664	0.3688	0.3995
Size	0.9849	0.9864	0.7664	1	0.7425	0.387
Wettability	0.7397	0.6713	0.3688	0.7425	1	−0.2723
Oil recovery	0.4262	0.479	0.3995	0.387	−0.2723	1

Table 3 Comparison of R-squared and mean square error (MSE) value for neuron 6

	Training		Validation		Testing	
Algorithm	R-square	MSE	R-square	MSE	R-square	MSE
Levenberg-Marquardt	1	9.16E−10	1	3.09E−09	1	1.06E−08
Bayesian Regulization	0.9883	0.1973	/	/	0.9602	0.5178
Scaled Conjugate Gradient	1	0.0009	1	0.0009	1	0.0005

correlation for nanofluid injection rate is 0.3995, for size is 0.387 and for wettability is −0.2723 between the oil recovery, respectively. The low correlation values infer that the data indicate the best performance for the network. The comparison of the R-squared and MSE between the actual and predicted values for training, validation and testing set for several algorithms was carried out as shown in Table 3.

The results from Table 3 shows that Levenberg-Marquardt and Scaled Conjugate Gradient (SCG) training algorithms have good values on Mean square Error (MSE) and coefficient of determination (R-square). The Bayesian Regulization algorithm has the lowest values for MSE and R-square. SCG algorithm is chosen for this study since it has a lower computational time which is (2 s) compared to LM algorithm that takes almost 39 s to complete the full calculation. Based on the results, an ANN model consisting of input, 6 neurons in the hidden layer and prediction layer gives the best performance. The mean square error (MSE) and coefficient of determination (R^2) between the actual and predicted values were determined as 0.0009 and 1.0000 during training, 0.0009 and 1.0000 for validation and 0.0005 and 1.0000 for testing.

Figure 3 shows the plots of the predicted against actual values based on the SCG algorithm for the training, validation, testing for all data sets. As shown in the figures, the predicted model is closed to the actual values.

3.1 Comparison Between Experimental Data and Neural Network Model

An ANN with the setup shown in Table 4 was identified for the study.

The ANN model is firstly used to predict the percentage of oil recovery. Figures 4 and 5 present the graph for actual and predicted results for 205 data sets and their relative errors respectively. It showed that the ANN model is able to predict well since most of the relative errors are less than 2%. The results indicate the potential of ANN for prediction of oil recovery when each of the parameters are varied.

The performance of the ANN with the empirical model for each of the parameters is shown in Fig. 6. From the graph, it can be seen that the ANN gives a good accuracy in predicting on oil recovery for the parameters.

The trend of how the oil recovery changes with each parameter is also studied. The parameter of interest is varied while keeping the other parameters constant. Figure 7

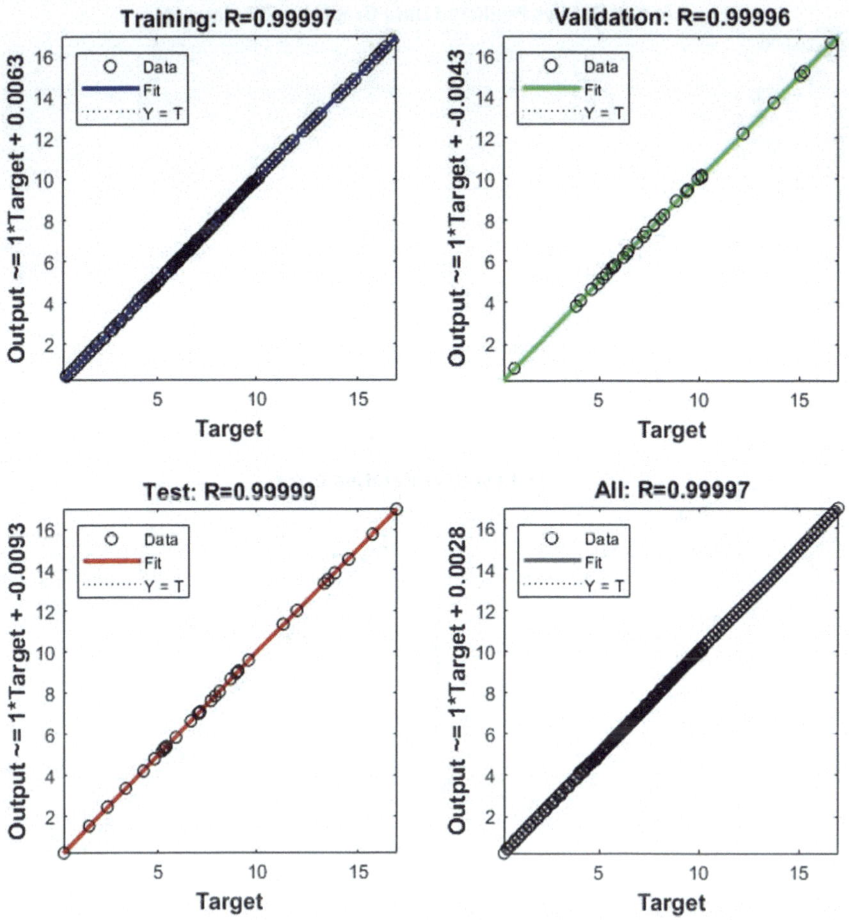

Fig. 3 The plots of actual against predicted values for training, validation and testing of data sets

Table 4 Parameter set up of ANN

Input	Size, permeability, wettability, nanofluid injection rate, temperature
Output	Oil recovery
Data distribution	205 datasets 143—training data (70%) 31—validation data (15%) 31—testing data (15%)
Neuron	6 hidden neurons
Algorithm	Scaled conjugate gradient

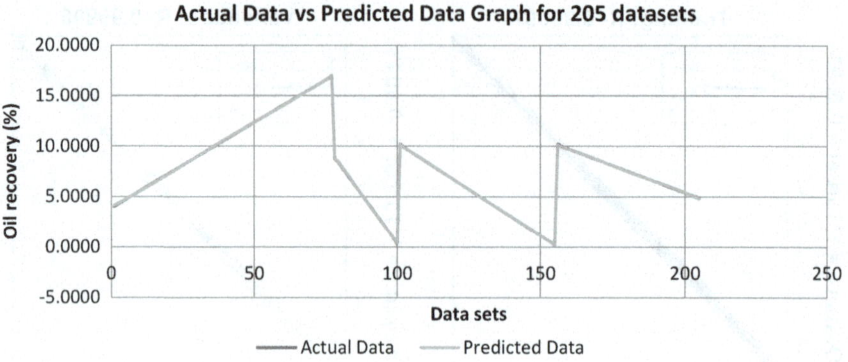

Fig. 4 Performance of ANN and actual

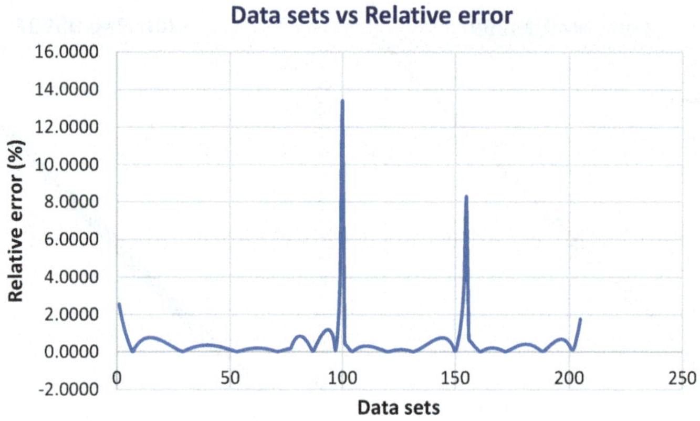

Fig. 5 Relative error between predicted and actual

shows the results when there is a change in size of nanoparticle. In all the simulations, the nanofluid injection is fixed at 0.2 cm^3 per minute and wettability is 0.8.

In this simulation, the range of the nanoparticle size considered is between 41 and 98 nm. The percentage of oil recovery does not show much increase when the nanoparticle size increases from 41 to 65 nm. However, there is a sharp increase in the percentage of oil recovery when the particle size increases from size 68 to 98 nm. The results seem to imply that as the size increases, the percentage of oil recovered also increases.

The change in temperature against percentage of oil recovery is shown in Fig. 8. the graph shows that the percentage of oil recovery increases as temperature increases which is from 81 to 100 °C. It can be inferred that higher temperature is able to reduce the viscosity of the oil, hence resulting in an increase in oil recovery.

The change in permeability against percentage oil recovery is shown in Fig. 9. Permeability is a measure of the ease of flow of a fluid through a porous solid.

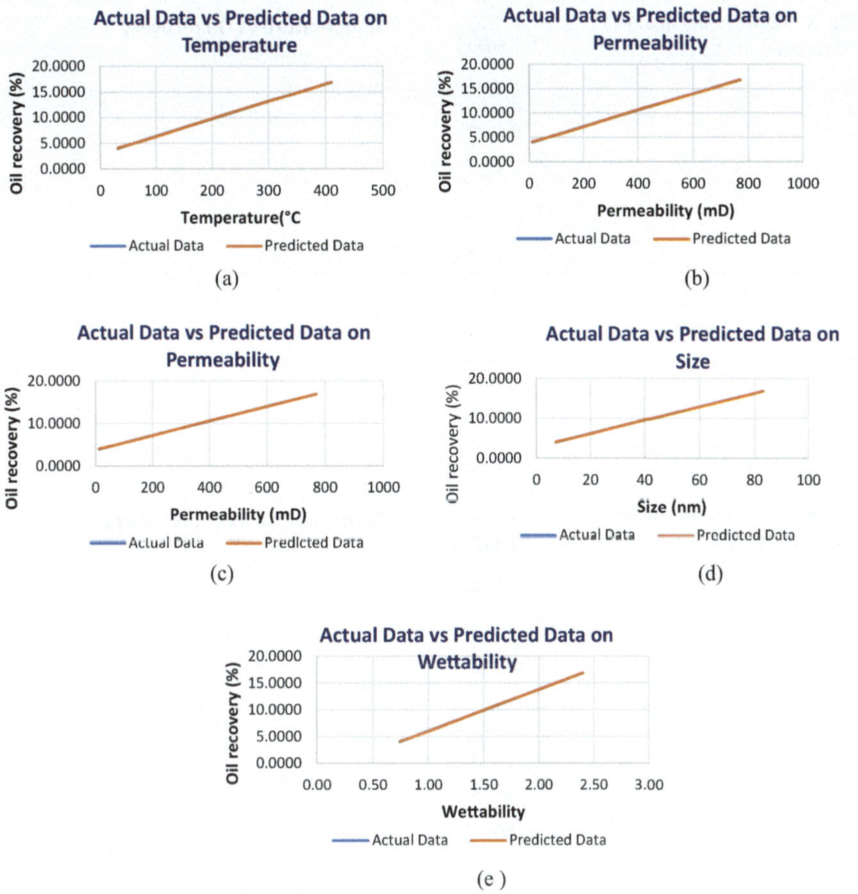

Fig. 6 Graph for actual versus predicted data on (**a**) temperature, (**b**) permeability, (**c**) size, (**d**) nanofluid injection rate, (**e**) wettability

Fig. 7 Size of nanoparticle against oil recovery (%)

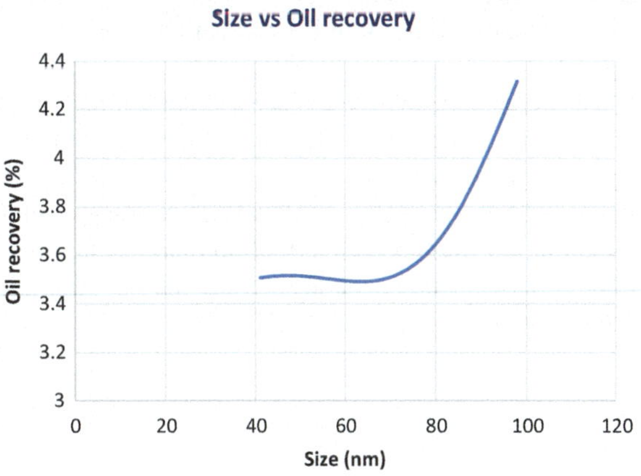

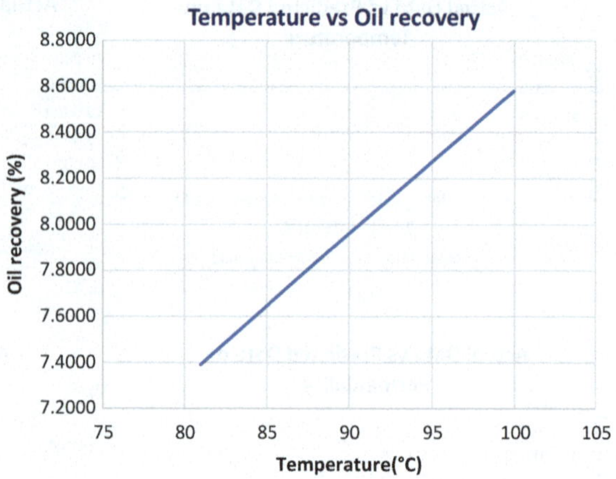

Fig. 8 Temperature against oil recovery (%)

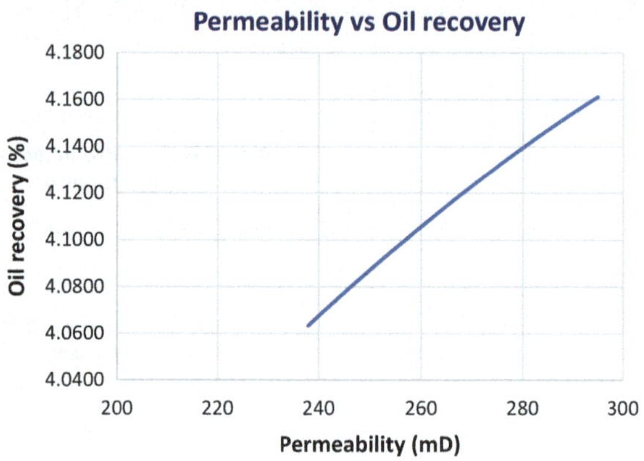

Fig. 9 Permeability against oil recovery (%)

From the graph, it shows that percentage of oil recovery increases, as permeability increases from 238 to 295 mD.

The results on the trend for injection rate against oil recovery (%) is given in Fig. 10.

The percentage of oil recovery was determined for injection rate between 0.85 and 1.04 cm^3/min. It can be observed that the percentage of oil recovery increases as the injection rate is increased from 0.85 to 1.00 cm^3/min. However, the oil percentage of recovery decreases as the injection rate increases from 1.03 cm^3/min onwards. From the graph, the highest recovery of oil obtained is 5% when the injection rate is about 1.00 cm^3/min. This result shows that, the injection rate of 1.00 cm^3/min will result in the maximum recovery of oil.

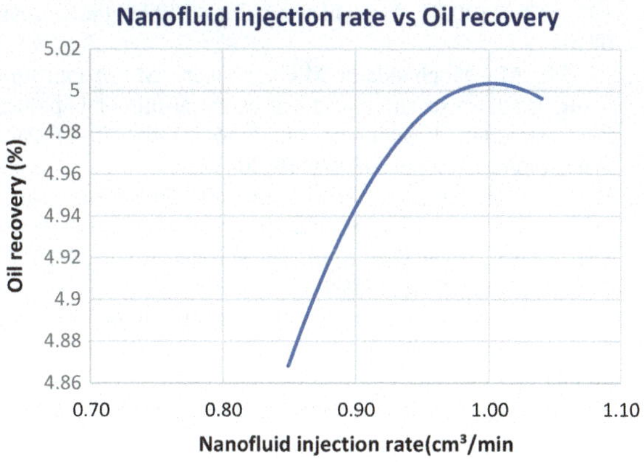

Fig. 10 Injection rate against oil recovery (%)

4 Summary

In this chapter, a neural network architecture-based model for oil recovery was presented. A neural network model with six neurons and one hidden layer, four input neurons, and one output neuron, and a scaled conjugate training algorithm were developed for oil recovery. Four parameters; wettability, permeability, temperature and injection rate were considered as the input. The results shows that the percentage of oil recovery will increase when permeability, and temperature is increased. The maximum percentage oil recovery is obtained at 1.00 cm^3/m. There is a significant increase in oil recovery percentage when the particle size is greater than 65 nm. Comparison between actual and predicted data shows no huge difference; hence the accuracy ANN training algorithm is promising. The lowest correlation between oil recovery is size with 0.387, and the highest correlation is permeability with 0.479. All the correlation between the oil recovery is less than 0.5. Low correlation between data gives the best performance for the network.

References

1. Esfandyari Bayat, A., Junin, R., Samsuri, A., Piroozian, A., Hokmabadi, M.: Impact of metal oxide nanoparticles on enhanced oil recovery from limestone media at several temperatures. Energy Fuels **28**(10), 6255–6266 (2014). Available from: https://doi.org/10.1021/ef5013616
2. Irfan, S.A., Shafie, A.: Artificial neural network modeling of nanoparticles assisted enhanced oil recovery. Stud. Syst. Decis. Control **320** (2021)
3. Irfan, S.A., Shafie, A., Yahya, N., Zainuddin, N.: Mathematical modeling and simulation of nanoparticle-assisted enhanced oil recovery—a review. Energies **12**(8), 1575 (2019). Available from: https://www.mdpi.com/1996-1073/12/8/1575
4. Yekeen, N., Padmanabhan, E., Abdulelah, H., Irfan, S.A., Okunade, O.A., Khan, J.A., et al.: CO_2/brine interfacial tension and rock wettability at reservoir conditions: a critical review of previous studies and case study of black shale from Malaysian formation. J. Pet. Sci. Eng.

196, 107673 (2021). Available from: https://linkinghub.elsevier.com/retrieve/pii/S0920410520307403
5. Khajeh, M., Moghaddam, M.G., Shakeri, M.: Application of artificial neural network in predicting the extraction yield of essential oils of Diplotaenia cachrydifolia by supercritical fluid extraction. J. Supercrit. Fluids **69**, 91–96 (2012). Available from: https://linkinghub.elsevier.com/retrieve/pii/S089684461200188X
6. Hendraningrat, L., Li, S., Torsæter, O.: Effect of some parameters influencing enhanced oil recovery process using silica nanoparticles: an experimental investigation. In: All Days. SPE (2013). Available from: https://onepetro.org/SPERCSC/proceedings/13RCSC/All-13RCSC/AbuDhabi,UAE/178041
7. Andrei, N.: Scaled conjugate gradient algorithms for unconstrained optimization. Comput. Optim. Appl. **38**(3):401–416 (2007). Available from: https://doi.org/10.1007/s10589-007-9055-7
8. Moller, M.F.: A scaled conjugate gradient algorithm for fast supervised learning. Neural Netw. **6**, 525–533 (1993)

Mathematical Study of a Fractional-Order General Pathogen Dynamic Model with Immune Impairment

Shaimaa Abdelhamed Azoz and Fatima Hussien

Abstract The global features of a fractional-order general pathogen infection model with immune deficiency and two types of target cells are examined in this chapter. Both cell-to-cell and pathogen-to-cell transmissions are presented. We demonstrate that the system solutions are positivity and confined, Lyapunov function is considered to show global asymptotic stability for disease-free equilibrium point of the model. Analytical expression for basic reproduction number $\mathcal{R}_0$ and the required condition under which disease-free and chronic equilibrium points are asymptotically stable are determined. We carried out numerical simulations to utilize it to document the analytical results.

Keywords Pathogen dynamic · Pathogen-to-cell transmission · Cell-to-cell transmission · Fractional-order differential equations · Immune impairment · Global asymptomatic stability

1 Introduction

The study of disease dynamics caused via parasites, bacteria, mosquitoes, and other pathogens has been the focus of attention in many types of research for many years due to the great role that these studies provide and this mathematical analysis in testing and presenting new insights and evaluating the parameters that control the infection process without the need to resort to clinical trials (see e.g. [1]–[11]). Some of these papers take into account the CTLs effect, CTLs are cytotoxic T lymphocytes cells that play a major role as part of the body's natural immunity to fight pathogens. So, CTLs are supposed to be the essential hostile immune agent deciding the pathogen load. Based on that, the study of pathogen infection dynamics with

S. A. Azoz (✉) · F. Hussien
Faculty of Science, Department of Mathematics, Assiut University, Assiut 71516, Egypt
e-mail: azoz@aun.edu.eg

F. Hussien
e-mail: fatma.hussin@sceince.aun.edu.eg

© Institute of Technology PETRONAS Sdn Bhd 2022
S. A. Abdul Karim and A. Shafie (eds.), *Towards Intelligent Systems Modeling and Simulation*, Studies in Systems, Decision and Control 383,
https://doi.org/10.1007/978-3-030-79606-8_24

CTL response has attracted the focus of attention for many studies in related fields [14]–[28]. Moreover, it was found that the pathogen occurs impairment in CTL function during the infection. Several papers have looked at pathogenic dynamics models with impaired CTL immunity (see e.g. [29]–[33]). These studies assumed that the uninfected cells become infected as a result of get in touch with pathogen but the pathogen can also be passed from an infected cell to an uninfected cell directly. Some papers discussed two types of pathogen transmissions, cell-to-cell and pathogen-to-cell ([34]–[41]) but these papers ignored the immune impairment effect. Two types of transmissions including pathogen-to-cell and cell-to-cell in addition to immune impairment presented in papers [42] and [43].

On the other hand, systems presented by integral operators and fractional-order derivative are more universal than the traditional deterministic systems. Fractional derivative is a appropriate implement to describe real phenomena which has a memory like most biological systems. This indicates that to calculate the fractional derivative at a specific point, we must consider the whole history from the beginning point to this point. For these reasons, using fractional derivative to model a real process has piqued the interest of a number of authors from different fields [45, 46] and [47]. The fractional derivative has been demonstrated to be helpful in biology for analyzing cell rheological properties. Moreover, the membranes of biological organism cells were reported to have an electrical conductance in fractional order. Fractional-order derivative or integral operators are commonly used to generate fractional-order differential equations. As its derivatives and integral operators have non-locally characteristic, fractional-order derivative and integral operators are dependent not only on their present condition but also on all their previous states, making them very useful and strong for evaluating the system's next state. As a result these operators are more effective than other traditional deterministic operators. The most well-known fractional-order operators and widely used in fractional-order differential equations applications are Riemann–Liouville, Grünwald–Letnikov, and Caputo operators.

All the pathogen infection models mentioned above examine the interaction of the pathogen with one target cell. Sometimes multiple cells are susceptible to a specific infection for example HCV, HIV and other viruses including rotaviruses, adenoviruses, and picornaviruses. Elaiw and his collaborators have examined global asymptotic stability of the stable HIV model states with two target cells classes as they considered the macrophages cell beside CD4+ T cells. In context of an entire live organism, the study of macrophagic biology offers unique opportunities to understand the dynamic cell subset contribution in the reaction to infections. These discoveries have sparked the development of novel therapeutic strategies for the treatment of infectious diseases. We need to develop two-compartment mathematical models to be able to study the interaction between pathogen and two target cells (see e.g. [12] and [13]).

During this study, we present a pathogen model with two target cells. In both, we considered the effect of two types of transmission cell-to-cell and pathogen-to-cell transmission represented by general incidence rates. We will show that the system solutions are positive and ultimately bounded which ascertains the well-posed of the model. Biological threshold parameter $\mathcal{R}_0$ has been deduced to prove the existence of equilibrium points and to study their stability. We use the Lyapunov method with LaSalle's invariance principle to investigate global stability of uninfected equilibrium point. At the end of the chapter, we will present numerical simulations through an example of the presented model for the investigation of the analytical results.

2 Two Target Cells and General Incidence Rate Model

This section addresses the pathogen dynamics model of two target cells with general incidence rate for pathogen-to-cell and cell-to-cell as follows:

$$D^\alpha S_1(t) = \Lambda_1 - \Upsilon_1 S_1(t) - \Big(\chi_1(P(t)) + \psi_1(I_1(t))\Big)F_1(S_1(t)), \tag{1}$$

$$D^\alpha I_1(t) = \Big(\chi_1(P(t)) + \psi_1(I_1(t))\Big)F_1(S_1(t)) - \Theta I_1(t) - \Delta I_1(t)C(t), \tag{2}$$

$$D^\alpha S_2(t) = \Lambda_2 - \Upsilon_2 S_2(t) - \Big(\chi_2(P(t)) + \psi_2(I_2(t))\Big)F_2(S_2(t)), \tag{3}$$

$$D^\alpha I_2(t) = \Big(\chi_2(P(t)) + \psi_2(I_2(t))\Big)F_2(S_2(t)) - \Theta I_2(t) - \Delta I_2(t)C(t), \tag{4}$$

$$D^\alpha P(t) = \Omega(I_1(t) + I_2(t)) - \Sigma P(t), \tag{5}$$

$$D^\alpha C(t) = \Psi(I_1(t) + I_2(t)) - \Gamma C(t) - \Phi(I_1(t) + I_2(t))C(t), \tag{6}$$

where $S_k(t)$ and $I_k(t)$ signify the densities of uninfected and infected cells for two target cells, respectively at time t, $k = 1, 2$. $P(t)$ and $C(t)$ signify the densities of pathogens and CTLs respectively at time t. Λ_k represent the restoration rate of uninfected cells, while Υ_k are the rate at which they die, k = 1, 2. The infected cells die at rate Θ and CTLs kill them at rate Δ each of the two targets. Pathogens multiply at rate Ω and at rate Σ, its die. CTLs multiply at rate Ψ and die at rate Γ, Φ measures the impairment of the CTLs. The uninfected cells become infected with general incidence rate $\Big(\chi_k(P(t)) + \psi_k(I_k(t))\Big)F_k(S_k(t))$. Functions χ_k, ψ_k and F_k are continuously differentiable and bounded with the following criteria:

(A1) $\chi_k(u) > 0$, $\psi_k(u) > 0$ and $F_k(u) > 0$ for all $u > 0$ and $\chi_k(0) = \psi_k(0) = F_k(0) = 0, k = 1, 2$.

(A2) $\chi_k'(u) > 0$, $\psi_k'(u) > 0$ and $F_k'(u) > 0$ for all $u \geq 0$.

(A3) $\left(\frac{F_k(S_k)}{S_k}\right)' \leq 0$, $\left(\frac{\chi_k(P)}{P}\right)' \leq 0$ and $\left(\frac{\psi_k(I_k)}{I_k}\right)' \leq 0$ for all $S_k > 0$, $I_k > 0$ and $P > 0, k = 1, 2$.

2.1 *Basic Properties*

Non-negativity and finiteness of system (1)–(6) solutions are discussed in the present subsection.

Lemma 1 *A non-negative invariant compact set corresponding to system (1)-(6) exists such that*

$$\Omega_1 = \left\{(S_1, I_1, S_2, I_2, P, C) \in \mathbb{R}^6_{\geq 0} : 0 \leq S_1, I_1, S_2, I_2 \leq m_1, 0 \leq P \leq m_2, 0 \leq C \leq m_3\right\}. \quad (7)$$

Proof It is clear that

$$\begin{aligned}
D^\alpha S_1|_{(S_1=0)} &= \Lambda_1 > 0,\\
D^\alpha I_1|_{(I_1=0)} &= (\chi_1(P(t))F_1(S_1(t)) \geq 0, \text{ for all } S_1, P \geq 0,\\
D^\alpha S_2|_{(S_2=0)} &= \Lambda_2 > 0,\\
D^\alpha I_2|_{(I_2=0)} &= (\chi_2(P(t))F_2(S_2(t)) \geq 0, \text{ for all } S_2, P \geq 0,\\
D^\alpha P|_{(P=0)} &= \Omega(I_1 + I_2) \geq 0, \text{ for all } I_1, I_2 \geq 0,\\
D^\alpha C|_{(C=0)} &= \Psi(I_1 + I_2) \geq 0, \text{ for all } I_1, I_2 \geq 0.
\end{aligned}$$

This proves the non-negative invariant property of $\mathbb{R}^6_{\geq 0}$ for the system (1)-(6). Let $Q = S_1 + I_1 + S_2 + I_2 + \frac{\Theta}{2\Omega}P + \frac{\Theta}{4\Psi}C$, then

$$\begin{aligned}
D^\alpha Q =&\Lambda_1 - \Upsilon_1 S_1(t) - \Big(\chi_1(P(t)) + \psi_1(I_1(t))\Big)F_1(S_1(t))\\
&+ \Big(\chi_1(P(t)) + \psi_1(I_1(t))\Big)F_1(S_1(t)) - \Theta I_1(t) - \Delta I_1(t)C(t)\\
&+ \Lambda_2 - \Upsilon_2 S_2(t) - \Big(\chi_2(P(t)) + \psi_2(I_2(t))\Big)F_2(S_2(t))\\
&+ \Big(\chi_2(P(t)) + \psi_2(I_2(t))\Big)F_2(S_2(t)) - \Theta I_2(t) - \Delta I_2(t)C(t)\\
&+ \frac{\Theta}{2\Omega}\Big(\Omega(I_1(t) + I_2(t)) - \Sigma P(t)\Big)\\
&+ \frac{\Theta}{4\Psi}\Big(\Psi(I_1(t) + I_2(t)) - \Gamma C(t) - \Phi(I_1(t) + I_2(t))C(t)\Big)\\
=&\Lambda_1 + \Lambda_2 - \Upsilon_1 S_1 - \Upsilon_2 S_2 - \frac{\Theta}{4}(I_1 + I_2)\\
&- \left(\Delta + \frac{\Theta\Phi}{4\Psi}\right)(I_1 + I_2)C - \frac{\Theta\Sigma}{2\Omega}P - \frac{\Theta\Gamma}{4\Psi}C\\
\leq&\Lambda_1 + \Lambda_2 - \Upsilon_1 S_1 - \Upsilon_2 S_2 - \frac{\Theta}{4}(I_1 + I_2) - \frac{\Theta\Sigma}{2\Omega}P - \frac{\Theta\Gamma}{4\Psi}C\\
\leq&\Lambda_1 + \Lambda_2 - \omega\left(S_1 + I_1 + S_2 + I_2 + \frac{\Theta}{2\Omega}P + \frac{\Theta}{4\Psi}C\right) = \Lambda_1 + \Lambda_2 - \omega Q,
\end{aligned}$$

where, $\omega = \min\{\Upsilon_1, \Upsilon_2, \frac{\Theta}{4}, \Sigma, \Gamma\}$. Then

$$Q(t) \leq e^{-\omega t}\left(Q(0) - \frac{(\Lambda_1 + \Lambda_2)}{\omega}\right) + \frac{(\Lambda_1 + \Lambda_2)}{\omega}.$$

This yields, $0 \le Q(t) \le m_1$ for all $t \ge 0$ if $Q(0) \le m_1$, where $m_1 = \frac{(\Lambda_1+\Lambda_2)}{\omega}$. It follows that $0 \le S_1(t), I_1(t), S_2(t), I_2(t) \le m_1, 0 \le P(t) \le m_2$ and $0 \le C(t) \le m_3$ for all $t \ge 0$ if $S_1(0) + I_1(0) + S_2(0) + I_2(0) + \frac{\Theta}{2\Omega}P(0) + \frac{\Theta}{4\Psi} C(0) \le m_1$, where $m_2 = \frac{2\Omega(\Lambda_1+\Lambda_2)}{\Theta\omega}$ and $m_3 = \frac{4\Psi(\Lambda_1+\Lambda_2)}{\Theta\omega}$. This ends the proof of the finiteness of S_1, I_1, S_2, I_2, P and C.

2.2 *Equilibrium Points and basic number of reproduction $\mathcal{R}_0$*

To compute the equilibrium points we let

$$0 = \Lambda_1 - \Upsilon_1 S_1 - \Big(\chi_1(P) + \psi_1(I_1)\Big)F_1(S_1), \tag{8}$$

$$0 = \Big(\chi_1(P) + \psi_1(I_1)\Big)F_1(S_1) - \Theta I_1 - \Delta I_1 C, \tag{9}$$

$$0 = \Lambda_2 - \Upsilon_2 S_2(t) - \Big(\chi_2(P) + \psi_2(I_2)\Big)F_2(S_2), \tag{10}$$

$$0 = \Big(\chi_2(P) + \psi_2(I_2)\Big)F_2(S_2) - \Theta I_2 - \Delta I_2 C, \tag{11}$$

$$0 = \Omega(I_1 + I_2) - \Sigma P, \tag{12}$$

$$0 = \Psi(I_1 + I_2) - \Gamma C - \Phi(I_1 + I_2)C. \tag{13}$$

From Equations (8)-(13), We find that we have two possibilities, first one if $I_1 = I_2 = 0$, we have the disease free equilibrium point $\Gamma_0 = (S_1^0, 0, S_2^0, 0, 0, 0)$, where $S_1^0 = \dfrac{\Lambda_1}{\Upsilon_1}$ and $S_2^0 = \dfrac{\Lambda_2}{\Upsilon_2}$. When $I_1, I_2 \neq 0$ another equilibrium point exists $\Gamma^* = (S_1^*, I_1^*, S_2^*, I_2^*, P^*, C^*)$ defined by infected equilibrium and satisfies the next equations

$$0 = \frac{\Big(\chi_1(P) + \psi_1(I_1)\Big)F_1(S_1)}{I_1} - \Theta - \Delta C,$$

$$0 = \frac{\Big(\chi_2(P) + \psi_2(I_2)\Big)F_2(S_2)}{I_2} - \Theta - \Delta C,$$

such that

$$P = \frac{\Omega(I_1 + I_2)}{\Sigma}, \tag{14}$$

$$C = \frac{\Psi(I_1 + I_2)}{\Phi(I_1 + I_2) + \Gamma}, \tag{15}$$

S_1, S_2 satisfy the following equations

$$0 = \Lambda_1 - \Upsilon_1 S_1 - \Big(\chi_1(P) + \psi_1(I_1)\Big)F_1(S_1),$$
$$0 = \Lambda_2 - \Upsilon_2 S_2 - \Big(\chi_2(P) + \psi_2(I_2)\Big)F_2(S_2),$$

define function G_1 on $[0, \infty)$ by

$$G_1(I_1, I_2) = \frac{\Big(\chi_1(P) + \psi_1(I_1)\Big)F_1(S_1)}{I_1} + \frac{\Big(\chi_2(P) + \psi_2(I_2)\Big)F_2(S_2)}{I_2} - \Theta - \Delta C,$$

from the finiteness of χ_1, ψ_1, χ_2 and ψ_2 and Eq. 14, we get $\lim_{I_1\to\infty}\frac{\chi_1(p)}{I_1} = \lim_{I_1\to\infty}\frac{\psi_1(I_1)}{I_1} = 0$ and $\lim_{I_2\to\infty}\frac{\chi_2(p)}{I_2} = \lim_{I_2\to\infty}\frac{\psi_2(I_2)}{I_2} = 0$.
Then $\lim_{(I_1,I_2)\to(\infty,\infty)} G_1 = -\Theta - \dfrac{\Psi\Delta}{\Phi} < 0$, $\lim_{(I_1,I_2)\to(0,0)} G_1 = \left(\frac{\Omega}{\Sigma}\chi_1'(0) + \psi_1'(0)\right) F_1(S_1^0) + \left(\frac{\Omega}{\Sigma}\chi_2'(0) + \psi_2'(0)\right) F_2(S_2^0) - \Theta > 0$. Therefore there exists $I_1, I_2 \in (0, \infty)$ and from Equations (14)-(15) we have $P^* = \dfrac{\Omega(I_1^* + I_2^*)}{\Sigma} > 0$ and $C^* = \dfrac{\Psi(I_1^* + I_2^*)}{\Phi(I_1^* + I_2^*) + \Gamma} > 0$ when $\Theta\left[\left(\frac{\Omega\chi_1'(0)}{\Theta\Sigma} + \frac{\psi_1'(0)}{\Theta}\right) F_1(S_1^0) + \left(\frac{\Omega\chi_2'(0)}{\Theta\Sigma} + \frac{\psi_2'(0)}{\Theta}\right) F_2(S_2^0) - 1\right] > 0$. Thus the chronic equilibrium $\Gamma^* = (S_1^*, I_1^*, S_2^*, I_2^*, P^*, C^*)$ exists when the condition $\Theta\left[\left(\frac{\Omega\chi_1'(0)}{\Theta\Sigma} + \frac{\psi_1'(0)}{\Theta}\right) F_1(S_1^0) + \left(\frac{\Omega\chi_2'(0)}{\Theta\Sigma} + \frac{\psi_2'(0)}{\Theta}\right) F_2(S_2^0) - 1\right] > 0$ is met.

Now, by using the method of next generation matrix, we evaluate the basic reproduction number $\mathcal{R}_0$ as follows: let $X = (I_1, I_2, P, C)^T$, then we write system (1)-(6) as

$$D^\alpha X = \aleph(X) - \hbar(X),$$

$$\aleph(X) = \begin{pmatrix} \Big(\chi_1(P) + \psi_1(I_1)\Big)F_1(S_1) \\ \Big(\chi_2(P) + \psi_2(I_2)\Big)F_2(S_2) \\ 0 \\ 0 \end{pmatrix}, \quad \hbar(X) = \begin{pmatrix} \Theta I_1 + \Delta I_1 C \\ \Theta I_2 + \Delta I_2 C \\ \Sigma P - \Omega(I_1 + I_2) \\ \Gamma C + \Phi(I_1 + I_2)C - \Psi(I_1 + I_2) \end{pmatrix},$$

The special matrices $F = (\dfrac{\partial \aleph_i}{\partial X_j})$ and $V = (\dfrac{\partial \hbar_i}{\partial X_j})$ for $1 \le i, j \le 4$ take the form

$$F = \begin{pmatrix} \psi_1'(I_1)F_1(S_1) & 0 & \chi_1'(P)F_1(S_1) & 0 \\ 0 & \psi_2'(I_2)F_2(S_2) & \chi_2'(P)F_2(S_2) & 0 \\ 0 & 0 & 0 & 0 \\ 0 & 0 & 0 & 0 \end{pmatrix},$$

$$V = \begin{pmatrix} \Theta + \Delta C & 0 & 0 & \Delta I_1 \\ 0 & \Theta + \Delta C & 0 & \Delta I_2 \\ -\Omega & -\Omega & \Sigma & 0 \\ \Phi C - \Psi & \Phi C - \Psi & 0 & \Gamma + \Phi(I_1 + I_2) \end{pmatrix},$$

then at the uninfected equilibrium point $\Gamma_0 = (S_1^0, 0, S_2^0, 0, 0, 0)$ of system (1)-(6) we get

$$FV^{-1} = \begin{pmatrix} \left(\frac{\Omega\chi_1'(0)}{\Theta\Sigma} + \frac{\psi_1'(0)}{\Theta}\right)F_1(S_1^0) & \frac{\Omega\chi_1'(0)}{\Theta\Sigma}F_1(S_1^0) & \frac{\chi_1'(0)}{\Sigma}F_1(S_1^0) & 0 \\ \frac{\Omega\chi_2'(0)}{\Theta\Sigma}F_2(S_2^0) & \left(\frac{\Omega\chi_2'(0)}{\Theta\Sigma} + \frac{\psi_2'(0)}{\Theta}\right)F_2(S_2^0) & \frac{\chi_2'(0)}{\Sigma}F_2(S_2^0) & 0 \\ 0 & 0 & 0 & 0 \\ 0 & 0 & 0 & 0 \end{pmatrix}. \tag{16}$$

Where $\mathcal{R}_0$ known as the matrix (FV^{-1}) eigenvalue that is dominant, therefore

$$\mathcal{R}_0 = \rho(FV^{-1}) = \left(\frac{\Omega\chi_1'(0)}{\Theta\Sigma} + \frac{\psi_1'(0)}{\Theta}\right)F_1(S_1^0) + \left(\frac{\Omega\chi_2'(0)}{\Theta\Sigma} + \frac{\psi_2'(0)}{\Theta}\right)F_2(S_2^0) \tag{17}$$

$$= \mathcal{R}_1 + \mathcal{R}_2.$$

where $\mathcal{R}_1$ and $\mathcal{R}_1$ represent the basic reproduction numbers for first and second target cells nope separately. Accordingly, we introduce the next lemma.

Lemma 2 *suppose that $\mathcal{R}_0 > 0$ of system (1)-(6) and let Conditions **A1–A3** are fulfilled, then*
(i) only one equilibrium point Γ^0 exists if $\mathcal{R}_0 \leq 1$,
(ii) two equilibrium points Γ^0 and Γ^* exist if $\mathcal{R}_0 > 1$.

2.3 Analysis of Stability

In the current subsection, we investigate local and global stability of the equilibrium point Γ^0 of system (1)–(6).

2.3.1 Local Characteristics

Theorem 1 *For the uninfected equilibrium point Γ^0 of system (1)-(6), if $\mathcal{R}_0 \leq 1$ then Γ^0 is locally asymptotically stable and if $\mathcal{R}_0 > 1$ is unstable.*

Proof. System (1)–(6) Jacobian matrix at $\Gamma^0 = (S_1^0, 0, S_2^0, 0, 0, 0)$ is

$$J(\Gamma^0) = \begin{pmatrix} -\Upsilon_1 & -\psi_1'(0)F_1(S_1^0) & 0 & 0 & -\chi_1'(0)F_1(S_1^0) & 0 \\ 0 & \psi_1'(0)F_1(S_1^0) - \Theta & 0 & 0 & \chi_1'(0)F_1(S_1^0) & 0 \\ 0 & 0 & -\Upsilon_2 & -\psi_2'(0)F_2(S_2^0) & -\chi_2'(0)F_2(S_2^0) & 0 \\ 0 & 0 & 0 & \psi_2'(0)F_2(S_2^0) - \Theta & \chi_2'(0)F_2(S_2^0) & 0 \\ 0 & \Omega & 0 & \Omega & -\Sigma & 0 \\ 0 & \Psi & 0 & \Psi & 0 & -\Gamma \end{pmatrix}.$$

Then, the characteristic equation corresponding to $J(\Gamma^0)$ is

$$(\Upsilon_1 + \lambda)(\Gamma + \lambda)(\Upsilon_2 + \lambda)(\lambda^3 + A\lambda^2 + B\lambda + D) = 0, \tag{18}$$

it is clear that Eq. (18) has three negative roots $\lambda_1 = -\Upsilon_1$, $\lambda_2 = -\Upsilon_2$ and $\lambda_3 = -\Gamma$, then the stability of Γ^0 depends on the roots of the sub-equation

$$P(\lambda) := \lambda^3 + A\lambda^2 + B\lambda + D = 0, \tag{19}$$

where,

$$\begin{aligned}
A &= 2\Theta + \Sigma - \psi_1'(0)F_1(S_1^0) - \psi_2'(0)F_2(S_2^0) \\
&= \frac{\Omega}{\Sigma}\Big(\chi_1'(0)F_1(S_1^0) + \chi_2'(0)F_2(S_2^0)\Big) + \Theta + \Sigma + \Theta(1 - \mathcal{R}_0), \\
B &= \Big(\Theta - \psi_1'(0)F_1(S_1^0)\Big)\Big(\Sigma + \Theta - \psi_2'(0)F_2(S_2^0)\Big) + \Sigma\Big(\Theta - \psi_2'(0)F_2(S_2^0)\Big) \\
&\quad - \Omega\chi_1'(0)F_1(S_1^0) = \frac{\Theta\Omega}{\Sigma}\Big(\chi_1'(0)F_1(S_1^0) + \chi_2'(0)F_2(S_2^0)\Big) + \Omega\chi_2'(0)F_2(S_2^0) \\
&\quad + \psi_1'(0)\psi_2'(0)F_1(S_1^0)F_2(S_2^0) + \Sigma\Theta + \Theta(\Theta + \Sigma)(1 - \mathcal{R}_0), \\
D &= \Sigma\Big(\Theta - \psi_2'(0)F_2(S_2^0)\Big)\Big(\Theta - \psi_1'(0)F_1(S_1^0)\Big) - \Omega\chi_2'(0)F_2(S_2^0)\Big(\Theta - \psi_1'(0)F_1(S_1^0)\Big) \\
&\quad + \Omega\chi_1'(0)F_1(S_1^0))\Big(\psi_2'(0)F_2(S_2^0) - \Theta\Big) \\
&= \Omega F_1(S_1^0)F_2(S_2^0)\Big(\psi_1'(0)\chi_2'(0) + \psi_2'(0)\chi_1'(0)\Big) + \Sigma\psi_1'(0)\psi_2'(0)F_1(S_1^0)F_2(S_2^0) \\
&\quad + \Sigma\Theta^2(1 - \mathcal{R}_0).
\end{aligned}$$

The discriminant $D(P)$ of a polynomial $P(\lambda)$ is:

$$\begin{aligned}
D(P) &= \begin{pmatrix} 1 & A & B & D & 0 \\ 0 & 1 & A & B & D \\ 3 & 2A & B & 0 & 0 \\ 0 & 3 & 2A & B & 0 \\ 0 & 0 & 3 & 2A & B \end{pmatrix} \\
&= 18ABD + (AB)^2 - 4D(A)^3 - 4(B)^3 - 27(D)^2.
\end{aligned}$$

If $\mathcal{R}_0 \leq 1$, then we have $A > 0, B > 0, D > 0$ and $AB - D > 0$. Based on Routh-Hurwitz standards, Γ^0 is locally asymptotically stable if either

$$(i) D(P) > 0 \text{ for } \alpha \in (0, 1] \quad or$$
$$(ii) D(P) < 0 \text{ for } \alpha \in (0, 2/3)$$

is satisfied, and this ends the proof.

2.3.2 Global Characteristics

In this subsection we are going to confirm the global stability of uninfected equilibrium point of model (1)–(6) by creating appropriate Lyapunov function.

Remark 1 From Assumption (A3) we have $\frac{\chi_1(P)}{P} \leq \lim_{P\to 0^+} \frac{\chi_1(P)}{P} = \chi_1'(0)$, $\frac{\chi_2(P)}{P} \leq \lim_{P\to 0^+} \frac{\chi_2(P)}{P} = \chi_2'(0)$, $\frac{\psi_1(I_1)}{I_1} < \lim_{I_1\to 0^+} \frac{\psi_1(I_1)}{I_1} = \psi_1'(0)$ and $\frac{\psi_2(I_2)}{I_2} \leq \lim_{I_2\to 0^+} \frac{\psi_2(I_2)}{I_2} = \psi_2'(0)$.

Theorem 2 *For model (1)-(6), if $\mathcal{R}_0 < 1$, then Γ^0 is globally asymptotically stable.*

Proof. Define $V_1(S_1, I_1, S_2, I_2, P, C)$ as:

$$V_1(S_1, I_1, S_2, I_2, P, C) = S_1 - S_1^0 - \int_{S_1^0}^{S_1} \frac{F_1(S_1^0)}{F_1(\zeta)} d\zeta + I_1 + S_2 - S_2^0 - \int_{S_2^0}^{S_2} \frac{F_2(S_2^0)}{F_2(\zeta)} d\zeta + I_2$$
$$+ \frac{F_1(S_1^0)\chi_1'(0) + F_2(S_2^0)\chi_2'(0)}{\Sigma} P + \frac{\Theta(1 - \mathcal{R}_0)}{\Psi} C.$$

Clearly, $V_1(S_1, I_1, S_2, I_2, P, C) > 0$ for all $S_1, I_1, S_2, I_2, P, C > 0$ and $V_1(S_1^0, 0, S_2^0, 0, 0, 0) = 0$. Calculating $D^\alpha V_1$ along the system (1)-(6), we get

$$\begin{aligned}
D^\alpha V_1 &= \left(1 - \frac{F_1(S_1^0)}{F_1(S_1)}\right)\left[\Lambda_1 - \Upsilon_1 S_1 - \left(\chi_1(P) + \psi_1(I_1)\right)F_1(S_1)\right] \\
&+ \left(\chi_1(P) + \psi_1(I_1)\right)F_1(S_1) - \Theta I_1 - \Delta I_1 C \\
&+ \left(1 - \frac{F_2(S_2^0)}{F_2(S_2)}\right)\left[\Lambda_2 - \Upsilon_2 S_2 - \left(\chi_2(P) + \psi_2(I_1)\right)F_2(S_2)\right] \\
&+ \left(\chi_2(P) + \psi_2(I_2)\right)F_2(S_2) - \Theta I_2 - \Delta I_2 C \\
&+ \frac{F_1(S_1^0)\chi_1'(0) + F_2(S_2^0)\chi_2'(0)}{\Sigma}\left[\Omega(I_1 + I_2) - \Sigma P\right] \\
&+ \frac{\Theta(1 - \mathcal{R}_0)}{\Psi}\left[\Psi(I_1 + I_2) - \Gamma C - \Phi(I_1 + I_2)C\right] \\
&= \left(1 - \frac{F_1(S_1^0)}{F_1(S_1)}\right)[\Lambda_1 - \Upsilon_1 S_1] + \left(\chi_1(P) + \psi_1(I_1)\right)F_1(S_1^0)
\end{aligned}$$

$$+\left(1-\frac{F_2(S_2^0)}{F_2(S_2)}\right)[\Lambda_2-\Upsilon_2 S_2]+\Big(\chi_2(P)+\psi_2(I_2)\Big)F_2(S_2^0)$$
$$-\Delta(I_1+I_2)C+\frac{F_1(S_1^0)\chi_1'(0)+F_2(S_2^0)\chi_2'(0)}{\Sigma}\Omega(I_1+I_2)$$
$$-\Big[F_1(S_1^0)\chi_1'(0)+F_2(S_2^0)\chi_2'(0)\Big]P-\Theta\mathcal{R}_0(I_1+I_2)$$
$$-\frac{\Theta(1-\mathcal{R}_0)}{\Psi}\Big[\Gamma C+\Phi(I_1+I_2)C\Big]$$

Using $\Lambda_1=\Upsilon_1 S_1^0$, $\Lambda_2=\Upsilon_2 S_2^0$ and from Remak 1 we get

$$D^\alpha V_1\le-\Upsilon_1 S_1^0\left(1-\frac{S_1}{S_1^0}\right)\left(1-\frac{F_1(S_1^0)}{F_1(S_1)}\right)-\Upsilon_2 S_2^0\left(1-\frac{S_2}{S_2^0}\right)\left(1-\frac{F_2(S_2^0)}{F_1(S_1)}\right)$$
$$+\psi_1'(0)I_1F_1(S_1^0)+\psi_2'(0)I_2F_2(S_2^0)-\Delta(I_1+I_2)C$$
$$+\frac{\Omega}{\Sigma}\Big[F_1(S_1^0)\chi_1'(0)+F_2(S_2^0)\chi_2'(0)\Big](I_1+I_2)-\Theta\Big(\frac{\Omega\chi_1'(0)}{\Theta\Sigma}+\frac{\psi_1'(0)}{\Theta}\Big)F_1(S_1^0)(I_1+I_2)$$
$$-\Theta\Big(\frac{\Omega\chi_2'(0)}{\Theta\Sigma}+\frac{\psi_2'(0)}{\Theta}\Big)F_2(S_2^0)(I_1+I_2)-\frac{\Theta(1-\mathcal{R}_0)}{\Psi}\Big[\Gamma C+\Phi(I_1+I_2)C\Big].$$

Finally, we obtain

$$D^\alpha V_1\le-\Upsilon_1 S_1^0\left(1-\frac{S_1}{S_1^0}\right)\left(1-\frac{F_1(S_1^0)}{F_1(S_1)}\right)-\Upsilon_2 S_2^0\left(1-\frac{S_2}{S_2^0}\right)\left(1-\frac{F_2(S_2^0)}{F_1(S_1)}\right)$$
$$-\psi_1'(0)I_2F_1(S_1^0)-\psi_2'(0)I_1F_2(S_2^0)-\Delta(I_1+I_2)C$$
$$-\frac{\Theta(1-\mathcal{R}_0)}{\Psi}\Big[\Gamma+\Phi(I_1+I_2)\Big]C.$$

According to **A2**, we find that $\left(1-\frac{S_1}{S_1^0}\right)\left(1-\frac{F_1(S_1^0)}{F_1(S_1)}\right)\le 0$ and $\left(1-\frac{S_2}{S_2^0}\right)\left(1-\frac{F_2(S_2^0)}{F_2(S_2)}\right)\le 0$, also $\psi_1'(0)>0$ and $\psi_2'(0)>0$. Obviously if $\mathcal{R}_0<1$, then $D^\alpha V_1\le 0$ for all $S_1, I_1, S_2, I_2, P, C>0$. Moreover $D^\alpha V_1=0$ only if $S_1(t)=S_1^0$, $S_2(t)=S_2^0$, $I_1(t)=0$, $I_2(t)=0$ and $C(t)=0$. Let $\mathcal{D}_0=\{(S_1,I_1,S_2,I_2,P,C): D^\alpha V_1=0\}$ and $\acute{\mathcal{D}}_0$ be largest invariant subset of $\mathcal{D}_0$. The solutions of the model (1)-(6) goes to $\acute{\mathcal{D}}_0$. For any item in $\acute{\mathcal{D}}_0$ when $S_1(t)=S_1^0, S_2(t)=S_2^0, I_1(t)=0, I_2(t)=0$ and $C(t)=0$. Thus Eq. (5) Leads to

$$D^\alpha P(t)=0=\Omega(I_1(t)+I_2(t))-\Sigma P(t),$$

as a result $P(t)=0$. Therefore, $\acute{\mathcal{D}}_0$ includes only one point, which is $(S_1^0, 0, S_2^0, 0, 0, 0)$. LaSalle's invariance principle (LIP) indicates that Γ^0 is GAS when $\mathcal{R}_0<1$.

3 Numerical Simulations and Dissection

We provide an example and conduct numerical simulations in this section to confirm our theoretical findings had shown in this chapter. All of numerical computations were conducted by MATLAB.

Example of the model (1)–(6):

To perform numerical simulations and demonstrate global asymptotic stability of equilibrium points of our model, let us choose the functions $\digamma_1(S_1(t)) = \frac{\sqrt{S_1}}{1+\alpha_3\sqrt{S_1}}$, $\digamma_2(S_2(t)) = \frac{\sqrt{S_2}}{1+\alpha_5\sqrt{S_2}}$, $\chi_1(P(t)) = \frac{\beta_1 P}{1+\alpha_1 P}$, $\chi_2(P(t)) = \frac{\beta_2 P}{1+\alpha_4 P}$, $\psi_1(I_1(t)) = \frac{\eta_1 I_1}{1+\alpha_2 I_1}$ and $\psi_2(I_2(t)) = \eta_2 I_2$, where $\alpha_1, \alpha_2, \alpha_3, \alpha_4$ and α_5 are positive constants. We introduce the following model as a special case of the model (1)–(6):

$$D^\alpha S_1(t) = \Lambda_1 - \Upsilon_1 S_1(t) - \left(\frac{\beta_1 P}{1+\alpha_1 P} + \frac{\eta_1 I_1}{1+\alpha_2 I_1}\right)\frac{\sqrt{S_1}}{1+\alpha_3\sqrt{S_1}}, \tag{20}$$

$$D^\alpha I_1(t) = \left(\frac{\beta_1 P}{1+\alpha_1 P} + \frac{\eta_1 I_1}{1+\alpha_2 I_1}\right)\frac{\sqrt{S_1}}{1+\alpha_3\sqrt{S_1}} - \Theta I_1(t) - \Delta I_1(t)C(t), \tag{21}$$

$$D^\alpha S_2(t) = \Lambda_2 - \Upsilon_2 S_2(t) - \left(\frac{\beta_2 P}{1+\alpha_4 P} + \eta_2 I_2\right)\frac{\sqrt{S_2}}{1+\alpha_5\sqrt{S_2}}, \tag{22}$$

$$D^\alpha I_2(t) = \left(\frac{\beta_2 P}{1+\alpha_4 P} + \eta_2 I_2\right)\frac{\sqrt{S_2}}{1+\alpha_5\sqrt{S_2}} - \Theta I_2(t) - \Delta I_2(t)C(t), \tag{23}$$

$$D^\alpha P(t) = \Omega(I_1(t) + I_2(t)) - \Sigma P(t), \tag{24}$$

$$D^\alpha C(t) = \Psi(I_1(t) + I_2(t)) - \Gamma C(t) - \Phi(I_1(t) + I_2(t))C(t), \tag{25}$$

We now check the conditions **(A1)**-**(A3)**:

(A1) It is obvious that $\digamma_1(S_1(t)) = \frac{\sqrt{S_1}}{1+\alpha_3\sqrt{S_1}} > 0$ as $S_1 > 0$, $\digamma_2(S_2(t)) = \frac{\sqrt{S_2}}{1+\alpha_5\sqrt{S_2}} > 0$ as $S_2 > 0$, $\chi_1(P(t)) = \frac{\beta_1 P}{1+\alpha_1 P} > 0$, $\chi_2(P(t)) = \frac{\beta_2 P}{1+\alpha_4 P} > 0$, as $P > 0$, $\psi_1(I_1(t)) = \frac{\eta_1 I_1}{1+\alpha_2 I_1} > 0$ as $I_1 > 0$ and $\psi_2(I_2(t)) = \eta_2 I_2 > 0$ as $I_2 > 0$, also $\digamma_k(0) = \chi_k(0) = \psi_k(0) = 0, k = 1, 2$.

(A2) $\digamma_1{}'(S_1) = \frac{1}{2\sqrt{S_1}(1+\alpha_3\sqrt{S_1})^2} > 0$, $\digamma_2{}'(S_2) = \frac{1}{2\sqrt{S_2}(1+\alpha_5\sqrt{S_2})^2} > 0$, $\chi_1'(P) = \frac{\beta_1}{(1+\alpha_1 P)^2} > 0$, $\chi_2'(P) = \frac{\beta_2}{(1+\alpha_4 P)^2} > 0$, $\psi_1'(I_1) = \frac{\eta_1}{(1+\alpha_2 I_1)^2}$ and $\psi_2'(I_2) = \eta_2 > 0$.

(A3) $\left(\frac{\digamma_1(S_1)}{S_1}\right)' = \frac{-\frac{1}{2}-\alpha_3\sqrt{S_1}}{(\sqrt{S_1})^3(1+\alpha_3\sqrt{S_1})^2} < 0$, $\left(\frac{\digamma_2(S_2)}{S_2}\right)' = \frac{-\frac{1}{2}-\alpha_5\sqrt{S_2}}{(\sqrt{S_2})^3(1+\alpha_5\sqrt{S_2})^2} < 0$, $\left(\frac{\chi_1(P)}{P}\right)' = \frac{-\beta_1\alpha_1}{(1+\alpha_1 P)^2} < 0$, $\left(\frac{\chi_2(P)}{P}\right)' = \frac{-\beta_2\alpha_4}{(1+\alpha_4 P)^2} < 0$, $\left(\frac{\psi_1(I_1)}{I_1}\right)' = \frac{-\eta_1\alpha_2}{(1+\alpha_2 I_1)^2} < 0$ and $\left(\frac{\psi_2(I_2)}{I_2}\right)' = 0$ for $S_k > 0$, $P > 0$ and $I_k > 0$ and $k = 1, 2$. For the previous functions the basic reproduction number is $\mathcal{R}_0 = \mathcal{R}_1 + \mathcal{R}_2 = \left(\frac{\Omega\beta_1}{\Theta\Sigma} + \frac{\eta_1}{\Theta}\right)\frac{\sqrt{S_1^0}}{1+\alpha_3\sqrt{S_1^0}} + \left(\frac{\Omega\beta_2}{\Theta\Sigma} + \frac{\eta_2}{\Theta}\right)\frac{\sqrt{S_2^0}}{1+\alpha_5\sqrt{S_2^0}}$.

Using the parameter values mentioned in Table 1, we will run numerical simulations for the system (20)-(25), we choose three initial conditions as:

IC1: $S_1(0) = 900, I_1(0) = 50, S_2(0) = 800, I_2(0) = 40, P(0) = 100, C(0) = 5.5$,

IC2: $S_1(0) = 600, I_1(0) = 30, S_2(0) = 500, I_2(0) = 20, P(0) = 50, C(0) = 4.5$, and

IC3: $S_1(0) = 300, I_1(0) = 20, S_2(0) = 200, I_2(0) = 10, P(0) = 20, C(0) = 3.5$.

3.1 *Effect of β_1 and β_2 on equilibrium points stability:*

Based on Table 1 and taking $\alpha = 0.97$ and β_1, β_2 are varied as:

(i) If we take $\beta_1 = 0.03$ and $\beta_2 = 0.02$, then we have $\mathcal{R}_0 = 0.7450 < 1$, Specifically, $\mathcal{R}_1 = 0.4535 < 1$, $\mathcal{R}_2 = 0.2916 < 1$. From Lemma 2, the system has only one equilibrium point $\Gamma^0 = (1300, 0, 1200, 0, 0, 0)$. In Fig. 1, we note that the concentrations of uninfected cells grow up to $S_1^0 = 1300$ and $S_2^0 = 1200$, while the concentration of infected cells, pathogenes and CTL cells decreases and approaches zero for IC1-IC3. It demonstrates that Γ^0 is GAS, which indicates the pathogenes are removed and therefore theorem 2 is supported.

(ii) Take $\beta_1 = 0.08$ and $\beta_2 = 0.02$. By computation $\mathcal{R}_0 = 1.4812 > 1$, Specifically, $\mathcal{R}_1 = 1.1896 > 1$, $\mathcal{R}_2 = 0.2916 < 1$. Figure 2 shows as β_1 is increased, the uninfected cells concentrations are decreased.

(iii) Take $\beta_1 = 0.03$ and $\beta_2 = 0.09$. With such choice we get $\mathcal{R}_0 = 1.7456 > 1$, Specifically, $\mathcal{R}_1 = 0.4535 < 1$, $\mathcal{R}_2 = 1.2921 > 1$. State trajectories of system (20)-(25) are shown in Fig. 3.

(iv) Take $\beta_1 = 0.09$ and $\beta_2 = 0.08$. By computation we get $\mathcal{R}_0 = 2.4860 > 1$, Specifically, $\mathcal{R}_1 = 1.3368 > 1$, $\mathcal{R}_2 = 1.1492 > 1$. Figure 4 shows state trajectories of system (20)-(25). Consequently, the numerical results confirm that the chronic equilibrium point Γ^* exists and is globally asymptomatically stable if and only if $\mathcal{R}_0 > 1$ and it must be $\mathcal{R}_1 > 1$ and $\mathcal{R}_2 > 1$.

Table 1 Parameters values of the model (20)–(23)

Parameter	Value	Parameter	Value	Parameter	Value	Parameter	Value
Λ_1	260	Λ_2	240	Θ	4.5	Σ	2
Υ_1	0.2	Υ_2	0.2	Δ	0.04	Ψ	0.5
η_1	0.002	η_2	0.001	Φ	0.1	Γ	2
Ω	5	β_1, β_2	*varied*	α_1	0.01	α_2	0.01
α_3	0.01	α_4	0.01	α_5	0.01		

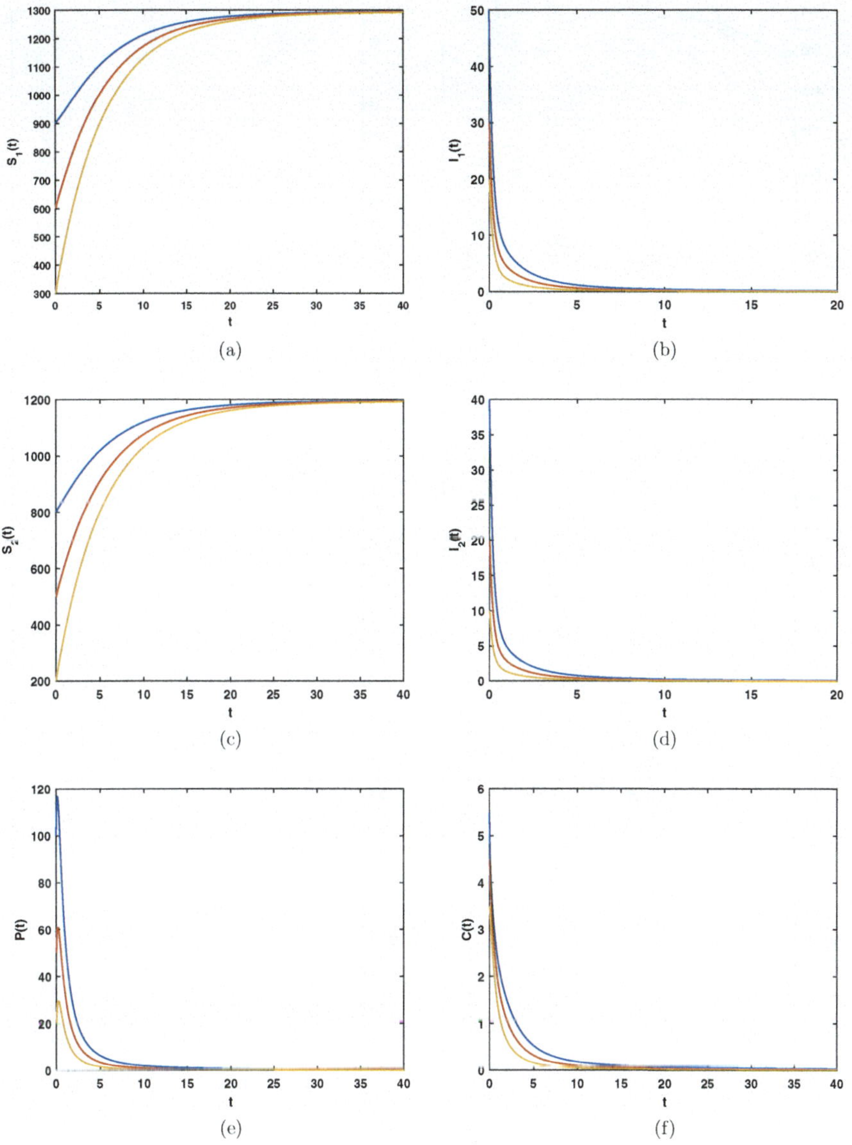

Fig. 1 State trajectories of system (20)–(25) with the initial conditions IC1-IC3 and $\mathcal{R}_0 = 0.7450 < 1$, $\mathcal{R}_1 = 0.4535 < 1$, $\mathcal{R}_2 = 0.2916 < 1$

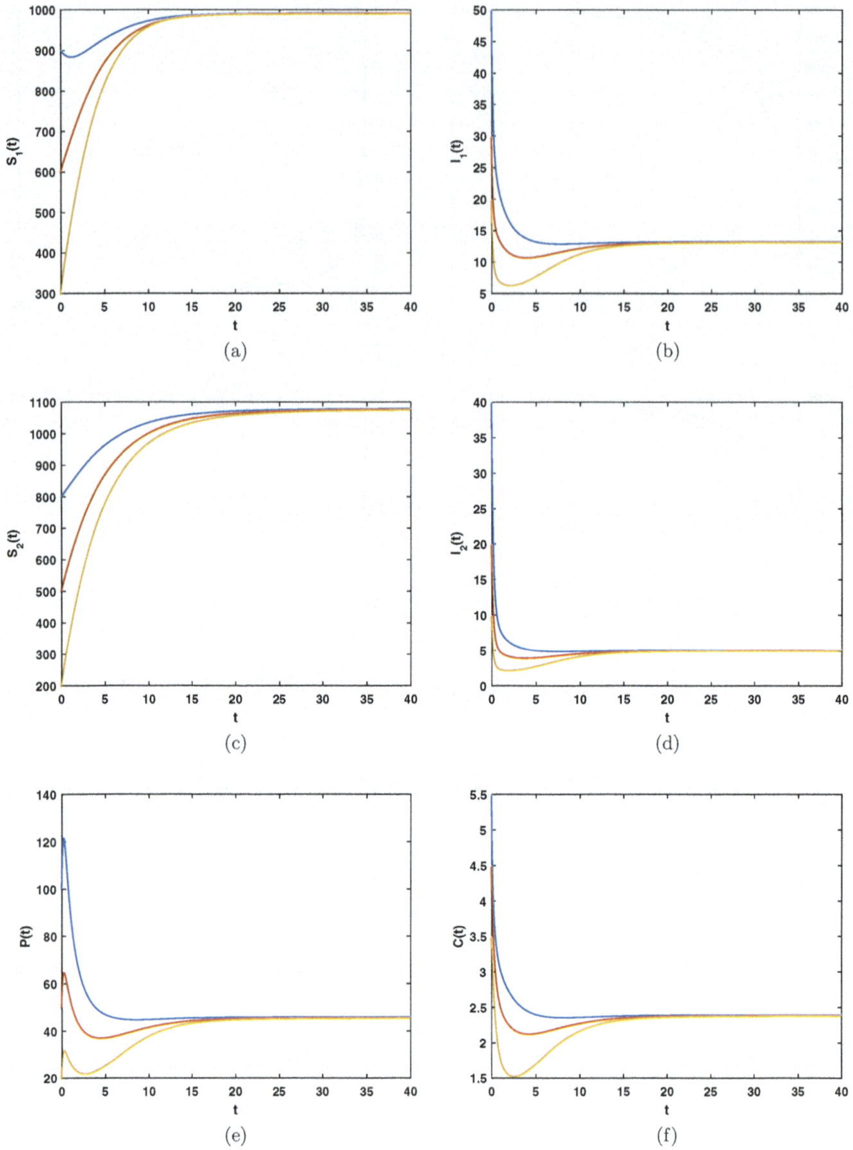

Fig. 2 State trajectories of system (20)–(25) with the initial conditions IC1-IC3 and $\mathcal{R}_0 = 1.4812 > 1$, $\mathcal{R}_1 = 1.1896 > 1$, $\mathcal{R}_2 = 0.2916 < 1$

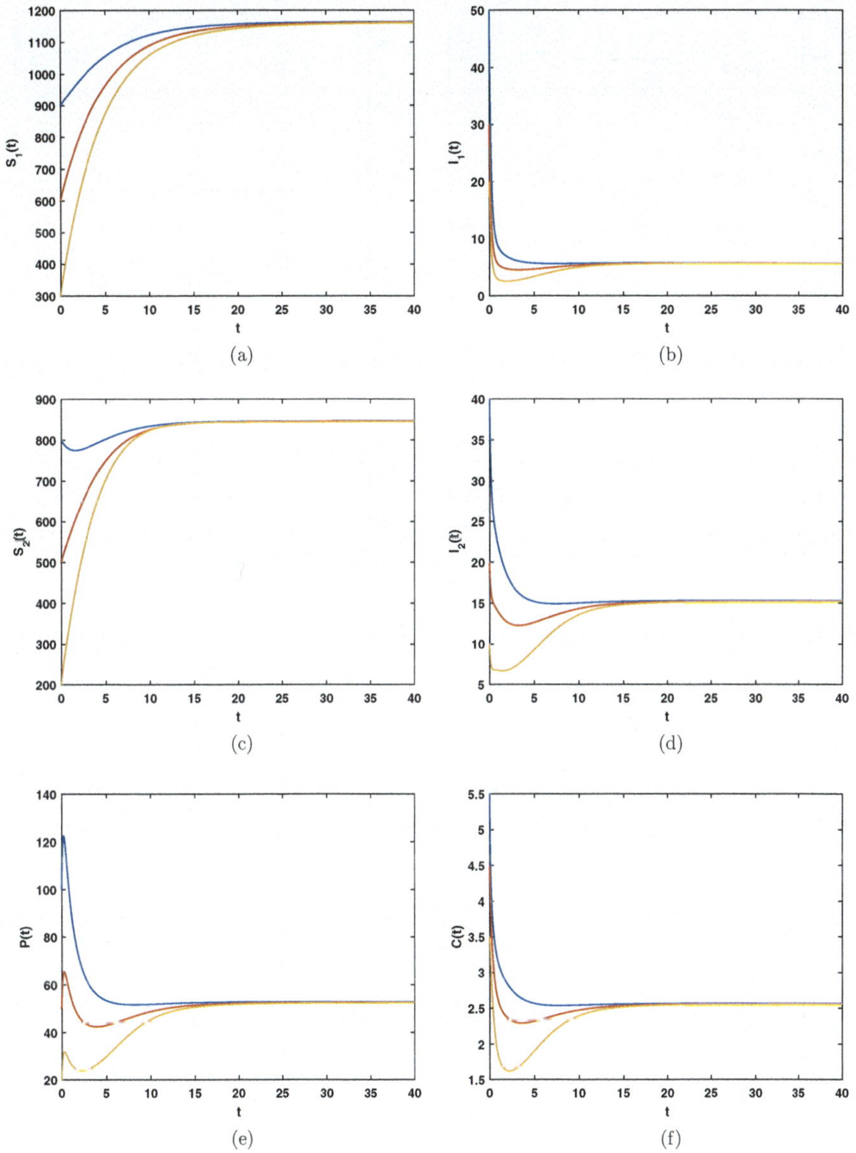

Fig. 3 State trajectories of system (20)–(25) with the initial conditions IC1-IC3 and $\mathcal{R}_0 = 1.7456 > 1$, $\mathcal{R}_1 = 0.4535 < 1$, $\mathcal{R}_2 = 1.2921 > 1$

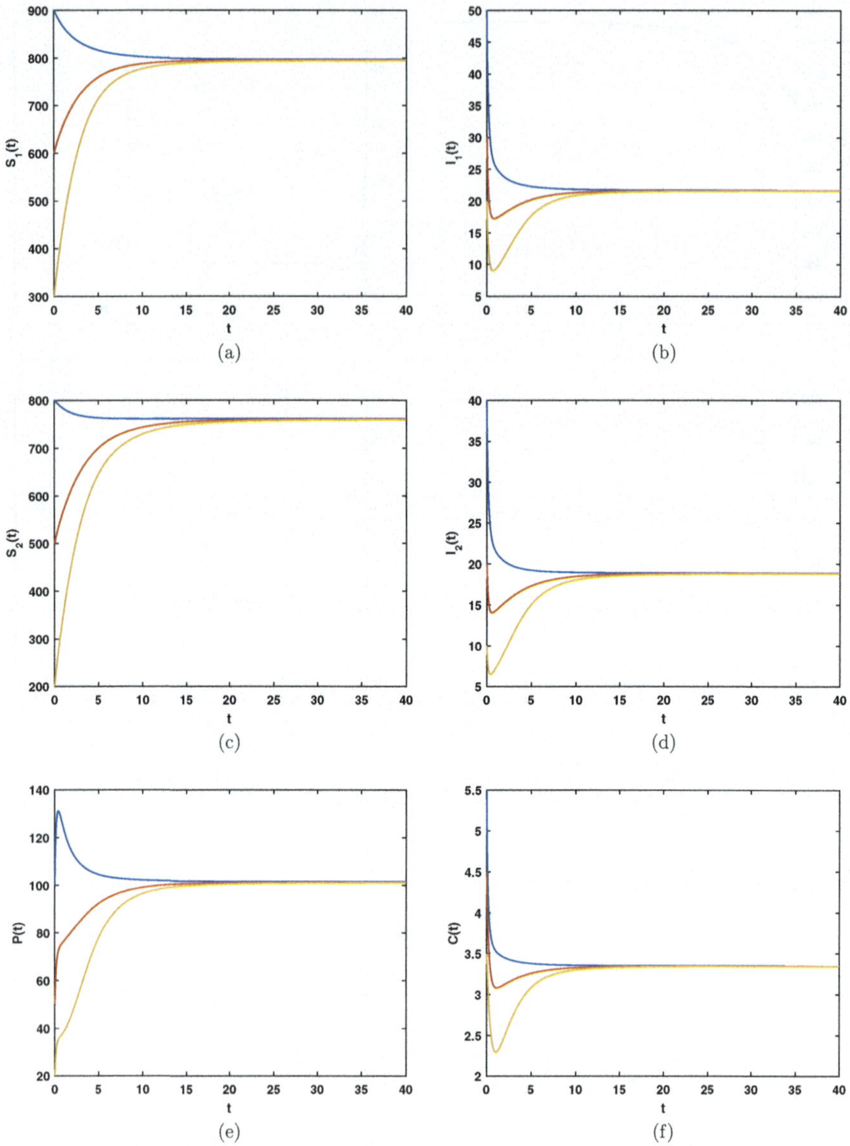

(a) (b)

(c) (d)

(e) (f)

Fig. 4 State trajectories of system (20)–(25) with the initial conditions IC1-IC3 and $\mathcal{R}_0 = 2.4860 > 1$, $\mathcal{R}_1 = 1.3368 > 1$, $\mathcal{R}_2 = 1.1492 > 1$

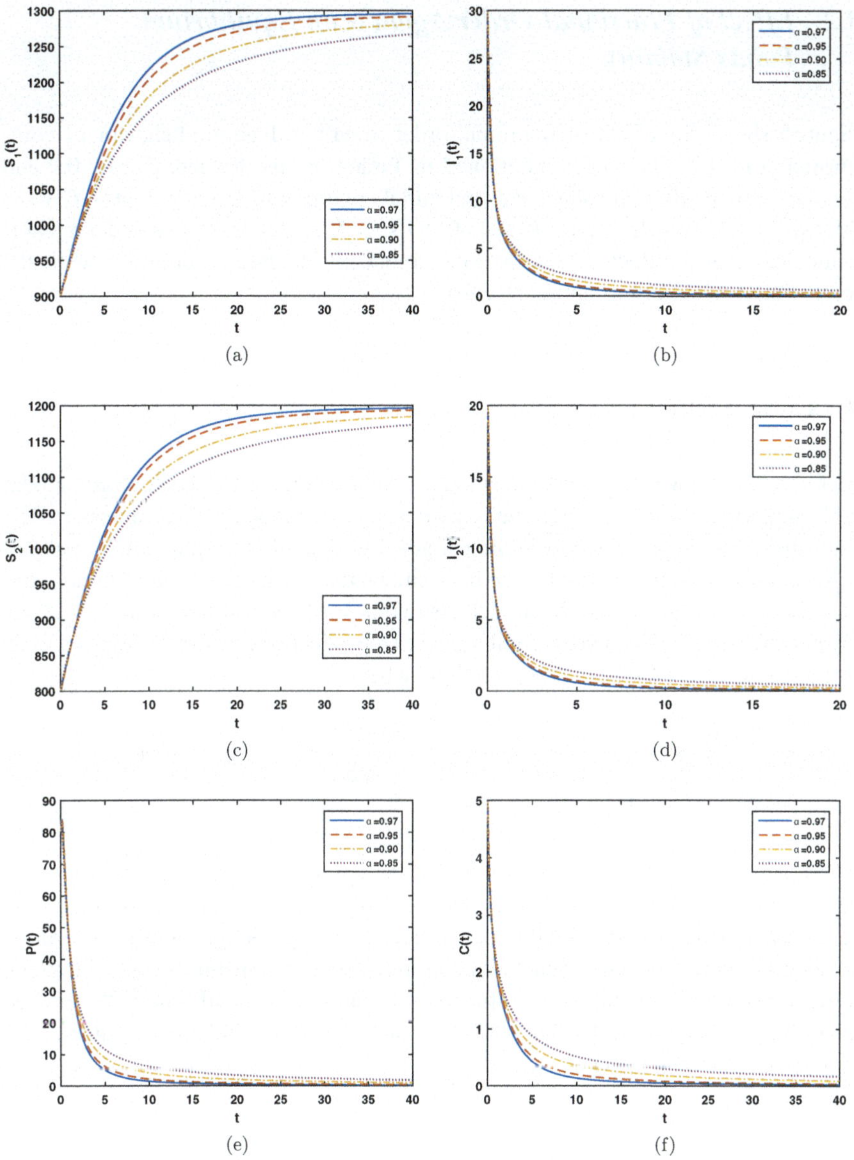

Fig. 5 State trajectories of system (20)–(25) with the initial conditions IC1 and different values of α

3.2 Effect of Fractional-Order Agent α on Equilibrium Points Stability

Figure 5 shows the effect of fractional-order α on the dynamic behavior of equilibrium point Γ^0. The values mentioned in Table 1 are used where $\beta_1 = 0.03$, and $\beta_2 = 0.02$ with different values of α and initial condition IC4: $S_1(0) = 900$, $I_1(0) = 30$, $S_2(0) = 800$, $I_2(0) = 20$, $P(0) = 80$, $C(0) = 5$. As the fractional-order agent α value increases, uninfected cells for both target cells increase accordingly, while the remaining concentrations approach zero.

4 Conclusion

In this document we suggested and studied a pathogen-to-cell and cell-to-cell model with impairment of CTL's immune reactions, and two modes of transmission. The transmission are represented by nonlinear general functions that generalize a number of special forms in the literature. We have shown that the model's solutions are non-negative and restrictive and the model has two possible equilibrium, uninfected one Γ^0 and infected Γ^*. The fundamental infection reproduction number $\mathcal{R}_0$ was obtained using the next-generation matrix technique where

$$\begin{aligned}\mathcal{R}_0 &= \frac{\Omega\chi_1'(0)}{\Theta\Sigma}F_1(S_1^0) + \frac{\psi_1'(0)}{\Theta}F_1(S_1^0) + \frac{\Omega\chi_2'(0)}{\Theta\Sigma}F_2(S_2^0) + \frac{\psi_2'(0)}{\Theta}F_2(S_2^0) \\ &= \mathcal{R}_1^P + \mathcal{R}_1^C + \mathcal{R}_2^P + \mathcal{R}_2^C.\end{aligned} \tag{26}$$

where $\mathcal{R}_1^P$, $\mathcal{R}_2^P$, $\mathcal{R}_1^C$ and $\mathcal{R}_2^C$ represent the basic reproduction numbers for two target cells due to the pathogen-to-cell and cell-to-cell transmissions separately which indicates that transmission from cell to cell affects the pathogen dynamics significantly.

We have built Lyapunov functions and used the invariance principle of LaSalle to demonstrate the global asymptotic stability uninfected equilibrium. We proved analytically and numerically that if $\mathcal{R} < 1$; then Γ^0 is GAS, and proved numerically that if $\mathcal{R} > 1$ then Γ^* is GAS.

References

1. Nowak, M.A., May, R.M.: Virus dynamics: Mathematical Principles of Immunology and Virology. Oxford University, Oxford (2000)
2. Wang, J., Pang, J., Kuniya, T.: A note on global stability for malaria infections model with latencies. Math. Biosci. Eng. **11**, 995–1001 (2014)
3. Callaway, D.S., Perelson, A.S.: HIV-1 infection and low steady state viral loads. Bull. Math. Biol. **64**, 29–64 (2002)

4. Elaiw, A.M., Azoz, S.A.: Global properties of a class of HIV infection models with Beddington-DeAngelis functional response. Math. Methods Appl. Sci. **36**, 383–394 (2013)
5. Elaiw, A.M.: Global properties of a class of HIV models. Nonlinear Anal. Real World Appl. **11**, 2253–2263 (2010)
6. Li, M.Y., Wang, L.: Backward bifurcation in a mathematical model for HIV infection in vivo with anti-retroviral treatment. Nonlinear Anal. Real World Appl. **17**, 147–160 (2014)
7. Wang, K., Fan, A., Torres, A.: Global properties of an improved hepatitis B virus model. Nonlinear Anal. Real World Appl. **11**, 3131–3138 (2010)
8. Zhang, F., Li, J., Zheng, C., Wang, L.: Dynamics of an HBV/HCV infection model with intracellular delay and cell proliferation. Commun. Nonlinear Sci. Numer. Simul. **42**, 464–476 (2017)
9. Neumann, A.U., Lam, N.P., Dahari, H., Gretch, D.R., Wiley, T.E., Layden, T,J., Perelson, A.S.: Hepatitis C viral dynamics in vivo and the antiviral efficacy of interferon-alpha therapy. Science **282**, 103–107 (1998)
10. Zhang, S., Xu, X.: Dynamic analysis and optimal control for a model of hepatitis C with treatment. Commun. Nonlinear Sci. Numer. Simul. **46**, 14–25 (2017)
11. Wang, L., Li, M.Y., Kirschner, D.: Mathematical analysis of the global dynamics of a model for HTLV-I infection and ATL progression. Math. Biosci. **179**, 207–217 (2002)
12. Elaiw, A.M.: Global dynamics of an HIV infection model with two classes of target cells and distributed delays. Discrete Dyn. Nature Soc. **2012** (2012). Article ID 253703
13. Elaiw, A.M., Hassanien, I.A., Azoz, S.A.: Global stability of HIV infection models with intracellular delays. J. Korean Math. Soc. **49**, 779–794 (2012)
14. Elaiw, A.M., Raezah, A.A., Hattaf, K.: Stability of HIV-1 infection with saturated virus-target and infected-target incidences and CTL immune response. Int. J. Biomathematics **5**, 1750070 (2017)
15. Azoz, S.A., Ibrahim, A.M.: Effect of cytotoxic T lymphocytes on HIV-1 dynamics. J. Comput. Anal. Appl. **25**(1), 111–125 (2018)
16. Aidoo, M., Udhayakumar, V.: Field studies of cytotoxic T lymphocytes in Malaria infections: implications for Malaria vaccine development. Parasitol. Today **16**, 50–56 (2000)
17. Nowak, M.A., Bangham, C.R.M.: Population dynamics of immune responses to persistent viruses. Science **272**, 74–79 (1996)
18. Li, M.Y., Shu, H.: Global dynamics of a mathematical model for HTLV-I infection of CD4+ T cells with delayed CTL response. Nonlinear Anal. Real World Appl. **13**, 1080–1092 (2012)
19. Lv, C., Huang, L., Yuan, Z.: Global stability for an HIV-1 infection model with Beddington-DeAngelis incidence rate and CTL immune response. Commun. Nonlinear Sci. Numer. Simula. **19**, 121–127 (2014)
20. Mondragon, E., Esteva, L.: On CTL response against mycobacterium tuberculosis. SIAM J. Appl. Math. **8**, 2383–2389 (2014)
21. Shu, H., Wang, L., Watmough, J.: Global stability of a nonlinear viral infection model with infinitely distributed intracellular delays and CTL imune responses. SIAM J. Appl. Math. **73**, 1280–1302 (2013)
22. Wang, X., Elaiw, A.M., Song, X.: Global properties of a delayed HIV infection model with CTL immune response. Appl. Math. Comput. **218**, 9405–9414 (2012)
23. Li, X., Fu, S.: Global stability of a virus dynamics model with intracellular delay and CTL immune response. Math. Methods Appl. Sci. **38**, 420–430 (2015)
24. Huang, D., Zhang, X., Guo, Y., Wang, H.: Analysis of an HIV infection model with treatments and delayed immune response. Appl. Math. Model. **40**, 3081–3089 (2016)
25. Pang, J., J.-An Cui, J., Hui, J.: The importance of immune responses in a model of hepatitis B virus. Nonlinear Dyn. **67**, 723–734 (2012)
26. Zhao, Y., Xu, Z.: Global dynamics for a delyed hepatitis C virus, infection model. Electron. J. Diff. Equ. **2014**, 1–18 (2014)
27. Regoes, R., Wodarz, D., Nowak, M.A.: Virus dynamics: the effect to target cell limitation and immune responses on virus evolution. J. Theor. Biol. **191**, 451–462 (1998)

28. Wang, J., Guo, M., Liu, X., Zhao, Z.: Threshold dynamics of HIV-1 virus model with cell-to-cell transmission, cell-mediated immune responses and distributed delay. Appl. Math. Comput. **291**, 149–161 (2016)
29. Wang, S., Song, X., Ge, Z.: Dynamics analysis of a delayed viral infection model with immune impairment. App. Math. Model. **35**(10), 4877–4885 (2011)
30. Krishnapriya, P., Pitchaimani, M.: Analysis of time delay in viral infection model with immune impairment. J. Appl. Math. Comput. **55**, 421–453 (2017)
31. Hu, Z., Zhang, J., Wang, H., Ma, W., Liao, F.: Dynamics analysis of a delayed viral infection model with logistic growth and immune impairment. Appl. Math. Model. **38**, 524–534 (2014)
32. Jia, J., Shi, X.: Analysis of a viral infection model with immune impairment and cure rate. J. Nonlinear Sci. Appl. **9**, 3287–3298 (2016)
33. Eric, A.V., Noe, C.C., Gerardo, G.A.: Analysis of a viral infection model with immune impairment, intracellular delay and general non-linear incidence rate. Chaos Solitons Fractals **69**, 1–9 (2014)
34. Culshaw, R.V., Ruan, S., Webb, G.: A mathematical model of cell-to-cell spread of HIV-1 that includes a time delay. J .Math. Biol. **46**, 425–444 (2003)
35. Pourbashash, H., Pilyugin, S.S., De Leenheer, P., McCluskey, C.: Global analysis of within host virus models with cell-to-cell viral transmission. Discrete Contin. Dyn. Syst. Ser. B **10**, 3341–3357 (2014)
36. Wang, J., Lang, J., Zou, X.: Analysis of an age structured HIV infection model with virus-to-cell infection and cell-to-cell transmission. Nonlinear Anal. Real World Appl. **34**, 75–96 (2017)
37. Chen, S.S., Cheng, C.Y., Takeuchi, Y.: Stability analysis in delayed within-host viral dynamics with both viral and cellular infections. J. Math. Anal. Appl. **442**, 642–672 (2016)
38. Elaiw, A.M., Raezah, A.A., Shehata, A.M.: Stability of general virus dynamics models with both cellular and viral infections. J. Nonlinear Sci. Appl. **10**, 1538–1560 (2017)
39. Lai, X., Zou, X.: Modeling cell-to-cell spread of HIV-1 with logistic target cell growth. J. Math. Anal. Appl. **426**, 563–584 (2015)
40. Lai, X., Zou, X.: Modelling HIV-1 virus dynamics with both virus-to-cell infection and cell-to-cell transmission. SIAM J. Appl. Math. **74**, 898–917 (2014)
41. Yang, Y., Zou, L., Ruan, S.: Global dynamics of a delayed within-host viral infection model with both virus-to-cell and cell-to-cell transmissions. Math. Biosci. **270**, 183–191 (2015)
42. Elaiw, A.M., Raezah, A.A., Alofi, B.S.: Stability of pathogen dynamics models with viral and cellular infections and immune impairment. J. Nonlinear Sci. Appl. **11**, 456–468 (2018)
43. Alofi, B.S., Azoz, S.A.: Stability of general pathogen dynamic models with two types of infectious transmission with immune impairment. AIMS Math. J. **6**(1), 114–140 (2021)
44. B. Buonomo, and C. Varglobally asymptotically stable-De-Le, *Global stability for an HIV-1 infection model including an eclipse stage of infected cells*, Journal of Mathematical Analysis and Applications, 385 (2012), 709–720
45. Pindo, C.M., Machado, J.A.: Fractional model for malaria transmission under control strategies. Comput. Math. Appl. J. **66**, 908–916 (2013)
46. Rida, S.Z., Farghaly, A.A., Azoz, S.A., Hussien, F.: Global stability of a delayed fractional-order SEI epidemic model with logistic growth. Appl. Math. Inf. Sci J. **15**, 31–42 (2021)
47. Ullah, S., Khan, M.A., Farooq, M.: A new fractional model for the dynamics of the hepatitis B virus using the Caputo-Fabrizio derivative. Eur. Phys. J. Plus **133**, 237 (2018)

Efficiency Evaluation of Half-Sweep Newton-EGSOR Method to Solve 1D Nonlinear Porous Medium Equations

Jackel Vui Lung Chew, Elayaraja Aruchunan, and Jumat Sulaiman

Abstract This work presents the efficiency evaluation of the Half-Sweep Newton-EGSOR method to compute the numerical solution of several one-dimensional nonlinear porous medium equations. The Half-Sweep finite difference method leads to a nonlinear approximation equation to the porous medium equation and sparse nonlinear systems. Newton method is used to solve the resulting systems by yielding the corresponding linear system. The linear system of equations is solved by using the developed EGSOR iterative method. Local truncation error and stability of the Half-Sweep finite difference scheme are discussed. The efficiency of the proposed Half-Sweep Newton-EGSOR method is evaluated based on the number of iterations, computation time and maximum absolute error. Some examples are delivered with a comparison to the existing Newton-Gauss–Seidel and the Newton-EGSOR methods. The computational complexity of the tested methods is discussed.

Keywords Porous medium equation · Finite difference method · Half-Sweep · Newton method · Successive over relaxation · Explicit group

1 Introduction

Nonlinear partial differential equations (NPDE) have an important role in the mathematical modelling of the complicated phenomenon in physics. One of the simplest

J. V. L. Chew (✉)
Faculty of Computing and Informatics, Universiti Malaysia Sabah Labuan International Campus, 87000 Labuan F.T., Malaysia
e-mail: jackelchew93@ums.edu.my

E. Aruchunan
Faculty of Science, Institute of Mathematical Sciences, University of Malaya, 50603 Kuala Lumpur, Malaysia

J. Sulaiman
Mathematics with Economics Programme, Faculty of Science and Natural Resources, Universiti Malaysia Sabah, Jalan UMS, Kota Kinabalu 88400, Sabah, Malaysia
e-mail: jumat@ums.edu.my

© Institute of Technology PETRONAS Sdn Bhd 2022
S. A. Abdul Karim and A. Shafie (eds.), *Towards Intelligent Systems Modeling and Simulation*, Studies in Systems, Decision and Control 383,
https://doi.org/10.1007/978-3-030-79606-8_25

examples of an NPDE is a porous medium equation (PME). It describes heat transfer, fluid flow and gas diffusion. Its theory and properties depart strongly from the classical heat equation attribute to the presence of nonlinear term that made the mathematical modelling more realistic [30]. Recently, this NPDE is applied in physics mathematical models such as the fire extinguishing in an aircraft engine nacelle [11], the counter-current imbibition phenomenon [17], the hydrodynamic for the slow reservoirs [4], and the instability phenomenon from the process of extracting crude oil from the field [5]. One-dimensional PME in a general form [24]:

$$u_t = \rho(u^m u_x)_x, \tag{1}$$

and can be further derived using the method of calculus into

$$u_t = \rho\left(mu^{m-1}(u_x)^2 + u^m u_{xx}\right). \tag{2}$$

where ρ and m are real numbers with $\rho > 0$. $u(x, t)$ may express the temperature distribution in a thin and long rod or the velocity of fluid flow (or gas-particle diffusion) in a long pipe with a small diameter. x represents the particular location in the rod or pipe at the time t. u^m is the applicable term that determines the transfer of heat, the flow of fluid, or the diffusion of gas particle.

For the past decades, researchers proposed different methods to solve PME. Cho [10], Monsaingeon [20], and Chew and Sulaiman [6, 7, 9] used the finite difference method (FDM) to approximate the solution of PME. Liu et al. [18] and Abedian et al. [2] developed a high order accurate weighted essentially non-oscillatory (WENO) FDM. Bessemoulin-Chatard and Filbet [3], and Yoshioka and Triadis [32] applied the finite volume method (FVM). Ngo and Huang [21, 22] utilised the finite element method (FEM). Saberi nik et al. [25] solved PME using the homotopy perturbation method. Last but not least, Duque et al. [12, 13] formulated the moving mesh method for the numerical solution of PME.

FDM is the most direct approach to discretising differential equations compared to FVM and FEM. By selecting a point in space and replaces the selected point with a set of finite difference equations, differential equations can be approximated. Since the finite difference method is typically defined on a regular framework of grid, such as rectangular or block-shaped models, this method is usually used to develop very efficient solution methods.

Different finite difference schemes have been introduced for solving PME. Liu et al. [18] designed sixth order and eighth order WENO finite difference schemes and demonstrated that these schemes could achieve high order accuracy. Cho [10] introduced adaptive time meshes in the finite difference construction to tackle the blow-up solution problem of PME. Abedian et al. [2] proposed a mixed WENO finite difference scheme and linear multi-step. Monsaingeon [20] applied an explicit scheme of FDM for PME modelled based on interface tracking and hole filling problem. Chew and Sulaiman [6, 9] initiated the application of the Half-Sweep finite difference scheme by Abdullah [1] in developing a family of the efficient iterative method. The Half-Sweep scheme was proven can reduce the computational complexity of solving a sparse and large-scaled linear system by considering the iteration process of half of the total grid points [19, 26, 28, 29].

This paper aims to develop and evaluate the efficiency of a Newton Explicit Group Successive Over Relaxation (EGSOR) iterative method based on the Half-Sweep scheme of FDM in solving PME. The approximation equation to PME is formulated by imposing the Half-Sweep FDM on Eq. (2). Local truncation error and stability of the scheme are discussed. Moreover, the reduction in the number of iteration and computation time is discussed numerically through the comparison between the proposed Half-Sweep Newton-EGSOR and the two existing methods: Newton-EGSOR and Newton-Gauss-Seidel. Some examples are delivered to show the efficiency evaluation of the Half-Sweep Newton-EGSOR method to solve PME.

2 Formulation of Half-Sweep Finite Difference Approximation

Generally speaking, the Half-Sweep finite difference scheme for approximating differential equations is similar to the standard finite difference scheme. The scheme depends on the transformation of a calculus problem into an algebraic problem in which the finite domain is discretised into a discrete difference grid point $U\left(x_p, t_n\right)$. The grid point is denoted as $U\left(x_p, t_n\right) \cong u(x,t) + \varepsilon$, where $u(x,t)$ and ε represent the exact solution and the approximation error, respectively. For simplicity, the approximate solution can be considered as $U_{p,n}$. The finite solution domain in space is partitioned into unit length, i.e., $x_p = ph$, $h = \frac{L}{M}$ with $M - 1$ grid points, and the time level is $t_n = nk$, $k = \frac{T}{100}$. From the formation of discrete difference grid points, we have numerous grid points that need to be computed, i.e., $U_{p,n}$ for $p = 1, 2, \ldots, M-2, M-1$ and for $n = 1, 2, \ldots, 99, 100$.

The implicit finite difference scheme for approximating Eq. (2) can be defined as follows [27]:

The backward difference at the time t_{n+1} for the time derivative u_t:

$$u_t = \frac{U_{p,n} - U_{p,n-1}}{k}. \tag{3}$$

The second-order central difference for space derivatives, i.e., u_x and u_{xx}:

$$u_x = \frac{U_{p+1,n} - U_{p-1,n}}{2h}, \tag{4}$$

and

$$u_{xx} = \frac{U_{p+1,n} - 2U_{p,n} + U_{p-1,n}}{h^2}. \tag{5}$$

The central difference schemes in space, as shown by Eqs. (4) and (5) can be known as the Full-Sweep scheme because the computation will take all available grid points in the domain simultaneously. Eventually, the computational cost will be high, especially when we attempt to increase the total of grid points for obtaining more accurate solutions.

In this paper, the central difference schemes (4) and (5) are extended using the Half-Sweep technique from Abdullah [1]. Based on the concept of the Half-Sweep finite difference scheme, the distance between any two consecutive grid points that are to be computed can be lengthened by $2h$ instead of h [1, 6, 9]. In other words, the computation will consider alternating grid points, which as a result, only half of the total available grid points is computed iteratively. The remaining grid points can be approximated by using the average point technique or interpolation approach. Thus, the Half-Sweep finite difference scheme for the first and second-order space derivatives is denoted as:

$$u_x = \frac{U_{p+2,n} - U_{p-2,n}}{4h}, \tag{6}$$

and

$$u_{xx} = \frac{U_{p+2,n} - 2U_{p,n} + U_{p-2,n}}{4h^2}. \tag{7}$$

The comparison between the Full-Sweep and Half-Sweep finite difference schemes, in terms of the distribution of grid points, can be referred to Figs. 1 and 2.

Based on Fig. 1, the framework contains full solid black grid points. Each grid point is calculated iteratively by forming the linear system that connects all the points. Based on Fig. 2, the framework consists of an equal number of solid black and non-solid black grid points. Solid black points are computed using the iterative method, whereas non-solid black grid points are approximated using the averaging technique or interpolation.

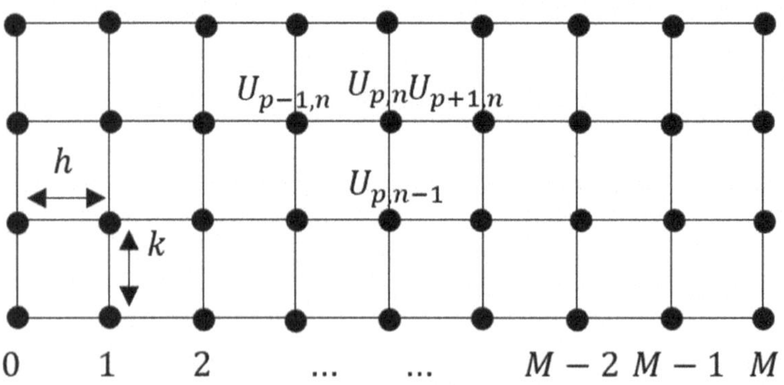

Fig. 1 Full-Sweep finite difference grid point

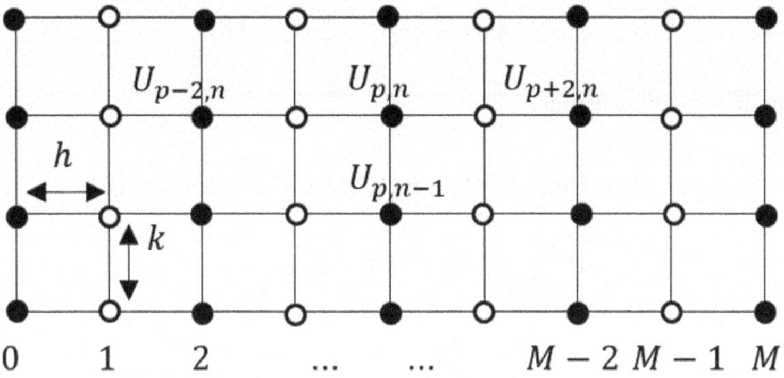

Fig. 2 Half-Sweep finite difference grid point

The Full-Sweep approximation to Eq. (2) can be formulated using the schemes as in Eqs. (3) to (5) as

$$\begin{aligned}{}^{F}F_{p,n} = U_{p,n} - \alpha_1 m U_{p,n}^{m-1} U_{p+1,n}^2 + 2\alpha_1 m U_{p,n}^{m-1} U_{p+1,n} U_{p-1,n} \\ - \alpha_1 m U_{p,n}^{m-1} U_{p-1,n}^2 - \beta_1 U_{p,n}^m U_{p+1,n} + 2\beta_1 U_{p,n}^{m+1} \\ - \beta_1 U_{p,n}^m U_{p-1,n} - U_{p,n-1},\end{aligned} \quad (8)$$

where $\alpha_1 = k/4h^2$ and $\beta_1 = k/h^2$. Meanwhile, the Half-Sweep approximation to Eq. (2) is formulated into

$$\begin{aligned}{}^{H}F_{p,n} = U_{p,n} - \alpha_2 m U_{p,n}^{m-1} U_{p+2,n}^2 + 2\alpha_2 m U_{p,n}^{m-1} U_{p+2,n} U_{p-2,n} \\ - \alpha_2 m U_{p,n}^{m-1} U_{p-2,n}^2 - \beta_2 U_{p,n}^m U_{p+2,n} + 2\beta_2 U_{p,n}^{m+1} \\ - \beta_2 U_{p,n}^m U_{p-2,n} - U_{p,n-1},\end{aligned} \quad (9)$$

where $\alpha_2 = k/16h^2$ and $\beta_2 = k/4h^2$. The local truncation error and the analysis of stability of the Half-Sweep finite difference scheme of the linear form of PME is considered as follows.

Theorem 1 The local truncation error of the Half-Sweep finite difference scheme to Eq. (1) is of $O(4h^2) + O(k)$.

Proof Assuming $U(x_p, t_n)$ exists within the region of $0 < x, t \leq 1$ for $p > 0$ and $n > 0$. Also, assuming u^m is bounded at some maximum value as

$$\mu = C|u^m|_{\max}. \quad (10)$$

The discrete approximation equation to the linear form of Eq. (1) based on the Half-Sweep scheme can be written as

$$\frac{U_{p,n}-U_{p,n-1}}{k}=\mu\left(\frac{U_{p+2,n}-2U_{p,n}+U_{p-2,n}}{4h^2}\right). \tag{11}$$

From Taylor's expansion,

$$U_{p,n-1}=U(ph,(n-1)k)=U(ph,nk)-k\frac{\partial}{\partial t}U(ph,nk)$$
$$+\frac{1}{2}k^2\frac{\partial^2}{\partial t^2}U(ph,nk)+O\left(k^2\right), \tag{12}$$

$$U_{p+2,n}=U((p+2)h,nk)=U(ph,nk)+2h\frac{\partial}{\partial x}U(ph,nk)$$
$$+2h^2\frac{\partial^2}{\partial x^2}U(ph,nk)+O\left(h^2\right), \tag{13}$$

and

$$U_{p-2,n}=U((p-2)h,nk)=U(ph,nk)-2h\frac{\partial}{\partial x}U(ph,nk)$$
$$+2h^2\frac{\partial^2}{\partial x^2}U(ph,nk)+O\left(h^2\right). \tag{14}$$

Substituting Eqs. (12)–(14) into Eq. (11) gives

$$\frac{U(ph,nk)-\left(U(ph,nk)-k\frac{\partial}{\partial t}U(ph,nk)+\frac{1}{2}k^2\frac{\partial^2}{\partial t^2}U(ph,nk)+O\left(k^2\right)\right)}{k}$$
$$=\frac{\mu\left(U(ph,nk)+2h\frac{\partial}{\partial x}U(ph,nk)+2h^2\frac{\partial^2}{\partial x^2}U(ph,nk)+O\left(4h^2\right)\right)}{4h^2}$$
$$-\frac{2\mu(U(ph,nk))}{4h^2}$$
$$+\frac{\mu\left(U(ph,nk)-2h\frac{\partial}{\partial x}U(ph,nk)+2h^2\frac{\partial^2}{\partial x^2}U(ph,nk)+O\left(4h^2\right)\right)}{4h^2}. \tag{15}$$

Since $U(ph,nk)$ is the exact solution of Eq. (2),

$$\frac{\partial}{\partial t}U(ph,nk)=\mu\frac{\partial^2}{\partial x^2}U(ph,nk). \tag{16}$$

Hence, from Eqs. (15) and (16), the local truncation error of the Half-Sweep finite difference scheme is of $O\left(4h^2\right)+O(k)$, and

$$\lim_{h,k\to 0}\left[O\left(4h^2\right)+O(k)\right]=0, \tag{17}$$

Theorem 2 The Half-sweep finite difference scheme for Eq. (1) is unconditionally stable.

Proof From the von Neumann method,

$$U_{p,n} = \xi^n e^{p\theta i}. \tag{18}$$

Using Eq. (18), the linear form of Eq. (1) can be derived as

$$\xi\left(1 - \lambda\left(e^{2\theta i} - 2 + e^{-2\theta i}\right)\right) = 1, \tag{19}$$

where $\lambda = \mu k/4h^2$.

Since $e^{2\theta i} - 2 + e^{-2\theta i} = -4\sin^2\theta$, Eq. (19) can be simplified as

$$\xi = \frac{1}{1+4\lambda\sin^2\theta}. \tag{20}$$

Hence, $0 < \xi \leq 1$ for $\lambda > 0$ and $\theta \in [0, 2\pi]$ and the Half-Sweep finite difference scheme to the linear form of PME is proven to be unconditionally stable.

3 Formulation of Newton-EGSOR Iterative Method

The formulation of the Newton-EGSOR iterative method begins with the following general expression:

$$\begin{aligned} F_{p,n} = {} & U_{p,n} - \alpha_2 m U_{p,n}^{m-1} U_{p+s,n}^2 + 2\alpha_2 m U_{p,n}^{m-1} U_{p+s,n} U_{p-s,n} \\ & - \alpha_2 m U_{p,n}^{m-1} U_{p-s,n}^2 - \beta_2 U_{p,n}^m U_{p+s,n} + 2\beta_2 U_{p,n}^{m+1} \\ & - \beta_2 U_{p,n}^m U_{p-s,n} - U_{p,n-1}, \end{aligned} \tag{21}$$

where $\alpha_s = k/4(sh)^2$, $\beta_s = k/(sh)^2$. $s = 1, 2$ indicating Full- and Half- Sweep, respectively. Based on the s-Sweep function as in Eq. (21), a general form of the nonlinear system has the form of

$$\underline{F}_n(\underline{U}) = 0, \tag{22}$$

where $\underline{F}_n(\underline{U}) = \left(F_{s,n}(\underline{U}), F_{2s,n}(\underline{U}), \ldots, F_{M-2s,n}(\underline{U}), F_{M-s,n}(\underline{U})\right)^T$ and $\underline{U} = (U_s, U_{2s}, \ldots, U_{M-2s}, U_{M-s})$.

Newton method is used to solve the nonlinear system (Eq. 22) by transforming it into a linear system as follows:

$$J_F\left(\underline{U}^{(l)}\right)\Delta\underline{U}^{(l)} = F\left(\underline{U}^{(l)}\right), \tag{23}$$

where

$$J_F\left(\underline{U}^{(l)}\right) = \begin{bmatrix} \frac{\partial F_{s,n}}{\partial U_{s,n}} & \frac{\partial F_{s,n}}{\partial U_{2s,n}} & \cdots & \frac{\partial F_{s,n}}{\partial U_{M-2s,n}} & \frac{\partial F_{s,n}}{\partial U_{M-s,n}} \\ \frac{\partial F_{2s,n}}{\partial U_{s,n}} & \frac{\partial F_{2s,n}}{\partial U_{2s,n}} & \cdots & \frac{\partial F_{2s,n}}{\partial U_{M-2s,n}} & \frac{\partial F_{2s,n}}{\partial U_{M-s,n}} \\ \vdots & \vdots & & \vdots & \vdots \\ \frac{\partial F_{M-2s,n}}{\partial U_{s,n}} & \frac{\partial F_{M-2s,n}}{\partial U_{2s,n}} & \cdots & \frac{\partial F_{M-2s,n}}{\partial U_{M-2s,n}} & \frac{\partial F_{M-2s,n}}{\partial U_{M-s,n}} \\ \frac{\partial F_{M-s,n}}{\partial U_{s,n}} & \frac{\partial F_{M-s,n}}{\partial U_{2s,n}} & \cdots & \frac{\partial F_{M-s,n}}{\partial U_{M-2s,n}} & \frac{\partial F_{M-s,n}}{\partial U_{M-s,n}} \end{bmatrix} \tag{24}$$

and

$$\Delta\underline{U}^{(l)} = \underline{U}^{(l)} - \underline{U}^{(l-1)}, l = 1, 2, \ldots. \tag{25}$$

When the total grid points to be computed in the linear system (Eq. 23) increases, the computational complexity also increases. Hence, to reduce the computational complexity, EGSOR iterative method is applied. This iterative method combines the block-partitioning technique from Evans [14] and the SOR iterative method [15, 16]. The block-partitioning technique uses small-sized groups of grid point strategy to evenly distribute the computation load. In this paper, the block of four points approach is introduced.

Taking the first four grid points to be computed from the linear system (Eq. 23), a system of four equations can be formed as

$$\begin{bmatrix} \mathcal{M}_{11} & \mathcal{M}_{12} & \mathcal{M}_{13} & \mathcal{M}_{14} \\ \mathcal{M}_{21} & \mathcal{M}_{22} & \mathcal{M}_{23} & \mathcal{M}_{24} \\ \mathcal{M}_{31} & \mathcal{M}_{32} & \mathcal{M}_{33} & \mathcal{M}_{34} \\ \mathcal{M}_{41} & \mathcal{M}_{42} & \mathcal{M}_{43} & \mathcal{M}_{44} \end{bmatrix} \begin{bmatrix} \Delta U_1 \\ \Delta U_2 \\ \Delta U_3 \\ \Delta U_4 \end{bmatrix} = \begin{bmatrix} F_1^* \\ F_2^* \\ F_3^* \\ F_4^* \end{bmatrix}. \tag{26}$$

The inverse of coefficient matrix A in Eq. (26) and adding a parameter, $1 < \omega < 2$, yields

$$\begin{bmatrix} \Delta U_1 \\ \Delta U_2 \\ \Delta U_3 \\ \Delta U_4 \end{bmatrix}^{(l)} = (1-\omega)\begin{bmatrix} \Delta U_1 \\ \Delta U_2 \\ \Delta U_3 \\ \Delta U_4 \end{bmatrix}^{(l-1)} + \omega \begin{bmatrix} \mathcal{M}_{11} & \mathcal{M}_{12} & \mathcal{M}_{13} & \mathcal{M}_{14} \\ \mathcal{M}_{21} & \mathcal{M}_{22} & \mathcal{M}_{23} & \mathcal{M}_{24} \\ \mathcal{M}_{31} & \mathcal{M}_{32} & \mathcal{M}_{33} & \mathcal{M}_{34} \\ \mathcal{M}_{41} & \mathcal{M}_{42} & \mathcal{M}_{43} & \mathcal{M}_{44} \end{bmatrix}^{-1} \begin{bmatrix} F_1^* \\ F_2^* \\ F_3^* \\ F_4^* \end{bmatrix}. \tag{27}$$

Since the s-Sweep function is developed based on an implicit finite difference scheme, the coefficient matrix A is a tridiagonal matrix. Hence, the Newton-EGSOR iterative method can be formulated in a general form as

$$\begin{bmatrix} \Delta U_p \\ \Delta U_{p+s} \\ \Delta U_{p+2s} \\ \Delta U_{p+3s} \end{bmatrix}^{(l)} = (1-\omega) \begin{bmatrix} \Delta U_p \\ \Delta U_{p+s} \\ \Delta U_{p+2s} \\ \Delta U_{p+3s} \end{bmatrix}^{(l-1)}$$
$$+\omega \begin{bmatrix} \mathcal{M}_{p,p} & \mathcal{M}_{p,p+s} & 0 & 0 \\ \mathcal{M}_{p+s,p} & \mathcal{M}_{p+s,p+s} & \mathcal{M}_{p+s,p+2s} & 0 \\ 0 & \mathcal{M}_{p+2s,p+s} & \mathcal{M}_{p+2s,p+2s} & \mathcal{M}_{p+2s,p+3s} \\ 0 & 0 & \mathcal{M}_{p+3s,p+2s} & \mathcal{M}_{p+3s,p+3s} \end{bmatrix}^{-1} \begin{bmatrix} F_p^* \\ F_{p+s}^* \\ F_{p+2s}^* \\ F_{p+3s}^* \end{bmatrix}, \tag{28}$$

where $p = s, 5s, \ldots, M - 5s, M - s$, $F_p^* = F_p - \mathcal{M}_{p,p-s} F_{p-s}$, $F_{p+s}^* = F_{p+s}$, $F_{p+2s}^* = F_{p+2s}$, and $F_{p+3s}^* = F_{p+3s} - \mathcal{M}_{p+3s,p+4s} F_{p+4s}$.

The computation algorithm of the Half-Sweep Newton-EGSOR iterative method is provided as follows:

Algorithm 1
Half-Sweep Newton-EGSOR

i. Set the initial iteration $\underline{U}^{(0)} = 1.0000$, the initial corrector $\Delta\underline{U}^{(0)} = 0$, the tolerance error $\varepsilon = 10^{-10}$ and the parameter $\omega = 1.00$,
ii. For $n = 1, 2, \ldots$, compute Eq. (28),
iii. Check the convergence of $\left|\Delta\underline{U}^{(\ell)} - \Delta\underline{U}^{(\ell-1)}\right| \leq \varepsilon$,
iv. If converges, compute $\underline{U}^{(l)} = \Delta\underline{U}^{(l)} + \underline{U}^{(l-1)}$,
v. Check the convergence of $\left|\underline{U}^{(\ell)} - \underline{U}^{(\ell-1)}\right| \leq \varepsilon$,
vi. If converges, go to $n + 1$,
vii. If $n > 100$, display output such as numerical solution, number of iterations, computation time and maximum absolute errors.

By referring to Algorithm 1, the optimal value of ω (± 0.01) is determined by running the developed C program a few times until the minimum number of iterations is obtained.

4 Results and Discussion

This section presents three different standard examples of one-dimensional PME to test the proposed Half-Sweep Newton-EGSOR (HSNEGSOR) method. The results are compared to the existing Newton-EGSOR (FSNEGSOR) method [8] and Newton-Gauss–Seidel (NGS) method [23].

Example 1 [24]

$$u_t = (uu_x)_x, \tag{29}$$

subjects to the initial condition, $u(x, 0) = x,\ 0 \leq x \leq 1$, and the boundary condition, $u(0, t) = t, u(1, t) = 1 + t,\ 0 \leq t \leq 1$. The exact solution is $u(x, t) = x + t$.

Example 2 [31]

$$u_t = \left(u^2 u_x\right)_x, \tag{30}$$

with the initial condition $(x + 1)/4, 0 \leq x \leq 1$, and the boundary $1/\left(2(4 - t)^{1/2}\right)$ and The exact solution for the numerical solution comparison purpose is $u(x, t) = (x + 1)/\left(2(4 - t)^{1/2}\right)$.

Example 3 [31]

$$u_t = 0.5\left(u^{-2} u_x\right)_x, \tag{31}$$

The initial condition is $(0.7x + 1.35)^{-1/2}$ for $0 \leq x \leq 1$, whereas the boundary condition is $(1.35 - 0.1225t)^{-1/2}$ and $(0.7 - 0.1225t + 1.35)^{-1/2}$, at $0 \leq t \leq 1$. The exact solution is $u(x, t) = (0.7x - 0.1225\mathrm{t} + 1.35)^{-1/2}$.

The efficiency of HSNEGSOR, FSNEGSOR and NGS methods is studied based on the maximum number of iterations (ℓ_{total}), the computation time (seconds) and the maximum error of absolute ($\varepsilon_{\max}$). The comparison between the three methods is made for different size of matrix, M, 256, 512, 1024, 2048, and 4096.

From the numerical results shown in Table 1 until Table 3, using NGS as a control method, both FSNEGSOR and HSNEGSOR have successfully reduced the number of

Table 1 Numerical result of solving Example 1

M	Method	ℓ_{total}	Seconds	$\varepsilon_{\max}$
256	NGS	48,395	19.31	5.33E−07
	FSNEGSOR	1010	1.28	6.90E−10
	HSNEGSOR	513	0.76	4.28E−11
512	NGS	169,693	133.84	2.10E−06
	FSNEGSOR	2027	2.02	7.69E−10
	HSNEGSOR	1010	0.81	6.90E−11
1024	NGS	587,031	919.22	7.62E−06
	FSNEGSOR	4072	9.70	1.54E−10
	HSNEGSOR	2027	3.92	7.69E−11
2048	NGS	1,993,096	6208.25	2.67E−05
	FSNEGSOR	8277	40.80	8.97E−09
	HSNEGSOR	4072	7.22	1.54E−10
4096	NGS	6,612,931	40,998.73	9.66E−05
	FSNEGSOR	15,761	156.87	1.36E−08
	HSNEGSOR	8277	45.08	8.97E−09

iterations and the computation time. The reason is due to the EGSOR iterative method, which uses the combination of small-sized groups of grid point strategy and a relaxation parameter to improve the efficiency in numerical computation. HSNEGSOR has shown its capability to reduce the number of iterations in computing the numerical solution by half than FSNEGSOR. This indicates that by applying the Half-Sweep scheme, the number of iterations required to obtain the final numerical solution is reduced by 50%. Consequently, the computation time becomes shorter. Besides that, the accuracy of the solution by HSNEGSOR is comparable to FSNEGSOR and NGS methods.

Based on the tabulated results, it can be seen that the HSNEGSOR method has superiority in terms of the number of iterations, the computation time and the maximum absolute errors over FSNEGSOR and NGS methods. The computational complexity by the HSNEGSOR method has been investigated based on the total arithmetic operations per iteration and compared against the two other methods.

Based on Table 4, the sizes of grid points computed by iteration by the three methods are different. After the boundary condition is imposed, the NGS method considers $M - 1$ grid points per iteration. NGS implemented two operations for Plus/Minus and three operations for Multiply/Divide. On the other hand, FSNEGSOR considers $M/4 - 1$ grid points and required 6 Plus/Minus operations and 10 Multiply/Divide operations. HSNEGSOR method is more efficient than both NGS and FSNEGSOR because HSNEGSOR computes only $M/8 - 1$ grid points per iteration. Hence, the HSNEGSOR method is more efficient in solving PME than

Table 2 Numerical result of solving Example 2

M	Method	ℓ_{total}	Seconds	$\varepsilon_{\max}$
256	NGS	17,308	10.88	8.39E−05
	FSNEGSOR	873	1.14	8.39E−05
	HSNEGSOR	437	0.69	8.39E−05
512	NGS	61,658	76.91	8.40E−05
	FSNEGSOR	1718	3.02	8.39E−05
	HSNEGSOR	873	1.72	8.39E−05
1024	NGS	218,147	557.97	8.43E−05
	FSNEGSOR	3344	11.45	8.39E−05
	HSNEGSOR	1718	3.80	8.39E−05
2048	NGS	763,998	3839.51	8.55E−05
	FSNEGSOR	6558	52.20	8.39E−05
	HSNEGSOR	3344	13.00	8.39E−05
4096	NGS	2,630,914	26,497.28	8.99E−05
	FSNEGSOR	12,780	166.32	8.39E−05
	HSNEGSOR	6558	47.88	8.39E−05

Table 3 Numerical result of solving Example 3

M	Method	ℓ_{total}	Seconds	$\varepsilon_{\max}$
256	NGS	24,325	16.70	2.71E−06
	FSNEGSOR	734	1.37	2.97E−06
	HSNEGSOR	380	0.69	2.96E−06
512	NGS	81,729	113.26	1.86E−06
	FSNEGSOR	1428	2.86	2.98E−06
	HSNEGSOR	734	2.07	2.97E−06
1024	NGS	265,698	767.23	3.33E−06
	FSNEGSOR	2881	10.35	2.97E−06
	HSNEGSOR	1428	3.72	2.95E−06
2048	NGS	882,282	5164.64	1.66E−05
	FSNEGSOR	6008	39.78	2.98E−06
	HSNEGSOR	2881	11.86	2.97E−06
4096	NGS	2,853,985	33,726.52	6.10E−05
	FSNEGSOR	11,891	153.55	2.98E−06
	HSNEGSOR	6008	44.47	2.98E−06

Table 4 Arithmetic operations per iteration by HSNEGSOR, FSNEGSOR and NGS

Method	Arithmetic operations per iteration	
	Plus/minus	Multiply/divide
NGS	$2(M-1)$	$3(M-1)$
FSNEGSOR	$6(\frac{M}{4}-1)+5$	$10(\frac{M}{4}-1)+8$
HSNEGSOR	$6(\frac{M}{8}-1)+5$	$10(\frac{M}{8}-1)+8$

FSNEGSOR and NGS methods because of a lower computational complexity, which attributes to the use of the Half-Sweep finite difference scheme.

5 Conclusion

This paper proposed a Half-Sweep Newton-EGSOR method to efficiently solve one-dimensional PME based on an unconditionally stable Half-Sweep finite difference scheme. It is also shown that the local truncation error of the scheme is first-order accurate in time and second-order accurate in space. The numerical results confirmed that the Half-Sweep Newton-EGSOR method is computationally efficient and lower complexity than the Full-Sweep Newton-EGSOR and Newton-Gauss–Seidel methods.

Acknowledgements The corresponding author acknowledges the financial support from Universiti Malaysia Sabah (UMS) under Acculturation Grant Scheme with the project code SGA0061-2019.

References

1. Abdullah, A.R.: The four point explicit decoupled group (EDG) Method: a fast poisson solver. Int. J. Comput. Math. **38**(1–2), 61–70 (1991). https://doi.org/10.1080/00207169108803958
2. Abedian, R., Adibi, H., Dehghan, M.: A high-order weighted essentially non-oscillatory (WENO) finite difference scheme for nonlinear degenerate parabolic equations. Comput. Phys. Commun. **184**(8), 1874–1888 (2013). https://doi.org/10.1016/j.cpc.2013.03.012
3. Bessemoulin-Chatard, M., Filbet, F.: A finite volume scheme for nonlinear degenerate parabolic equations. SIAM J. Sci. Comput. **34**(5), B559–B583 (2012). https://doi.org/10.1137/110853807
4. Bonorino, L., de Paula, R., Gonçalves, P., Neumann, A.: Hydrodynamics of porous medium model with slow reservoirs. J. Stat. Phys. **179**(3), 748–788 (2020). https://doi.org/10.1007/s10955-020-02550-y
5. Borana, R., Pradhan, V., Mehta, M.: Numerical solution of instability phenomenon arising in double phase flow through inclined homogeneous porous media. Perspect. Sci. **8**, 225–227 (2016). https://doi.org/10.1016/j.pisc.2016.04.033
6. Chew, J.V.L., Sulaiman, J.: Half-Sweep Newton-Gauss-seidel for implicit finite difference solution of 1D nonlinear porous medium equations. Glob. J. Pure Appl. Math. **12**(3), 2745–2752 (2016)
7. Chew, J.V.L., Sulaiman, J.: Application of MSOR iteration with Newton scheme for solutions of 1D nonlinear porous medium equations. In: AIP Conference Proceedings, 2016, p. 020017. https://doi.org/10.1063/1.4952497
8. Chew, J.V.L., Sulaiman, J.: Application of four-point Newton-EGSOR iteration for the numerical solution of 2D porous medium equations. J. Phys.: Conf. Ser. **890**, 012075 (2017). https://doi.org/10.1088/1742-6596/890/1/012075
9. Chew, J.V.L., Sulaiman, J.: Implicit finite difference solution of one-dimensional porous medium equations using Half-Sweep Newton-explicit group iterative method. J. Eng. Appl. Sci. **13**(5), 1286–1290 (2018). https://doi.org/10.3923/jeasci.2018.1286.1290
10. Cho, C.-H.: On the finite difference approximation for blow-up solutions of the porous medium equation with a source. Appl. Numer. Math. **65**, 1–26 (2013). https://doi.org/10.1016/j.apnum.2012.11.001
11. Díaz, J.L.: Modeling of an aircraft fire extinguishing process with a porous medium equation. SN Appl. Sci. **2**(12), 2108 (2020). https://doi.org/10.1007/s42452-020-03891-9
12. Duque, J.C.M., Almeida, R.M.P., Antontsev, S.N.: Numerical study of the porous medium equation with absorption, variable exponents of nonlinearity and free boundary. Appl. Math. Comput. **235**, 137–147 (2014). https://doi.org/10.1016/j.amc.2014.02.100
13. Duque, J.C.M., Almeida, R.M.P., Antontsev, S.N.: Application of the moving mesh method to the porous medium equation with variable exponent. Math. Comput. Simul. **118**, 177–185 (2015). https://doi.org/10.1016/j.matcom.2014.11.025
14. Evans, D.J.: Parallel S.O.R. iterative methods. Parallel Comput. **1**(1), 3–18 (1984). https://doi.org/10.1016/S0167-8191(84)90380-6.
15. Hadjidimos, A.: Successive overrelaxation (SOR) and related methods. J. Comput. Appl. Math. **123**(1–2), 177–199 (2000). https://doi.org/10.1016/S0377-0427(00)00403-9
16. Hageman, L.A., Young, D.M.: Iterative solution of large linear systems. The Am. Math. Monthly **80**(1), 92 (1973). https://doi.org/10.2307/2319285.
17. Kesarwani, J., Meher, R.: Modeling of an imbibition phenomenon in a heterogeneous cracked porous medium on small inclination. Spec. Top. Rev. Porous Media: An Int. J. **12**(1), 27–52 (2021). https://doi.org/10.1615/SpecialTopicsRevPorousMedia.2020030994

18. Liu, Y., Shu, C.-W., Zhang, M.: High order finite difference WENO schemes for nonlinear degenerate parabolic equations. SIAM J. Sci. Comput. **33**(2), 939–965 (2011). https://doi.org/10.1137/100791002
19. Mat Ali, N.A., Rahman, R., Sulaiman, J., Ghazali, K.: Solutions of reaction-diffusion equations using similarity reduction and HSSOR iteration. Indonesian J. Electr. Eng. Comput. Sci. **16**(3), 1430–1438 (2019). https://doi.org/10.11591/ijeecs.v16.i3.pp1430-1438
20. L. Monsaingeon, An explicit finite-difference scheme for one-dimensional Generalized Porous Medium Equations: Interface tracking and the hole filling problem, *ESAIM: Mathematical Modelling and Numerical Analysis*, vol. 50, no. 4, pp. 1011–1033, Jul. 2016, doi: https://doi.org/10.1051/m2an/2015063.
21. Ngo, C., Huang, W.: A study on moving mesh finite element solution of the porous medium equation. J. Comput. Phys. **331**, 357–380 (2017). https://doi.org/10.1016/j.jcp.2016.11.045
22. Ngo, C., Huang, W.: Adaptive finite element solution of the porous medium equation in pressure formulation. Numerical Methods for Partial Differential Equations **35**(3), 1224–1242 (2019). https://doi.org/10.1002/num.22347
23. Ortega, J.M., Rheinboldt, W.C.: Monotone Iterations for Nonlinear Equations with Application to Gauss-Seidel Methods. SIAM J. Numer. Anal. **4**(2), 171–190 (1967). https://doi.org/10.1137/0704017
24. Polyanin, A.D., Zaitsev, V.F.: Handbook of nonlinear partial differential equations. Choice Reviews Online (2004). https://doi.org/10.5860/CHOICE.41-5960
25. H. Saberi nik, M. S. Zahedi, R. Buzhabadi, and S. Effati, Homotopy perturbation method and He's polynomials for solving the porous media equation, *Computational Mathematics and Modeling*, vol. 24, no. 2, pp. 279–292, Apr. 2013, doi: https://doi.org/10.1007/s10598-013-9177-y.
26. Saudi, A., Sulaiman, J.: An Efficient Two-Stage Half-Sweep Modified Arithmetic Mean (HSMAM) Method for the Solution of 2D Elliptic Equation. Adv. Sci. Lett. **24**(3), 1917–1921 (2018). https://doi.org/10.1166/asl.2018.11188
27. Strikwerda, J.C.: Finite Difference Schemes and Partial Differential Equations, 2nd edn. Society for Industrial and Applied Mathematics, Philadelphia (2004)
28. Sunarto, A., Sulaiman, J.: Performance Numerical Method Half-Sweep Preconditioned Gauss-Seidel for Solving Fractional Diffusion Equation. Mathematical Modelling of Engineering Problems **7**(2), 201–204 (2020). https://doi.org/10.18280/mmep.070205
29. Suparmin, S., Saudi, A.: Path planning of indoor mobile robot using harmonic potentials via half-sweep modified SOR method (2017). https://doi.org/10.1109/I2CACIS.2017.8239031
30. Vazquez, J.L.: The Porous Medium Equation. Oxford University Press, Oxford (2006)
31. Wazwaz, A.-M.: The variational iteration method: A powerful scheme for handling linear and nonlinear diffusion equations. Comput. Math. Appl. **54**(7–8), 933–939 (2007). https://doi.org/10.1016/j.camwa.2006.12.039
32. H. Yoshioka and D. Triadis, A Regularized Finite Volume Numerical Method for the Extended Porous Medium Equation Relevant to Moisture Dynamics with Evaporation in Non-woven Fibrous Sheets, in *Communications in Computer and Information Science*, 2016, pp. 3–16.

Examination of Half-Sweep Closed Newton–Cotes Quadrature Schemes in Solving Dense System

Elayaraja Aruchunan, Mohana Sundaram Muthuvalu, Zailan Siri, Sachin Sharma Ashok Kumar, Jumat Sulaiman, Jackel Vui Lung Chew, and Majid Khan Majahar Ali

Abstract The main objective of this research is to apply and analyse the performance of the half-sweep iteration concept to the low order to high order Newton–Cotes and finite difference schemes via the Fredholm integro-differential equations to form a system of linear equations. Then generated linear systems will be computed by half-sweep Conjugate Gradient Normal Equation (HSCGNR) iterative method. The fundamental designs and formulations of full- and half-sweep Newton–Cotes and finite difference schemes in combined with the full- and half-sweep Conjugate Gradient Normal Equations methods are also presented. Analysis of the computational complexity and reduction in computational amount are also included to show

E. Aruchunan (✉) · Z. Siri
Institute of Mathematical Sciences, Faculty of Science, University of Malaya, 50603 Kuala Lumpur, Malaysia
e-mail: elayarajah@um.edu.my

Z. Siri
e-mail: zailansiri@um.edu.my

M. S. Muthuvalu
Fundamental and Applied Sciences Department, Universiti Teknologi PETRONAS, 32610 Seri Iskandar Perak, Malaysia

S. S. A. Kumar
Department of Physics, Faculty of Science, Centre for Ionic Universiti Malaya, 50603 Kuala Lumpur, Malaysia

J. Sulaiman
Mathematics with Economics Programme, Faculty of Science and Natural Resources, Universiti Malaysia Sabah, Jalan UMS, Kota Kinabalu 88400, Sabah, Malaysia
e-mail: jumat@ums.edu.my

J. V. L. Chew
Faculty of Computing and Informatics, Universiti Malaysia Sabah Labuan International Campus, 87000 Labuan F.T., Malaysia
e-mail: jackelchew93@ums.edu.my

M. K. M. Ali
School of Mathematical Sciences, Universiti Sains Malaysia, 11800 Penang, Malaysia
e-mail: majidkhanmahajarali@usm.my

© Institute of Technology PETRONAS Sdn Bhd 2022
S. A. Abdul Karim and A. Shafie (eds.), *Towards Intelligent Systems Modeling and Simulation*, Studies in Systems, Decision and Control 383,
https://doi.org/10.1007/978-3-030-79606-8_26

that the combination of the HSCGNR iterative method with high order discretisation schemes is superior compared with other low order schemes with full-sweep or standard Conjugate Gradient Normal Equation method via some examples.

Keywords Fredholm · Integro-differential equations · Conjugate Gradients Normal Equations · Half-sweep iteration · Newton–Cotes Schemes

The main research objective aimed to apply and analyze the half-sweep iteration performance concept from low to high-order Newton–Cotes and finite difference schemes using the Fredholm integro-differential equations (IDEs) for a system of linear equations. The generated linear systems would then be computed by a half-sweep Conjugate Gradient Normal Equation (HSCGNR) iterative method. The fundamental designs and formulations of full and half-sweep Newton–Cotes and finite difference schemes combined with the full and half-sweep Conjugate Gradient Normal Equation methods were also illustrated. Furthermore, analyses on the computational complexity and reduction in computational amount were presented to demonstrate that the integration of the HSCGNR iterative method with high-order discretization schemes was superior to other low-order schemes with full-sweep or standard Conjugate Gradient Normal Equation methods based on specific examples.

1 Introduction

The IDEs have become significantly crucial in representing most of the engineering and industrial challenges in the areas of mechanics, physics, chemistry, astronomy, biology, potential theory, electrostatics, nano-hydrodynamics, and glass-forming processes [1–4]. Given that IDE solutions failed to attain initial or boundary value problems using a closed-form, the problems must be resolved with various approximate and numerical methods. In this study, two value problem types initial and boundary value problems of linear Fredholm integro-differential equations (LFIDEs)] from the first and second orders are formulated as follows:

$$\lambda y'(x) = p(x)y(x) + g(x) + \xi \int_a^b K(x,t)y(t)dt, \quad x,t \in (a,b] \tag{1}$$

with

$$y(a) = A_1,$$

and

$$\lambda y''(x) = q(x)y'(x) + p(x)y(x) + g(x) + \xi \int_a^b K(x,t)y(t)dt, \; x, t \in (a,b) \tag{2}$$

with

$$y(a) = A_0 \text{ and } y(b) = B_0$$

Notably, $K(x,t)$, $g(x)$, $q(x)$, and $p(x)$ indicated known functions, ξ denoted a real parameter, and $y(x)$ implied the unknown function to be identified. Various numerical approaches were proposed (to discretize the LFIDEs) in generating linear algebraic equation systems to obtain the favored approximate solutions over several decades. Some of the conventional methods include the Taylor collocation [5], quadrature-difference [1, 6, 7], Lagrange interpolation [8, 9], Taylor polynomial [10, 11], Wavelet-Galerkin [12], rationalized Haar functions [13], Tau [14], Gauss–Seidel [15, 16], and Generalized Minimal Residual [17]. Regardless, the methods were fundamentally based on standard or full-sweep (FS) iterations, wherein the computational complexities in each nodal point per iteration were extremely high. Resultantly, the computational time increased dramatically with the parallel increase in mesh size. In minimizing the aforementioned complexities, advanced mathematical approaches were required to simplify the computation, specifically in a large and dense system.

The fundamental concept of the half-sweep (HS) iterative method was introduced by Abdullah [18] through the Explicit Decoupled Group (EDG) approach to solve the two-dimensional Poisson equation. In this vein, the primary objective of the HS iteration concept served to minimize computational complexities during the iterative process. By implementing the aforementioned concept, only the unknown value involving half of the nodes in the solution domain required calculation. Various investigations were subsequently performed to validate the HS iteration scheme with one-stage iterative methods [15, 19–23].

This study examined two applications concerning the HS iteration technique combined with high-order Newton–Cotes [20] and finite difference discretization schemes to solve problems (1) and (2). First, the discretized equation systems for both problems were derived. The approximation equations were then resolved with the HSCGNR iterative method. Specifically, the remaining study components are organized as follows: Sect. 2 provides brief explanations and formulations of the full and HS iteration concepts with high-order Newton–Cotes and finite difference schemes to discretize LFIDEs; Sect. 3 presents formulations for the standard or FS conjugate gradient normal equation (FSCGNR) and HSCGNR methods to solve problems (1) and (2) with algorithms; Sect. 4 revealed some numerical results and validated the effectiveness of high-order Newton–Cotes and finite difference schemes; Sect. 5 presents the computational complexity and percentage reduction analyses; Sect. 6 outlined the summary of the study conclusion.

2 Half-Sweep Iteration Concept

In numerically resolving problems (1) and (2), Fig. 1 indicates the finite grid networks that established the full and HS Newton–Cotes and finite difference schemes approximation equations.

Following Fig. 1, the full and HS iterative methods computed approximate values at nodal points of type ● only until the convergence criterion was attained. Other approximate solutions at the remaining points of type ○ could then be directly computed as mentioned in [15, 19].

The formulation of direct method calculation is as follows:

$$y_i = \frac{y_{i-1} + y_{i+1}}{2}, \quad i = 1, 3, 5, \cdots, n-1. \tag{3}$$

2.1 *Derivation of the Half-Sweep Newton–Cotes Schemes*

Based on Sect. 1, the family of Newton–Cotes schemes was analyzed as a means of solving problems (1) and (2). In discretizing LFIDEs, composite Simpson's $\frac{1}{3}$ (CS1), composite Simpson's $\frac{3}{8}$ (CS2), and composite Boole (CB) were duly integrated with high-order finite difference schemes. Thus, the Newton–Cotes schemes could generally be expressed as follows:

$$\int_a^b y(t)dt = \sum_{j=0}^{n} \chi_j y(x_j) + \varepsilon_n(y) \tag{4}$$

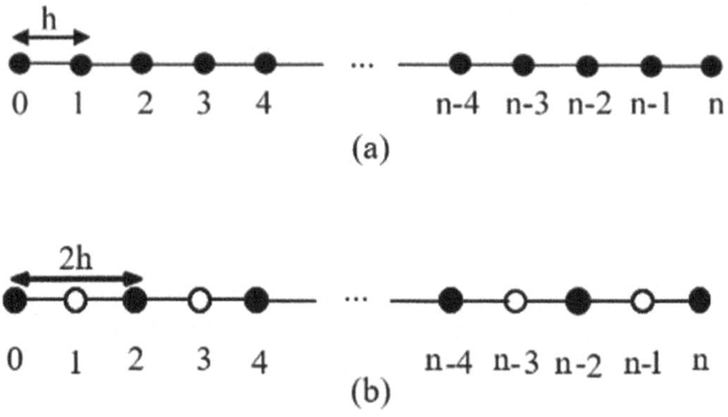

Fig. 1 **a** and **b** indicates uniform nodal point distributions for FS and HS cases

Notably, $\chi_j (j = 0, 1, \ldots, n)$ were the abscissas of the partition points concerning the integration interval $[a, b]$, $\chi_j (j = 0, 1, \ldots, n)$ were numerical coefficients that were independent of the $y(x)$ function, and $\varepsilon_n(y)$ denoted the truncation error of (4). In expressing the full and HS approximation equations for problems (1) and (2), further discussions were confined to CS1, CS2, and CB schemes following interpolation formulas with equally-spaced data. Based on the CS1, CS2, and CB schemes, the numerical coefficients A_j in Eq. (4) were fulfilled based on the following relationships:

1. HSCS1 $\frac{1}{3}$

$$\chi_j = \begin{cases} \frac{1}{3}\Delta x, & j = 0, n \\ \frac{4}{3}\Delta x, & j = p, 3p, 5p, \cdots, n - p \\ \frac{2}{3}\Delta x, & j = 2p, 4p, 6p, \cdots, n - 2p \end{cases}$$

2. HSCS2 $\frac{3}{8}$

$$\chi_j = \begin{cases} \frac{3}{8}\Delta x, & j = 0, n \\ \frac{6}{8}\Delta x, & j = 3p, 6p, 9p \cdots, n - 3p \\ \frac{9}{8}\Delta x, & j = p, 2p, 4p, 5p \cdots, n - p, n - 2p \end{cases}$$

3. HSCB

$$\chi_j = \begin{cases} \frac{14}{45}\Delta x, & j = 0, n \\ \frac{24}{45}\Delta x, & j = 2p, 6p, 10p \cdots, n - 2p \\ \frac{28}{45}\Delta x, & j = 4p, 8p, 12p \cdots, n - 4p \\ \frac{64}{45}\Delta x & otherwise. \end{cases}$$

2.2 Derivation of the HS Finite Difference

High-order finite difference methods were utilized to discretize the differential components in problems (1) and (2). The backward and central difference schemes concerning second-order accuracy were also regarded. Specifically, the standard finite difference schemes were modified by applying the HS iteration concept to implement the discretization process. The HS finite difference schemes in discretization problem (1) are formulated as follows:

$$y'(x_i) \cong \frac{y(x_{i+p}) - y(x_{i-p})}{2\Delta x} + O(h^2), \tag{5}$$

for $i = p, 2p, 3p, \cdots, n - p$ and at point $i = n$,

$$y'(x_n) \cong \frac{3y(x_n) - 4y(x_{n-p}) + y(x_{n-2p})}{2\Delta x} + O(h^2). \tag{6}$$

In solving problem (2), the second derivative of HS central difference schemes is formulated as follows:

$$y''(x_i) = \frac{y(x_{i+p}) - 2y(x_i) + y(x_{i-p})}{(\Delta x)^2} + O(h^2). \tag{7}$$

where the value of p, represents 1 and 2 for full and HS cases, respectively. Based on the HS Newton–Cotes and finite-difference formulation in subsections *A* and *B*, the constant step size, h is illustrated as follows:

$$h = \Delta x = \frac{b - a}{n} \tag{8}$$

Notably, n denoted the number of subintervals in the $[a, b]$ interval. The combination of discretization schemes, namely, Finite Difference and Composite Simpson's $\frac{1}{3}$ (FD-CS1) schemes, Finite Difference and Composite Simpson's $\frac{3}{8}$ (FD-CS2) schemes, and Finite Difference-Composite Boole (FD-CB) schemes were outlined for full and HS cases.

2.3 *First-Order Linear Fredholm IDE Approximations*

In solving the first-order LFIDEs in problem (1), a combination of discretization schemes (4), (5), and (6) were applied to the differential and integral parts in formulating approximation equations. Generally, the formulation of full and HS approximation equations following the incorporation of Newton–Cotes and finite difference schemes for first-order LFIDE was represented as follows:

$$\left.\begin{array}{l} \dfrac{y_{i+p}-y_{i-p}}{2\Delta x}=g_i+P_iy_i+\xi\displaystyle\sum_{j=p,2p}^{n-1}\chi_jy(x_j) \quad \text{for} \quad i=p,\,2p,\,\ldots n-p \\ \dfrac{3y_i-4y_{i-p}-y_{i-2p}}{2\Delta x}=g_i+P_iy_i+\xi\displaystyle\sum_{i=n}^{n}\chi_jy(x_j), \quad \text{at note point } i=n\,. \end{array}\right\} \tag{9}$$

The linear system generated by the full or HS approximation equation could be simplified as follows:

$$E\,\underset{\sim}{y}=\underset{\sim}{f}, \tag{10}$$

wherein

$$E=\begin{bmatrix} a_{p,p} & b_{p,2p} & d_{p,3p} & d_{p,4p} & \cdots & d_{p,n} \\ c_{2p,p} & a_{2p,2p} & b_{2p,3p} & d_{2p,4p} & \cdots & d_{2p,n} \\ d_{3p,p} & c_{3p,2p} & a_{3p,3p} & b_{3p,4p} & \cdots & d_{3p,n} \\ \tau_{4p,p} & d_{4p,2p} & c_{4p,3p} & a_{4p,4p} & \cdots & d_{4p,n} \\ \vdots & \vdots & \vdots & \vdots & \ddots & \vdots \\ d_{n,p} & d_{n,2p} & d_{n,3p} & d_{n,4p} & \cdots & a_{n,n} \end{bmatrix}$$

$$\underset{\sim}{f}=\begin{bmatrix} 2hg_p+\left(2h\chi_pK_{p,0}+1\right)y_0 \\ 2hg_{2p}+\left(2h\chi_pK_{2p,0}\right)y_0 \\ 2hg_{3p}+\left(2h\chi_pK_{3p,0}\right)y_0 \\ \vdots \\ 2hg_{n-2p}+\left(2h\chi_pK_{n-2p,0}\right)y_0 \\ 2hg_{n-p}+\left(2h\chi_pK_{n-p,0}\right)y_0 \\ 2hg_N+\left(2h\chi_pK_{n,0}\right)y_0 \end{bmatrix} \text{ and } \underset{\sim}{y_n}=\begin{bmatrix} y_n(x_p) \\ y_n(x_{2p}) \\ y_n(x_{3p}) \\ \vdots \\ y_n(x_{n-2p}) \\ y_n(x_{n-p}) \\ y_n(x_n) \end{bmatrix}$$

in which $a_{i,i}=-2hP_i-2h\chi_iK_{i,i}$, $b_{i,j}=1-2h\chi_jK_{i,j}$, $c_{i,j}=-1-2h\chi_jK_{i,j}$, $d_{i,j}=-2h\chi_jK_{i,j}$,$e_{i,j}=-4-2h\chi_jK_{i,j}$ and $\hbar_{i,i}=-3-2hP_i-2h\chi_iK_{i,i}$.

Notably, E denoted a dense and non-symmetrical coefficient matrix, $\underset{\sim}{f}$ was the known vector, and $\underset{\sim}{y_n}$ implied the unknown values to be identified.

2.4 Second-Order Linear Fredholm IDE Approximations

Following the integration of modified Newton–Cotes and finite difference schemes, the second-order LFIDE in problem (2) was discretized as follows:

$$\frac{y_{i+p} - 2y_i + y_{i-p}}{(ph)^2} + \frac{y_{i+p} - y_{i-p}}{2ph} = g_i + P_i y_i + \xi \sum_{j=0,p,2p}^{n} \chi_j y(x_j) \tag{11}$$

For $i = p, 2p, \ldots n - p$.

Specifically, the generated linear system for both full and HS approximations could be expressed by

$$G \underset{\sim}{y_n} = \underset{\sim}{\ell}, \tag{12}$$

wherein

$$G = \begin{bmatrix} \sigma_{p,p} & \varsigma_{p,2p} & \tau_{p,3p} & \tau_{p,4p} & \cdots & \tau_{p,n-p} \\ \varsigma_{2p,p} & \sigma_{2p,2p} & \varsigma_{2p,3p} & \tau_{p,4p} & \cdots & \tau_{2p,n-p} \\ \tau_{3p,p} & \varsigma_{3p,2p} & \sigma_{3p,3p} & \varsigma_{3p,4p} & \cdots & \tau_{3p,n-p} \\ \tau_{4p,p} & \tau_{4p,2p} & \varsigma_{4p,3p} & \sigma_{4p,4p} & \cdots & \tau_{4p,n-p} \\ \vdots & \vdots & \vdots & \vdots & \ddots & \vdots \\ \tau_{n-p,p} & \tau_{n-p,2p} & \tau_{n-p,3p} & \tau_{n-p,4p} & \cdots & \sigma_{n-p,n-p} \end{bmatrix}$$

$$\underset{\sim}{\ell} = \begin{bmatrix} h^2 g_p + \left(1 + h^2 \chi_p K_{p,0}\right) y_0 + \left(h^2 \chi_n K_{p,n}\right) y_n \\ h^2 g_{2p} + \left(h^2 \chi_p K_{2p,0}\right) y_0 + \left(h^2 \chi_n K_{2p,n}\right) y_n \\ h^2 g_{3p} + \left(h^2 \chi_p K_{3p,0}\right) y_0 + \left(h^2 \chi_n K_{3p,n}\right) y_n \\ \vdots \\ h^2 g_{n-3p} + \left(h^2 \chi_p K_{n-3,0}\right) y_0 + \left(h^2 \chi_n K_{n-3p,n}\right) y_n \\ h^2 g_{n-2p} + \left(h^2 \chi_p K_{n-2,0}\right) y_0 + \left(h^2 \chi_n K_{n-2p,0}\right) y_n \\ h^2 g_{n-p} + \left(h^2 \chi_p K_{n-p,0}\right) y_0 + \left(-1 + h^2 \chi_n K_{n-p,0}\right) y_n \end{bmatrix}$$

and

$$\underset{\sim}{y_n} = \begin{bmatrix} y_n(x_p) \\ y_n(x_{2p}) \\ y_n(x_{3p}) \\ \vdots \\ y_n(x_{n-2p}) \\ y_n(x_{n-p}) \\ y_n(x_n) \end{bmatrix}.$$

in which $\sigma_{i,i} = -2 - h^2 P_i - h^2 \chi_i K_{i,i}$, $\varsigma_{i,j} = 1 - h^2 \chi_j K_{i,j}$ and $\tau_{i,j} = -h^2 \chi_j K_{i,j}$.

Notably, G implied a dense and non-symmetrical coefficient matrix, $\underset{\sim}{\ell}$ was the known vector, and $\underset{\sim}{y_n}$ denoted the unknown values to be identified.

3 Formulation of HSCGNR

As the coefficient matrices for approximation Eqs. (9) and (11) were non-symmetrical and dense, the Conjugate Gradient (CG) iterative method failed to function due to convergence conditions. Thus, the CGNR algorithm was best known for solving non-symmetrical, positive, and definite linear systems [24]. In identifying a solution for the generated linear systems, Eqs. (2.0) and (2.2) were multiplied by the transposition of coefficient matrices to form symmetrical, positive, and definite linear systems.

Accordingly, the linear system (10) was multiplied by E^T on both sides to demonstrate CGNR as follows,

$$E^T E \underset{\sim}{y} = E^T \underset{\sim}{f}, \tag{13}$$

wherein the new coefficient matrix $E^T E$ denoted a symmetrical and positive definite. Consequently, the linear system (13) was developed to apply a standard or FS conjugate gradient on normal equations (FSCGNR) as represented by

$$E^* \underset{\sim}{y} = \underset{\sim}{f^*}, \tag{14}$$

wherein $E^* = E^T E$ and $E^* = E^T \underset{\sim}{f}$.

The same approach was performed for Eq. (11) to solve problem (2). Notably, the standard FSCGNR was incorporated with the HS iteration concept in forming a HSCGNR iterative method to solve the generated linear system by (14). The implementation of FSCGNR and HSCGNR iterative methods were depicted in Algorithm 1 for problem (1).

Algorithm 1: FSCGNR and HSCGNR methods

1 Initialize parameters $r, d, \alpha, \beta, h, E^*, K, f$

set $k = 0$

2 for $i = p, 2p, \ldots, n-2p, n-p, n$

Compute

i. $r_i^{(k)} = f_i - E_{i,j}^* y_i$ and $d_i^{(k)} = r_i^{(k)}$

ii. $$\alpha^{(k)} = \frac{\left(r_i^{(k)}\right)^T r_i^{(k)}}{\left(d_i^{(k)}\right)^T \sum_{j=p,2p,3p}^{n} E_{i,j}^* d_i^{(k)}}$$

iii. $y_i^{(k+1)} = y_i^{(k)} + \alpha^{(k)} d_i^k$

iv. $$r_i^{(k+1)} = r_i^{(k)} - \alpha^{(k)} \sum_{j=p,2p,3p}^{n} E_{i,j}^* d_i^{(k)}$$

v. $$\beta^{(k+1)} = \frac{\left(r_i^{k+1}\right)^T r_i^{(k+1)}}{\left(r_i^k\right)^T r_i^{(k)}}$$

vi. $d_i^{(k+1)} = r_i^{(k+1)} + \beta^{(k+1)} d_i^{(k)}$

Monitor the convergence. If the converge criterion is satisfied, namely, $\left\| y_i^{(k+1)} - y_i^{(k)} \right\| \leq \varepsilon$, go to Step 3.

Otherwise, repeat steps iv to vi until the solution converges.

3 End

4 Numerical Treatments

The performance of the Newton–Cotes and finite difference combination schemes using the HS iteration concept was measured with the HSCGNR iterative method. For the comparison analysis, the standard or FS Composite Trapezoidal (FS-CT) with the FS Gauss Seidel iterative method (FSGS) was employed as a control method. The tolerance error $\varepsilon = 10^{-10}$ was also considered for the convergence criterion. In the numerical treatments, two well-posed problems from the first and second-order LFIDEs were duly examined.

Problem 1 [25]: First-order LFIDE:

$$y'(x) = 1 - \frac{1}{3}x + \int_0^x xy(s)\,ds, \quad 0 < x \le 1 \tag{15}$$

subject to an initial value problem with condition $y(0) = 0$. The precise solution for the aforementioned problem is $y(x) = x$.

Problem 2 [26]: Second-order LFIDE.

$$y''(x) = x - 2 + 60\int_0^x (x - s)y(s)\,ds, \quad 0 < x < 1 \tag{16}$$

subject to two boundary value problems with conditions $y(0) = 0$ and $y(1) = 1$. The accurate solution is $y(x) = x$.

Three performance parameters (number of iterations, computational time, and maximum absolute errors) were regarded in assessing the subjected numerical schemes. All the simulations were conducted in the same computer with processor Intel(R) Core(TM) i5-4460 CPU @ 3.20 GHz. Additionally, the algorithm codes were written in the C programming language. The mesh sizes n considered in this study were based on the common complete loop which was fulfilled with each discretization scheme. The outcomes of the numerical simulations for problems (1) and (2) are presented in Tables 1 and 2, respectively.

Specifically, Tables 1 and 2 illustrate the performance of HSCGNR over FSCGNR and FSGS in three different parameters. Each iterative method was computed with FD-CT, FD-CS1, FD-CS2, and FD-CB to solve problems (1) and (2). The results revealed that the number of iterations and computational time were significantly reduced by HSCGNR compared to the FSCGNR and FSGS iterative methods. Meanwhile, the accuracy of HSCGNR with FD-CB was stronger than the other three tested discretization schemes.

5 Percentage Reduction and Computational Complexity Analysis

The method performance was measured based on the percentage of reduction concerning the number of iterations and execution time for full and HS iteration cases. The results of the FSGS and FSCGNR iterative methods were subsequently utilized as a benchmark to illustrate the efficiency of the HSCGNR iterative method. The percentage reduction involving the number of iterations and execution time are summarized in Table 3.

Computational complexity was an essential element in numerical computation to determine method efficiency [27]. As such, the computation time required for the FSGS, FSCGNR, and HSCGNR iterative methods were calculated to identify the computational cost of arithmetic operations performed per iteration. Arithmetic operations, such as addition or subtractions (ADD/SUM) and multiplications or divisions (MUL/DIV) for first and second-order LFIDEs are summarized in Tables 4 and 5, respectively.

6 Summary

In the study context, the HS iteration concept was applied to high-order Newton–Cotes and finite difference schemes, namely, FD-SC1, FD-CS2, and FD-CB using the CGNR iterative method to solve first and second-order LFIDEs. Following the results in Tables 1 and 2, the implementation of high-order discretization schemes improved the solution accuracy for first and second-order LFIDEs. Moreover, the implementation of HS iteration significantly reduced the number of iterations and execution time compared to the standard form. Resultantly, HS high-order Newton–Cotes and finite difference with HSCGNR proved to be the best work for all parameters excluding the accuracy in solving LFIDEs. Regardless, the precision could be enhanced by implementing second-order Lagrange interpolation techniques for the remaining calculation points.

Acknowledgements The authors would like to thank Professor Praveen Agarwal for his constructive comments.

Appendix

Table 1 Comparison in terms of the number of iterations, execution time and maximum absolute error for problem 1

Mesh size	Methods	Schemes	Number of iterations	Execution time	Maximum absolute error
24	FSGS	FD-CT	502	0.22	4.65E−4
		FD-CS1	497	0.29	2.41E−6
		FD-CS2	497	0.31	5.43E−6
		FD-CB	497	0.34	2.37E−9
	FSCGNR	FD-CT	31	0.01	4.65E−4
		FD-CS1	32	0.01	2.41E−6
		FD-CS2	31	0.01	5.43E−6
		FD-CB	31	0.01	3.68E−11
	HSCGNR	FD-CT	11	0.01	2.88E−3

(continued)

Table 1 (continued)

Mesh size	Methods	Schemes	Number of iterations	Execution time	Maximum absolute error
		FD-CS1	11	0.01	3.24E−5
		FD-CS2	11	0.01	3.23E−5
		FD-CB	11	0.01	2.17E−10
48	FSGS	FD-CT	2101	0.49	1.16E−4
		FD-CS1	2097	0.50	1.38E−8
		FD-CS2	2097	0.57	3.27E−8
		FD-CB	2097	0.59	1.24E−8
	FSCGNR	FD-CT	90	0.02	1.16E−4
		FD-CS1	90	0.02	1.50E−7
		FD-CS2	90	0.02	3.40E−7
		FD-CB	90	0.02	5.47E−10
	HSCGNR	FD-CT	31	0.01	7.91E−4
		FD-CS1	32	0.01	8.40E−6
		FD-CS2	31	0.01	8.40E−6
		FD-CB	31	0.01	2.17E−9
72	FSGS	FD-CT	4628	1.17	5.17E−5
		FD-CS1	4625	1.18	8.89E−8
		FD-CS2	4625	1.20	3.83E−8
		FD-CB	4625	1.38	2.92E−8
	FSCGNR	FD-CT	176	0.15	5.17E−5
		FD-CS1	171	0.16	3.07E−8
		FD-CS2	184	0.19	6.77E−8
		FD-CB	185	0.20	5.23E−10
	HSCGNR	FD-CT	59	0.01	3.62E−4
		FD-CS1	59	0.01	3.77E−7
		FD-CS2	60	0.01	3.77E−7
		FD-CB	59	0.01	8.06E−8
96	FSGS	FD-CT	8034	2.30	5.17E−5
		FD-CS1	8032	2.34	8.89E−8
		FD-CS2	8032	2.69	3.83E−8
		FD-CB	8032	2.78	5.28E−8
	FSCGNR	FD-CT	291	0.25	2.91E−5
		FD-CS1	292	0.26	1.20E−8
		FD-CS2	291	0.27	2.58E−8
		FD-CB	292	0.28	4.23E−9
	HSCGNR	FD-CT	90	0.02	2.07E−4

(continued)

Table 1 (continued)

Mesh size	Methods	Schemes	Number of iterations	Execution time	Maximum absolute error
		FD-CS1	90	0.02	2.13E−7
		FD-CS2	90	0.02	2.13E−7
		FD-CB	90	0.03	3.39E−7
120	FSGS	FD-CT	12,278	4.37	1.85E−5
		FD-CS1	12,276	4.10	7.96E−8
		FD-CS2	12,276	4.77	7.48E−8
		FD-CB	12,276	4.98	8.35E−8
	FSCGNR	FD-CT	436	0.39	1.86E−5
		FD-CS1	448	0.45	6.21E−9
		FD-CS2	435	0.42	1.65E−8
		FD-CB	436	0.47	9.51E−9
	HSCGNR	FD-CT	137	0.06	1.33E−4
		FD-CS1	132	0.17	1.37E−7
		FD-CS2	132	0.22	1.37E−7
		FD-CB	132	0.34	1.02E−7

Table 2 Comparison in terms of the number of iterations, execution time and maximum absolute error for problem 2

Mesh Size	Methods	Schemes	Number of Iterations	Execution time	Maximum absolute error
24	FSGS	FD-CT	7814	9.38	1.65E−4
		FD-CS1	7964	10.25	4.76E−8
		FD-CS2	7911	10.98	1.62E−8
		FD-CB	8006	11.74	4.78E−8
	FSCGNR	FD-CT	26	0.01	1.65E−4
		FD-CS1	26	0.01	6.59E−14
		FD-CS2	26	0.01	3.10E−14
		FD-CB	26	0.01	2.57E−14
	HSCGNR	FD-CT	13	0.01	6.62E−4
		FD-CS1	13	0.01	2.44E−15
		FD-CS2	13	0.01	1.45E−14
		FD-CB	13	0.01	1.42E−14
48	FSGS	FD-CT	23,006	130.27	4.12E−5
		FD-CS1	23,428	131.81	1.51E−8
		FD-CS2	23,504	134.87	1.48E−8
		FD-CB	23,543	135.60	1.52E−7

(continued)

Table 2 (continued)

Mesh Size	Methods	Schemes	Number of Iterations	Execution time	Maximum absolute error
	FSCGNR	FD-CT	52	0.01	4.13E−5
		FD-CS1	52	0.01	8.99E−14
		FD-CS2	52	0.01	1.35E−13
		FD-CB	52	0.02	1.17E−13
	HSCGNR	FD-CT	26	0.01	1.65E−4
		FD-CS1	26	0.01	6.59E−14
		FD-CS2	26	0.01	3.10E−14
		FD-CB	26	0.01	2.57E−14
72	FSGS	FD-CT	45,002	801.24	1.80E−5
		FD-CS1	45,756	804.56	3.12E−8
		FD-CS2	45,901	807.84	3.06E−8
		FD-CB	45,964	812.73	3.13E−7
	FSCGNR	FD-CT	79	0.80	1.83E−5
		FD-CS1	78	0.95	1.79E−13
		FD-CS2	78	0.96	1.74E−13
		FD-CB	78	0.97	1.27E−13
	HSCGNR	FD-CT	39	0.01	7.34E−5
		FD-CS1	39	0.01	2.85E−14
		FD-CS2	39	0.01	4.37E−14
		FD-CB	39	0.01	1.60E−14
96	FSGS	FD-CT	73,430	2927.12	9.82E−6
		FD-CS1	74,614	2934.98	5.29E−8
		FD-CS2	74,728	2940.54	5.19E−8
		FD-CB	74,934	2947.05	5.31E−7
	FSCGNR	FD-CT	105	1.41	1.03E−5
		FD-CS1	105	1.45	1.03E−13
		FD-CS2	104	1.46	9.62E−13
		FD-CB	105	1.47	3.31E−13
	HSCGNR	FD-CT	52	0.02	4.13E−5
		FD-CS1	52	0.03	8.99E−14
		FD-CS2	52	0.03	1.35E−13
		FD-CB	52	0.03	1.17E−13
120	FSGS	FD-CT	107,988	7514.36	3.50E−6
		FD-CS1	109,685	7636.10	1.23E−8

(continued)

Table 2 (continued)

Mesh Size	Methods	Schemes	Number of Iterations	Execution time	Maximum absolute error
		FD-CS2	109,987	7654.89	7.88E−8
		FD-CB	110,144	7758.67	8.06E−7
	FSCGNR	FD-CT	131	1.85	6.61E−6
		FD-CS1	131	1.93	1.55E−13
		FD-CS2	131	1.92	2.64E−14
		FD-CB	131	1.94	7.68E−13
	HSCGNR	FD-CT	65	0.73	2.64E−5
		FD-CS1	65	0.77	8.68E−14
		FD-CS2	65	0.78	4.97E−14
		FD-CB	65	0.81	5.30E−14

Table 3 Percentage reductions of number of iterations and execution time of HSCGNR method relative to the FSGS and FSCGNR methods in solving problems 1 and 2 respectively

Problems	Relative	Methods	Number of iterations (%)	Execution time (%)
1	FSGS	HSCGNR	99.90–99.99	99.99–99.99
	FSCGNR	HSCGNR	49.23–50.38	71.42–87.72
2	FSGS	HSCGNR	98.58–99.02	96.25–98.90
	FSCGNR	HSCGNR	65.15–74.55	76.19–92.52

Table 4 Number of arithmetic operations required per iterations for the FSGS, FSCGNR and HSCGNR methods for first order LFIDEs

	Arithmetic Operations Per Node	
	ADD/SUB	*MUL/DIV*
FSGS	$(n-1)^2$	n^2-1
FSCGNR	$2n^2+2n-9$	$2n^2+3n+-3$
HSCGNR	$2\left(\frac{n}{2}\right)^2+2\left(\frac{n}{2}\right)-9$	$2\left(\frac{n}{2}\right)^2+3\left(\frac{n}{2}\right)-3$

Table 5 Number of arithmetic operations required per iterations for the FSGS, FSCGNR and HSCGNR methods for second order LFIDEs

	Arithmetic Operations Per Node	
	ADD/SUB	*MUL/DIV*
FSGS	$n(n-1)$	$n(n+1)$
FSCGNR	$2n^2+6n-4$	$2n^2+7n+2$
HSCGNR	$2\left(\frac{n}{2}\right)^2+6\left(\frac{n}{2}\right)-4$	$2\left(\frac{n}{2}\right)^2+7\left(\frac{n}{2}\right)+2$

References

1. Aruchunan, E., Muthuvalu, M.S., Sulaiman, J.: Quarter-sweep iteration concept on conjugate gradient normal residual method via second order quadrature-finite difference schemes for solving Fredholm Integro-differential equations. Sains Malaysiana **44**(1), 139–146 (2015)
2. Kurt, N., Sezer, M.: Polynomial solution of high-order linear Fredholm integro-differential equations with constant coefficients. J. Franklin Inst. **345**, 839–850 (2008)
3. Maleknejad, K., Mahmoudi, Y.: Numerical solution of linear Fredholm integral equation by using hybrid Taylor and block-pulse functions. Appl. Math. Comput. **149**, 799–806 (2004)
4. Yousif, W.S., Evans, D.J.: Explicit de-coupled group iterative methods and their implementations. Parallel Algorithms Appl. **7**, 53–71 (1995)
5. Karamete, A., Sezer, M.: A Taylor collocation method for the solution of linear Integro-differential equations. Int. J. Comput. Math. **79**(9), 987–1000 (2002)
6. Fedotov, A.I.: Quadrature-difference methods for solving linear and nonlinear singular integro-differential equations. Nonlinear Anal. **71**, 303–308 (2009)
7. Aruchunan, E., Sulaiman, J.: Half-sweep quadrature-difference schemes with iterative method in solving linear Fredholm Integro-differential equations. Prog. Appl. Math. **5**(1), 11–21 (2013)
8. Rashed, M.T.: Lagrange interpolation to compute the numerical solutions differential and integro-differential equations. Appl. Math. Comput. **151**, 869–878 (2003)
9. Shahsavaran, A.: On the convergence of Lagrange interpolation to solve special type of second kind Fredholm Integro differential equations. Appl. Math. Sci. **6**, 343–348 (2012)
10. Wang, W., Lin, C.: A new algorithm for integral of trigonometric functions with mechanization. Appl. Math. Comput. **164**, 71–82 (2005)
11. Yalcinbas, S., Sezer, M.: The approximate solution of high-order linear Volterra-Fredholm integro-differential equations in terms of Taylor polynomials. Appl. Math. Comput. **112**, 291–308 (2000)
12. Avudainayagam, A., Vani, C.: Wavelet-Galerkin method for integro-differential equations. Appl. Numer. Math. **32**, 247–254 (2000)
13. Maleknejad, K., Mirzaee, F., Abbasbandy, S.: Solving linear integro-differential equations system by using rationalized Haar functions method. Appl. Math. Comput. **155**, 317–328 (2004)
14. Hosseini, S.M., Shahmorad, S.: Numerical solution of a class of integro-differential equations by the Tau method with error estimation. Appl. Math. Comput. **136**, 559–570 (2003)
15. Aruchunan, E., Sulaiman, J.: Application of the central-difference scheme with half-sweep Gauss-Seidel method for solving first order linear Fredholm integro-differential equations. World Acad. Sci. Eng. Technol. **6**(8), 335–339 (2012)
16. Aruchunan, E., Muthuvalu, M.S., Sulaiman, J., Koh, W.S., Akhir, M.K.M.: An iterative solution for second order linear Fredholm Integro-differential equations. Malays. J. Math. Sci. **8**(2), 158–170 (2014)
17. Aruchunan, E., Sulaiman, J.: Numerical solution of second-order linear Fredholm Integro-differential equation using generalized minimal residual (GMRES) method. Am. J. Appl. Sci. **7**(6), 780–783 (2010)
18. Abdullah, A.R.: The four-point explicit decoupled group (EDG) method: A fast Poisson solver. Int. J. Comput. Math. **38**, 61–70 (1991)
19. Aruchunan, E., Sulaiman, J.: HS conjugate gradient method for solving first order linear Fredholm Integro-differential equations. Aust. J. Basic Appl. Sci. **5**, 38–43 (2011)
20. Aruchunan, E., Sulaiman, J.: Comparison of closed repeated Newton-Cotes quadrature schemes with half-sweep iteration concept in solving linear Fredholm Integro-differential equations. Int. J. Sci. Eng. Invest. **1**(8), 296–300 (2012)
21. Muthuvalu, M.S., Sulaiman, J.: Half-sweep arithmetic mean method with composite trapezoidal scheme for solving linear Fredholm integral equations. Appl. Math. Comput. **217**, 5442–5448 (2011)

22. Sulaiman, J., Hasan, M.K., Othman, M.: Red-Black half-sweep iterative method using triangle finite element approximation for 2D Poisson equations. In: Computational Science, Lecture Notes in Computer Science (LNCS 4487, pp. 326–333). Springer- Verlag, Berlin (2007)
23. Yalcinbas, S.: Taylor polynomial solution of nonlinear Volterra-Fredholm integral equations. Appl. Math. Comput. 127, 195–206 (2002)
24. Barrett, R., Berry, M., Chan, M.T.F., Demmel, J., Donato, J., Dongarra, J., Eijkhou, V., Pozo, R., Romine, C., & van der Vorst: Templates for the solution of linear systems. In: Building Blocks for Iterative Methods. Society for Industrial and Applied Mathematics, Philadelphia (1993)
25. Darania, P., Ebadia, A.: A method for numerical solution of integro-differential equations. Appl. Math. Comput. **188**, 657–668 (2007)
26. Delves, L.M., Mohamed, J.L.: Computational Methods for Integral Equations. Clarendon Press, Oxford, London (1985)
27. Hasan, M.K., Sulaiman, J., Karim, S.A.A., Othman, M.: Development of some numerical methods applying complexity reduction approach for solving scientific problem. J. Appl. Sci. **11**(7), 1255–1260 (2011)
28. Ren, Y., Zhang, B., Qiao, H.: A simple Taylor-series expansion method for a class of second kind integral equations. J. Comput. Appl. Math. **110**, 15–24 (1999)

Index

A
Absolute error, 71
Airflow, 153
Air mass, 154
Algorithm, 423
Alternative to existing multiple elimination, 283
Alzheimer's disease, 250
Ambiguity function, 96
Amplitude distribution, 330
Amyloidogenic processing, 251
Amyloid precursor protein, 250
Analysis, 422
Analyzing, 71
Antenna, 124
Approximate, 416
Approximate equation, 65
Approximate solution, 69, 71
Approximation order, 2
Approximations, 108
A quaternion finite measure, 80
Arithmetic operations, 424
Artificial neural network, 367
A sufficient condition of weak convergence, 87
Attenuating, 146
Autoregressive, 165
Autoregressive model, 173
Azimuths, 131

B
Bandlimited quaternion function, 81
Basic reproduction number, 379
Biopolymer, 359
Block backward differentiation formula, 108
Block-partitioning technique, 406
Bochner-Minlos theorem, 80
Boundary value problem, 63

C
CD4+ T cells, 380
Channels, 255
Characterization, 360
Chebyshev, 65
CIGRE standard, 220
Cleavage, 252
Closed-loop, 283
Cluster, 34
Coating, 358
Coefficient, 51, 421
Coefficient matrix, 69
Compact support, 11
Complexity, 59
Complex overburden, 330
Compositionality, 254
Computation, 49
Computational complexity, 406, 423
Computational cost, the, 424
Computational error, 116
Computationally efficient, 410
Computational strategies, 64
Computational time, 54
Computer aided geometric designs, 3
Computing flow, 54
Condensation, 362
Confident interval, 172
Conjugate function property, The, 101
Conjugate gradient, 421
Connects, 66
Conquering equations, 40

© Institute of Technology PETRONAS Sdn Bhd 2022

S. A. Abdul Karim and A. Shafie (eds.), *Towards Intelligent Systems Modeling and Simulation*, Studies in Systems, Decision and Control 383,
https://doi.org/10.1007/978-3-030-79606-8

Continuous, 9
Controlled source electromagnetic, 135, 236
Control source electromagnetic, 123
Convection, 43, 48
Convergence, 42
Convergence criterion, 416, 422
Converging, 158
Convexity, 1
Correlation, 371
Correlation analysis, 165
Correlation coefficient, 167
Coupling agents, 358
Covariance matrices, 241
C programming, 423
CPU time, 158
Current maximum capacity, 217
Curved antenna, 125
Cytotoxic T lymphocytes cells, 379

D
Decomposed, 51
Density, 112
Dependent, 165
Derivatives, 67
Diagonal matrix, 69
Differential, 417, 418
Direct method, 66, 69
Discretization, 63, 418
Discretization scheme, 423
Discretized, 71
Disease, 31
Disease dynamics, 379
Dissolved, 113
Distribution, 48
2D response, 132
3D visualization, 133

E
Economic, 35
Effectiveness, 43
Electrical efficiencies, 154
Electric field, 125, 141, 142, 240
Electromagnetic, 135
Electromagnetic response, 132
Electrostatic IA stable and unstable soliton, 278
Elliptic partial differential equations, 48
Empirical model, 372
Emulsification, 112, 114
Energy, 152
Enhanced oil recovery, 358, 368
Epidemic, 27
Equally-spaced, 417
Equilibrium points, 383
Error, 167, 422
Exponentially, 146

F
Faster rate of convergence, 283
Field data of Malay Basin, 283
Finite measure, 79
Finite difference approach, 65
Finite difference approximation, 49
Finite difference method, 42, 400
Finite difference schemes, 414
Finite element, 238
First order, 171
Fit, 258
Flatten the curve, 33
Flowrate, 154
Fluid, 39
Focal beam, 330
Forecast, 117
Forward modelling, 330
Fourier Series Model, 221
Fractional-order differential equations, 380
Fredholm integro-differential equations, 414
Frequency, 136

G
Gaussian function, 104
Gaussian process regression, 237
GDP, 35
Geological layers, 137
Global asymptotic stability, 379
Grafting, 361
Graph theory, 66
Grid size, 57
Growth rate, 30

H
Half-sweep iteration, 414
Half-Sweep Newton-EGSOR, 399
Heat, 48
Heat balance, 151
Historical data, 168
Holder exponent, 10
Horizontal electric dipole, 124
Hosoya approximation, 63
Hybrid technology, 155
Hydrocarbon, 127
Hydrocarbon exploration, 133

I

Identification, 164
IEEE Standard 738, 219
Illumination factor, 330
Illumination methods, 329
Immune impairment, 380
Immunized, 31
Improving seismic acquisition design, 330
Increases, 58
Incremental search, 158
Industrial Revolution 4.0, 48
Integral, 418
Integrated method, 330
Interpolating, 3
Ion Acoustic Waves (IAWs), 266
Iteration, 42, 54, 158, 255, 424
Iterative method, 158, 405, 421

K

Kernel function, 225
Korteweg–de Vries equation (KdVE), 266

L

LaSalle's invariance principle, 388
Laurent Polynomial, 4
LFM signal, 105
Likelihood, 164, 173
Limit curve, 21
Limited reference, 25
Line ampacity, 221
Linear canonical ambiguity function, 94, 97
Linear Canonical Transform (lCT), 93
Linear FM signal, 94, 103
Linear relationship, 167
Linear systems, 421
Llinear, 157
Local learners, 224
Local truncation error, 401
Lower triangular matrix, 69
Low frequency, 124
Lyapunov function, 379

M

Magnetic, 48
Magnetohydrodynamics, 40
Marine controlled source electromagnetic, 136
Mathematical analysis, 379
Mathematical induction, 1
Mathematical correlation, 31
Mathematical model, 25, 149, 154
Mathematics, 2
Matlab, 160
Matrix, 51, 69
Maximum error, 59
Maxwell equations, 125
Modeling, 252
Modified KdV Equation (mKdVE), 266
Movement control order, 33
Moyal's formula, 94
Multi-component plasmas, 265
Multiple difference, 42
Multiple removal, 283

N

Nanocomposite, 360
Nanofluid, 358
Nanofluid injection rate, 367
Nanoparticle, 371
Nanotechnology, 359
Negative-log marginal likelihood, 241
Network, 66
Neural networks, 229
Neurodegeneration, 250
New ordering, 49
Newton's method, 158
Newton-Cotes, 414
Newton-EGSOR, 401
Newton-Gauss-Seidel, 401
Next generation matrix, 182, 384
Nodal points, 416
Nondimensional, 41
Non-linear, 157
Non-linear algebraic, 162
Nonlinear Evolution Equations (NLEEs), 266
Nonlinearity, 266
Nonlinear partial differential equations, 399
Nonlinear Schrödinger Equation (NLSE), 266
Non-standard finite difference method, 39
Nonsteady equation, 41
Non-steady MHD, 43
Novel coronavirus (2019-nCoV), 25
Numerical simulations, 389
Numerical analysis, 158
Numerically, 67
Numerical schemes, 43
Numerical system, 47
Numerical weather prediction, 220

O

Occurrence, 164, 172, 173

Ode45, 108
Offset, 142
Oil dissolution, 112
Oil evaporated, 113
Oil recovery, 367
Oil spill, 108
Oil-weathering, 108
Optimization, 371
Outbreak, 25, 26
Overhead transmission lines, 217
Over relaxation methods, 59

P
Pandemics, 26
Parameter, 41, 52, 167
Parameter μ, 13
Parametric range, 21
Parseval's formula, 85
Partial differential equation, 47
Particle size, 367
Partition points, 417
Pathogen-to-cell, 380
Pearson Correlation, 172
Pentasection, 158
Percentage difference, 132
Permeability, 367
Phenomenon, 145
Photovoltaic/Thermal, 150
6-point, 3
Poisson's equations, 47
Poor images, 329
Porous medium equation, 399
Posterior distribution, 240
Predict, 110
Prediction, 164
Prediction system, 27
Probability, 255
Propagate, 146
P-values, 172

Q
Quaternion Fourier transform of finite measure, the, 80, 83
Quaternion Fourier transform, the, 79
Quarter sweep reduction, 59

R
Randomness, 254
Rate of spillage, 111
Rational Finite Difference schemes, 65
Ray hit count, 330
Ray interpolation, 329
Recovered, 34
Recover the primaries, 283
Recovery, 31
Recovery rate, 30
Rectangular grid, 49
Reduced, 59
Reductive Perturbation Technique (RPT), the, 266
Regression analysis, 169
Relaxation parameter, 409
Renewable energy, 149
Reservoir, 137
Reservoir characterization, 330
Resistivity, 240
Response, 142
Retarded functional differential equations, 181
Risk, 173
Risk analysis, 163
Road, 164
Road accidents, 163, 171
Robust, 158
Root finding, 158
Rotated accelerated over-relaxation, 52
Rotated strategy, 59
Routh-Hurwitz standards, 387

S
Scaled conjugate gradient, 371
Schemes, 63
Schizophyllan, 357
Scientific phenomena, 47
Seabed logging, 123, 136, 236
Sea surface, 111
Secant, 158
Second-order, 167
Secretases, 252
Semi-circle antenna, 125
Severity, 30
Shallow and complex geological area, 283
Shallow gas cloud, 329
Shape control parameters, 21
Shape-preserving, 2
Significant, 169
Simplify, 68
Simulation, 108, 116, 237, 255
SiO_2, 359
Solar energy, 149
Sol-gel, 360
Solve, 69
Solving, 43

SORLA, 250
Sorting protein-related receptor, 250
Space and astrophysical environments, 266
Sparsity constraint, 283
Speed, 49
Speed-up, 58
SPiM, 254
Sporadic, 35
Stability, 401
Standard Gardner Equation (SGE), 266
Steady temperature distribution, 48
Step size, 110
Stochastic Process Algebra, 252
Stopping criteria, 157
Strategy, 49
Strictly, 69
Strictly convex data, 18
Strong positive, 172
Subdivision, 16
Subdivision schemes, 3
Subintervals, 418
Successive over-relaxation, 52, 401
Surface modification, 358
Surface related multiple elimination, 283
Symmetric, 11
Synthetic, 283
System, 66
System collector, 150
System of linear equations, 69

T
Target depths, 238
Tedious mathematical procedures, 266
Temperature, 39, 113
Tension parameter μ, 1
Ternary, 1
Theoretical model equations, the, 266
Thermal efficiency, 152
Third order, 110, 167
Time, 42, 116
Time series model, 222
Tolerance, 422
Tolerance range, 158
Traffic fatalities, 164
Transition rate, 255
Transmembrane glycoprotein, 250
Transmitter, 238
Transportation, 164, 170
Trisection, 158
Two-dimensional, 47
Two-point, 63, 108
Two-point boundary value problems, 65
Two target cells, 380

U
Uniform measure, 84
Upper triangular matrix, 69

V
Value at risk, 165, 173
Variable, 165
Vertices, 66
Viscosity, 112

W
Wave field distortion, 330
Wave propagation, 329
Weather parameters, 221
Weighted, 52
Weiner index graph, 64
Wettability, 367
Wigner-Ville distribution, 94

The manufacturer's authorised representative in the EU is Springer Nature Customer Service Centre GmbH, Europaplatz 3, 69115 Heidelberg, Germany. If you have any concerns regarding our products, please contact ProductSafety@springernature.com

Printed and bound by CPI Group (UK) Ltd, Croydon, CR0 4YY
05/10/2026
02228339-0006